Tsunamis: Geology, Hazards and Risks

Geological Society books refereeing procedures

The Society makes every effort to ensure that the scientific and production quality of its books matches that of its journals. Since 1997, all book proposals have been refereed by specialist reviewers as well as by the Society's Books Editorial Committee. If the referees identify weaknesses in the proposal, these must be addressed before the proposal is accepted.

Once the book is accepted, the Society Book Editors ensure that the volume editors follow strict guidelines on refereeing and quality control. We insist that individual papers can only be accepted after satisfactory review by two independent referees. The questions on the review forms are similar to those for *Journal of the Geological Society*. The referees' forms and comments must be available to the Society's Book Editors on request.

Although many of the books result from meetings, the editors are expected to commission papers that were not presented at the meeting to ensure that the book provides a balanced coverage of the subject. Being accepted for presentation at the meeting does not guarantee inclusion in the book.

More information about submitting a proposal and producing a book for the Society can be found on its website: www.geolsoc.org.uk.

It is recommended that reference to all or part of this book should be made in one of the following ways:

Scourse, E. M., Chapman, N. A., Tappin, D. R. & Wallis, S. R. (eds) 2018. *Tsunamis: Geology, Hazards and Risks*. Geological Society, London, Special Publications, **456**.

Wallis, S. R., Fujiwara, O. & Goto, K. 2018. Geological studies in tsunami research since the 2011 Tohoku earthquake. *In*: Scourse, E. M., Chapman, N. A., Tappin, D. R. & Wallis, S. R. (eds) *Tsunamis: Geology, Hazards and Risks*. Geological Society, London, Special Publications, **456**, 39–53. First published online July 18, 2017, https://doi.org/10.1144/SP456.12

GEOLOGICAL SOCIETY SPECIAL PUBLICATION NO. 456

Tsunamis: Geology, Hazards and Risks

EDITED BY

E. M. SCOURSE
MCM Environmental Services Ltd, UK

N. A. CHAPMAN
University of Sheffield, UK

D. R. TAPPIN
British Geological Survey, UK

and

S. R. WALLIS
Nagoya University, Japan
University of Tokyo, Japan

2018
Published by
The Geological Society
London

THE GEOLOGICAL SOCIETY

The Geological Society of London (GSL) was founded in 1807. It is the oldest national geological society in the world and the largest in Europe. It was incorporated under Royal Charter in 1825 and is Registered Charity 210161.

The Society is the UK national learned and professional society for geology with a worldwide Fellowship (FGS) of over 10 000. The Society has the power to confer Chartered status on suitably qualified Fellows, and about 2000 of the Fellowship carry the title (CGeol). Chartered Geologists may also obtain the equivalent European title, European Geologist (EurGeol). One fifth of the Society's fellowship resides outside the UK. To find out more about the Society, log on to www.geolsoc.org.uk.

The Geological Society Publishing House (Bath, UK) produces the Society's international journals and books, and acts as European distributor for selected publications of the American Association of Petroleum Geologists (AAPG), the Indonesian Petroleum Association (IPA), the Geological Society of America (GSA), the Society for Sedimentary Geology (SEPM) and the Geologists' Association (GA). Joint marketing agreements ensure that GSL Fellows may purchase these societies' publications at a discount. The Society's online bookshop (accessible from www.geolsoc.org.uk) offers secure book purchasing with your credit or debit card.

To find out about joining the Society and benefiting from substantial discounts on publications of GSL and other societies worldwide, consult www.geolsoc.org.uk, or contact the Fellowship Department at: The Geological Society, Burlington House, Piccadilly, London W1J 0BG: Tel. +44 (0)20 7434 9944; Fax +44 (0)20 7439 8975; E-mail: enquiries@geolsoc.org.uk.

For information about the Society's meetings, consult *Events* on www.geolsoc.org.uk. To find out more about the Society's Corporate Affiliates Scheme, write to enquiries@geolsoc.org.uk.

Published by The Geological Society from:
The Geological Society Publishing House, Unit 7, Brassmill Enterprise Centre, Brassmill Lane, Bath BA1 3JN, UK

The Lyell Collection: www.lyellcollection.org
Online bookshop: www.geolsoc.org.uk/bookshop
Orders: Tel. +44 (0)1225 445046, Fax +44 (0)1225 442836

British Library Cataloguing in Publication Data

A catalogue record for this book is available from the British Library.
ISBN 978-1-78620-318-2
ISSN 0305-8719

Distributors
For details of international agents and distributors see:
www.geolsoc.org.uk/agentsdistributors

Typeset by Nova Techset Private Limited, Bengaluru & Chennai, India
Printed and bound by CPI Group (UK) Ltd, Croydon CR0 4YY

Contents

Foreword: Geological Society of Japan

We are delighted that this Special Publication on the link between geology and tsunamis has been published by the Geological Society of London. This volume brings together a collection of the latest knowledge on tsunami deposits from various locations around the globe and of various geological ages with the results of risk modelling. We believe that this publication will not only deepen the understanding of tsunami and high-energy (e.g. storm and swell) deposits by Earth scientists but also provide basic knowledge on tsunami impact, which can benefit all those working in the fields of disaster prevention and mitigation.

It was in the afternoon of 11 March 2011 when a large earthquake struck eastern Japan. This earthquake, later called 'the 2011 earthquake off the Pacific coast of Tohoku', generated a series of large tsunami waves, which repeatedly struck the coastal region of eastern Japan. As a direct result of this disaster, more than 20 000 people lost their lives or are listed as missing. Many Earth scientists in Japan felt a great sense of remorse and responsibility that they were unable to do more to protect the many victims, and to reduce the huge damage to property and infrastructure. This sense of regret was especially poignant, because at the time of the 2011 tsunami some Earth scientists had already identified the presence of several tsunami-deposit layers showing that the coastal region of NE Japan had been repeatedly struck by large tsunamis historically. But, unfortunately, this knowledge was insufficiently reflected in either social or industrial planning. In the Japanese research community there is a heightened desire to reflect deeply on the shortcomings of the past, and to be more active in suggesting new measures and directions that can lead to geohazard prevention and mitigation. More than 6 years after the disaster in NE Japan, this thought has not changed, and it will surely help guide our research for many years to come.

The events of 2011 formed the background for the signing of a memorandum of understanding (MoU) between the Geological Societies of Japan and London. As our first collaboration, our Societies agreed to hold a pair of symposia on tsunamis and geology, the first in Kagoshima in 2014 and the second in London in 2015, each associated with a field excursion. The outcomes of the first symposium were published in a thematic section, 'Geological records of storms, tsunamis and other extreme events', of *Island Arc* (Volume 25, Issue 5, 2016), the official English language journal of the Geological Society of Japan. This new Special Publication collects outcomes of the second symposium in London, and other contributions to the study of tsunamis and their deposits.

We express our heartfelt gratitude to the editors of this Special Publication: Ellie Scourse, Neil Chapman, David Tappin and Simon Wallis. We also thank the Great Britain Sasakawa Foundation for their support of this project, and Professor David Cope for his assistance. We are also grateful to the Japanese Embassy for hosting a post-meeting reception in London.

Yoshio Watanabe
President of the Geological Society of Japan
(June 2016–present)
Yasufumi Iryu
President of the Geological Society of Japan
(June 2014–May 2016)

Foreword: Geological Society of London

With Earth's growing population clustered increasingly on coastlines, tsunami hazards are of concern worldwide. This publication stems from two linked symposia on tsunami hazards that were organized in 2014 and 2015 in Japan and the UK. The significance of these symposia is that they marked the first outcomes of a bilateral co-operation agreement between the Geological Societies of London and Japan, initiated in 2013.

The 2014 Symposium (*Tsunami and their Geological Evidence*: 14 September 2014, Kagoshima, Japan) was held as part of the Annual Meeting of the Geological Society of Japan and focused principally on the geological evidence left by tsunamis on the Pacific Rim, whilst the second Symposium (*Tsunami Hazards and Risks: Using the Geological Record*: Arthur Holmes Meeting of the Geological Society of London, 25 September 2015, London, UK) focused more on the North Atlantic and Mediterranean regions, on tsunami modelling, and on hazard and risk assessment.

The symposia were not only valuable in establishing approaches to co-operation between the two Geological Societies, but also in bringing together geoscientists and risk assessors to assess tsunami hazards in an integrated manner, with a view to facilitating more quantitative and evidence-based evaluation of their scale, nature, location and timescales.

It was a great pleasure for the Geological Society of London to welcome the President of the Geological Society of Japan and other colleagues from Japan to the London symposium in 2015. The importance of the collaboration was recognized in a post-symposium reception that was kindly hosted by the Embassy of Japan and addressed by the Ambassador.

Our considerable thanks go to the Great Britain Sasakawa Foundation for facilitating contacts in both Japan and the UK, in particular to Professor David Cope for his help and enthusiasm in moving the project forward. The Foundation provided a multi-year grant, which not only helped support travel for some participants in the two symposia and the associated field excursions in central Japan and the Shetland Islands, but also extended to other joint activities of our two Societies. I would also like to thank the Lighthill Risk Network for additional financial support for the London symposium.

The linked symposia provide a model for what we hope will be many more years of fruitful collaboration between the Geological Societies of London and Japan.

Malcolm Brown
President, Geological Society of London

Tsunamis: geology, hazards and risks – introduction

ELLIE M. SCOURSE[1]*, NEIL A. CHAPMAN[2], DAVID R. TAPPIN[3] & SIMON R. WALLIS[4,5]

[1]*MCM Environmental Services Ltd, 1 Temple Way, Bristol BS2 0BY, UK*

[2]*Department of Materials Science and Engineering, University of Sheffield, Sir Robert Hadfield Building, Mappin Street, Sheffield S1 3JD, UK*

[3]*British Geological Survey, Keyworth, Nottingham NG12 5GG, UK*

[4]*Department of Earth and Planetary Sciences, Nagoya University, Furo-cho, Chikusa-ku, Nagoya 464-8601, Japan*

[5]*Department of Earth and Planetary Science, University of Tokyo, 7-3-1 Hongo, Bunkyo-ku, Tokyo 113-0033, Japan*

**Correspondence: ellie.scourse@mcmenvironmental.co.uk*

A decade or so ago, if you had asked almost anyone in Europe or North America, they might not have recognized the word 'tsunami'. The enormous and tragic event that swept across the shores of the Indian Ocean on 26 December 2004, followed only a few years later by the devastating tsunami caused by the March 2011 Great Tohoku earthquake off Japan, both with appalling loss of life, changed all that. Today, the words 'tsunami warning issued' seem to appear frequently on international 'breaking news', showing the extent to which we have become sensitized to the triggers that launch these deadly, but terrifyingly spectacular, natural events. Yet, great tsunamis and the tectonic events that cause them have not suddenly become more frequent. The historical records of old civilizations contain accounts of major inundations reaching back hundreds or thousands of years and sometimes even warnings to future generations – valuable, if they are heeded. What has changed, and has consequently raised the profile of tsunamis, is the exponential growth in world population over the last few 100 years, the great majority of whom live in coastal areas and are consequently exposed to hazard, along with instant global communication, which brings every large earthquake on Earth's plate margins directly and immediately onto our screens.

The consequence of recent events and our response to the compelling images that record these catastrophes is a global recognition of the potential hazards to life and infrastructure associated with tsunamis. For planners and decision-makers at all levels, what is demanded from scientists is more information: where can this happen; how big can it be; when will the next event occur and will there be any warning? Being able to address these questions and provide clear answers is a major responsibility for geoscientists, and a real challenge. The root of our knowledge here is a mechanistic understanding of the tectonic (and other) processes that cause tsunamis, and of how tsunamis evolve, propagate and fade. This understanding helps to start answering the 'where', 'how big' and 'how fast' questions, but in order to address the critical 'when' question, and obtain more focus on 'how big', requires seeking forensic geological evidence of the traces left by past tsunamis.

All of this information must then be integrated in a way that identifies and quantifies the inevitable uncertainties, before scientists can begin to discuss probabilities in a useful fashion for specific regions.

Academic societies have a role to play in promoting such research efforts and the Geological Society of London has collaborated with its partner, the Geological Society of Japan, in holding two symposia: one in 2014 in Kagoshima, and another – the Arthur Holmes Meeting in 2015 – in London. Both these symposia focused on 'Tsunami Hazards and Risks: Using the Geological Record' and emphasized the importance of geological studies in developing a better understanding of tsunamis. The results of the first meeting were summarized in a special issue of the journal *Island Arc* (September 2016). The results of the second are presented in this Special Publication.

The papers explore the sedimentological and dynamic traces of recent and prehistoric tsunamis globally – from Europe to the Pacific – as well as looking at historical records and how the information can be used to characterize the scale of impacts and areas that are most susceptible to tsunami

From: Scourse, E. M., Chapman, N. A., Tappin, D. R. & Wallis, S. R. (eds) 2018. *Tsunamis: Geology, Hazards and Risks*. Geological Society, London, Special Publications, **456**, 1–3.
First published online September 28, 2017, https://doi.org/10.1144/SP456.13

hazards. Armed with this information, scientists can begin to quantify risks – both to populations and in economic terms.

The papers are arranged into two sections: one on tsunami hazards globally, focusing on specific historical records, and another on risk modelling.

In the first section, **Tappin (2017)** presents a history of geological involvement in tsunami science, and its importance in advancing understanding of the extent, magnitude and nature of the hazard from tsunamis.

There are three chapters relating to tsunami studies in Japan: **Wallis *et al.* (2017)** investigate geological studies in tsunami research since the 2011 Tohoku earthquake; **Goda *et al.* (2017)** present tsunami simulations of a wide range of realistic slip distribution and kinematic rupture processes, reflecting the current best understanding of what might happen due to a future mega-earthquake at the Nankai–Tonankai Trough; and **Ikehara *et al.* (2017)** give an overview of spatial variability in sediment lithology and sedimentary processes along the Japan Trench, which can be used to reconstruct the recurrence history of large earthquakes and tsunamis.

Moving on from Japan, **Lindholm *et al.* (2017)** focus on tsunami hazards in Central America, where the Pacific subduction zone, coupled with the growing coastal tourist industry in this area, has the potential for creating large tsunami risk.

Focusing on Europe, **Boulton & Whitworth (2017)** investigate block and boulder accumulations on the southern coast of Crete, and interpret what these can tell us about tsunami propagation in the Mediterranean; **Mottershead *et al.* (2017)** also focus on the Mediterranean, specifically on the Maltese Islands, and on evidence for tsunami landfall, allowing for critical reassessment of the exposure of Malta to tsunami hazard. **Costa *et al.* (2017)** investigate the microtextural and heavy mineral analysis of sandy storm and tsunami deposits from Portugal, Scotland, Indonesia and the USA, considering inundation events of different chronologies and sources that affect contrasting coastal and hinterland settings with different regional oceanographic conditions.

In contrast, **Long (2017)** gives an overview of cataloguing tsunami events in the UK – located well away from any subduction zone, and so not exposed to as high a risk as the other locations in this section. Detailed examinations of the impact of three tsunamis on the UK coast are discussed as examples of events triggered by seismicity, submarine landslides and coastal landslides.

The second section focuses on risk modelling, with a comprehensive overview by **Woo (2017)** on risk-informed tsunami warnings presenting a probabilistic approach to assessing the tsunamigenicity of fault rupture, using both the geological record and historical catalogues. **Power *et al.* (2017)** present the New Zealand Probabilistic Tsunami Hazard Model (NZPTHM), for use as an effective quantitative estimate of tsunami hazard using the Monte Carlo method, and suggest detailed inundation modelling of a small set of scenarios in New Zealand. **Davies *et al.* (2017)** discuss a global-scale probabilistic tsunami hazard assessment (PTHA), extending previous global-scale assessments based largely on scenario analysis.

It can be seen from the wide range of contributions included in this volume that the field of tsunami research and understanding is diverse and highly dependent on the identification and interpretation of the geological traces left by past tsunamis. There is still some way to go in fully understanding and answering the 'where', 'how big', 'how fast' and 'when' questions. However, knowledge and techniques are developing rapidly, owing to the increased interest globally in this topical area of natural hazard research, and there is an expectation that the risks associated with tsunamis along overpopulated coastlines can, to some extent, be mitigated.

The editors would like to thank all of the contributors and reviewers for their patience and help in putting together this book, which we hope will prove a valuable resource for colleagues working in this field.

DRT publishes with the consent of the CEO of British Geological Survey.

References

BOULTON, S.J. & WHITWORTH, M.R.Z. 2017. Block and boulder accumulations on the southern coast of Crete (Greece): evidence for the 365 CE tsunami in the Eastern Mediterranean. *In*: SCOURSE, E.M., CHAPMAN, N.A., TAPPIN, D.R. & WALLIS, S.R. (eds) *Tsunamis: Geology, Hazards and Risks*. Geological Society, London, Special Publications, **456**. First published online February 9, 2017, https://doi.org/10.1144/SP456.4

COSTA, P.J.M., GELFENBAUM, G. *ET AL.* 2017. The application of microtextural and heavy mineral analysis to discriminate between storm and tsunami deposits. *In*: SCOURSE, E.M., CHAPMAN, N.A., TAPPIN, D.R. & WALLIS, S.R. (eds) *Tsunamis: Geology, Hazards and Risks*. Geological Society, London, Special Publications, **456**. First published online February 23, 2017, https://doi.org/10.1144/SP456.7

DAVIES, G., GRIFFIN, J. *ET AL.* 2017. A global probabilistic tsunami hazard assessment from earthquake sources. *In*: SCOURSE, E.M., CHAPMAN, N.A., TAPPIN, D.R. & WALLIS, S.R. (eds) *Tsunamis: Geology, Hazards and Risks*. Geological Society, London, Special Publications, **456**. First published online 23 February 2017, updated 2 March 2017, https://doi.org/10.1144/SP456.5

GODA, K., YASUDA, T., MAI, P.M., MARUYAMA, T. & MORI, N. 2017. Tsunami simulations of mega-thrust earthquakes in the Nankai–Tonankai Trough (Japan) based on stochastic rupture scenarios. *In*: SCOURSE, E.M., CHAPMAN, N.A., TAPPIN, D.R. & WALLIS, S.R. (eds) *Tsunamis: Geology, Hazards and Risks*. Geological Society, London, Special Publications, **456**. First published online February 22, 2017, https://doi.org/10.1144/SP456.1

IKEHARA, K., USAMI, K., KANAMATSU, T., ARAI, K., YAMAGUCHI, A. & FUKUCHI, R. 2017. Spatial variability in sediment lithology and sedimentary processes along the Japan Trench: use of deep-sea turbidite records to reconstruct past large earthquakes. *In*: SCOURSE, E.M., CHAPMAN, N.A., TAPPIN, D.R. & WALLIS, S.R. (eds) *Tsunamis: Geology, Hazards and Risks*. Geological Society, London, Special Publications, **456**. First published online March 3, 2017, https://doi.org/10.1144/SP456.9

LINDHOLM, C., STRAUCH, W. & FERNÁNDEZ, M. 2017. Tsunami hazard in Central America; history and future. *In*: SCOURSE, E.M., CHAPMAN, N.A., TAPPIN, D.R. & WALLIS, S.R. (eds) *Tsunamis: Geology, Hazards and Risks*. Geological Society, London, Special Publications, **456**. First published online February 23, 2017, https://doi.org/10.1144/SP456.2

LONG, D. 2017. Cataloguing tsunami events in the UK. *In*: SCOURSE, E.M., CHAPMAN, N.A., TAPPIN, D.R. & WALLIS, S.R. (eds) *Tsunamis: Geology, Hazards and Risks*. Geological Society, London, Special Publications, **456**. First published online June 29, 2017, https://doi.org/10.1144/SP456.10

MOTTERSHEAD, D.N., BRAY, M.J. & SOAR, P.J. 2017. Tsunami landfalls in the Maltese archipelago: reconciling the historical record with geomorphological evidence. *In*: SCOURSE, E.M., CHAPMAN, N.A., TAPPIN, D.R. & WALLIS, S.R. (eds) *Tsunamis: Geology, Hazards and Risks*. Geological Society, London, Special Publications, **456**. First published online February 23, 2017, updated 3 March 2017, https://doi.org/10.1144/SP456.8

POWER, W., WANG, X., WALLACE, L., CLARK, K. & MUELLER, C. 2017. The New Zealand Probabilistic Tsunami Hazard Model: development and implementation of a methodology for estimating tsunami hazard nationwide. *In*: SCOURSE, E.M., CHAPMAN, N.A., TAPPIN, D.R. & WALLIS, S.R. (eds) *Tsunamis: Geology, Hazards and Risks*. Geological Society, London, Special Publications, **456**. First published online March 3, 2017, https://doi.org/10.1144/SP456.6

TAPPIN, D.R. 2017. The importance of geologists and geology in tsunami science and tsunami hazard. *In*: SCOURSE, E.M., CHAPMAN, N.A., TAPPIN, D.R. & WALLIS, S.R. (eds) *Tsunamis: Geology, Hazards and Risks*. Geological Society, London, Special Publications, **456**. First published online June 28, 2017, https://doi.org/10.1144/SP456.11

WALLIS, S.R., FUJIWARA, O. & GOTO, K. 2017. Geological studies in tsunami research since the 2011 Tohoku earthquake. *In*: SCOURSE, E.M., CHAPMAN, N.A., TAPPIN, D.R. & WALLIS, S.R. (eds) *Tsunamis: Geology, Hazards and Risks*. Geological Society, London, Special Publications, **456**. First published online July 18, 2017, https://doi.org/10.1144/SP456.12

WOO, G. 2017. Risk-informed tsunami warnings. *In*: SCOURSE, E.M., CHAPMAN, N.A., TAPPIN, D.R. & WALLIS, S.R. (eds) *Tsunamis: Geology, Hazards and Risks*. Geological Society, London, Special Publications, **456**. First published online January 23, 2017, https://doi.org/10.1144/SP456.3

The importance of geologists and geology in tsunami science and tsunami hazard

DAVID R. TAPPIN

British Geological Survey, Keyworth, Nottingham, NG12 5GG, UK, drta@bgs.ac.uk

Abstract: Up until the late 1980s geology contributed very little to the study of tsunamis because most were generated by earthquakes which were mainly the domain of seismologists. In 1987–88 however, sediments deposited as tsunamis flooded land were discovered. Subsequently they began to be widely used to identify prehistorical tsunami events, providing a longer-term record than previously available from historical accounts. The sediments offered an opportunity to better define tsunami frequency that could underpin improved risk assessment. When over 2200 people died from a catastrophic tsunami in Papua New Guinea (PNG) in 1998, and a submarine landslide was controversially proven to be the mechanism, marine geologists provided the leadership that led to the identification of this previously unrecognized danger. The catastrophic tsunami in the Indian Ocean in 2004 confirmed the critical importance of sedimentological research in understanding tsunamis. In 2011, the Japan earthquake and tsunami further confirmed the importance of both sediments in tsunami hazard mitigation and the dangers from seabed sediment failures in tsunami generation. Here we recount the history of geological involvement in tsunami science and its importance in advancing understanding of the extent, magnitude and nature of the hazard from tsunamis.

Until the late 1980s tsunami science was mostly the province of seismologists, numerical modellers, geophysicists and historians; tsunamis received little attention from geologists (e.g. Bailey & Weir 1933; Coleman 1968, 1978). Earthquakes were considered the main, if not the only, mechanism that could generate elevated tsunami waves that were very destructive at the coast. Other tsunami-generation mechanisms, such as submarine landslides, were not considered a major hazard despite evidence to the contrary such as from the Grand Banks event of 1929 (Bardet *et al.* 2003). Numerical tsunami modelling of submarine landslides was in large part theoretical (Jiang & LeBlond 1992, 1994). The slow landslide failure velocities were perceived as inhibiting tsunami generation, in contrast to earthquake-generated tsunamis where rupture velocities of kilometres per second were regarded as instantaneous in the context of the relatively slow (metres per second) velocity of tsunami wave generation (Geist 2000; Ward 2001). Historians provided evidence on older tsunamis, hopefully for use in estimating tsunami frequency–magnitude relationships for future tsunami risk, although the limitations of historical data were recognized (Ambraseys & Jackson 1990). When a paper (rarely) considered the geology of tsunamis, it was on the sediments deposited from inundation and authored by seismologists (e.g. Wright & Mella 1963).

The involvement of geologists in tsunami research began in the early 1980s, with the first papers on deep-sea deposits in the Aegean Sea. Here, unusual seabed sediments, termed homogenites, were proposed as deposited from a tsunami generated by the Late Bronze Age eruption of Santorini (Kastens & Cita 1981; Cita *et al.* 1984). In the Hawaiian Islands, boulders and coarse-grained gravels preserved at high elevations (hundreds of metres) were considered to result from tsunamis generated by large-scale volcanic collapse (Moore & Moore 1984). The early results from the Mediterranean and Pacific were controversial due to the uncertainty over whether there were tsunamis generated at these locations. In addition, this approach of using sediments to identify tsunamis from their sedimentary record had not been attempted before. The tsunami from the Late Bronze Age (LBA) eruption of Santorini had been a major controversy for decades (Marinatos 1939). Sand had been found in Minoan buildings on the coast and used as evidence for tsunami inundation; its origin was however disputed as it was located at sea level and the sand could have been deposited from storms or was present for religious purposes. Subsequent validation has been impossible because the sand deposits were destroyed during subsequent excavations. The Aegean Sea homogenites provided the first geological evidence in support of the tsunami hypothesis but, even today, the origin of the

From: Scourse, E. M., Chapman, N. A., Tappin, D. R. & Wallis, S. R. (eds) 2018. *Tsunamis: Geology, Hazards and Risks*. Geological Society, London, Special Publications, **456**, 5–38.
First published online June 23, 2017, https://doi.org/10.1144/SP456.11

homogenites remains controversial (see Pareschi *et al.* 2006; Shanmugam 2006; Weiss 2008; Polonia *et al.* 2013). The suggestion that the Hawaiian deposits were from tsunamis was groundbreaking because of their use in identifying tsunamis from volcanic collapse. Previously, the deposits were interpreted as laid down during sea-level highstands (Stearns 1978). Their origin is also still disputed (Stearns 1978; Rubin *et al.* 2000; McMurtry *et al.* 2004*a*).

In 1987 and 1988, two groundbreaking papers published on tsunami sediments demonstrated their potential in evaluating tsunami hazard. The first (Atwater 1987) was on prehistoric sediments in Cascadia, which identified a sequence of earthquakes and their associated tsunamis that extended back in time to over 10 ka before present (BP). The second paper (Dawson *et al.* 1988) on sedimentary deposits in Scotland identified a major prehistoric tsunami from a massive submarine landslide (Storegga) located off Norway (Bugge 1983; Bugge *et al.* 1988). These sediments motivated the first attempt at numerical modelling of a tsunami from a submarine landslide mechanism (Harbitz 1992). In 1998, a devastating tsunami struck the north coast of Papua New Guinea (PNG) killing over 2200 people (Kawata *et al.* 1999). The associated M_w 7 magnitude earthquake was too small to generate all of the elevated local run-ups of 15 m. Amid confusion and controversy (e.g. Geist 2000), marine surveys organized in response to the disaster acquired hydroacoustic and sample data offshore of the impacted area. These surveys identified an offshore slump, which preliminary numerical simulations demonstrated to be the local tsunami mechanism (Tappin *et al.* 1999). The PNG event was seminal in identifying the major hazard from submarine landslides in tsunami generation. It was the first tsunami to be studied from responsive marine surveys led by geologists. Submarine landslides were well researched previously, but rarely in the context of tsunami generation and not by collaboration between geologists and numerical modellers (e.g. Grand Banks in 1929).

The 1987–88 research on tsunami sediments in the USA and Europe, together with the landslide-generated PNG tsunami, resulted in the major involvement of geologists in tsunami science and the recognition of the contribution which geology could make to an improved understanding of tsunamis and their hazard. Demonstrating the mechanism of the PNG tsunami was seminal in confirming the tsunami hazard from submarine landslides. It was based on marine surveys carried out by geologists, with the geological interpretations underpinning the numerical tsunami models. Much later, the PNG tsunami was the first attempt to use inverse and forward modelling of tsunami sediments to determine tsunami characteristics (Jaffe & Gelfenbuam 2007). More recent, devastating, events in the Indian Ocean (2004) and Japan (2011) have expanded this geological involvement in tsunami sediment characterization (e.g. Paris *et al.* 2007), in inverse modelling of tsunami-generation mechanisms (Spiske *et al.* 2010; Sugawara *et al.* 2014) and in research on submarine landslides in tsunami generation (Tappin *et al.* 2007, 2014). Storegga, PNG and most recently Japan have led to an increased realization of the tsunami hazard from submarine landslides.

Here it is recounted how over the past *c.* 30 years geologists became increasingly involved in tsunami science and how they have contributed to an improved understanding of tsunami mechanisms and their hazard. Although there were earlier, isolated precursors to the main 'catalyst' events identified above, it was during the 1980s that 'geology' became significant in advancing tsunami science; this advancement was initially from the application of tsunami sediments, followed by an improved understanding of tsunami frequency, hazard and risk, and more recently, with the Japan tsunami, from the use of sediments as a basis for inverse numerical models of tsunami-generation mechanisms (e.g. Sugawara *et al.* 2014). The motivations for this paper were the two meetings held in Japan and the United Kingdom in 2014 and 2015 on the theme of 'Tsunami Hazards and Risks: Using the Geological Record'. The focus of the Japan meeting was the use of tsunami sediments in the mitigation of tsunami hazard. The aim of the UK meeting was to bring together geologists and hazard and risk modellers. In this paper the use of the geological record in contributing to tsunami hazard is extended by the addition of how the identification of submarine landslides have led to their recognition as a major hazard in tsunami generation, a hazard previously overlooked (Bardet *et al.* 2003; Løvholt *et al.* 2015). To underpin realistic modelling of tsunamis generated from submarine landslides requires their architectures to be determined from hydroacoustic data, including multibeam echosounders (MBES) and seismics. Mapping of seabed failures has led to the transition from theoretical numerical models, which suggested that landslides were ineffective in tsunami generation (LeBlond & Jones 1995), to realistic, event-based models that prove otherwise (Harbitz 1992; Tappin *et al.* 1999). The use of the term 'geology' is interpretative in the sense that sub-seabed structure can be determined from remote data such as multibeam echosoundings and seismics. It is also recognized that the characterization of tsunami sediments and their discrimination from other depositional mechanisms is still undergoing review (see Shanmugam 2012), so here the focus is on well-studied examples from well-established mechanisms.

Tsunami sediments

Sediments deposited from tsunamis are mainly recognized on land. Those described from the seabed are rare and their identification and application in tsunami hazard assessment are more controversial (see Shanmugam 2006; Dawson & Stewart 2007), but they are covered here briefly for completeness (Fig. 1). Seabed tsunami sediments are found in deeper, oceanic water depths, include the homogenites in the eastern Mediterranean noted above. There are also shallow-water shelf deposits resulting from tsunami backwash (or backflow). Turbidites in the deep ocean have been well studied (e.g. Heezen *et al.* 1954; Kuenen 1957; Piper *et al.* 1988). Triggered by earthquakes, and in combination with onshore sediments, turbidites can be proxies for large-scale prehistoric earthquake events. Cascadia, on the west coast of the USA, is the best-studied area (Goldfinger *et al.* 2012). Here, turbidites along the oceanic margin were first researched in the 1970s (Griggs & Kulm 1970), with the first attempt to use these to date earthquake cycles in 1990 (Adams 1990) and subsequent research developed by Goldfinger *et al.* (2012). As the distal parts of submarine landslides, turbidites can be used in dating these (e.g. Normark *et al.* 2004). High-resolution age dating of submarine landslides is important in establishing relationships to climate change, which is a major control on sediment failure (Maslin *et al.* 2004) although still poorly understood (Urlaub *et al.* 2013). Backwash flow follows the maximum inundation of tsunami waves, after which the water recedes seawards. Backwash deposits on land are well described (Dawson 1994; Paris *et al.* 2007); however, in the ocean they are poorly researched. Interpretations are speculative because there are no reliable measurements of this process and few recent examples (Fig. 1; Dawson & Stewart 2007; Ikehara *et al.* 2014). The best evidence for backwash flow is from video footage of sediment plumes moving offshore from the Indian Ocean and Japan events (see Tappin *et al.* 2012). The backwash sediment flushed seawards has rarely been studied, but from the few case histories published on shelf deposits its long-term preservation potential is probably low because of reworking by longshore currents (Tappin *et al.* 2012) and storms (Noda *et al.* 2007; Sakuna *et al.* 2012; Feldens *et al.* 2012). Reworking compromises discrimination between tsunamis and storm deposits. On the outer shelf of Japan, the preservation of backwash deposits from the 2011 tsunami is considered likely only over short timescales (Ikehara *et al.* 2014). Nearer shore, the sediment flushed seawards was soon eroded by longshore drift and redeposited on the adjacent coast, where it repaired major coastal breaches (Tappin *et al.* 2012). The use of shallow-water tsunami deposits on the continental shelf in hazard assessment at present is therefore considered too poorly understood to be considered further.

Our imperative here is therefore on coastal deposits, particularly those from recent, historical and Quaternary events. Onshore sedimentary deposits from the older (pre-Quaternary) geological record have been attributed to tsunamis (e.g. Le Roux & Vargas 2005). Where the tsunami mechanism can be established, for example Chicxulub (e.g. Goto *et al.* 2008*b*), deposit origins can be validated; where the mechanism is more uncertain, as for many of the deposits on the west coast of South America, there may be considerable uncertainty (e.g. Bailey & Weir 1933; Spiske *et al.*

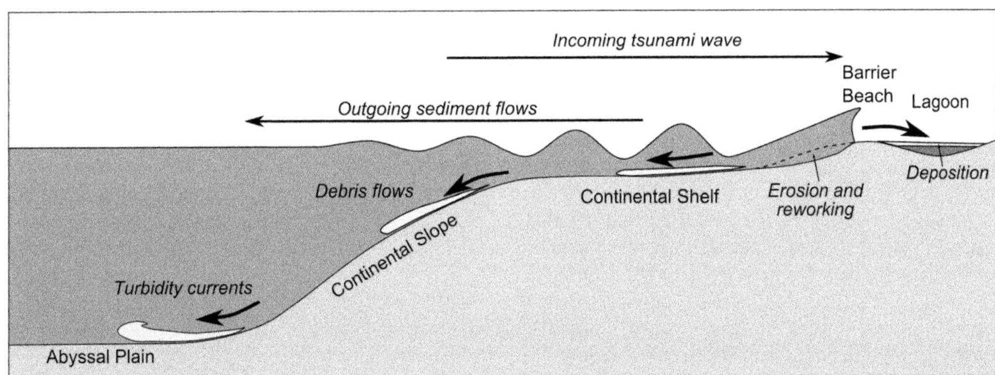

Fig. 1. Schematic illustration of principal pathways of tsunami sediment transport and deposition (Reproduced from MikeNorton; Wikipedia).

2014). The focus here is on those mainly recent, on-land deposits which can improve hazard assessment through: (1) better understanding of tsunami frequency and impact; (2) the validation of numerical models; and (3) as a basis for inverse modelling of tsunami inundation and tsunami-generation mechanisms (Huntington *et al.* 2007). These deposits include both fine-grained sediments and coarse-grained gravel/boulder deposits.

Early history of research (pre-1980)

Written observations of tsunami sediment deposition date back to 1868 from the Arica earthquake off Chile, where the US Postal Steamer *Wateree* (Fig. 2) was carried inland 430 m by a wave 18 m high at the coast and buried under a mass of sand and water (recorded in Myles 1985). The first scientific publication to suggest that tsunamis might be responsible for sediment deposition is Bailey & Weir (1933) on sediments of Jurassic age located on the west coast of Scotland. No further studies of tsunami inundation and sedimentation from historical or recent large earthquakes were published until the 1946 Aleutian tsunami, which struck Hawaii (Shepard *et al.* 1950). The first observational evidence for tsunamis transporting sediment was a series of photographs also from Hawaii from the 1957 earthquake in the Aleutians (referred to in Bourgeois 2009). Observations were reported of sediment deposition from the Chile tsunami of 1960 (Wright & Mella 1963) and the Alaska earthquake of 1964 (Reimnitz & Marshall 1965), but no detailed interpretations made. The 1960 Chile event left a 1–2 cm veneer of sand over the coastal lowlands (Wright & Mella 1963). There are numerous descriptions from Japan (e.g. Kon'no 1961), but these publications are mostly in Japanese so less accessible (Bourgeois 2009). Descriptions of the Suva earthquake and tsunami of 1953 were perhaps some of the first of the modern era to suggest an associated submarine landslide-generated tsunami, which deposited coral boulders on the adjacent reef (Houtz 1962). Probably the first modern 'geologically' focused paper that described tsunami sedimentation was by Coleman (1968).

Seminal events of the 1980–90s

The first geological evidence for sediments deposited from prehistoric tsunamis, which identified their potential for use in mitigation, was from

Fig. 2. USS *Wateree* (1863) beached at Arica, Chile, 430 yards above the usual high water mark, after she was deposited there by a tidal wave on 13 August 1868. Note the tsunami sand in the foreground (U.S. Navy photograph).

North America (Atwater 1987). The sediments were up to 10 ka old and preserved in sand sheets interbedded with marsh muds from the outer coast of Washington State. They were interpreted as deposited from earthquake-generated tsunamis. Recognition of the tsunami origin of the sediments led to the identification of earthquakes and tsunamis of much earlier age than that provided by historical records (up to 200 years old). A year later Dawson *et al.* (1988) described an unusual sand deposit (Fig. 3) within uplifted coastal sediment sequences in Scotland, which they attributed to a tsunami generated from the Storegga submarine landslide off the Norwegian coast (Jansen *et al.* 1987; Bondevik *et al.* 2005*a*). Deposits from Storegga had been described from the Shetland Islands by Birnie (1981), but had not been identified as originating from a tsunami. Along the Washington coast, the only explanation for the sediments was by rapid coastal subsidence from local earthquakes that generated tsunamis. In Scotland, radiocarbon dating of the sediments and Storegga landslide proved a close coeval correspondence. The identification of the deposits in Scotland motivated Harbitz (1992) to simulate a tsunami from a submarine landslide event, the first time that numerical modelling of a recorded submarine landside tsunami had been attempted.

The evidence for the origin of the sands in Cascadia and Scotland was circumstantial because their deposition from tsunamis had not been observed. For the first time the research was also by sedimentologists, unlike observations made at earlier historical events noted above which were by non-experts. Confirmatory evidence for the Cascadia research on the tsunami origin of the sands came first from a historical event in Japan. On the east coast of Honshu Island, sands preserved in lake sediments were also identified as laid down by a tsunami, generated from the 1983 Sea of Japan Earthquake (Minoura & Nakaya 1991). Older, underlying sands were dated as far back as 2700 years BP, confirming the association between tsunami inundation and earthquake subsidence as observed previously in Cascadia. Further confirmation of the Cascadia and Scottish interpretations of tsunami sand deposition was lacking because of an absence of post-1987 tsunami events. A devastating tsunami struck Flores Island in December 1992 killing 2190 people, over half of

Fig. 3. Tsunami sediments (pale grey by the handle of the Nejiri-gama) from the east coast of Scotland generated by the Storegga tsunami of 8.2 ka BP. Nejiri-gama 23 cm long (Photograph, D.R. Tappin).

whom died in the tsunami. The Flores tsunami was one of the first investigated by specifically organized, responsive, field surveys (Yeh *et al.* 1993). It was the first of the modern age to include geologists in the field team (Shi *et al.* 1995). The 1991 Japanese research by Minoura and Nakaya was based solely on core samples, but at Flores field surveys reported on both the geomorphology of tsunami impact and the sediments deposited (Shi *et al.* 1995; Minoura *et al.* 1997). For the first time, there was a direct correlation between an observed tsunami and the sediments deposited. Although the Flores surveys were aimed at understanding the sedimentary processes associated with tsunami inundation, they were also planned as the first attempt to use a recent event to improve the identification and interpretation of older, palaeotsunami deposits (Shi *et al.* 1995).

Scientific objectives

Two primary objectives of sedimentological research emerged from the early tsunami studies. The first was to establish unequivocal sedimentary characteristics of tsunami deposits that would allow their discrimination from other, high-energy depositional mechanisms such as storms (e.g. Shi *et al.* 1995; Dominey-Howes *et al.* 2006). The second was to use these sedimentary characteristics to identify older, prehistoric sediments. Identification of ancient, prehistoric tsunami sediments as a record of older events, together with their reliable age dating, would allow the quantification of tsunami recurrence intervals, improving tsunami hazard and risk assessment. Initially, discrimination of fine-grained tsunami sediments was based on simple criteria such as the extent of deposits, landwards grain-size fining and deposit thinning (fine-grained tsunami sediments were normally graded and storm deposits laminated), and rip-up clasts were significant (Morton *et al.* 2007). Discrimination of coarser-grained, boulder deposits was from imbrication and boulder orientation. There were also the first attempts at mathematical modelling of boulder transport (Moore & Moore 1988; Nott 1997, 2003; Weiss 2012).

Based on the analysis of sediments from more recent events, such as PNG and the Indian Ocean, objectives that are more ambitious were identified. Detailed grain size analysis of tsunami deposits from PNG for the first time were used to model onshore flow depth and speed, from which tsunami size could be quantified (Gelfenbaum & Jaffe 2003). The results were to provide the key for long-term hazard assessments based on tsunami source mechanisms (e.g. earthquake fault slip or submarine landslides) inverted from calculated tsunami wave characteristics. After the Indian Ocean tsunami of 2004, based on the new and developing quantitative approaches developed first for PNG Huntington *et al.* (2007) identified two further key challenges: (1) closing the knowledge gap in linking modern events to their deposits with an improved understanding of tsunami sediment transport; and (2) adapting this relationship to better interpret the geologic record.

Early advances (1990s–2004)

From the early 1990s there was a steady increase in research on tsunami deposits as reflected in the number of peer-reviewed papers published (see Bourgeois 2009, fig. 3.2). The research was from responsive field surveys, which became the norm after the Nicaragua tsunami of 1992 (Satake *et al.* 1993), but also because of continued work on seminal events such as Storegga and in the Hawaiian Islands. The focus was mainly on the convergent margins of the NW Pacific and Japan, and the north Atlantic passive margin. Research in other areas where the hazard is significant, such as New Zealand, was constrained because of the challenges in proving the tsunami sediment provenance (Goff *et al.* 2001). The experience gained from the early studies on Cascadia and Storegga and more recent events (e.g. Flores in 1992) was used to research older historical events including Lisbon in 1755 (Dawson *et al.* 1995), Grand Banks in 1929 (Tuttle *et al.* 2004; Moore *et al.* 2007) and events in New Zealand (Goff *et al.* 2001). In Cascadia, further research identified six major earthquakes over the past 7 ka, three of which at least were associated with tsunamis. During the past 3.5 ka, there were seven extensive tsunamis (Atwater & Hemphill-Haley 1997; Peters *et al.* 2007). At some locations, the record extended back to 14 ka BP, with three events older than 3.5 ka (Peters *et al.* 2007). Research on the Storegga tsunami confirmed previously established relationships between the landslide and the sediments in Scotland. This was based on additional evidence from sediments preserved in elevated lakes in Norway (Bondevik *et al.* 1997*a*, 2005*a*) and in the Faeroe Islands (Grauert *et al.* 2001). At the end of the period 1990–2004, the study of tsunami sediments had expanded from a few publications based on circumstantial relationships between cause and effect (as with Cascadia and Storegga) to the reporting of recent events validated by observations of tsunami inundation. The tsunami record at some locations now extended back beyond historical reporting. Preliminary identification of simple sediment characteristics led to optimism that, with time, absolute criteria would be identified for discrimination of tsunami deposits from other high-energy mechanisms of deposition.

Considering the limited published research, there were a large number of review papers (Bourgeois 2009). New terms, such as tsunamites, were introduced to describe the sediments, although these were later subject to some controversy over their definition (Shanmugam 2006). There were several special journal issues produced (Einsele 1996; Shiki *et al.* 2000), and an edited book was in preparation (later published as Shiki *et al.* 2008). Then the most catastrophic tsunami ever generated struck in the Indian Ocean.

Catastrophic events, advances and controversy: post-2004

The Indian Ocean tsunami led to major advances in tsunami sediment science because it generated a major surge in research. Observations of the tsunami flooding the land confirmed the sedimentary evidence for the tsunami source of the coastal sediments deposited. The Indian Ocean tsunami was the largest recorded event where there was a positive association between inundation and deposition. The responsive tsunami surveys carried out since 1992 (on Nicaragua) provided the basis for the management of the Indian Ocean field surveys which, carried out over the whole region impacted, provided a comprehensive database of the tsunami impact. An even greater surprise than the Indian Ocean tsunami was the Japanese event of 2011. Japan had the most sophisticated mitigation and response strategies of any country in the world in contrast to the Indian Ocean where there was no warning system, yet 19 000 people died. The research on the Indian Ocean tsunami resulted in improved understanding of qualitative depositional mechanisms, whereas the advances from Japan were quantitative. Although the advances from these events (and other smaller events) were significant; the original objectives identified above – tsunami deposit discrimination and its use in identifying older deposits – were not quite as successful as envisaged. Sedimentary characteristics have been identified, but absolute characterization of tsunami deposits has not been achieved (Shanmugam 2012). Coarser-grained (boulder) deposits have been especially difficult to characterize (Morton *et al.* 2006).

Sedimentary characteristics of fine-grained, tsunami sediments are now being used more effectively to estimate flow velocities from which wave magnitudes are derived. Preliminary studies (e.g. Jaffe *et al.* 2011; Sugawara *et al.* 2012; Tang & Weiss 2015) show that inverse modelling of these characteristics can be used to identify tsunami-generation mechanisms (MacInnes *et al.* 2010; Sugawara *et al.* 2012). However, palaeotsunami magnitudes and inflow characteristics derived from inverse numerical models, such as TsuSedMod (Jaffe & Gelfenbuam 2007; Spiske *et al.* 2010), are dependent on the successful discrimination of inflow from backflow deposits (Bahlburg & Spiske 2012). Most studies are of tsunami deposits from earthquakes; there are still only a few descriptions of deposits from submarine landslides (Dawson *et al.* 1988; Bondevik *et al.* 2003; Gelfenbaum & Jaffe 2003). Submarine landslide deposit research is still focused on Storegga and their sediments deposited on the coastlines of the north Atlantic (e.g. Bondevik *et al.* 2005*a*).

Intensive studies of the catastrophic Indian Ocean, 2004 and Japan, 2011 events show that previously proposed 'simple' discriminants described above are not necessarily unique to tsunami sediments; they may also be found in storm deposits. The assumption that tsunami sediments are only deposited from suspension settling (giving fining-upward beds) has given way to the recognition that they also result from bedload (traction) transport (giving laminated sediments), which compromises previous discrimination (Paris *et al.* 2007). Commonly deposited sediments during both the Indian Ocean and Japan events were laminated (from traction currents; Paris *et al.* 2007; Szczuciński *et al.* 2012). In addition, marine microfossils (diatoms and foraminifera), previously considered characteristic of tsunami sediments as in the Indian Ocean (Sawai *et al.* 2009), were rarely found in sediments from the Japan 2011 tsunami (Szczuciński *et al.* 2012). Without observational evidence of tsunami inundation, the identification in isolation of 'absolute' sedimentological criteria that allow identification of a palaeotsunami sediment does not yet seem possible.

Despite the recognition that simple discriminatory criteria are not as 'absolute' as previously proposed (e.g. Morton *et al.* 2007), there have been major advances in characterizing fine-grained sediments and the controls on deposition. Tsunami deposits vary because controls on sedimentation are richly variable. These controls include: (1) coastal and nearshore morphology; (2) the elevation of tsunami waves at the coast; (3) run-up (maximum inland elevation reached by the inundation); (4) the nature and amount of the sediment available for erosion; and (5) the preservation potential of the sedimentary imprint laid down during inundation. As a result, tsunami deposits are complex. There is positive news, however. Whereas there may not be absolute distinguishing sedimentological criteria that can be extrapolated between different locations, at any single location it is possible to discriminate between tsunami and storm deposits where both are present (e.g. Nanayama *et al.* 2000; Goff *et al.* 2004; Tuttle *et al.* 2004; Switzer *et al.* 2005). Research on the sediment history of tsunamis at individual locations can be used in improved understanding of event recurrence and frequency, as in Cascadia and Japan (Atwater & Hemphill-Haley 1997; Minoura *et al.* 2001; Ishimura & Miyauchi 2015). The surveys focused on the catastrophic events in the Indian Ocean and Japan yielded major advances in understanding of the (mainly) fine-grained sediments deposited, but also boulders. With the Indian Ocean tsunami, understanding was improved on characterizing the sediments and investigating their preservation and alteration over time. With Japan, the advances were in the geochemical characterization of the sediments and their use in numerical models (see 'Tsunami sediments in numerical modelling' below). In both instances, the focus was also on characterizing the deposits to provide valid diagnostic criteria to help identify palaeotsunami deposits at specific locations, from which the tsunami hazard at the location is better understood (Jankaew *et al.* 2008; Sugawara *et al.* 2012).

Analysis: new technological developments

Sedimentological analysis of deposits has advanced considerably, but new analytical techniques provide additional methodologies to support classical approaches. The use of geochemical profiling ('toolkits') of recent deposits has advanced considerably (Chagué-Goff *et al.* 2011, 2012). For example, geochemical analysis of soil profiles landwards of the limit of sand deposition now allow the maximum extent of tsunami inundation to be identified (e.g. Goto *et al.* 2011; Chagué-Goff *et al.* 2012). There are constraints with older deposits, however, where the use of geochemistry may be limited because of poor sediment preservation and post-depositional alteration (taphonomy) and reworking by rainfall, wind action, bioturbation and human activity (Szczuciński 2011, 2013). Other developing branches of study on tsunami sediments include the use of heavy minerals in their characterization, anisotropy of magnetic susceptibility (AMS) and X-ray tomography. Analysis of heavy minerals is now being used to discriminate tsunami deposits from other high-energy depositional mechanisms such as storms (e.g. Costa *et al.* 2017). AMS provides the sedimentary fabric of the tsunami deposits from which flow directions are identified. There are still too few events studied, but this technique has been used successfully on sandy deposits in northern Sumatra from the Indian Ocean tsunami (Wassmer *et al.* 2010) and pyroclastic volcanic deposits from the Krakatau eruption of 1883 (Paris *et al.* 2014). X-ray tomography is the most recent development, but has only been used on sandy deposits from the Lisbon tsunami of 1775 (Falvard & Paris 2017). It allows the characterization of grain-size distribution, structures, component analysis and sedimentary fabric of fine-grained unconsolidated tsunami deposits at resolutions down to particle scale. The results are validated by comparing to data obtained using other techniques such as laser diffraction, AMS and X-ray microfluorescence.

Numerical modelling

The study of deposits from the PNG, the Indian Ocean and Japan tsunamis has resulted in improved insights into the processes and forces acting during tsunami inundation based on the hydrodynamic characteristics of the tsunami that control sediment deposition (Jaffe & Gelfenbaum 2002; Spiske *et al.* 2008; Witter *et al.* 2012). Numerical inverse modelling, however, is still at an early stage of development and a more general physical understanding of hydrodynamic processes and their interplay with sediment is required to advance this approach (Cheng & Weiss 2013). For instance, studies of the deposits may be used to assess the tsunami flow velocity and the depth of tsunami inundations. As noted above, however, inflow and backflow deposits have to be reliably identified for the numerical models to be valid. Notwithstanding, tsunami sediments have been used to identify their earthquake source mechanisms. On the west coast of Kamchatka, alongshore distribution of tsunami sediments was used to discriminate between two earthquakes that took place in 1969 and 1971 (Martin *et al.* 2008). A study of the Kamchatka earthquake tsunami of 1952, based on the variable distribution of tsunami deposits, resulted in new models of earthquake slip (MacInnes *et al.* 2010).

The major scientific response to the Indian Ocean tsunami resulted in impacted coastlines being researched for tsunami sediments (e.g. Borrero 2005; Kench *et al.* 2006; Paris *et al.* 2007; Goto *et al.* 2008*a*; Morton *et al.* 2008). The devastating Japan tsunami of 2011 resulted in the acquisition of an extensive and comprehensive dataset of the sediments deposited. It provided another major opportunity to improve understanding of tsunami sedimentation and its use in mitigation (e.g. Goto *et al.* 2011; Szczuciński *et al.* 2012). The response to the coastal inundation was the most intensive of any previous event, with both national and international teams involved in field studies (e.g. Goto *et al.* 2011; Mori *et al.* 2011). From the Japanese field surveys, there were a number of new insights into tsunami sediment deposition: (1) tsunami inundation much greater than the depositional limit of sand; (2) geochemical analysis of muddy sediments that identify tsunami inundation beyond the limit of sand deposition; (3) a minor component of marine microfossils in tsunami sediment; (4) coarse gravel deposits thicker than sands; and (5) varied beach erosion, limited in some areas but intense in others

(Goto *et al.* 2011). These new observations formed the basis for improved understanding of older deposits preserved in the region of the 2011 inundation. One of the most devastating older historic events was the Jogan tsunami of 869. Comparison between the sediments from the two events (Minoura *et al.* 2001) showed the inland inundation of Jogan to be far greater than previously recognized. In addition, from inverse modelling of the deposits the 869 earthquake magnitude was re-evaluated and found to be much greater than previously estimated (Sugawara *et al.* 2013). The result was a major advance in understanding earthquake frequencies. The Japan, 2011 event therefore offered an important opportunity to improve: (1) how inland inundation can be identified; (2) the methodology of using inversion of sedimentary data to establish the 2011 earthquake rupture magnitude and extent; and (3) existing models of older, comparable, events in the same region (Sawai *et al.* 2012; Sugawara *et al.* 2013). The scientific advances were both qualitative (e.g. Goto *et al.* 2011) and quantitative (Sugawara & Goto 2012; Sugawara *et al.* 2014).

Boulder/gravel deposits

Tsunamis deposit individual boulder trains and boulder groups (Goto *et al.* 2007; Ramalho *et al.* 2015) as well as boulders entrained in finer-grained sediment (Paris *et al.* 2004; Yamada *et al.* 2014). The first attempts to understand the hydrodynamic mechanisms of deposition of both types of deposit were by Nott (1997); there are still uncertainties however because boulders are moved differently by storms and tsunamis, resulting in different deposit characteristics (Weiss 2012). Discrimination of tsunami boulder deposits from storm deposits is problematic (Felton 2002; Hall *et al.* 2006; Morton *et al.* 2006), but recent work suggests that tsunamis produce disorganized boulder deposits whereas, storms organize boulders along lines and clusters (Goto *et al.* 2010*b*; Weiss 2012). Individual boulders cast ashore have been used to identify tsunami impact (Goto *et al.* 2010*b*) as well as in validating tsunami-generation mechanisms and their magnitude (Moore *et al.* 1995; McMurtry *et al.* 2004*a*). As with fine-grained sediments, discrimination is possible where both storm and tsunami deposits are preserved at the same location (Goto *et al.* 2010*b*).

Boulder deposits of tsunami origin first described as far back as 1933 from eastern Scotland are coarse-grained conglomerates attributed to earthquake-triggered landslides, which were reworked by an ensuing tsunami (Bailey & Weir 1933). Subsequent research by Pickering (1984) proposed only a fault-scarp origin for the deposits, but surprisingly did not cite the earlier 1933 paper. This is unfortunate, because the opportunity to reconsider the previous interpretation of a tsunami origin in the context of improved sedimentological understanding was missed. Interpretation of other boulder deposits has also been controversial, for example discrimination between recent storm and boulder deposits on the Aruba, Bonaire and Curaçao (ABC) islands in the southern Caribbean (see Morton *et al.* 2006).

Some of the most controversial coarse-grained tsunami deposits are in the Hawaiian Islands. Here, boulders and marine gravels up to 200 m above present sea level (Fig. 4) are attributed to tsunamis from volcanic flank collapse (Moore 1964; Moore & Moore 1984, 1988; Moore *et al.* 1989). The deposits were first interpreted as marine highstands (Stearns & Macdonald 1946; Stearns 1978), then as tsunami deposits (in the Moore papers) and then again as highstands (Rubin *et al.* 2000). The best-preserved deposits are on Lanai Island, and recent interpretations attribute their origin to the last two interglacial highstands at 120 and 240 ka BP (Grigg & Jones 1997; Felton *et al.* 2000; Rubin *et al.* 2000). A problem in interpreting the origin of the deposits is the similarity between the ages of the highstands and the triggering of the landslides. The submarine landslides from volcanic collapse are probably triggered at the end of glaciations, at which time sea levels were rising rapidly (McMurtry *et al.* 1999; Quidelleur *et al.* 2008). The origin of the deposits are therefore quite complex. The importance in establishing the origin of the Hawaiian gravels is two-fold: (1) to prove (or not) the tsunami hazard from the volcanic collapses mapped offshore the islands (Fig. 5); and (2) to validate tsunami run-up elevations from the numerical models of volcanic collapse (McMurtry *et al.* 2004*b*, fig. 3). The controversy over the origin of the deposits (from highstands or tsunamis) compromises their use in identifying the tsunami-generating potential from volcanic collapse.

The basis of the controversy over the interpretation of the Hawaiian deposits is the uncertainty in the elevations of the islands when the sediments were deposited. The flank collapses (and sediments) are thousands to hundreds of thousands years old. The elevations of the Hawaiian Islands have changed since these deposits took place; some have been uplifted whereas others have subsided. Only the Big Island has a well-established vertical tectonic history, which shows subsidence for hundreds of thousands of years (Ludwig *et al.* 1991). The elevation of the sediments at their time of deposition is calculated from their age, present elevation and the known subsidence rate of the island. In the north of the Big Island at Kohala, coralliferous gravels are now at sea level. These deposits

Fig. 4. Tsunami sediments from Lanai Island, Hawaii deposited from large-scale volcanic flank collapse. Note the two coarsening-upwards cycles. Geological hammer 40 cm (Photograph, D.R. Tappin).

are dated at 120 ka BP, the same age as the Alika 2 landslide just offshore (Fig. 5). Alika 2 is the most likely source of the tsunami which laid the deposits down (McMurtry *et al.* 2004*a*). Since deposition 120 ka ago, Hawaii has been subsiding at a rate of 3.4 mm a^{-1}. The elevation of the sediments at time of deposition was therefore ~400 m above present sea level. Numerical tsunami modelling of the failure of the Alika phase-2 giant submarine landslide results in tsunami run-ups of hundreds of metres on Hawaii and Lanai (McMurtry *et al.* 2004*b*, fig. 3). The most recent evidence from the Big Island therefore confirms that the deposits are indeed from a tsunami hundreds of metres in elevation. The fringing coastal (? tsunami) deposits on both Hawaii and Lanai are compositionally very similar. It therefore seems likely that the explanation for the gravels at both locations (The Big Island and Lanai), at least for the youngest deposits dated at 120 ka BP, is a tsunami generated during postglacial volcanic collapse.

Individual boulders without associated finer-grained sediment are common along many shorelines, and have the potential to improve tsunami hazard assessment. As with fine-grained tsunami deposits, however, there is considerable controversy over their discrimination from the other potential depositional mechanisms such as storms, especially where both storms and tsunamis affect the same coast (Hearty 1997; Noormets *et al.* 2002; Mastronuzzi & Sanso 2004; Mylroie 2008; Switzer & Burston 2010; Weiss & Diplas 2015). The elevation of the deposits is an important discriminant (Ramalho *et al.* 2015) but local tectonic history needs to be established to avoid conflict, such as in Hawaii (Rubin *et al.* 2000; McMurtry *et al.* 2004*a*). Apart from boulder elevations, there are geomorphological and sedimentological methodologies proposed that might discriminate between the different mechanisms (e.g. Morton *et al.* 2006; Spiske *et al.* 2008; Goto *et al.* 2010*a*). Unfortunately, there are still too few case studies based on a systematic sedimentological approach to deposit analysis to formulate robust criteria for distinguishing between coarse-clast storm and tsunami deposits (Morton *et al.* 2006).

Submarine landslide tsunami

The potential for submarine landslides to generate tsunamis has been known for over 100 years (Milne 1898; Montessus de Ballore 1907; Gutenberg 1939), yet most research has been on

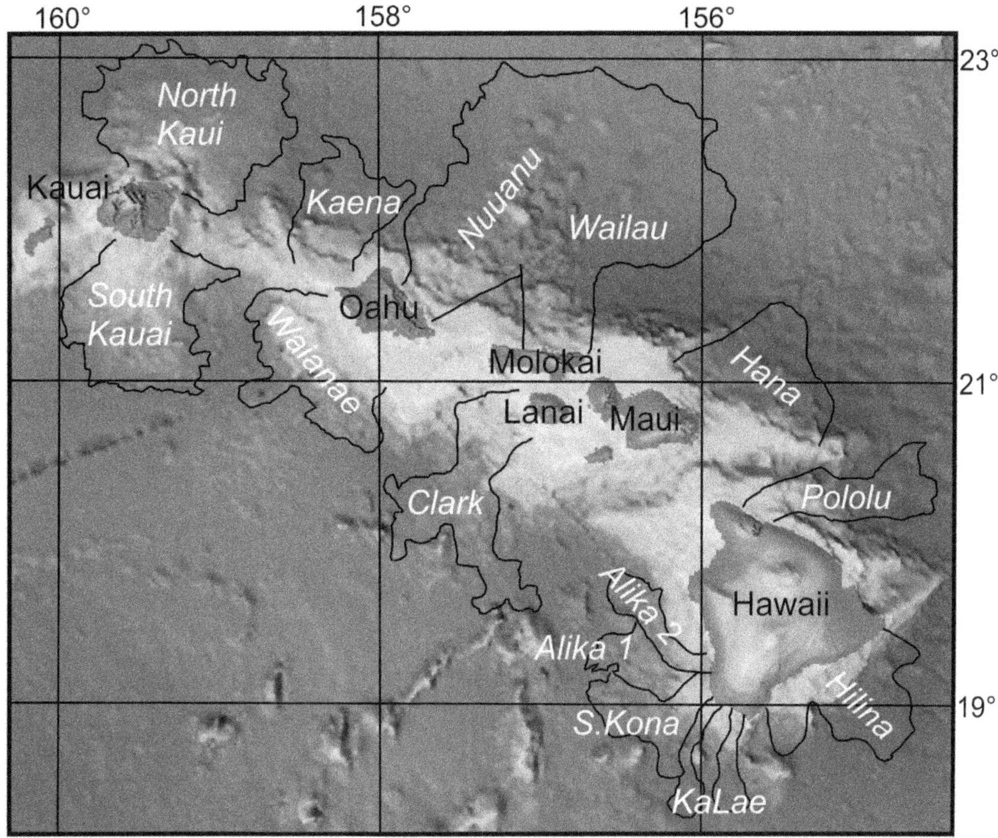

Fig. 5. Hawaiian seabed morphology and locations and extents of the Giant Submarine Landslides as mapped by Moore *et al.* (1989) (From Tappin, 2010*a*).

mechanisms of failure (Piper *et al.* 1988; Talling *et al.* 2007; Masson *et al.* 2009; Talling 2014) and very little published on their potential to generate tsunamis (Masson *et al.* 2006). Significant proven historical submarine landslide-generated tsunamis before 1998 include those of the Grand Banks in 1929 (Heezen & Ewing 1952), Nice in 1979 (Assier-Rzadkiewicz *et al.* 2000; Dan *et al.* 2007) and Skagway in 1994 (Kulikov *et al.* 1996; Rabinovich *et al.* 1999). Both Nice and Skagway landslides were probably triggered by human impact, although the tsunami mechanisms are controversial. Both these events came sharply into focus after the PNG tsunami (Synolakis & Bernard 2006). Other controversial events include the tsunami generated by the earthquake of 1946 in the Aleutians, where the 40 m local run-ups were most probably from a local submarine landslide triggered by the earthquake (Fryer *et al.* 2004; Okal & Herbert 2007; von Huene *et al.* 2014). The 1945 tsunami in the Indian Ocean off Pakistan is a similar event to the Aleutians in 1946 (e.g. Heidarzadeh *et al.* 2008), but no submarine landslide has been found or, indeed, even looked for. The tsunamis generated by the earthquake of 17 August 1999 at Izmit, Turkey are also probably associated with submarine landslides (Altinok *et al.* 2001).

The best-known historical submarine landslide tsunami before PNG was in 1929 on the Grand Banks, Canada but surprisingly, considering the importance of the event, most research has been on the landslide (e.g. Piper & Asku 1987) and earthquake (e.g. Hasegawa & Kanamori 1987) and not the tsunami. Only one paper is published on numerical tsunami modelling of the submarine landslide (Fine *et al.* 2005), but this is based on a theoretical mechanism. The most significant prehistorical tsunami from a submarine landslide is Storegga which, because of the discovery of the Ormen Lange Gasfield, is also the best studied. Research on tsunamis generated from volcanic collapse has identified another submarine landslide mechanism, analogous to clastic sedimentary events. However, volcanic collapse tsunamis, even those in the Hawaiian Islands, have not generated as much interest in their hazard to the same degree by which PNG raised the profile for non-volcanic submarine landslides. The exceptions are the Canary Islands landslides perhaps, which have had a consistently high media profile because of their potential to generate high-elevation, 'megatsunamis' in the far field, which could be a significant hazard to the east coast of the USA (Ward & Day 2001; Gisler *et al.* 2006; Løvholt *et al.* 2008; Hunt *et al.* 2013; Tehranirad *et al.* 2015). A point to note is that the Hawaiian collapses are submarine, whereas those in the Canary Islands are partly sub-aerial. Because of their retrogressive failure mechanisms however, which initiate on the seabed, the Canary Island collapses are included here. The Canary Island collapses are analogous to those of the Hawaiian volcanoes but smaller in volume which, together with their multistage, bottom-up collapse mechanisms proposed for the largest events such as those on Tenerife (Hunt *et al.* 2011), suggest that elevated tsunami wave heights could be more localized and concentrated near to source. Recent numerical tsunami modelling indeed suggests that, in the far field, frictional effects of propogation over the wide US coast shelf significantly reduces their onshore impact, although the volumes used in tsunami generation may be small (Tehranirad *et al.* 2015) and possibly underestimated (Simon Day, pers. comm., 2016).

The impact of Papua New Guinea, 1998

It was not until 1998 in PNG, when over 2200 people died in a tsunami now recognized as generated from an offshore seabed sediment failure (Kawata *et al.* 1999; Tappin *et al.* 1999), that the hazard from submarine landslide tsunamis was fully recognized (Bardet *et al.* 2003; Løvholt *et al.* 2015). The PNG event was the first investigated by a responsive programme of marine surveys (Fig. 6), and scepticism greeted the initial results (e.g. Geist 2000). The landslide architecture was constructed from the marine dataset, and for the first time used as the basis for realistic numerical models of tsunami generation (Tappin *et al.* 1999, 2001, 2008). Although the hazard from submarine landslides is now more generally recognized, the marine surveys required to map the submarine landslide hazard are expensive so only a few countries have so far carried these out to the degree necessary to fully identify the hazard (e.g. Grilli *et al.* 2009; ten Brink 2009; Clarke *et al.* 2014).

How submarine landslides generate tsunamis

Before PNG, simulations of tsunami generation were mainly confined to earthquake sources (Satake *et al.* 1996). Numerical earthquake tsunami-generation

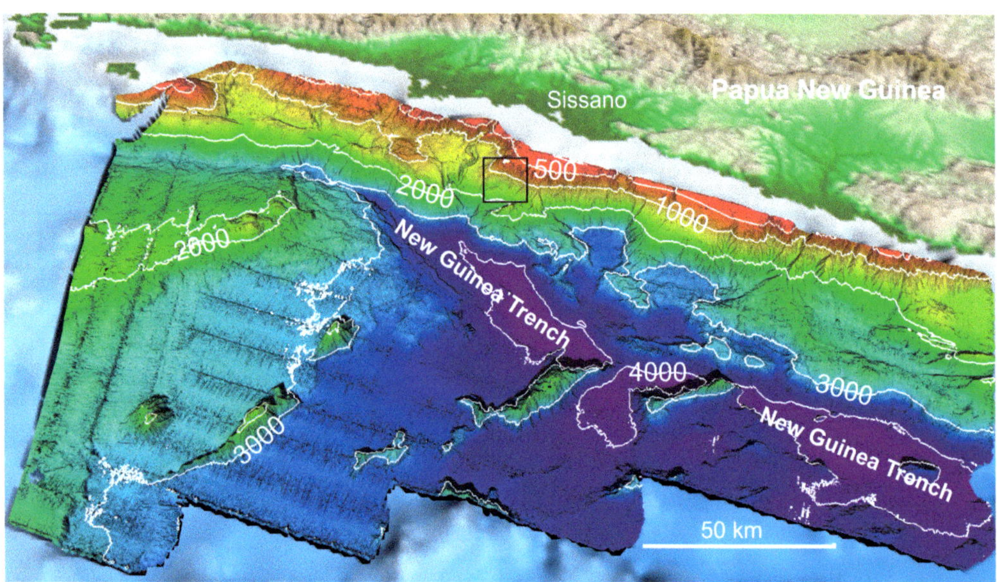

Fig. 6. Papua New Guinea regional bathymetry viewed from the north, showing the convergent margin in the foreground and the Papua land mass to the rear. Depths in metres. Black square the location of Figure 7 (From Tappin, 2010*a*).

models are based on an assumption that the initial water surface deformation is instantaneous and equal to that at the seabed. For the rise time of most earthquakes, the long-wave phase velocity in the ocean is slow enough so that displacement can be considered instantaneous. There are slight modifications to the tsunami wave field for earthquakes of slow rupture duration (tsunami earthquakes). Seabed deformation is calculated from earthquake fault parameters using theoretical deformation models such as Okada (1985). Underwater landslides were considered to be ineffective at generating significant tsunamis because of their longer source-generation times, smaller areas of seabed disturbance (compared to earthquakes) and the directivity of the tsunami produced (Hammack 1973; LeBlond & Jones 1995; Geist 2000). Before the PNG event, one of the major challenges was in understanding how the relatively slow-moving submarine landslides generate tsunamis.

A further complication in understanding tsunamis generated by submarine landslides is the range of failure mechanisms which vary according to morphology, sediment type and/or kinematics (see Hampton *et al.* 1996; Turner & Schuster 1996; Keating & McGuire 2000; O'Grady *et al.* 2000). Theoretical numerical modelling of submarine landslides was visualized as a Bingham-type fluid flow, analogous to a translational mechanism, where large blocks disintegrated on travelling downslope to form turbidites (Hampton 1972; Geist 2000). Modelling of solid block landslides at the time of the PNG tsunami was in its infancy (Watts 1998). Submarine landslide failure is dependent mainly on sediment composition, which controls landslide morphology and kinematics. Numerical tsunami-generation models were initially based on depth-averaged wave equations that represented immiscible liquids or water as a Bingham plastic (e.g. Jiang & LeBlond 1992, 1994). While depth-averaging accurately applies to tsunami generation from earthquakes, it is questionable when applied to landslide tsunamis because it does not allow for vertical fluid accelerations, important during submarine landslide motion and tsunami generation (Grilli *et al.* 2002). In 1998, landslide constitutive equations used in modelling were largely untested by laboratory experiments or case studies (Tappin *et al.* 2008). Submarine landslide models were idealized, and not based on geological data. There was no established method of merging geological data with numerical landslide models. In total, there was little appreciation of the complexity of modelling tsunamis generated by the different submarine landslide mechanisms. All this was to change because of two major events: one prehistoric and thousands of years old, Storegga; and the other recent and devastating, Papua New Guinea in 1998.

Submarine landslide numerical models

Storegga (8.5 ka BP). The first realistic attempt at numerically modelling a submarine landslide based on a slide architecture from seabed morphology was Storegga (Harbitz 1992). Validation of the tsunami generated was from run-up recorded in coastal sediments deposited on the east coast of Scotland (Fig. 3; Dawson *et al.* 1988; Long *et al.* 1989) and uplifted lake sediments in the west of Norway (Svendsen & Mangerud 1990). The landslide modelled was translational, moving at velocities of 20–35 m s^{-1} which were taken from measurements of the Grand Banks tsunami of 1929 (Heezen & Ewing 1952). Three major slide events were modelled: the first and third as partially liquefied debris flows; and the second retrogressive failing from the bottom upwards. Individual slide volumes were between 1700 and 3880 km^3. Average slide thicknesses used were between 88 and 114 m. The recorded run-up heights of 4 m on the east coast of Scotland were best reproduced by slide velocities of 35 m s^{-1}. The 1992 paper was a benchmark in tsunami numerical models as it was based on both a realistic landslide model and run-up data measured from sediments deposited on land. The numerical model was a major advance at the time, because only a few earthquake tsunamis had been simulated and the controls on tsunami generation by landslide architecture were not recognized. Submarine landslide tsunamis were only identified along two ocean margins: Storegga (Norway) and Grand Banks (Canada). In 1992, there were no numerical models of the Grand Banks tsunami.

Subsequent numerical models of Storegga post-dated the PNG event (Bondevik *et al.* 2005*a*; Hill *et al.* 2014) and made significant improvements on the 1992 results of Harbitz as they were based on a more comprehensive dataset of geophysics and coring of the landslide (Bryn *et al.* 2005). The motivation to investigate Storegga was not the 8.2 ka BP tsunami, but to ensure that exploitation of the underlying Ormen Lange gas field would not create another hazardous submarine landslide similar to the 8.2 ka BP event (Bryn *et al.* 2005). Validation of later numerical models was based on a more extensive dataset of tsunami run-up data from Norway (Bondevik *et al.* 1997*a*, *b*), Faroe Islands (Grauert *et al.* 2001), Shetland Islands (Bondevik *et al.* 2003, 2005*b*) and mainland Scotland (Smith *et al.* 2004, 2007). With the later numerical models based on improved landslide architecture, maximum tsunami run-ups on the Shetlands increased to 20 m which agreed with new studies on the elevations of tsunami sediments on the islands (Bondevik *et al.* 2003). Sea levels were much lower when the Storegga landslide took place (8.2 ka BP) than at present. The sea-level curve for the Shetlands is still uncertain, so these run-up elevations are most probably still underestimated.

The later numerical models of the Storegga tsunami also used a more comprehensive geotechnical and morphological dataset from the landslide (Forsberg 2002; Haflidason *et al.* 2005). These confirm that the Storegga failure mechanism was retrogressive, with large block failure initiated at the base of the slide at a water depth of *c.* 1000 m. The time lag between the individual block failures as the slide retreated landwards is a major control on tsunami generation (Bondevik *et al.* 2005*a*; Løvholt *et al.* 2005). Tsunami run-up elevations from sediments preserved at the different locations constrains the timing of block slide development. The shape and volume of the modelled slide used by Harbitz (1992) were adjusted until they fitted the new and much more detailed slide reconstruction. Modelling of Storegga was not just to simulate the tsunami, but also to establish the slide mechanics and triggering (Kvalstad *et al.* 2005). In the revised slide model the maximum thickness (400 m) of the slide is near the upper headwall, gradually becoming thinner towards the slide front in the offshore direction. The slide volume generating the tsunami is estimated to be 2400 km^3. The Storegga slide is probably the best studied of any tsunami from a translational-type failure mechanism.

Papua New Guinea tsunami (1998). The PNG tsunami struck during the initial development of the Ormen Lange gas field in 1998. PNG was an important precursor to the Indian Ocean tsunami of 2004 because the loss of life was greater than from any previous recent event. It resulted in a major revision in understanding of tsunami-generation mechanisms from submarine landslides; in this context it is second only in importance, although the generation mechanisms are different. Identifying the mechanism of the PNG tsunami required marine geophysical and geological data and, as with tsunami sediments, it utilized the expertise of geologists (Tappin *et al.* 1999, 2001, 2008). PNG also provided an important context to the identification of tsunamis from tsunami earthquakes (Kanamori 1972), where the associated tsunami is much larger than expected from the earthquake magnitude. It showed that although an earthquake might be small it might not be slow, and the tsunami mechanism could alternatively be from a submarine landslide. The discrimination between tsunamis generated by 'tsunami' earthquakes and submarine landslides is still uncertain, as initially suggested by Kanamori (1972) and further supported by events such as Java in 2006 (e.g. Kanamori 1972; Fritz *et al.* 2007).

The earthquake magnitude of the PNG event was small in comparison to the 10–15 m high tsunami that devastated the coast around Sissano Lagoon. It was not a 'tsunami' earthquake based on Newman & Okal's (1998*a*, *b*) discriminant of E/M0 (E/M0 is the ratio between high-frequency energy E and low-frequency seismic moment M0), because it did not have the required 'slow' source characteristics. Several other aspects of the event also suggested the earthquake was not responsible for the tsunami. The aftershock distribution indicated a shallow-dipping thrust (McCue 1998; Hurukawa *et al.* 2003) rather than a steeply dipping event necessary to generate the recorded tsunami. The peaked run-up distribution along the coast suggested a local focused source rather than the broad source usually associated with an earthquake tsunami. The 20 minute time lag between the felt earthquake and the tsunami striking the coast did not agree with an epicentre located just offshore. The tsunami in the far field was very small and undoubtedly generated from the earthquake (Kikuchi *et al.* 1999), but it seemed unlikely that the local tsunami was from this mechanism.

All evidence therefore initially converged on a submarine landslide mechanism located close offshore. This conclusion was highly controversial (e.g. Geist 2000), so marine hydroacoustic and sampling surveys were required for confirmation. For the first time after a major tsunami, responsive marine surveys funded by Japan and the USA acquired a comprehensive hydroacoustic dataset with some sediment sampling off northern PNG (Tappin *et al.* 1999, 2001; Sweet & Silver 2003). The hydroacoustic data comprised over 19 000 km^2 of multibeam bathymetry (Fig. 6), 4.2 kHz high-resolution, sub-bottom seismic lines (SBSL), and both single (SCS) and multichannel seismic (MCS) data. In the region of the landslide, four 7 m long sediment piston cores were recovered together with numerous shallow (30 cm) push cores of sediment, rock samples and marine organisms. Still and video photography of the seabed were acquired from a tethered remotely operated vehicle (ROV) and a manned submersible (MS).

Before the surveys, a translational landslide was proposed as the most likely tsunami mechanism (e.g. Geist 2000) and the first published numerical models were based on this mechanism (e.g. Heinrich *et al.* 2000). These numerical models successfully reproduced the recorded on-land tsunami elevation data, with the translational landslide treated as a fluid-like flow of a cohesionless granular material (Heinrich *et al.* 2000). From the marine surveys, however, this translational mechanism was shown to be unrealistic because seabed morphology and seismic data revealed a rotational slump (Fig. 7; Tappin *et al.* 1999). Theoretically, slumps by volume generate the largest tsunami (Tappin *et al.* 2008). At the time of the PNG tsunami, theoretical

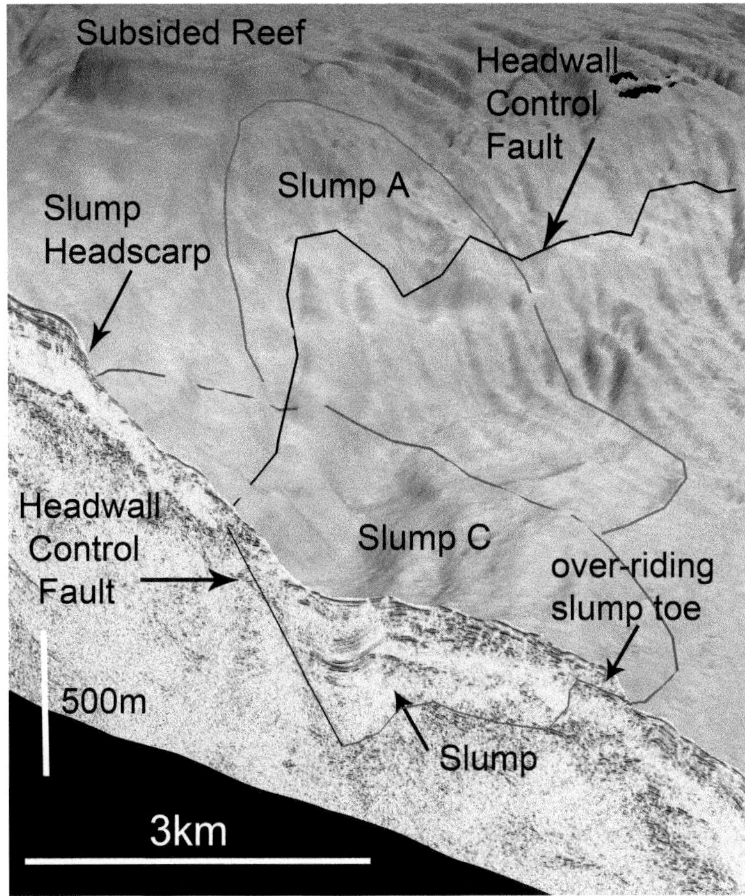

Fig. 7. 3D cutaway image of the PNG slump C that generated the 1998 tsunami, including a seismic section, viewed from the NE. Vertical exaggeration ×3. Location shown in Figure 6 (from Tappin *et al.* 2008).

numerical models of tsunamis were based on: (1) sliding blocks (e.g. Watts 1997; Grilli & Watts 1999); (2) finite volume discretization–volume of fluids (VOF) of the Navier–Stokes equations (Heinrich 1992); or (3) deformable landslides, with the generated waves governed by the finite volume discretization (Jiang & LeBlond 1992, 1994).

To model the tsunami from the rotational slump, new numerical models were developed based on rotational failure with travel distance being limited (800 m) (Watts *et al.* 2003). The results showed that non-linear and dispersive tsunami propagation models were necessary to model submarine landslide tsunamis, with the shape and motion of a realistic submarine landslide wavemaker defined from marine survey data. This is unlike earthquake tsunamis where numerical wavemaker models are based on ground deformation from fault slip, derived from inversion of seismological, geodetic or tsunami observations (Okada 1985). To identify the landslide mechanism at PNG (see Tappin *et al.* 2001) multibeam echosounder (MBES) technology was used to map the detailed seabed morphology, the first time this had been attempted. The MBES data, combined with sub-seabed seismic, led to the construction of the submarine landslide architecture (Fig. 7) that underpinned the numerical wavemaker models. In addition, whereas earthquake rupture models are mainly defined by strike, dip and rake, the failure mechanisms of submarine landslides are many and varied as they are dependent on the nature of the sediment. As well as the differences in initial tsunami wave generation between earthquakes and submarine landslides, there are also significant differences in how their tsunamis propagate. Coseismic displacement from vertical seafloor deformation usually generates tsunamis with longer wavelengths and periods than those generated by landslides, because of their larger source area (Hammack 1973; Watts 1998, 2000). Coseismic displacement generates tsunami amplitudes that correlate with earthquake magnitude (Hammack 1973; Geist 1998) except for tsunami earthquakes; submarine landslides produce tsunamis with amplitudes limited only by the vertical extent of centre of mass motion or the water depth (Murty 1979; Watts 1998).

The first numerical simulation of the PNG slump was devised at sea during the first survey in 1999; it was rudimentary, with many assumptions not validated. Initial tsunami-generation estimates were computed by hand, based on published (or soon to be published) literature (Watts 1998, 2000; Grilli & Watts 1999). The slump architecture was provisional because it was based only on the bathymetric data acquired during the survey (Tappin *et al.* 1999). The tsunami source used was a solid block two-dimensional (2D) underwater landslide (Grilli & Watts 1999). It did not use depth-averaging; instead, it solved fully non-linear potential flow (FNPF) equations which allowed for vertical water acceleration. However, the tsunami propagation simulation used linear shallow-water wave equations. The maximum wave height of 6 m generated by the model was located offshore of the sand spit at the 10 m water depth contour. Although approximating the relative distribution of run-up along the coast, the maximum offshore water height was not of the same magnitude as run-up measured on land by the responsive surveys. Despite these shortcomings, the results were a major advance; not only was it the first time that a landslide-generated tsunami was modelled from a real event, but the modelling was based on MBES data. Comparison with the alternative earthquake-generation mechanism, which gave a maximum wave height at the shore of 2 m based on a shallow-dipping rupture mechanism, demonstrated that it was the slump rather than the earthquake that generated most of the local tsunami.

Subsequently, numerical models improved with the definition of slump architecture (Fig. 8) (Watts *et al.* 1999; Tappin *et al.* 2001). Most recent numerical modelling (Tappin *et al.* 2008) is from a slump modified from the marine survey data that included seismics (Fig. 7). Numerical modelling used an initial condition (wavemaker) using 'Tsunami open and progressive initial conditions system' (TOPICS) software that provides the vertical landslide displacements as outputs, as well as a characteristic tsunami wavelength λ_0 and a characteristic tsunami period T_0. To account for the dispersive nature of landslide tsunamis, the Boussinesq propagation models GEOWAVE (Watts *et al.* 2003) and the later development FUNWAVE (Tappin *et al.* 2008) were used. The initial numerical models provided tsunami wave elevations offshore, not on-land run-ups, and the later modelling provided tsunami wave elevations at the coast. This was a significant improvement over earlier simulations using tsunami source and (non-dispersive) shallow-water, wave tsunami propagation models (see discussion in Tappin *et al.* 2008).

Raised awareness after PNG 1998 event

The great loss of life from the PNG tsunami and the demonstration of the non-earthquake mechanism from numerical models were a catalyst to an intense period of research on how submarine landslides generate tsunamis (Synolakis & Bernard 2006). Recognition of the submarine landslide hazard in tsunami generation resulted in increased awareness of these events but also that there were no comprehensive models that covered all aspects of landslide-induced tsunamis from source mechanism, through propagation to coastal inundation. Before the identification of the submarine landslide mechanism of the PNG tsunami, understanding of the mechanics of landslide tsunamis was lacking. After the PNG event, there was a re-evaluation in the USA of other possible submarine landslide events such as Palos Verdes, California (Fig. 9) where the under-prediction of the height of the leading wave led to the dismissal of the local tsunami hazard. PNG also led to the resolution of the dispute concerning the landslide trigger of the 1994 Skagway, Alaska tsunami (Synolakis & Bernard 2006). Marine geophysical data now reveal that submarine landslides are common along most continental margins (Fig. 9), especially those of California, Oregon and the east Coast of the USA. As a result, the level of hazard posed by relatively moderate earthquakes and submarine landslides was re-examined (e.g. Borrero *et al.* 2001, 2004). There was an increased recognition, by both the scientific research and tsunami forecasting communities, that earthquakes affecting oceanic margins frequently trigger submarine landslides (e.g. Geist *et al.* 2009; Grilli *et al.* 2015). Although many of these might not be tsunamigenic in the far field, they had the potential to generate significant local tsunamis, even if the earthquake magnitude was small and not of a sufficient magnitude to generate a significant coseismic event.

Once the landslide mechanism became evident, one of the first responses to the PNG tsunami was the workshop on landslide tsunamis sponsored by the National Science Foundation at the University of Southern California on 10–11 March 2000. The resulting special publication (Bardet *et al.* 2003) recognized that landslide tsunamis require multidisciplinary studies that build upon experience in engineering seismology, geotechnical engineering, marine geology, modelling of sediment deposition

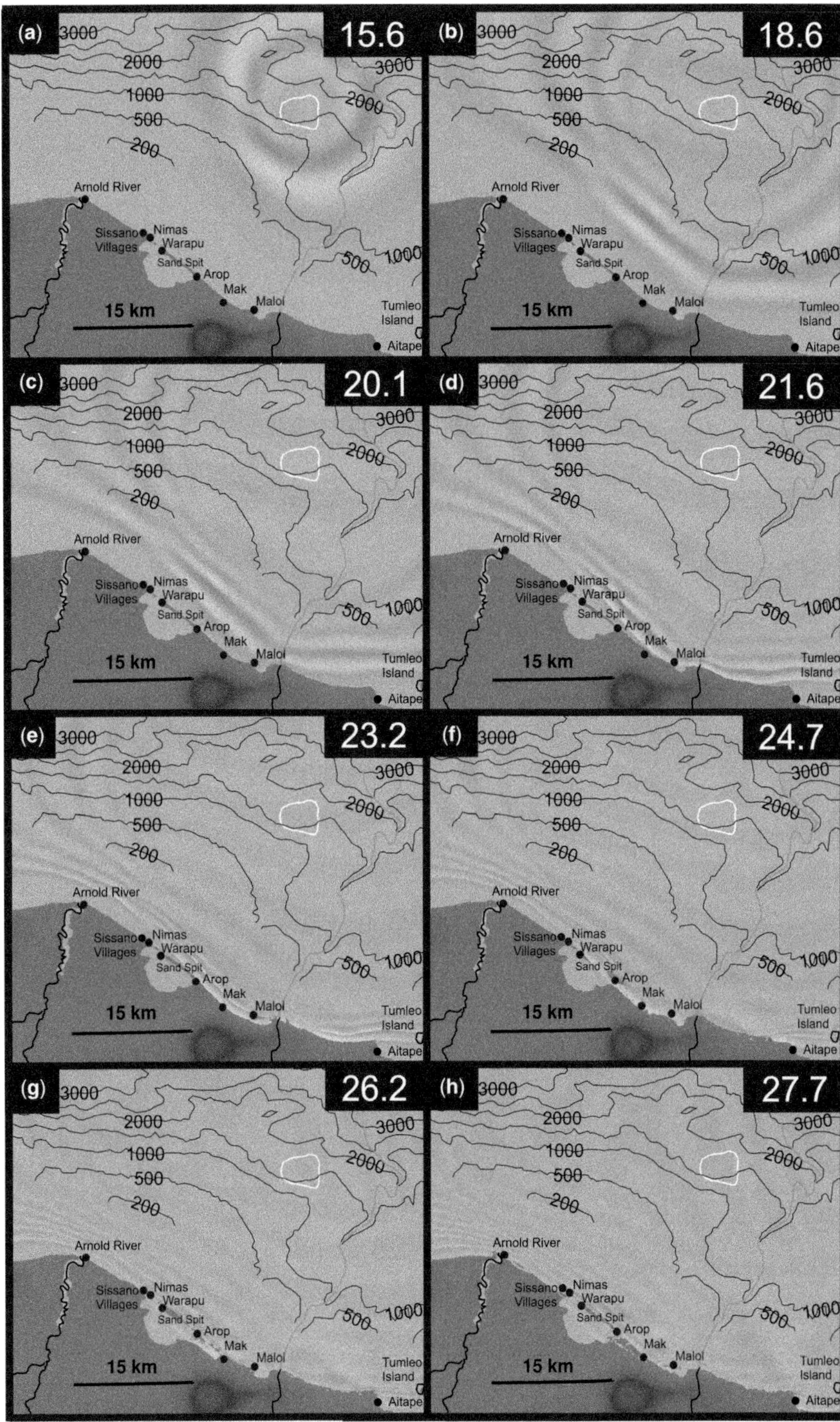

Fig. 8. Eight snapshots of Papua New Guinea, 1998 tsunami propagation and inundation from a slump source. Light blue are elevation waves and dark blue are depression waves. Numbers in the top right of each image are the tsunami propagation times after the main earthquake shock trigger. The slump location is in yellow (from Tappin *et al.* 2008).

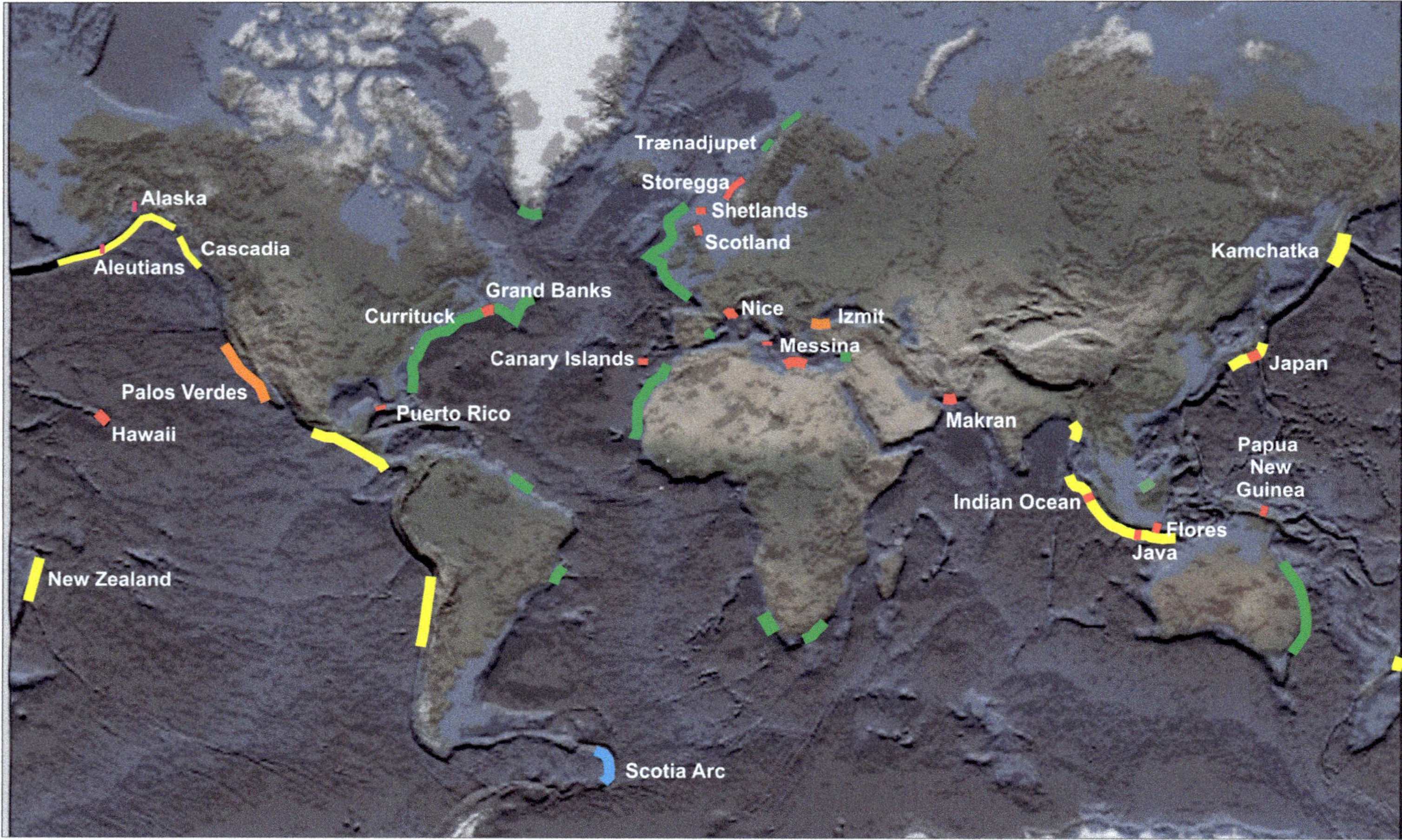

Fig. 9. Global distribution of mapped submarine landslides. Green lines: landslides on continental shelves and fan systems. Yellow lines: landslides located along convergent margins. Red lines: locations of landslide-sourced tsunamis, or where there may be a landslide contribution. Pale blue lines: volcanoes. Orange lines: landslides along strike-slip margins. Pale yellow lines: landslides along fjord margins. Significant landslides named.

and run-off, and hydrodynamics. Another special publication (Tappin 2007) focused on tsunami sediments. In approximately 35% of all tsunami events studied, nearshore waves from landslide tsunamis could exceed those resulting solely from coseismic ground motions (Watts 2004). In the USA, submarine landslide research programmes resulted in a special issue of the journal *Marine Geology* focused on the tsunami hazard along the east coast of the US (Chaytor *et al.* 2009; Geist & Parsons 2009; Grilli *et al.* 2009; Lee 2009; ten Brink 2009; ten Brink *et al.* 2009; Twichell *et al.* 2009).

Later tsunami landslide research made some significant counterintuitive discoveries. One of the most fascinating was that along the passive margins of the North Atlantic (Fig. 9) the majority of potentially tsunamigenic submarine landslides were found to occur on seabed slope angles of less than 5°, with some of the largest slides failing on slopes of less than 1° (Hühnerbach *et al.* 2004). This evidence on slope failure resulted in considerable conjecture on how submarine landslides fail (Canals *et al.* 2004; Smith *et al.* 2013; Talling *et al.* 2014). Subsequently, there has been a large amount of physical and numerical modelling work devoted to studying tsunamis generated by submarine landslides (e.g. Heinrich 1992; Grilli & Watts 1999, 2001, 2005; Watts 2000; Tinti *et al.* 2001; Ward 2001; Grilli *et al.* 2002; Lynett & Liu 2002; Enet *et al.* 2003; Watts *et al.* 2003, 2005; Locat *et al.* 2004; Enet & Grilli 2005, 2007; Fine *et al.* 2005; Haugen *et al.* 2005; Liu *et al.* 2005; Abadie *et al.* 2012; Ma *et al.* 2013, 2015; Løvholt *et al.* 2015; Smith *et al.* 2016).

Improved understanding after Japan 2011 event

Up until March 2011, PNG remained the only proven catastrophic tsunami generated by a submarine landslide. There were suspicions of a submarine landslide contribution to other events including Java, 2006 (Fritz *et al.* 2007) and the highly focused tsunami at Riangkroko, Flores Islands in 1992, where coastal (sub-aerial) landslides were the probable cause (Yeh *et al.* 1993). Without marine surveys at these locations, their proposed submarine landslide tsunami-generation mechanisms remain unresolved. It was not until the Japan tsunami of 11 March 2011 that exceptionally high (40 m) and focused run-ups along the Sanriku coast on northern Honshu Island, north of the main earthquake rupture, suggested another submarine landslide event in addition to the magnitude M_w 9 earthquake (Tappin *et al.* 2014). Two important aspects of this event suggested a second tsunami mechanism in addition to the earthquake. Firstly, the numerical tsunami simulations from earthquake mechanisms could not reproduce the elevated (40 m) wave elevations recorded along the coast between latitudes 39° 30′ and 40° 15′ N (e.g. Fujii *et al.* 2011). Secondly, even when inverting tsunami waveforms and using dispersive-wave Green's functions, the simulations could not satisfactorily reproduce the timing and high-frequency content of tsunami waveforms recorded at the nearshore GPS buoys located in this area, nor the timing and dispersive-wave train at the Deep-Ocean Assessment and Reporting of Tsunamis (DART) buoy #21418 located 600 km off the coast (Gusman *et al.* 2012; Iinuma *et al.* 2012; Løvholt *et al.* 2012; Romano *et al.* 2012; Grilli *et al.* 2013; Satake *et al.* 2013; Yamazaki *et al.* 2013). These deficiencies were identified by comparison of the tsunami mechanisms from ten earthquake source models, obtained by inverting seismic and geodetic data and tsunami waveforms (see MacInnes *et al.* 2013, fig. 4). This comparison found that none of the mechanisms satisfactorily reproduced the elevations of the recorded run-ups on the Sanriku coast north of latitude 39° 00′ N.

Marine hydroacoustic data of the type previously used to identify seabed failure elsewhere (such as PNG) were available in the region of the Japan earthquake; the area off Honshu Island had been mapped by MBES both before and after the tsunami. From these data, and a limited number of seismic lines, the presence of submarine landslides offshore of the elevated Sanriku run-ups was confirmed (Fig. 10). In addition, in the region of the earthquake there were offshore bottom sensors which recorded the frequency content of the tsunami waveforms. This was the first time that these marine seabed data were available to record a tsunami generated by a large-magnitude earthquake. Analysis of these data provided the potential to discriminate between different tsunami mechanisms, because the wave frequency content of tsunamis from earthquakes and submarine landslides are quite different. Since the wave frequency content of tsunamis from earthquakes is much lower than from submarine landslides, it could be used to identify the submarine landslide location and to validate the numerical tsunami models.

Although a submarine landslide was the most likely second tsunami mechanism, an alternative was out-of-sequence (splay) faulting. There was no evidence for the splay faulting in the published seismic profiling data for the source region (Tsuru *et al.* 2002); however, there was considerable evidence supporting a submarine landslide. Large slumps had been found on the margin of the accretionary prism off the Sanriku coast (Cadet *et al.* 1987; von Huene *et al.* 1994; Tsuru *et al.* 2002). Data from two ocean bottom pressure gauge stations

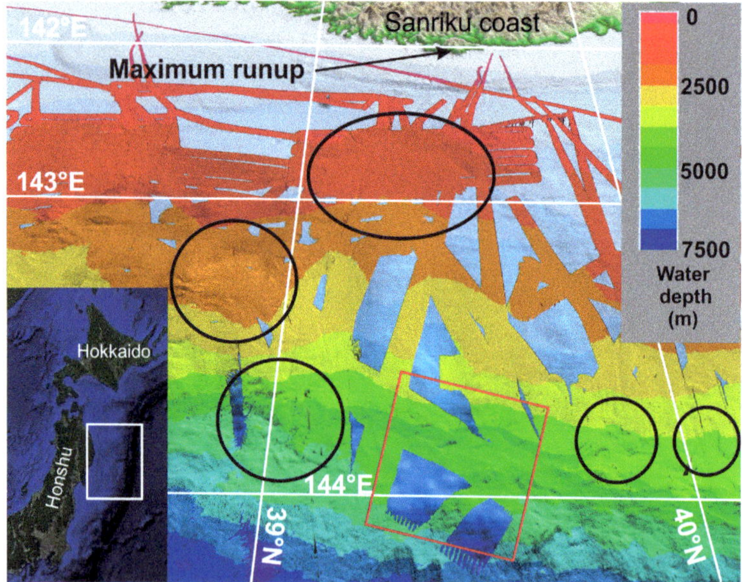

Fig. 10. Japan, 2011. Submarine landslides (black circles/ellipses) in the region off northeastern Honshu Island from bathymetry, viewed from the east. Red square: the location of the landslide triggered by the March 2011 earthquake. Highest elevation observed tsunami run-up/inundation (around 39.5° N) along the Sanriku coast is also marked. Approximate location of this figure shown in inset (From Tappin *et al.* 2014).

TM1 and TM2 off Kamaishi (latitude 39° 12′ N) indicated that at least part of the tsunami source in this region is a narrow area in the deeper part of the Japan Trench (Maeda *et al.* 2011). Seabed movement had been identified in the area of earthquake rupture from before and after bathymetry (Kawamura *et al.* 2012), but this was south of the region of elevated onshore run-ups. Backward ray tracing using the higher-frequency, leading elevation wave from the tsunami recorded by the seabed buoys located offshore of the region of high run-ups identified the most likely location of a submarine landslide that could have generated these. Within this area, north of that identified by Kawamura *et al.* (2012), a number of submarine landslides were identified from bathymetry acquired from before and after the earthquake (Tappin *et al.* 2014). There were few seismic data in this region and none at the location identified by the ray tracing as the potential additionary tsunami source, so sub-seabed structure was not available to confirm this interpretation. A slope stability analysis confirmed that earthquake shaking could have triggered the landslide. Numerical modelling of a dual earthquake and submarine landslide tsunami mechanism (Fig. 11) demonstrated that in combination they reproduced the waves recorded along the Honshu coast, especially in the north of the inundated area in the Sanriku region (Tappin *et al.* 2014).

Undefined hazard

Extensive mapping of continental shelves reveals the common presence of submarine landslides, although large regions remain unmapped (Fig. 9). At present however, only four significant submarine landslide tsunamis have been mapped, modelled and validated: Storegga, Grand Banks, PNG and Japan (two of which have led to a significant loss of life). For volcanoes, only the flank collapse of Alika 2 on the Big Island of Hawaii is well studied; the tsunamis from the Canary Island volcanoes remain controversial.

Regarding passive margins, the best-studied region is the North Atlantic (Fig. 9). Here, the general controls on landslide failure are related to climate-controlled influences on sedimentation during glacial and interglacial periods (e.g. Lee 2009; Talling *et al.* 2014), with earthquakes being the most likely triggering mechanism. The specific relationships between climate and landslide failure are still far from clear, however (Urlaub *et al.* 2013; Talling *et al.* 2014). Storegga is well studied in the NE Atlantic, and the associated tsunami was elevated along the surrounding coasts and geographically extensive. There is still no evidence for tsunamis from the other large-volume landslides off Norway (such as Trænadjupet; Fig. 9) located north of Storegga. Further south in the Shetlands

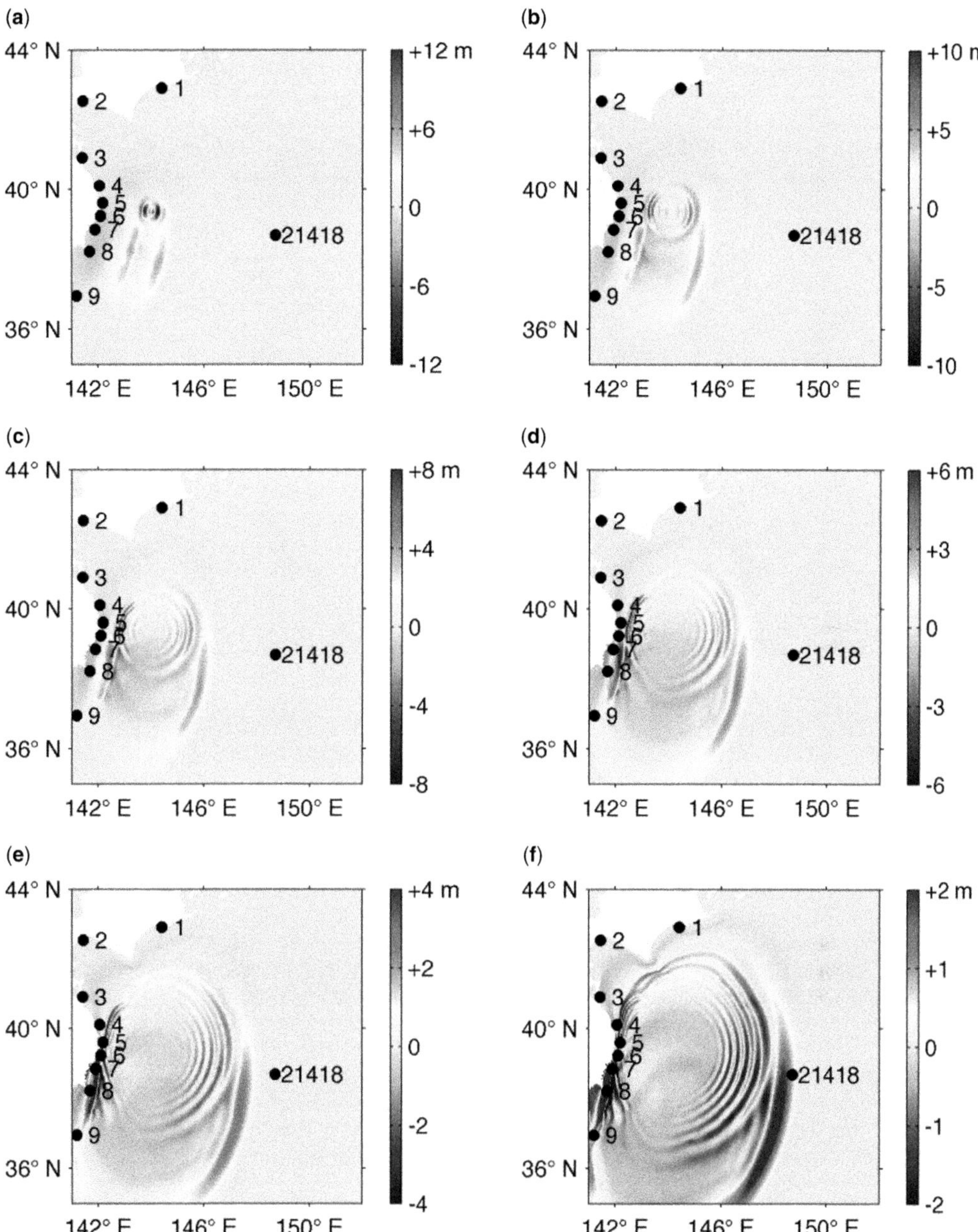

Fig. 11. Japan tsunami, 2011. Numerical simulation of the 2011 tsunami using a dual earthquake and submarine landslide source mechanism, showing instantaneous surface elevations at time (**a**) 5, (**b**) 10, (**c**) 15, (**d**) 20, (**e**) 25 and (**f**) 30 min. Labelled black dots mark the locations of GPS buoys and of DART buoy #21418. Note the highly dispersive nature of waves generated by the SMF source to the north, as compared to the longer-wavelength, long-crested, non-dispersive earthquake-generated tsunami waves to the south (From Tappin *et al.* 2014).

(Fig. 9) there are tsunami sediments much younger than Storegga, dated at 1.5 and 5 ka BP (Bondevik *et al.* 2005*b*), but there is no obvious tsunami mechanism for these events except the possibility of submarine landslides. The most likely location for the landslides is off Norway, for example Trænadjupet where two events are recognized, but the ages of these were dated at 3–5 ka BP and 19–22 ka BP

(Laberg & Vorren 2000), that is, different from the Shetland events. The mechanisms for the Shetland sediments therefore seem to be more local, but are yet to be found.

To address the submarine landslide hazard in the North Atlantic a major UK initiative was funded by the Natural Environment Research Council, with the focus on high-latitude ocean warming and its potential consequences on seabed slope stability. A marine programme acquiring hydroacoustic and sediment core data on the landslides off of Norway reappraised slope failure mechanisms and their ages, addressing some of the deficiencies identified by Urlaub *et al.* (2013) in the relationships between slope failures and their climate controls. Based on new data and numerical models, a further result of the project is a new numerical tsunami model of Storegga by Hill *et al.* (2014) which is the first to address the palaeobathymetric effects of lowered sea level at time of failure.

In the USA, scientific programmes to determine the tsunami hazard from submarine landslides focus on the passive margin of the east coast and Gulf of Mexico where there are large population centres and nuclear power plants (see special issue in *Marine Geology*, eight articles of which present new research; ten Brink 2009; ten Brink *et al.* 2009). Numerous submarine landslides have been mapped off the east coast of the USA (Fig. 9), but few have been well studied (Lee 2009). One of the best-researched and largest slide is Currituck, dated at 25–50 ka although it could be much younger (J. Chaytor pers. comm., 2017), which has been numerically modelled for tsunami generation (Grilli *et al.* 2015). Based on a slide volume of 134 km^3 and a rigid block failure, failure of Currituck would generate a local tsunami up to 5–6 m high, highlighting the hazard along this coast. It is likely that many of the passive margin landslides mapped on Figure 9 generated tsunamis, but there is no historical or geological evidence preserved that records their impact. Even without this evidence, the presence and distribution of such a large number of submarine landslides, some of significant volume, identifies a potential risk from their associated tsunamis (although statistical analysis suggests the risk is small; Grilli *et al.* 2009). Further research is required because the coast is densely populated and nuclear power plants are present (ten Brink *et al.* 2014). The tsunami hazard is mainly from submarine landslides rather than earthquakes (ten Brink 2009; ten Brink *et al.* 2014).

The only other country where there has been a concerted marine programme on a passive margin is Australia where, off the east coast, MBES mapping reveals a large number of landslide scars offshore of significant concentrations of population (Fig. 9) (Clarke *et al.* 2014). Numerical modelling suggests that the hazard here may be limited, however (Webster *et al.* 2016).

Along convergent margins, the hazard from submarine landslide tsunamis may be much greater than along passive margins. PNG is the best-known and most comprehensively researched submarine landslide tsunami along a convergent margin. Japan follows as a close second, although further confirmation is required to locate the exact position of the submarine landslide. Other convergent margin tsunamis where there is a possible landslide component include those of Messina (1908), Makran (1945), Aleutians (1946), Alaska (1964), Puerto Rico (1918), Flores Islands (1992) and Java (2006) (Fig. 9). One of the most devastating historical, convergent-margin tsunamis was Messina, 1908 (Fig. 9). A total of 50 000 people died in the earthquake from collapsed buildings in Messina and Calabria with a large, but uncertain, number drowned in the ensuing tsunami. Over 600 people died in the Java tsunami and 1000 at Flores, where the highly focused tsunami flooded up to 25 m above sea level. There is a hazard programme off NW USA similar to that on the east coast, established for dual earthquake and submarine landslide-generated tsunami along convergent margins. This programme recently re-evaluated the Aleutian tsunami of 1946 because of the controversy over the submarine landslide mechanism of the local 40 m high tsunami (Fryer *et al.* 2004; López & Okal 2006; Locat *et al.* 2009). Re-evaluation based on MBES and seismic data (von Huene *et al.* 2014) now identifies the landslide location previously proposed by Fryer *et al.* (2004), but this has yet to be numerically modelled. For most of the convergent-margin tsunamis identified above, the submarine landslide contribution remains uncertain because not all have marine survey data on which to evaluate the seabed for landslides that can underpin numerical tsunami models. Hydroacoustic data have been used to identify submarine landslides for the Aleutians, Alaska, Messina and Puerto Rico events (Fig. 9) and re-evaluate their tsunami hazard.

Discussion

The numerous tsunamis experienced since 1992 suggest that we are living in a period where these events may be more frequent than previously. Recognition of this (apparent) high frequency and concomitant hazard has resulted in a greater awareness of the tsunami impact, which requires better understanding so the hazard and risk can be fully addressed on a sound scientific basis. An improved understanding of the hazard and risk will necessarily require a continued geological input. A major requirement in hazard mitigation is for longer-term records so that the frequency of tsunami events can be better understood. This longer-term record can only be gained from the sediments laid down as tsunamis flood the coast. Geologists, as the arbiters of this record, can contribute based on major advances resulting from the recent catastrophic events of PNG (1998), the Indian Ocean (2004) and Japan (2011). Tsunami sediments provide a longer-term record for tsunami impact. Together with improved understanding of inundation limits from new methodologies (such as geochemistry), the impacts of older historical and prehistorical tsunamis and their generation mechanisms can be revised as in Japan with the Jogan (869) event. Based on recent events, where tsunami hydrodynamics are recorded and the origin of the sediments established, tsunami sediment can again be used through novel inverse modelling methodologies of tsunami flow speeds and depths to better understand older events at the same locations. This methodology also has potential in regions where there are no observed recent events, but where there is a historical or prehistorical record. Inverse modelling of tsunami sediment inundation provides new insight into generation mechanisms and magnitudes.

Although earthquakes are undoubtedly the most frequent mechanism of tsunami generation, two events (PNG and Storegga) indicate that submarine landslides are a secondary but important hazard. These latter events were recognized and researched over a similar period of time in the late 1990s to early 2000s; one (Storegga) was prehistoric so had little human impact, and the other (PNG) was very recent, killing over 2200 people. The recognition of the Storegga tsunami was from a coeval relationship between the tsunami sediments discovered in Scotland and a submarine landslide off the Norwegian coast, with the identification of the sediments motivating the first numerical modelling. The discovery of the Ormen Lange Gas field in 1997 led to a major investigation into slope stability and tsunami generation. When the PNG tsunami struck in 1998 and the landslide mechanism was identified, the Storegga gas field developers recognized its importance. Storegga and PNG remain the best-studied and validated examples of landslide-generated tsunamis. The reasons for their study are entirely different. At Ormen Lange, the imperative was to prove that exploitation of the gas field would not trigger another tsunami. With PNG, the scale of the loss of life and initial uncertainty in the generation mechanism dictated that the cause of the tsunami had to be understood. Unlike Storegga, where oil money was available to fund a comprehensive investigation of the landslide, with PNG there was no obvious donor to fund the marine surveys which were organized on humanitarian grounds. In both instances, geologists made major contributions both to understanding tsunami generation and validation of numerical models. PNG was the first recent event where there was a focused post-event marine survey organized and managed by geologists. For Storegga, the numerical models were based on marine data and validated by sediments deposited by the tsunami on adjacent coastlines. For PNG, although the tsunami sediments were analysed and in fact used as a basis for the first inverse modelling, the numerical models were validated from tsunami run-up elevations acquired during post-event surveys.

Although only four (five with Alika 2) significant tsunamis are positively identified as generated by submarine landslides, seabed mapping of continental shelves reveals their ubiquity (Fig. 9). Too few landslides are yet dated, and even fewer studied to a degree necessary to understand the controls on their generation. These controls on sediment failure and landslide frequency are therefore poorly understood, so the hazard from submarine landslides remains undefined. Recent studies suggest that submarine landslide failure is random (e.g. Urlaub *et al.* 2013), but there are still too few events studied in sufficient detail or accurately dated to be certain of this. Urlaub *et al.* (2013) analysed 68 events, 50% of which they were unable to date to sufficient resolution to relate to potential landslide controlling factors. Because the controls on failure are complex (Tappin 2009, 2010*b*; Talling 2014) more events require researching before reliable conclusions can be drawn on the most important failure mechanisms. Recent overviews (ten Brink *et al.* 2016) suggest that as more events are studied, improved understanding of landslide failure and triggering mechanisms, particularly in specific tectonic environments, will result. Although earthquakes are still the most likely landslide triggers, these may not only be 'tectonic' in the context of convergent margin environments. There are also strong climate controls on earthquake rupture, for example in high latitudes due to glacioisostatic processes, such as established from research on Storegga (e.g. Bungum *et al.* 2005). In addition, new research suggests there may be triggering relationships between earthquakes and continental shelf loading from sea-level rise not previously recognized (Brothers *et al.* 2013; Smith *et al.* 2013). Regarding tsunami generation, most submarine landslides have the potential to generate hazardous events if of sufficient volume. The preservation potential of tsunami sediment is low; so, even though evidence for tsunamis associated with the landslides may be absent, this may not necessarily discount tsunami generation.

The Japanese tsunami of March 2011 led to the confirmation that tsunami sediments were critical in identifying inundation limits. It revealed that submarine landslides are an additional, yet unforeseen, major hazard even where a large-magnitude earthquake generates a devastating tsunami. It also demonstrated the importance of new technologies, such as seabed pressure sensors and improved geological (geochemical) methodologies, in the identification and discrimination between tsunami mechanisms and their coastal impact.

Challenges remain in improving our understanding of tsunamis and their hazard, to which geologists can contribute. The disintegration of translational landslides, and how this controls tsunami generation, has yet to be fully addressed by numerical models, which need to be more complex to be realistic. Geologists can contribute in developing more realistic landslide models from marine data and in their validation from sediment data. A major challenge is in determining the hazard from tsunamis generated from volcanic eruption, which is hardly researched. There have been some theoretical studies (e.g. Pareschi *et al.* 2006; Novikova *et al.* 2011), but these have not been validated. Even validated studies leave uncertainties over mechanisms (e.g. Ulvrova *et al.* 2016), however. There are numerical models of small-scale events, such as from Montserrat (Pelinovsky 2004), but their relevance to large-scale eruptions is uncertain. It is almost certain that, as with recent events such as the Indian Ocean, Japan and PNG where the seminal research has resulted from a catastrophe, the challenge of eruption tsunami mechanisms will be met only in response to a future major event with significant loss of life and/or major economic impact. The most recent (and only) catastrophic eruption tsunami in 1883 was when Krakatau in the Java Strait exploded, devastating surrounding coastlines and killing 36 000 people. To understand the eruption tsunami mechanism demands validated numerical models and, fortunately, Krakatau was subject to the first post-event tsunami survey ever carried out (Verbeek 1885). Although there was an immediate response to the tsunami impact, and the volcanic eruption has been well studied (e.g. Self & Rampino 1981), there is still uncertainty over the final cataclysmic explosion during which the devastating tsunami was generated. Alternative possible tsunami mechanisms include pyroclastic flows, caldera collapse or both as a dual mechanism (Francis 1985). There is only one comprehensive numerical modelling study, and this supports the entry of pyroclastic flows into the sea (Maeno & Imamura 2011). The results are however questionable because of the validation used. Globally, there are 42 volcanoes similar to Krakatau which could erupt with similar consequences. The uncertainty over the tsunami mechanisms of eruption therefore remains an important issue that needs addressing if appropriate mitigation and response strategies for eruption tsunamis are to be developed, similar to those for earthquakes and submarine landslides.

Conclusions

Over the past 30 years, there have been major advances in understanding the mechanisms of tsunami generation and tsunami impact. Improvements in technology, documentation of events and ability to model them have largely been because of devastating events. For the past 20 of these 30 years many of the advances have been by contributions from geology and by geologists. Most now-accepted ideas, such as that tsunamis lay down sediment when they flood the land and that submarine landslides generate hazardous tsunamis, were at first controversial and considered unlikely. Now, post-event surveys acquire geological data as a matter of course. Earthquakes are undoubtedly the primary tsunami mechanism, but improved understanding of earthquake mechanisms, magnitudes and frequencies in tsunami generation result from research on their associated tsunami deposits. Studies of prehistoric tsunami sediments have resulted in timescales now extended back in time for thousands of years, beyond historical records. These extended records have improved the understanding of the frequencies of these events that together allow improved mitigation and response strategies. Comparison of sediments from recent tsunamis with those preserved in geological records at the same locations offers the opportunity to better understand past events. Recent major advances in inverse modelling of earthquake tsunami magnitudes from sediments indicate their potential in this field.

Understanding tsunami generation from landslides, especially submarine events, is underpinned by geological research in mapping the locations and architectures to identify failure mechanisms and then using these mechanisms for realistic numerical simulations. Whereas the hazard from submarine landslide tsunamis is now recognized, the extent and risk from these events is still uncertain. Submarine landslides are present along the margins of most continents; many of these margins remain largely unmapped however, so their hazard is not known.

Eruption-generated tsunamis remain poorly researched, and their mechanisms are not well understood. Their global extent requires more focused programmes of research to address their hazard and risk. Based on the most recent history of scientific advance for earthquake and landslide tsunami, it may well be that advances in understanding eruption events await the next catastrophic event. Nevertheless, continued research on tsunami sediments, submarine landslides and volcanic eruptions promises to better define the global tsunami hazard from all generation mechanisms. Geology and geologists will continue to make essential contributions to this better understanding.

Many thanks to both Jon Hill and the indefatigable Simon Wallis for in-depth and constructive reviews that made a major contribution to the final text, and to Ellie Scourse for management of the review process. This paper is published with the permission of the CEO of the British Geological Survey, Natural Environmental Research Council, United Kingdom.

References

ABADIE, S.M., HARRIS, J.C., GRILLI, S.T. & FABRE, R. 2012. Numerical modeling of tsunami waves generated by the flank collapse of the Cumbre Vieja Volcano (La Palma, Canary Islands): tsunami source and near field effects. *Journal of Geophysical Research*, **117**(C5), C0503.

ADAMS, J. 1990. Paleoseismicity of the Cascadia subduction zone: evidence from turbidites off the Oregon-Washington margin. *Tectonics*, **9**, 569–583.

ALTINOK, Y., TINTI, S., ALPAR, B., YALÇINER, A.C., ERSOY, Ş., BORTOLUCCI, E. & ARMIGLIATO, A. 2001. The Tsunami of August 17, 1999 in Izmit Bay, Turkey. *Natural Hazards*, **24**, 133–146.

AMBRASEYS, N.N. & JACKSON, J.A. 1990. Seismicity and associated strain of central Greece between 1890 and 1988. *Geophysical Journal International*, **101**, 663–708.

ASSIER-RZADKIEWICZ, S., HEINRICH, P., SABATIER, P.C., SAVOYE, B. & BOURILLET, J.-F. 2000. Numerical modelling of a landslide-generated tsunami: the 1979 Nice event. *Pure and Applied Geophysics*, **157**, 1707–1727.

ATWATER, B.A. 1987. Evidence for great Holocene earthquakes along the outer coast of Washington State. *Science*, **236**, 942–944.

ATWATER, B.F. & HEMPHILL-HALEY, E. 1997. Recurrence intervals for great earthquakes of the past 3,500 years at northeastern Willapa Bay, Washington. USGS Professional Paper 1576.

BAHLBURG, H. & SPISKE, M. 2012. Sedimentology of tsunami inflow and backflow deposits: key differences revealed in a modern example. *Sedimentology*, **59**, 1063–1086.

BAILEY, E.B. & WEIR, J. 1933. XIV. Submarine Faulting in Kimmeridgian Times: East Sutherland. *Earth and Environmental Science Transactions of the Royal Society of Edinburgh*, **57**, 429–467.

BARDET, J.-P., SYNOLAKIS, C.E., DAVIES, H.L., IMAMURA, F. & OKAL, E.A. 2003. Landslide tsunamis: recent findings and research directions. *Pure and Applied Geophysics*, **160**, 1793–1809.

BIRNIE, J. 1981. *Environmental changes in Shetland since the end of the last glaciation*. Unpublished PhD thesis. University of Aberdeen.

BONDEVIK, S., SVENDSEN, J.I., JOHNSEN, G., MANGERUD, J.A.N. & KALAND, P.E. 1997*a*. The Storegga tsunami along the Norwegian coast, its age and run up. *Boreas*, **26**, 29–53.

BONDEVIK, S., SVENDSEN, J.I. & MANGERUD, J.A.N. 1997*b*. Tsunami sedimentary facies deposited by the Storegga tsunami in shallow marine basins and coastal lakes, western Norway. *Sedimentology*, **44**, 1115–1131.

BONDEVIK, S., MANGERUD, J., DAWSON, S., DAWSON, A. & LOHNE, Ø. 2003. Record-breaking height for 8000-year-old tsunami in the North Atlantic. *EOS, Transactions of American Geophysical Union*, **84**, 289–293.

BONDEVIK, S., LØVHOLT, F., HARBITZ, C., MANGERUD, J., DAWSON, A. & SVENDSEN, J.I. 2005*a*. The Storegga Slide tsunami – comparing field observations with numerical simulations. *Marine and Petroleum Geology*, **22**, 195–208.

BONDEVIK, S., MANGERUD, J., DAWSON, S., DAWSON, A.R. & LOHNE, Ø. 2005*b*. Evidence for three North Sea tsunamis at the Shetland Islands between 8000 and 1500 years ago. *Quaternary Science Reviews*, **24**, 1757–1775.

BORRERO, J.C. 2005. Field data and satellite imagery of tsunami effects in Banda Aceh. *Science*, **308**, 1596.

BORRERO, J.C., DOLAN, J.F. & SYNOLAKIS, C.E. 2001. Tsunamis within the Eastern Santa Barbara Channel. *Geophysical Research Letters*, **28**, 643–646.

BORRERO, J., LEGG, M.R. & SYNOKALIS, C.E. 2004. Tsunami sources in the southern California Bight. *Geophysical Research Letters*, **31**, https://doi.org/10.1029/2004GL020078

BOURGEOIS, J. 2009. Geologic effects and records of tsunamis. *In*: BERBARD, E.N. & ROBINSON, A.R. (eds) *The Sea, Tsunamis*. 15, Harvard University Press: Cambridge, Massachusetts, 53–91.

BROTHERS, D.S., LUTTRELL, K.M. & CHAYTOR, J.D. 2013. Sea-level-induced seismicity and submarine landslide occurrence. *Geology*, **41**, 979–982.

BRYN, P., BERG, K., FORSBERG, C.F., SOLHEIM, A. & LIEN, R. 2005. Explaining the Storegga Slide. *Marine and Petroleum Geology*, **22**, 11–19.

BUGGE, T. 1983. Submarine slides on the Norwegian continental margin, with special emphasis on the Storegga area. *IKU Report*, **110**, 1–152.

BUGGE, T., BELDERSON, R.H. & KENYON, N.H. 1988. The Storegga Slide. *Philosophical Transactions of the Royal Society of London. Series A, Mathematical and Physical Sciences*, **325**, 357–388.

BUNGUM, H., LINDHOLM, C. & FALEIDE, J.I. 2005. Post-glacial seismicity offshore mid-Norway with emphasis on spatio-temporal-magnitudal variations. *Marine and Petroleum Geology*, **22**, 137–148.

CADET, J.P., KOBAUASHI, K. ET AL. 1987. The Japan Trench and its juncture with the Kuril trench: cruise results of the Kaiko project, Leg 3. *Earth and Planetary Science Letters*, **83**, 267–285.

CANALS, M., LASTRAS, G. ET AL. 2004. Slope failure dynamics and impacts from seafloor and shallow sub-seafloor geophysical data: case studies from the COSTA project. *Marine Geology*, **213**, 9–72.

CHAGUÉ-GOFF, C., SCHNEIDER, J.-L., GOFF, J.R., DOMINEY-HOWES, D. & STROTZ, L. 2011. Expanding the proxy toolkit to help identify past events – Lessons from the 2004 Indian Ocean Tsunami and the 2009 South Pacific Tsunami. *Earth-Science Reviews*, **107**, 107–122.

CHAGUÉ-GOFF, C., ANDREW, A., SZCZUCIŃSKI, W., GOFF, J. & NISHIMURA, Y. 2012. Geochemical signatures up to the maximum inundation of the 2011 Tohoku-oki tsunami – Implications for the 869 AD Jogan and other palaeotsunamis. *Sedimentary Geology*, **282**, 65–77.

CHAYTOR, J.D., TEN BRINK, U.S., SOLOW, A.R. & ANDREWS, B.D. 2009. Size distribution of submarine landslides along the U.S. *Atlantic margin. Marine Geology*, **264**, 16–27.

CHENG, W. & WEISS, R. 2013. On sediment extent and runup of tsunami waves. *Earth and Planetary Science Letters*, **362**, 305–309.

CITA, M.B., BEGHI, C. ET AL. 1984. Turbidites and mega-turbidites from the Herodotus abyssal plain (eastern Mediterranean) unrelated to seismic events. *Marine Geology*, **55**, 79–101.

CLARKE, S., HUBBLE, T. & AIREY, D. 2014. Morphology of Australia's Eastern continental slope and related tsunami hazard. *In*: KRASTEL, S. & BEHRMANN, J.-H. ET AL. (eds) *Submarine Mass Movements and Their Consequences*. Springer, 529–538.

COLEMAN, P.J. 1968. Tsunamis as geological agents. *Journal of Geological Society of Australia*, **15**, 267–273.

COLEMAN, P.J. 1978. Tsunami sedimentation. *In*: FAIRBRIDGE, R.W. & BOURGEOIS, J. (eds) *Encyclopedia of Sedimentology*. Dowden, Hutchinson and Ross, Stroudsbourg, PA, 828–832.

COSTA, P.J.M., GELFENBAUM, G. ET AL. 2017. The application of microtextural and heavy mineral analysis to discriminate between storm and tsunami deposits. *In*: SCOURSE, E.M., CHAPMAN, N.A., TAPPIN, D.R. & WALLIS, S.R. (eds) *Tsunamis: Geology, Hazards and Risks*. Geological Society, London, Special Publications, **456**. First published online February 23, 2017, doi:10.1144/SP456.7

DAN, G., SULTAN, N. & SAVOYE, B. 2007. The 1979 Nice harbour catastrophe revisited: trigger mechanism inferred from geotechnical measurements and numerical modelling. *Marine Geology*, **245**, 40–64.

DAWSON, A.G. 1994. Geomorphological effects of tsunami run-up and backwash. *Geomorphology*, **10**, 83–94.

DAWSON, A.G. & STEWART, I. 2007. Tsunami deposits in the geological record. *Sedimentary Geology*, **200**, 166–183.

DAWSON, A.G., LONG, D. & SMITH, D.E. 1988. The Storegga Slides: evidence from eastern Scotland for a possible tsunami. *Marine Geology*, **82**, 271–276.

DAWSON, A.G., HINDSON, R., ANDRADE, C., FREITAS, C., PARISH, R. & BATEMAN, M. 1995. Tsunami sedimentation associated with the Lisbon earthquake of 1 November AD 1755: Boca do Rio, Algarve, Portugal. *The Holocene*, **5**, 209–215.

DOMINEY-HOWES, D.T.M., HUMPHREYS, G.S. & HESSE, P.P. 2006. Tsunami and palaeotsunami depositional signatures and their potential value in understanding the late-Holocene tsunami record. *The Holocene*, **16**, 1095.

EINSELE, G. (ed) 1996. Marine sedimentary events and their records. *Sedimentary Geology*, **104**, 1–257

ENET, F. & GRILLI, S.T. 2005. Tsunami landslide generation: Modelling and experiments. *Proceedings of the 5th International Conference on OceanWave Measurement and Analysis (WAVES 2005), Madrid, Spain, July 2005*, ASCE Publication, paper 88.

ENET, F. & GRILLI, S.T. 2007. Experimental study of tsunami generation by three-dimensional rigid underwater landslides. *Journal of Waterway, Port, Coastal, and Ocean Engineering*, **133**, 442–454.

ENET, F., GRILLI, S.T. & WATTS, P. 2003. Laboratory experiments for tsunamis generated by underwater landslides: comparison with numerical modeling. *Proceedings of the 13th International Offshore and Polar Engineering Conference, ISOPE03*, Honolulu, Hawaii, **3**, 372–379.

FALVARD, S. & PARIS, R. 2017. X-ray tomography of tsunami deposits: towards a new depositional model of tsunami deposits. *Sedimentology*, **64**, 453–477.

FELDENS, P., SCHWARZER, K., SAKUNA, D., SZCZUCIŃSKI, W. & SOMPONGCHAIYAKUL, P. 2012. Sediment distribution on the inner continental shelf off Khao Lak (Thailand) after the 2004 Indian Ocean tsunami. *Earth Planets Space*, **64**, 875–887.

FELTON, E.A. 2002. Sedimentology of rocky shorelines: 1. A review of the problem, with analytical methods, and insights gained from the Hulopoe Gravel and the modern rocky shoreline of Lanai, Hawaii. *Sedimentary Geology*, **152**, 221–245.

FELTON, E.A., CROOK, K.A.W. & KEATING, B.H. 2000. The Hulopoe gravel, Lanai, Hawaii: new sedimentological data and their bearing on the 'giant wave' (mega-tsunami) emplacement hypothesis. *Pure and Applied Geophysics*, **157**, 1257–1284.

FINE, I.V., RABINOVICH, A.B., BORNHOLD, B.D., THOMSON, R.E. & KULIKOV, E.A. 2005. The Grand Banks landslide-generated tsunami of November 18, 1929: preliminary analysis and numerical modeling. *Marine Geology*, **215**, 45–57.

FORSBERG, C.F. 2002. *Reconstruction of the Pre-Storegga Slide Stratigaphy*. Norsk Hydro Report 37-00NH-X15-00040.

FRANCIS, P.W. 1985. The origin of the 1883 Krakatau tsunamis. *Journal of Volcanology and Geothermal Research*, **25**, 349–363.

FRITZ, H.M., KONGKO, W. ET AL. 2007. Extreme runup from the 17 July 2006 Java tsunami. *Geophysical Research Letters*, **34**, https://doi.org/10.1029/2007GL029404

FRYER, G.J., WATTS, P. & PRATSON, L.F. 2004. Source of the great tsunami of 1 April 1946: a landslide in the upper Aleutian forearc. *Marine Geology*, **203**, 201–218.

FUJII, Y., SATAKE, K., SAKAI, S.I., SHINOHARA, M. & KANAZAWA, T. 2011. Tsunami source of the 2011 off the Pacific coast of Tohoku Earthquake. *Earth Planets Space*, **63**, 815–820.

GEIST, E.L. 1998. Local tsunamis and earthquake source parameters. *Advances in Geophysics*, **39**, 117–209.

GEIST, E.L. 2000. Origin of the 17 July, 1998 Papua New Guinea tsunami: Earthquake or landslide? *Seismological Research Letters*, **71**, 344–351.

GEIST, E.L. & PARSONS, T. 2009. Assessment of source probabilities for potential tsunamis affecting the U.S. Atlantic coast. *Marine Geology*, **264**, 98–108.

GEIST, E.L., LYNETT, P.J. & CHAYTOR, J.D. 2009. Hydrodynamic modeling of tsunamis from the Currituck landslide. *Marine Geology*, **264**, 41–52.

GELFENBAUM, G. & JAFFE, B. 2003. Erosion and sedimentation from the 17 July, 1998 Papua New Guinea tsunami. *Pure and Applied Geophysics*, **160**, 1969–1999.

GISLER, G., WEAVER, R. & GITTINGS, M. 2006. SAGE calculations of the tsunami threat from La Palma. *Science of Tsunami Hazards*, **24**, 288–301.

GOFF, J., CHAGUÉ-GOFF, C. & NICHOL, S. 2001. Paleotsunami deposits: a New Zealand perspective. *Sedimentary Geology*, **143**, 1–6.

GOFF, J., MCFADGEN, B.G. & CHAGUÉ-GOFF, C. 2004. Sedimentary differences between the 2002 Easter storm and the 15th Century Okoropunga tsunami, southeastern North Island, New Zealand. *Marine Geology*, **204**, 235–250.

GOLDFINGER, C., NELSON, C.H. *ET AL.* 2012. Turbidite event history – Methods and implications for Holocene paleoseismicity of the Cascadia subduction zone. U.S. Geological Survey Professional Paper 1661–F, 170.

GOTO, K., CHAVANICH, S.A. *ET AL.* 2007. Distribution, origin and transport process of boulders deposited by the 2004 Indian Ocean tsunami at Pakarang Cape, Thailand. *Sedimentary Geology*, **202**, 821–837.

GOTO, K., IMAMURA, F. *ET AL.* 2008*a*. *Distribution and Significance of the 2004 Indian Ocean Tsunami Deposits: Initial Results from Thailand and Sri Lanka, Tsunamiites*. Elsevier, Amsterdam.

GOTO, K., TADA, R. *ET AL.* 2008*b*. Lateral lithological and compositional variations of the Cretaceous/Tertiary deep-sea tsunami deposits in northwestern Cuba. *Cretaceous Research*, **29**, 217–236.

GOTO, K., KAWANA, T. & IMAMURA, F. 2010*a*. Historical and geological evidence of boulders deposited by tsunamis, southern Ryukyu Islands, Japan. *Earth-Science Reviews*, **102**, 77–99.

GOTO, K., MIYAGI, K., KAWAMATA, H. & IMAMURA, F. 2010*b*. Discrimination of boulders deposited by tsunamis and storm waves at Ishigaki Island, Japan. *Marine Geology*, **269**, 34–45.

GOTO, K., CHAGUÉ-GOFF, C. *ET AL.* 2011. New insights of tsunami hazard from the 2011 Tohoku-oki event. *Marine Geology*, **290**, 46–50.

GRAUERT, M., BJÖRCK, S. & BONDEVIK, S. 2001. Storegga tsunami deposits in a coastal lake on Suouroy, the Faroe Islands. *Boreas*, **30**, 263–271.

GRIGG, R.W. & JONES, A.T. 1997. Uplift caused by lithospheric flexure in the Hawaiian Archipelago as revealed by elevated coral deposits. *Marine Geology*, **141**, 11–25.

GRIGGS, G.B. & KULM, L.D. 1970. Sedimentation in Cascadia deep-sea Channel. *Geological Society of America Bulletin*, **81**, 1361–1384.

GRILLI, S.T. & WATTS, P. 1999. Modelling of waves generated by a moving submerged body: applications to underwater landslides. *Engineering Analysis with Boundary Elements*, **23**, 645–656.

GRILLI, S.T. & WATTS, P. 2001. Modeling of tsunami generation by an underwater landslide in a 3D-NWT. *Proceedings of the 11th Offshore and Polar Engineering Conference (ISOPE01)*, Stavanger, Norway, June 2001, **III**, 132–139.

GRILLI, S.T. & WATTS, P. 2005. Tsunami generation by submarine mass failure part I: modeling, experimental validation, and sensitivity analyses. *Journal of Waterway, Port, Coastal, and Ocean Engineering*, **131**, 283–297.

GRILLI, S.T., VOGELMANN, S. & WATTS, P. 2002. Development of a 3D numerical wave tank for modelling tsunami generation by underwater landslides. *Engineering Analysis with Boundary Elements*, **26**, 301–313.

GRILLI, S.T., TAYLOR, O.-D.S., BAXTER, C.D.P. & MARETZKI, S. 2009. A probabilistic approach for determining submarine landslide tsunami hazard along the upper east coast of the United States. *Marine Geology*, **264**, 74–97.

GRILLI, S.T., HARRIS, J.C., TAJALLI BAKHSH, T.S., MASTERLARK, T.L., KYRIAKOPOULOS, C., KIRBY, J.T. & SHI, F. 2013. Numerical simulation of the 2011 Tohoku Tsunami based on a new transient FEM co-seismic source: comparison to far- and near-field observations. *Pure and Applied Geophysics*, **170**, 1333–1359.

GRILLI, S., O'REILLY, C. *ET AL.* 2015. Modeling of SMF tsunami hazard along the upper US East Coast: detailed impact around Ocean City, MD. *Natural Hazards*, **76**, 705–746.

GUSMAN, A.R., TANIOKA, Y., SAKAI, S. & TSUSHIMA, H. 2012. Source model of the great 2011 Tohoku earthquake estimated from tsunami waveforms and crustal deformation data. *Earth and Planetary Science Letters*, **341-344**, 234–242.

GUTENBERG, B. 1939. Tsunamis and earthquakes. *Bulletin of the Seismological Society of America*, **29**, 517–526.

HAFLIDASON, H., LIEN, R., SEJRUP, H.P., FORSBERG, C.F. & BRYN, P. 2005. The dating and morphometry of the Storegga Slide. *Marine and Petroleum Geology*, 123–136.

HALL, A.M., HANSOM, J.D., WILLIAMS, D.M. & JARVIS, J. 2006. Distribution, geomorphology and lithofacies of cliff-top storm deposits: examples from the high-energy coasts of Scotland and Ireland. *Marine Geology*, **232**, 131–155.

HAMMACK, J.L. 1973. A note on tsunamis: their generation and propagation in an ocean of uniform depth. *Journal of Fluid Mechanics*, **60**, 769–799.

HAMPTON, M. 1972. The role of subaqueous debris flow in generating turbidity currents. *Journal of Sedimentary Research*, **42**, 775–993.

HAMPTON, M.A., LEE, H.J. & LOCAT, J. 1996. Submarine landslides. *Reviews of Geophysics*, **34**, 33–59.

HARBITZ, C.B. 1992. Model simulation of tsunamis generated by the Storegga Slides. *Marine Geology*, **105**, 1–21.

HASEGAWA, H.S. & KANAMORI, H. 1987. Source mechanism of the magnitude 7.2 Grand Banks earthquake of November 1929: double couple or submarine landslide? *Bulletin of the Seismological Society of America*, **77**, 1984–2004.

HAUGEN, K.B., LØVHOLT, F. & HARBITZ, C.B. 2005. Fundamental mechanisms for tsunami generation by submarine mass flows in idealised geometries. *Marine and Petroleum Geology*, **22**, 209–217.

HEARTY, J.P. 1997. Boulder deposits from large waves during the last interglaciation on North Eleuthera island, Bahamas. *Quaternary Research*, **48**, 326–338.

HEEZEN, B.C. & EWING, M. 1952. Turbidity currents and submarine slumps, and the 1929 Grand Banks Earthquake. *American Journal of Science*, **250**, 849–873.

HEEZEN, B.C., ERICSSON, D.B. & EWING, M. 1954. Further evidence of a turbidity current following the 1929 Grand Banks earthquake. *Deep Sea Research*, **1**, 193–202.

Heidarzadeh, M., Pirooz, M.D., Zaker, N.H., Yalciner, A.C., Mokhtari, M. & Esmaeily, A. 2008. Historical tsunami in the Makran Subduction Zone off the southern coasts of Iran and Pakistan and results of numerical modeling. *Ocean Engineering*, **35**, 774–786.

Heinrich, P. 1992. Nonlinear water waves generated by submarine and aerial landslides. *Journal of Waterways, Port, Coast, Ocean Engineering*, **118**, 249–266.

Heinrich, P., Piatanesi, A., Okal, E.A. & Hébert, H. 2000. Near-field modelling of the July 17, 1998 event in Papua New Guinea. *Geophysical Research Letters*, **27**, 3037–3040.

Hill, J., Collins, G.S., Avdis, A., Kramer, S.C. & Piggott, M.D. 2014. How does multiscale modelling and inclusion of realistic palaeobathymetry affect numerical simulation of the Storegga Slide tsunami? *Ocean Modelling*, **83**, 11–25.

Houtz, R.E. 1962. The 1953 Suva earthquake and tsunami. *Bulletin of the Seismological Society of America*, **52**, 1–12.

Hühnerbach, V., Masson, D.G. & COSTA project partners 2004. Landslides in the north Atlantic and its adjacent seas: an analysis of their morphology, setting and behaviour. *Marine Geology*, **213**, 343–362.

Hunt, J.E., Wynn, R.B., Masson, D.G., Talling, P.J. & Teagle, D.A.H. 2011. Sedimentological and geochemical evidence for multistage failure of volcanic island landslides: a case study from Icod landslide on north Tenerife, Canary Islands. *Geochemistry, Geophysics, Geosystems*, **12**, Q12007.

Hunt, J.E., Wynn, R.B., Talling, P.J. & Masson, D.G. 2013. Multistage collapse of eight western Canary Island landslides in the last 1.5 Ma: sedimentological and geochemical evidence from subunits in submarine flow deposits. *Geochemistry, Geophysics, Geosystems*, **14**, 2159–2181.

Huntington, K., Bourgeois, J., Gelfenbaum, G., Lynett, P., Jaffe, B., Yeh, H. & Weiss, R. 2007. Sandy signs of a tsunami's onshore depth and speed. *Eos*, **88**, 577–578.

Hurukawa, N., Tsuji, Y. & Waluyo, B. 2003. The 1998 Papua New Guinea earthquake and its fault plane estimated from relocated aftershocks. *Pure and Applied Geophysics*, **160**, 1829–1841.

Iinuma, T., Hino, R. et al. 2012. Coseismic slip distribution of the 2011 off the Pacific Coast of Tohoku Earthquake (M9.0) refined by means of seafloor geodetic data. *Journal of Geophysical Research: Solid Earth*, **117**, B07409.

Ikehara, K., Irino, T., Usami, K., Jenkins, R., Omura, A. & Ashi, J. 2014. Possible submarine tsunami deposits on the outer shelf of Sendai Bay, Japan resulting from the 2011 earthquake and tsunami off the Pacific coast of Tohoku. *Marine Geology*, **358**, 120–127.

Ishimura, D. & Miyauchi, T. 2015. Historical and paleotsunami deposits during the last 4000 years and their correlations with historical tsunami events in Koyadori on the Sanriku Coast, northeastern Japan. *Progress in Earth and Planetary Science*, **2, 16**.

Jaffe, B.E. & Gelfenbaum, G. 2002. Using tsunami deposits to improve assessment of tsunami risk. *Solutions to Coastal Disasters '02, Conference Proceedings, ASCE*, February 24–27, 2002. American Society of Civil Engineers, San Diego, California, United States, 836–847.

Jaffe, B.E. & Gelfenbuam, G. 2007. A simple model for calculating tsunami flow speed from tsunami deposits. *Sedimentary Geology*, **200**, 347–361.

Jaffe, B., Buckley, M. et al. 2011. Flow speed estimated by inverse modeling of sandy sediment deposited by the 29 September 2009 tsunami near Satitoa, east Upolu, Samoa. *Earth-Science Reviews*, **107**, 23–37.

Jankaew, K., Atwater, B.F., Sawai, Y., Choowong, M., Charoentitirat, T., Martin, M.E. & Prendergast, A. 2008. Medieval forewarning of the 2004 Indian Ocean tsunami in Thailand. *Nature*, **455**, 1228–1231.

Jansen, E., Befring, S., Bugge, T., Eidvin, T., Holtedahl, H. & Sejrup, H.P. 1987. Large submarine slides on the Norwegian continental margin: sediments, transport and timing. *Marine Geology*, **78**, 77–107.

Jiang, L. & LeBlond, P.H. 1992. The coupling of a submarine slide and the surface wave which it generates. *Journal of Geophysical Research*, **97**, 12731–12744.

Jiang, L. & LeBlond, P.H. 1994. Three dimensional modelling of tsunami generation due to submarine mudslide. *Journal of Physical Oceanography*, **24**, 559–573.

Kanamori, H. 1972. Mechanisms of tsunami earthquakes. *Physics of the Earth and Planetary Interiors*, **6**, 346–359.

Kastens, K.A. & Cita, M.B. 1981. Tsunami-induced sediment transport in the abyssal Mediterranean Sea. *Geological Society of America Bulletin*, **92**, 591–604.

Kawamura, K., Sasaki, T., Kanamatsu, T., Sakaguchi, A. & Ogawa, Y. 2012. Large submarine landslides in the Japan Trench: a new scenario for additional tsunami generation. *Geophysical Research Letters*, **39**, L05308.

Kawata, Y., Benson, B.C. et al. 1999. Tsunami in Papua New Guinea was as intense as first thought. *Eos, Transactions of the American Geophysical Union*, **80**, 101, 104–105.

Keating, H.B. & McGuire, J.W. 2000. Island edifice failures and associated tsunami hazards. *Pure and Applied Geophysics*, **157**, 899–955.

Kench, P.S., McLean, R.F. et al. 2006. Geological effects of tsunami on mid-ocean atoll islands: the Maldives before and after the Sumatran tsunami. *Geology*, **34**, 177–180.

Kikuchi, M., Yamanaka, Y., Abe, K. & Morita, Y. 1999. Source rupture process of the Papua New Guinea earthquake of July 17th, 1998 inferred from teleseismic body waves. *Earth Planets Space*, **51**, 1319–1324.

Kon'no, E.E. 1961. Geological observations of the Sanriku coastal region damaged by tsunami due to the Chile earthquake in 1960, Contributions to the Institute of Geology and Paleontology, Tohoku University, 1–40.

Kuenen, P.H. 1957. Sole markings of graded greywacke beds. *Journal of Geology*, **65**, 231–258.

Kulikov, E.A., Rabinovich, A.B., Thomson, R.E. & Bornhold, B.D. 1996. The landslide tsunami of November 3, 1994, Skagway Harbor, Alaska. *Journal of Geophysical Research: Oceans*, **101**, 6609–6615.

KVALSTAD, T.J., ANDRESEN, L., FORSBERG, C.F., BERG, K., BRYN, P. & WANGEN, M. 2005. The Storegga Slide: evaluation of triggering sources and slide mechanics. *Marine and Petroleum Geology*, **22**, 245–256.

LABERG, J.S. & VORREN, T.O. 2000. The Trænadjupet Slide, offshore Norway – morphology, evacuation and triggering mechanisms. *Marine Geology*, **171**, 95–114.

LEBLOND, P.H. & JONES, A. 1995. Underwater landslides ineffective at tsunami generation. *Science of Tsunami Hazards*, **13**, 25–26.

LEE, H.J. 2009. Timing of occurrence of large submarine landslides on the Atlantic Ocean margin. *Marine Geology*, **264**, 53–64.

LE ROUX, J.P. & VARGAS, G. 2005. Structure and depositional processes of a gravelly tsunami deposit in a shallow marine setting: Lower Cretaceous Miyako Group, Japan – discussion. *Sedimentary Geology*, **201**, 485–487.

LIU, P.L.-F., WU, T.-R., RAICHLEN, F., SYNOLAKIS, C.E. & BORRERO, J.C. 2005. Runup and rundown generated by three-dimensional sliding masses. *Journal of Fluid Mechanics*, **536**, 107–144.

LOCAT, J., LOCAT, P., LEE, H.J. & IMRAN, J. 2004. Numerical analysis of the mobility of the Palos Verdes debris avalanche, California, and its implication for the generation of tsunamis. *Marine Geology*, **20**, 269–280.

LOCAT, J., LEE, H., TEN BRINK, U.S., TWICHELL, D., GEIST, E. & SANSOUCY, M. 2009. Geomorphology, stability and mobility of the Currituck slide. *Marine Geology*, **264**, 28–40.

LONG, D., SMITH, D.E. & DAWSON, A.G. 1989. A Holocene tsunami deposit in eastern Scotland. *Journal of Quaternary Science*, **4**, 61–66.

LÓPEZ, A.M. & OKAL, E.A. 2006. A seismological reassessment of the source of the 1946 Aleutian 'tsunami' earthquake. *Geophysics Journal International*, **165**, 835–849.

LØVHOLT, F., HARBITZ, C.B. & HAUGEN, K.B. 2005. A parametric study of tsunamis generated by submarine slides in the Ormen Lange/Storegga area off western Norway. *Marine and Petroleum, Geology*, **22**, 219–231.

LØVHOLT, F., PEDERSEN, G. & GISLER, G. 2008. Oceanic propagation of a potential tsunami from the La Palma Island. *Journal of Geophysical Research*, **113**, https://doi.org/10.1029/2007JC004603

LØVHOLT, F., KAISER, G., GLIMSDAL, S., SCHEELE, L., HARBITZ, C.B. & PEDERSEN, G. 2012. Modeling propagation and inundation of the 11 March 2011 Tohoku tsunami. *Natural Hazards and Earth Systems Science*, **12**, 1017–1028.

LØVHOLT, F., PEDERSEN, G., HARBITZ, C.B., GLIMSDAL, S. & KIM, J. 2015. On the characteristics of landslide tsunamis. *Philosophical Transactions of the Royal Society A: Mathematical, Physical and Engineering Science*, **373**, https://doi.org/10.1098/rsta.2014.0376

LUDWIG, K.R., SZABO, B.J., MOORE, J.G. & SIMMONS, K.R. 1991. Crustal subsidence rate of Hawaii determined from 234U/238U ages of drowned coral reefs. *Geology*, **19**, 171–174.

LYNETT, P. & LIU, P.L. 2002. A numerical study of submarine landslide generated waves and runup. *Proceedings of the Royal Society of London, A*, **458**, 2885–2910.

MA, G., KIRBY, J.T. & SHI, F. 2013. Numerical simulation of tsunami waves generated by deformable submarine landslides. *Ocean Modelling*, **69**, 146–165.

MA, G., KIRBY, J.T., HSU, T.-J. & SHI, F. 2015. A two-layer granular landslide model for tsunami wave generation: theory and computation. *Ocean Modelling*, **93**, 40–55.

MACINNES, B.T., WEISS, R., BOURGEOIS, J. & PINEGINA, T.K. 2010. Slip distribution of the 1952 Kamchatka Great Earthquake based on near-field tsunami deposits and historical records. *Bulletin of the Seismological Society of America*, **100**, 1695–1709.

MACINNES, B.T., GUSMAN, A.R., LEVEQUE, R.J. & TANIOKA, Y. 2013. Comparison of earthquake source models for the 2011 Tohoku event using tsunami simulations and near-field observations. *Bulletin of the Seismological Society of America*, **103**, 1256–1274.

MAEDA, T., FURUMURA, T., SAKAI, S.I. & SHINOHARA, M. 2011. Significant tsunami observed at ocean-bottom pressure gauges during the 2011 off the Pacific coast of Tohoku Earthquake. *Earth, Planets and Space*, **63**, 803–808.

MAENO, F. & IMAMURA, F. 2011. Tsunami generation by a rapid entrance of pyroclastic flow into the sea during the 1883 Krakatau eruption, Indonesia. *Journal of Geophysical Research*, **116**, B09205.

MARINATOS, S. 1939. The volcanic destruction of Minoan Crete. *Antiquity*, **13**, 425–439.

MARTIN, M.E., WEISS, R., BOURGEOIS, J., PINEGINA, T.K., HOUSTON, H. & TITOV, V.V. 2008. Combining constraints from tsunami modeling and sedimentology to untangle the 1969 Ozernoi and 1971 Kamchatskii tsunamis. *Geophysical Research Letters*, **35**, L01610.

MASLIN, M., OWEN, M., DAY, S. & LONG, D. 2004. Linking continental-slope failures and climate change: testing the clathrate gun hypothesis. *Geology*, **32**, 53–56.

MASSON, D.G., HARBITZ, C.B., WYNN, R.B., PEDERSEN, G. & LØVHOLT, F. 2006. Submarine landslides: processes, triggers and hazard prediction. *Philosophical Transactions of the Royal Society, A*, **364**, 2009–2039.

MASSON, D.G., WYNN, R.B. & TALLING, P.J. 2009. Large landslides on passive continental margins: processes, hypotheses and outstanding questions. *In*: MOSHER, D.C., SHIPP, R.C., MOSCARDILLI, L., CHAYTOR, J.D., BAXTER, C.D.P., LEE, H.J. & URGELES, R. (eds) *Submarine Mass Movements and Their Consequences*. Springer Science, Dordrecht, Heidelberg, London, New York, 153–165.

MASTRONUZZI, G. & SANSO, P. 2004. Large boulder accumulations by extreme waves along the Adriatic coast of southern Apulia (Italy). *Quaternary International*, **120**, 1173–1184.

MCCUE, K.F. 1998. An AGSO perspective of PNG's tsunamagenic earthquake of 17 July 1998. *AusGeo International*, **9**, 1–2.

MCMURTRY, G.M., HERRERO-BERVERA, E., CREMER, M., RESIG, J., SHERMAN, C., SMITH, J.R. & TORRESAN, M.E. 1999. Stratigraphic constraints on the timing and emplacement of the Alika 2 giant Hawaiian submarine landslide. *Journal of Volcanology and Geothermal Research*, **94**, 35–58.

MCMURTRY, G.M., FRYER, G.J. ET AL. 2004*a*. Megatsunami deposits on Kohala volcano, Hawaii, from flank collapse of Mauna Loa. *Geology*, **32**, 741–744.

McMurtry, G.M., Watts, P., Fryer, G.J., Smith, J.R. & Imamura, F. 2004*b*. Giant landslides, mega-tsunamis, and paleo-sea level in the Hawaiian Islands. *Marine Geology*, **203**, 219–233.
Milne, J. 1898. *Earthquakes and Other Earth Movements*. Paul, Trench, Trübner & Co., London, UK.
Minoura, K. & Nakaya, S. 1991. Traces of tsunami preserved in inter-tidal lacustrine and marsh deposits: some examples from Northeast Japan. *Journal of Geology*, **99**, 265–287.
Minoura, K., Imamura, F., Takahashi, T. & Shuto, N. 1997. Sequence of sedimentation processes caused by the 1992 Flores tsunami: evidence from Babi Island. *Geology*, **25**, 523–526.
Minoura, K., Imamura, F., Sugawara, D., Kono, Y. & Iwashita, T. 2001. The 869 Jogan tsunami deposit and recurrence interval of large-scale tsunami on the Pacific coast of northeast Japan. *Journal of Natural Disaster Science*, **23**, 83–88.
Montessus de Ballore, F. 1907. *La Science Séismologique. A.* Colin, Paris, France.
Moore, A.L., McAdoo, B.G. & Ruffman, A. 2007. Landward fining from multiple sources in a sand sheet deposited by the 1929 Grand Banks tsunami, Newfoundland. *Sedimentary Geology*, **200**, 336–346.
Moore, G.W. & Moore, J.G. 1988. Large-scale bedforms in boulder gravel produced by giant waves in Hawaii: sedimentologic consequences of convulsive geologic events. *Special Papers of the Geological Society of America*, 101–110.
Moore, J.G. 1964. Giant submarine landslides on the Hawaiian Ridge. United States Geological Survey Professional Paper 501-D, 95–98.
Moore, J.G. & Moore, G.W. 1984. Deposit from a giant wave on the island of Lanai, Hawaii. *Science*, **226**, 1312–1315.
Moore, J.G., Clague, D.A., Holcomb, R.T., Lipman, P.W., Normark, W.R. & Torresan, M.E. 1989. Prodigious submarine landslides on the Hawaiian Ridge. *Journal of Geophysical Research*, **94**, 17465–17484.
Moore, J.G., Bryan, W.B., Beeson, M.H. & Normark, W.R. 1995. Giant blocks in the South Kona landslide, Hawaii. *Geology*, **23**, 125–128.
Mori, N., Takahashi, T., Yasuda, T. & Yanagisawa, H. 2011. Survey of 2011 Tohoku earthquake tsunami inundation and run-up. *Geophysical Research Letters*, **38**, L00G14.
Morton, R.A., Richmond, B.M., Jaffe, B.E. & Gelfenbaum, G. 2006. *Reconnaissance Investigation of Caribbean Extreme Wave Deposits – Preliminary Observations, Interpretations, and Research Directions*. Open-File Report 2006-1293, US Department of the Interior, US Geological Survey.
Morton, R.A., Gelfenbaum, G. & Jaffe, B.E. 2007. Physical criteria for distinguishing sandy tsunami and storm deposits using modern examples. *Sedimentary Geology*, **200**, 184–207.
Morton, R.A., Goff, J.R. & Nichol, S.L. 2008. Hydrodynamic implications of textural trends in sand deposits of the 2004 tsunami in Sri Lanka. *Sedimentary Geology*, **207**, 56–64.
Murty, T.S. 1979. Submarine slide-generated water waves in Kitimat Inlet, British Columbia. *Journal of Geophysical Research*, **84**, 7777–7779.
Myles, D. 1985. *The Great Waves*. Robert Hale Ltd., London.
Mylroie, J.E. 2008. Late Quaternary sea-level position: evidence from Bahamian carbonate deposition and dissolution cycles. *Quaternary International*, **183**, 61–75.
Nanayama, F., Shigeno, K., Satake, K., Shimokawa, K., Koitabashi, S., Mayasaka, S. & Ishi, M. 2000. Sedimentary differences between 1993 Hokkaido-Nansei-Oki tsunami and 1959 Miyakijima typhoon at Tasai, southwestern Hokkaido, northern Japan. *Sedimentary Geology*, **135**, 255–264.
Newman, A.V. & Okal, E.A. 1998*a*. Moderately slow character of the July 17, 1998 Sandaun earthquake as studied by teleseismic energy estimates (abs.). *Eos Transactions, American Geophysical Union*, **79**, Fall Meeting, Supplement F564.
Newman, A.V. & Okal, E.A. 1998*b*. Teleseismic estimates of radiated seismic energy: the E/M0 discriminant for tsunami earthquakes. *Journal of Geophysical Research*, **103**, 26885–26826, 26898.
Noda, A., Katayama, H. *et al.* 2007. Evaluation of tsunami impacts on shallow marine sediments: an example from the tsunami caused by the 2003 Tokachi-oki earthquake, northern Japan. *Sedimentary Geology*, **200**, 314–327.
Noormets, R., Felton, E.A. & Crook, K.A.W. 2002. Sedimentology of rocky shorelines: 2. Shoreline megaclasts on the north shore of Oahu, Hawaii: origins and history. *Sedimentary Geology*, **150**, 31–45.
Normark, W.R., McGann, M. & Sliter, R. 2004. Age of Palos Verdes submarine debris avalanche, southern California. *Marine Geology*, **203**, 247–259.
Nott, J. 1997. Extremely high-energy wave deposits inside the Great Barrier Reef, Australia, determining the cause: tsunami or tropical cyclone. *Marine Geology*, **141**, 193–207.
Nott, J. 2003. Tsunami or stormwaves? Determining the origin of a spectacular field of wave emplaced boulders using numerical storm, surge and wave models and hydrodynamic transport equations. *Journal of Coastal Research*, **19**, 348–356.
Novikova, T., Papadopoulos, G.A. & McCoy, F.W. 2011. Modelling of tsunami generated by the giant Late Bronze Age eruption of Thera, South Aegean Sea, Greece. *Geophysical Journal International*, **186**, 665–680.
O'Grady, D.B., Syvitski, J.P.M., Pratson, L.F. & Sarg, J.F. 2000. Categorizing the morphologic variability of siliciclastic passive continental margins. *Geology*, **28**, 207–210.
Okada, Y. 1985. Surface deformation due to shear and tensile faults in a half-space. *Bulletin of the Seismological Society of America*, **75**, 1135–1154.
Okal, E. & Herbert, H. 2007. Far-field simulation of the 1946 Aleutian tsunami. *Geophysical Journal International*, **169**, 129–1238.
Pareschi, M.T., Favalli, M. & Boschi, E. 2006. Impact of the Minoan tsunami of Santorini: simulated scenarios in the eastern Mediterranean. *Geophysical Research Letters*, **33**, https://doi.org/10.1029/2006GL027205
Paris, R., Pérez Torrado, F.J., Cabrera, M.D.C., Wassmer, P., Schneider, J.L. & Carracedo, J.C.

2004. Tsunami-induced conglomerates and debris-flow deposits on the Western Coast of Gran Canaria (Canary Islands). *Acta Vulcanalogica*, **16, 133–136**.

PARIS, R., LAVIGNE, F., WASSMER, P. & SARTOHADI, J. 2007. Coastal sedimentation associated with the December 26, 2004 tsunami in Lhok Nga, west Banda Aceh (Sumatra, Indonesia). *Marine Geology*, **238**, 93–106.

PARIS, R., WASSMER, P. ET AL. 2014. Coupling eruption and tsunami records: the Krakatau 1883 case study, Indonesia. *Bulletin of Volcanology*, **76**, 1–23.

PELINOVSKY, E. 2004. Tsunami gnerated by the volcano eruption on July 12–13, 2003T Montserrat, Lesser Antilles. *Science of Tsunami Hazards*, **22**, 44–45.

PETERS, R., JAFFE, B. & GELFENBAUM, G. 2007. Distribution and sedimentary characteristics of tsunami deposits along the Cascadia margin of western North America. *Sedimentary Geology*, **200**, 372–386.

PICKERING, K.T. 1984. The Upper Jurassic 'Boulder Beds' and related deposits: a fault-controlled submarine slope, NE Scotland. *Journal of the Geological Society of London*, **141**, 357–374.

PIPER, D.J.W. & ASKU, A.E. 1987. The source and origin of the 1929 Grand Banks turbidity current inferred from sediment budgets. *Geo Marine Letters*, **7**, 177–182.

PIPER, D.J.W., SHOR, A.N. & CLARKE, J.E.H. 1988. The 1929 Grand Banks earthquake, slump and turbidity current. *Geological Society of America Special Paper*, **229**, 77–92.

POLONIA, A., BONATTI, E., CAMERLENGHI, A., LUCCH, R.G., PANIERI, G. & GASPERINI, L. 2013. Mediterranean megaturbidite triggered by the AD 365 Crete earthquake and tsunami. *Scientific Reports*, **3**, 1–12.

QUIDELLEUR, X., HILDENBRAND, A. & SAMPER, A. 2008. Causal link between Quaternary paleoclimatic changes and volcanic islands evolution. *Geophysical Research Letters*, **35**, L02303.

RABINOVICH, A.B., THOMSON, R.E., KULIKOV, E.A., BORNHOLD, B.D. & FINE, I.V. 1999. The landslide-generated tsunami of November 3, 1994 in Skagway Harbor, Alaska: a case study. *Geophysical Research Letters*, **26**, 3009–3012.

RAMALHO, R.S., WINCKLER, G. ET AL. 2015. Hazard potential of volcanic flank collapses raised by new megatsunami evidence. *Science Advances*, **1**, https://doi.org/10.1126/sciadv.1500456

REIMNITZ, E. & MARSHALL, N.F. 1965. Effects of the Alaska earthquake and tsunami on recent deltaic sediments. *Journal of Geophysical Research*, **70**, 2363–2376.

ROMANO, F., PIATANESI, A. ET AL. 2012. Clues from joint inversion of tsunami and geodetic data of the 2011 Tohoku-oki earthquake. *Scientific Reports*, **2**, https://doi.org/10.1038/srep00385

RUBIN, K.H., FLETCHER, C.H., III & SHERMAN, C. 2000. Fossiliferous Lanai deposits formed by multiple events rather than a single giant tsunami. *Nature*, **408**, 675–681.

SAKUNA, D., SZCZUCIŃSKI, W., FELDENS, P., SCHWARZER, K. & KHOKIATTIWONG, S. 2012. Sedimentary deposits left by the 2004 Indian Ocean tsunami on the inner continental shelf offshore of Khao Lak, Andaman Sea (Thailand). *Earth, Planets and Space*, **64**, 931–943.

SATAKE, K., BOURGEOIS, J. ET AL. 1993. Tsunami field survey of the 1992 Nicaragua earthquake. *Eos, Transactions of the American Geophysical Union*, **74**, 145–157.

SATAKE, K., SHIMAZAKI, K., TSUJI, Y. & UEDA, K. 1996. Time and size of a giant earthquake in Cascadia inferred from Japanese tsunami records of January 1700. *Nature*, **379**, 246–249.

SATAKE, K., FUJII, Y., HARADA, T. & NAMEGAYA, Y. 2013. Time and space distribution of coseismic slip of the 2011 Tohoku earthquake as inferred from tsunami waveform data. *Bulletin of the Seismological Society of America*, **103**, 1473–1492.

SAWAI, Y., JANKAEW, K., MARTIN, M.E., PRENDERGAST, A., CHOOWONG, M. & CHAROENTITIRAT, T. 2009. Diatom assemblages in tsunami deposits associated with the 2004 Indian Ocean tsunami at Phra Thong Island, Thailand. *Marine Micropaleontology*, **73**, 70–79.

SAWAI, Y., NAMEGAYA, Y., OKAMURA, Y., SATAKE, K. & SHISHIKURA, M. 2012. Challenges of anticipating the 2011 Tohoku earthquake and tsunami using coastal geology. *Geophysical Research Letters*, **39**, L21309.

SELF, S. & RAMPINO, M.R. 1981. The 1883 eruption of Krakatau. *Nature*, **294**, 699–704.

SHANMUGAM, G. 2006. The tsunamite problem. *Journal of Sedimentary Research*, **76**, 718–730.

SHANMUGAM, G. 2012. Process-sedimentological challenges in distinguishing paleo-tsunami deposits. *Natural Hazards*, 1–26.

SHEPARD, F.P., MACDONALD, G.A. & COX, D.C. 1950. The tsunami of April 1st 1946 (Hawaii). *California University, Scripps Institute Oceanography Bulletin*, **5**, 391–528.

SHI, S., DAWSON, A.G. & SMITH, D.E. 1995. Coastal sedimentation associated with the December 12th, 1992 Tsunami in Flores, Indonesia. *PAGEOPH*, **144**, 525–536.

SHIKI, T., CITA, M.B. & GORSLINE, D.S. 2000. Sedimentary features of seismites, seismo-turbidites and tsunamiites – an introduction. *Sedimentary Geology*, **135**, vii–ix.

SHIKI, T., TSUJI, Y., YAMAZAKI, T. & MINOURA, K. 2008. *Tsunamiites*. Elsevier, Amsterdam.

SMITH, D.E., SHI, S. ET AL. 2004. The Holocene Storegga Slide tsunami in the United Kingdom. *Quaternary Science Reviews*, **23**, 2291–2321.

SMITH, D.E., FOSTER, I.D.L., LONG, D. & SHI, S. 2007. Reconstructing the pattern and depth of flow onshore in a palaeotsunami from associated deposits. *Sedimentary Geology*, **200**, 362–371.

SMITH, D.E., HARRISON, S. & JORDAN, J.T. 2013. Sea level rise and submarine mass failures on open continental margins. *Quaternary Science Reviews*, **82**, 93–103.

SMITH, R.C., HILL, J., COLLINS, G.S., PIGGOTT, M.D., KRAMER, S.C., PARKINSON, S.D. & WILSON, C. 2016. Comparing approaches for numerical modelling of tsunami generation by deformable submarine slides. *Ocean Modelling*, **100**, 125–140.

SPISKE, M., BÖRÖCZ, Z. & BAHLBURG, H. 2008. The role of porosity in discriminating between tsunami and hurricane emplacement of boulders – a case study from the Lesser Antilles, southern Caribbean. *Earth and Planetary Science Letters*, **268**, 384–396.

Spiske, M., Weiss, R., Bahlburg, H., Roskosch, J. & Amijaya, H. 2010. The TsuSedMod inversion model applied to the deposits of the 2004 Sumatra and 2006 Java tsunami and implications for estimating flow parameters of palaeo-tsunami. *Sedimentary Geology*, **224**, 29–37.

Spiske, M., Bahlburg, H. & Weiss, R. 2014. Pliocene mass failure deposits mistaken as submarine tsunami backwash sediments – An example from Hornitos, northern Chile. *Sedimentary Geology*, **305**, 69–82.

Stearns, H.T. 1978. Quaternary shorelines in the Hawaiian Islands. *Bernice P. Bishop Museum Bulletin*, **237**, 57.

Stearns, H.T. & Macdonald, G.A. 1946. Geology and groundwater resources of the Island of Hawaii. *Hawaii Division of Hydrography Bulletin*, **9, 393**.

Sugawara, D. & Goto, K. 2012. Numerical modeling of the 2011 Tohoku-oki tsunami in the offshore and onshore of Sendai Plain, Japan. *Sedimentary Geology*, **282**, 110–123.

Sugawara, D., Goto, K., Imamura, F., Matsumoto, H. & Minoura, K. 2012. Assessing the magnitude of the 869 Jogan tsunami using sedimentary deposits: prediction and consequence of the 2011 Tohoku-oki tsunami. *Sedimentary Geology*, **282**, 14–26.

Sugawara, D., Imamura, F., Goto, K., Matsumoto, H. & Minoura, K. 2013. The 2011 Tohoku-oki Earthquake Tsunami: similarities and differences to the 869 Jogan Tsunami on the Sendai Plain. *Pure and Applied Geophysics*, 170: 831, https://doi.org/10.1007/s00024-012-0460-1

Sugawara, D., Goto, K. & Jaffe, B.E. 2014. Numerical models of tsunami sediment transport – current understanding and future directions. *Marine Geology*, **352**, 295–320.

Svendsen, J.l. & Mangerud, J. 1990. Sea-level changes and pollen stratigraphy on the outer coast of Sunnmøre, western Norway. *Norsk Geologisk Tidsskrift*, **70**, 111–134.

Sweet, S. & Silver, E.A. 2003. Tectonics and slumping in the source region of the 1998 Papua New Guinea tsunami from seismic reflection images. *Pure and Applied Geophysics*, **160**, 1945–1968.

Switzer, A.D. & Burston, J.M. 2010. Competing mechanisms for boulder deposition on the southeast Australian coast. *Geomorphology*, **114**, 42–54.

Switzer, A.D., Pucillo, K., Haredy, R.A., Jones, B.G. & Bryant, E.A. 2005. Sea level, storm, or tsunami: enigmatic sand sheet deposits in a sheltered coastal embayment from southeastern New South Wales, Australia. *Journal of Coastal Research*, **21**, 655–663.

Synolakis, C.E. & Bernard, E.N. 2006. Tsunami science before and beyond Boxing Day 2004. *Philosophical Transactions of the Royal Society of London A: Mathematical, Physical and Engineering Sciences*, **364**, 2231–2265.

Szczuciński, W. 2011. The post-depositional changes of the onshore 2004 tsunami deposits on the Andaman Sea coast of Thailand. *Natural Hazards*, **60**, 115–133.

Szczuciński, W. 2013. Limitations of tsunami deposits identification – problem of sediment sources, sedimentary environments and processes, and post-depositional changes. *4th International INQUA Meeting on Paleoseismology, Active Tectonics and Archeoseismology (PATA)*. INQUA Focus Group on Paleoseismology and Active Tectonics, Aachen, Germany.

Szczuciński, W., Kokociński, M., Rzeszewski, M., Chagué-Goff, C., Cachão, M., Goto, K. & Sugawara, D. 2012. Sediment sources and sedimentation processes of 2011 Tohoku-oki tsunami deposits on the Sendai Plain, Japan – insights from diatoms, nannoliths and grain size distribution. *Sedimentary Geology*, **282**, 40–56.

Talling, P.J. 2014. On the triggers, resulting flow types and frequencies of subaqueous sediment density flows in different settings. *Marine Geology*, **352**, 155–182.

Talling, P.J., Wynn, R.B. *et al.* 2007. Onset of submarine debris flow deposition far from original giant landslide. *Nature*, **450**, 541–544.

Talling, P.J., Clare, M., Urlaub, M., Pope, E., Hunt, J.E. & Watt, S.F.L. 2014. Large submarine landslides on continental slopes: geohazards, methane release, and climate change. *Oceanography*, **27**, 32–45.

Tang, H. & Weiss, R. 2015. A model for tsunami flow inversion from deposits (TSUFLIND). *Marine Geology*, **370**, 55–62.

Tappin, D.R. 2007. Sedimentary features of tsunami deposits – their origin, recognition and discrimination: an introduction. *Sedimentary Geology*, **200**, 151–154.

Tappin, D.R. 2009. Mass transport events and their tsunami hazard. *In*: Mosher, D.C., Shipp, R.C., Moscardilli, L., Chaytor, J.D., Baxter, C.D.P., Lee, H.J. & Urgeles, R. (ed.) *Submarine Mass Movements and Their Consequences*. Springer Science + Business Media, Dordrecht, Heidelberg, London, New York, 667–684.

Tappin, D.R. 2010*a*. Digital elevation models in the marine domain: investigating the offshore tsunami hazard from submarine landslides. *In*: Fleming, C., Marsh, S.H. & Giles, J.R.A. (eds) *Elevation Models for Geoscience*. Geological Society, London, Special Publications, London, 81–101, https://doi.org/10.1144/SP345.10

Tappin, D.R. 2010*b*. Submarine mass failures as tsunami sources – their climate control. *Philosophical Transactions of the Royal Society A*, **368**, 2317–2368.

Tappin, D.R., Matsumoto, T. *et al.* 1999. Sediment slump likely caused 1998 Papua New Guinea Tsunami. *EOS, Transactions of the American Geophysical Union*, **80**, 329, 334, 340.

Tappin, D.R., Watts, P., McMurtry, G.M., Lafoy, Y. & Matsumoto, T. 2001. The Sissano Papua New Guinea tsunami of July 1998 – offshore evidence on the source mechanism. *Marine Geology*, **175**, 1–23.

Tappin, D.R., McNeil, L., Henstock, T. & Mosher, D. 2007. Mass wasting processes – offshore Sumatra. *In*: Lykousis, V., Sakellarious, D. & Locat, J. (eds) *Submarine Mass Movements and Their Consequences*. Springer, Dordrecht, 327–336.

Tappin, D.R., Watts, P. & Grilli, S.T. 2008. The Papua New Guinea tsunami of 17 July 1998: anatomy of a catastrophic event. *Natural Hazards and Earth System Sciences*, **8**, 243–266.

Tappin, D.R., Evans, H.M., Richmond, B., Sugawara, D. & Goto, K. 2012. Coastal changes in the Sendai area from the impact of the 2011 Tōhoku-oki tsunami: interpretations of time series satellite images, helicopter-borne video footage and field observations. *Sedimentary Geology*, **282**, 151–174.

TAPPIN, D.R., GRILLI, S.T. *ET AL.* 2014. Did a submarine landslide contribute to the 2011 Tohoku tsunami? *Marine Geology*, **357**, 344–361.

TEHRANIRAD, B., HARRIS, J., GRILLI, A., GRILLI, S., ABADIE, S., KIRBY, J. & SHI, F. 2015. Far-field tsunami impact in the North Atlantic Basin from large scale flank collapses of the Cumbre Vieja Volcano, La Palma. *Pure and Applied Geophysics*, **172**, 3589–3616.

TEN BRINK, U. 2009. Tsunami hazard along the U.S. Atlantic coast. *Marine Geology*, **264**, 1–3.

TEN BRINK, U.S., LEE, H.J., GEIST, E.L. & TWICHELL, D. 2009. Assessment of tsunami hazard to the U.S. East Coast using relationships between submarine landslides and earthquakes. *Marine Geology*, **264**, 65–73.

TEN BRINK, U.S., CHAYTOR, J.D., GEIST, E.L., BROTHERS, D.S. & ANDREWS, B.D. 2014. Assessment of tsunami hazard to the U.S. Atlantic margin. *Marine Geology*, **353**, 31–54.

TEN BRINK, U.S., ANDREWS, B.D. & MILLER, N.C. 2016. Seismicity and sedimentation rate effects on submarine slope stability. *Geology*, **44**, 563–566.

TINTI, S., BORTOLUCCI, E. & CHIAVETTIERI, C. 2001. Tsunami excitation by submarine slides in shallow-water approximation. *Pure and Applied Geophysics*, **158**, 759–797.

TSURU, T., PARK, J.-O., MIURA, S., KODAIRA, S., KIDO, Y. & HAYASHI, T. 2002. Along-arc structural variation of the plate boundary at the Japan Trench margin: implication of interplate coupling. *Journal of Geophysical Research*, **107**, 2357.

TURNER, A.K. & SCHUSTER, R.L. 1996. *Landslides: Investigation and Mitigation.* Special Report 247, Transportation Research Board, National Academy Press, Washington, DC.

TUTTLE, M.P., RUFFMAN, A., ANDERSON, T. & JETER, H. 2004. Distinguishing tsunami from storm deposits in eastern North America: the 1929 Grand Banks tsunami v. the 1991 Halloween storm. *Seismological Research Letters*, **75**, 117–131.

TWICHELL, D.C., CHAYTOR, J.D., TEN BRINK, U.S. & BUCZKOWSKI, B. 2009. Morphology of late Quaternary submarine landslides along the U.S. Atlantic continental margin. *Marine Geology*, **264**, 4–15.

ULVROVA, M., PARIS, R., NOMIKOU, P., KELFOUN, K., LEIBRANDT, S., TAPPIN, D.R. & MCCOY, F.W. 2016. Source of the tsunami generated by the 1650 AD eruption of Kolumbo submarine volcano (Aegean Sea, Greece). *Journal of Volcanology and Geothermal Research*, **321**, 125–139.

URLAUB, M., TALLING, P.J. & MASSON, D.G. 2013. Timing and frequency of large submarine landslides: implications for understanding triggers and future geohazard. *Quaternary Science Reviews*, **72**, 63–82.

VERBEEK, R.D.M. 1885. *Krakatau.* Government Press, Batavia.

VON HUENE, R., KLAESCHEN, D., CROPP, B. & MILLER, J. 1994. Tectonic structure across the accretionary and erosional parts of the Japan Trench margin. *Journal of Geophysical Research*, **99**, 22349–22361.

VON HUENE, R., KIRBY, S., MILLER, J. & DARTNELL, P. 2014. The destructive 1946 Unimak near-field tsunami: new evidence for a submarine slide source from reprocessed marine geophysical data. *Geophysical Research Letters*, **41, 6811–6818**.

WARD, S.N. 2001. Landslide tsunami. *Journal of Geophysical Research*, **106**, 11201–11215.

WARD, S.N. & DAY, S. 2001. Cumbre Vieja volcano – potential collapse and tsunami at La Palma, Canary Islands. *Geophysical Research Letters*, **28**, 3397–3400.

WASSMER, P., SCHNEIDER, J.-L., FONFRÈGE, A.-V., LAVIGNE, F., PARIS, R. & GOMEZ, C. 2010. Use of anisotropy of magnetic susceptibility (AMS) in the study of tsunami deposits: application to the 2004 deposits on the eastern coast of Banda Aceh, North Sumatra, Indonesia. *Marine Geology*, **275**, 255–272.

WATTS, P. 1997. *Water waves generated by underwater landslides.* Unpublished PhD thesis. California Institute of Technology, Pasadena, CA.

WATTS, P. 1998. Wavemaker curves for tsunamis generated by underwater landslides. *Journal of Waterway, Port, Coastal, and Ocean Engineering*, **124**, 127–137.

WATTS, P. 2000. Tsunami features of solid block underwater landslides. *Journal of Waterway, Port, Coastal, and Ocean Engineering*, **126**, https://doi.org/10.1061/(ASCE)0733-950X(2000)126:3(144)

WATTS, P. 2004. Probabilistic predictions of landslide tsunamis off Southern California. *Marine Geology*, **203**, 281–301.

WATTS, P., BORRERO, J.C., TAPPIN, D.R., BARDET, J.-P., GRILLI, S.T. & SYNOLAKIS, C.E. 1999. Novel simulation technique employed on the 1998 Papua New Guinea tsunami (abs). *Proceedings of 22nd General Assembly IUGG*, Birmingham, UK, JSS42.

WATTS, P., GRILLI, S.T., KIRBY, J.T., FRYER, G.J. & TAPPIN, D.R. 2003. Landslide tsunami case studies using a Boussinesq model and a fully nonlinear tsunami generation model. *Natural Hazards and Earth System Science*, **3**, 391–402.

WATTS, P., GRILLI, S.T., TAPPIN, D.R. & FRYER, G.J. 2005. Tsunami generation by submarine mass failure part II: predictive equations and case studies. *Journal of Waterway, Port, Coastal, and Ocean Engineering*, **131**, 298–310.

WEBSTER, J.M., George, N.P.J. *ET AL.* 2016. Submarine landslides on the Great Barrier Reef shelf edge and upper slope: a mechanism for generating tsunamis on the north-east Australian coast? *Marine Geology*, **371**, 120–129.

WEISS, R. 2008. Sediment grains moved by passing tsunami waves: tsunami deposits in deep water. *Marine Geology*, **250**, 251–257.

WEISS, R. 2012. The mystery of boulders moved by tsunamis and storms. *Marine Geology*, **295-298**, 28–33.

WEISS, R. & DIPLAS, P. 2015. Untangling boulder dislodgement in storms and tsunamis: is it possible with simple theories? *Geochemistry, Geophysics, Geosystems*, **16**, 890–898.

WITTER, R., JAFFE, B., ZHANG, Y. & PRIEST, G. 2012. Reconstructing hydrodynamic flow parameters of the 1700 tsunami at Cannon Beach, Oregon, USA. *Natural Hazards*, **63**, 223–240.

WRIGHT, C. & MELLA, A. 1963. Modifications to the soil pattern of South-Central Chile resulting from seismic and associated phenomenona during the period May to August 1960. *Bulletin of the Seismological Society of America*, **53**, 1367–1402.

YAMADA, M., FUJINO, S. & GOTO, K. 2014. Deposition of sediments of diverse sizes by the 2011 Tohoku-oki tsunami at Miyako City, Japan. *Marine Geology*, **358**, 67–78.

YAMAZAKI, Y., CHEUNG, K.F. & LAY, T. 2013. Modeling of the 2011 Tohoku Near-Field Tsunami from Finite-Fault Inversion of Seismic Waves. *Bulletin of the Seismological Society of America*, **103**, 1444–1455.

YEH, H., IMAMURA, F., SYNOLAKIS, C., TSUJI, Y., LIU, P. & SHI, S.Z. 1993. The Flores Island Tsunamis. *EOS, Transactions of the American Geophysical Union*, **74**, 369, 371–373.

Geological studies in tsunami research since the 2011 Tohoku earthquake

SIMON R. WALLIS[1]*, OSAMU FUJIWARA[2] & KAZUHISA GOTO[3]

[1]*Department of Earth and Planetary Sciences, Graduate School of Environmental Studies, Nagoya University, Nagoya, 464-8601, Japan*

[2]*Geological Survey of Japan, National Institute of Advanced Industrial Science and Technology (AIST), Tsukuba 305-8567, Japan*

[3]*International Research Institute of Disaster Science, Tohoku University, Sendai 980-0845, Japan*

**Correspondence: swallis@eps.nagoyau.ac.jp*

Abstract: The 2011 Tohoku earthquake, tsunami and resulting damage is often referred to as 3.11 after the date on which it took place. Leading to almost 20 000 people dead or missing, a major nuclear disaster and severe economic damage, 3.11 represents the biggest challenge faced by Japan since the end of World War II. Before 3.11, the possibility of a Mw 9 earthquake in this area was not generally recognized, highlighting the need to reassess seismic risk in NE Japan. The large amount of new quantitative data covering a range of disciplines and from onshore and offshore studies makes 3.11 an important case study that can contribute to improving our understanding of tsunamis, including their formation, their effects on coastal regions and the effectiveness of defensive measures old and new. Geological studies have a key role to play in this new phase of tsunami studies and this is the only method available for determining recurrence intervals over timescales of thousands of years. Data from 3.11 have improved our ability to identify sedimentary records of tsunami events and to estimate tsunami size from geological data. More complete databases will provide invaluable information for long-term planning of coastal regions in convergent plate margins.

The 2011 Mw 9.0–9.1 earthquake of NE Japan is commonly referred to as the Tohoku event. Its associated destruction has been given several different names, but commonly it is simply referred to as 3.11 after the date on which the earthquake occurred. This seems appropriate: it is date that will be the focus for study in many different fields for years to come and a date that will be marked in Japanese history; it has been identified by Prime Minister Kan as the biggest challenge for Japanese society since World War II.

The National Police Agency of Japan gives the number of dead and missing from the 3.11 earthquake and tsunami as 18 457 (NPA 2016) with close to 400 000 buildings destroyed. More than 90 per cent of the deaths are thought to have been caused by drowning related to the tsunami (Nakahara & Ichikawa 2013), but the low temperatures and lack of sufficient medical supplies were clearly also a factor in the immediate aftermath (Asahi Newspaper 2011). Large areas of residential and industrial buildings were damaged and many completely obliterated. The wreckage has been cleared, but there is only limited reconstruction in many of the most affected areas. The number of internal refugees reached a peak of about 470 000 (MIAC 2011) and still remained at around 200 000 almost five years after the event (NPA 2016). The long-lasting effects mark this out as one of the most destructive earthquake disasters ever recorded.

This is that fourth largest recorded earthquake – exceeded only by the Sumatra 2004 (Mw 9.2), Alaska 1964 (Mw 9.2) and Chile 1960 (Mw 9.5) events – and the resulting tsunamis affected many countries outside Japan. The global reach of the 3.11 Tohoku earthquake has also been recognized in many ways: a resulting reduction in the Earth's spin by a few microseconds and the calving of kilometre-sized icebergs in the Sulzberger ice shelf of the Antarctic – an event that had not happened for 50 years (Brunt *et al.* 2011). But one area where the Tohoku earthquake exceeds all others is the economic cost of more than 23 trillion yen (*c.* 220 billion USD) (Mahul & White 2012), making it the costliest natural disaster ever. This unparalleled natural disaster has also provided an unparalleled opportunity for scientists to obtain detailed real-time information about large earthquakes and related tsunami phenomena. Since 1993, Japan has invested heavily in developing a network of

From: Scourse, E. M., Chapman, N. A., Tappin, D. R. & Wallis, S. R. (eds) 2018. *Tsunamis: Geology, Hazards and Risks*. Geological Society, London, Special Publications, **456**, 39–53.
First published online July 18, 2017, https://doi.org/10.1144/SP456.12

GNSS-based control stations that provide continuous observation of crustal movements on land. A submarine network known as The Dense Ocean Floor Network System for Earthquakes and Tsunamis (DONET) can measure tilt and vertical displacement of the sea floor, but only became operational in August 2011, after 3.11. Data from the dense on-land geodesic networks allowed detailed analysis of the 3D movement of the Earth's surface before, during and after the 3.11 earthquake, providing valuable information that enabled fault slip and geometry to be constrained. There are also good records from surveys of the land before and after the tsunami allowing us to examine in detail the effects of the rise and fall of water on buildings, plants and soil and to some extent changes in the seafloor.

The 3.11 event followed only six years after the M_W 9.1–9.3 2004 Indian Ocean earthquake and tsunami, the latter claiming the lives of as many as 300 000 people in coastal communities of 12 different countries (e.g. Tohoku University 2005). The great human and economic cost of the two magnitude 9 tsunami earthquakes of the twenty-first century have provided a clear incentive to study tsunamis and their effects more closely. The resulting increase of associated research is clearly reflected in the numbers of published papers in earth sciences as well as other fields (Fig. 1). Academic societies have played their part in promoting the research efforts. One example is the pair of meetings held jointly by the Geological Societies of London and Japan: *Tsunami Hazards and Risks: Using the Geological Record* held in 2014 in Kagoshima, and the *Arthur Holmes Meeting – Tsunami Hazards and Risks: Using the Geological Record* held in 2015 in London. Both these emphasized the importance of geological studies in developing a better understanding of tsunamis. The results of the first meeting were summarized in a special issue of the journal *Island Arc* (2016).

Here we focus on the significance of the 3.11 tsunami for the geological community, in particular in Japan, and how Japanese government policies have attempted to incorporate a new recognition of what can happen and of how geological studies can be used to inform emergency planning scenarios. There have been 22 historical tsunamis in NE Japan, nine of which were associated with fatalities (e.g. Shuto *et al.* 2007; Goto *et al.* 2012) including the major Sanriku-Keicho 1611 and Jogan 869 events (Keicho and Jogan are taken from the names for the emperors' reigns during which the events occurred). Of these older events only the Jogan event is thought to be equivalent to 3.11 in both maximum height of the tsunami and regional distribution of the destruction (Goto *et al.* 2012). There have also been large tsunamis associated with volcanic eruptions such as the 1792 Shimabara

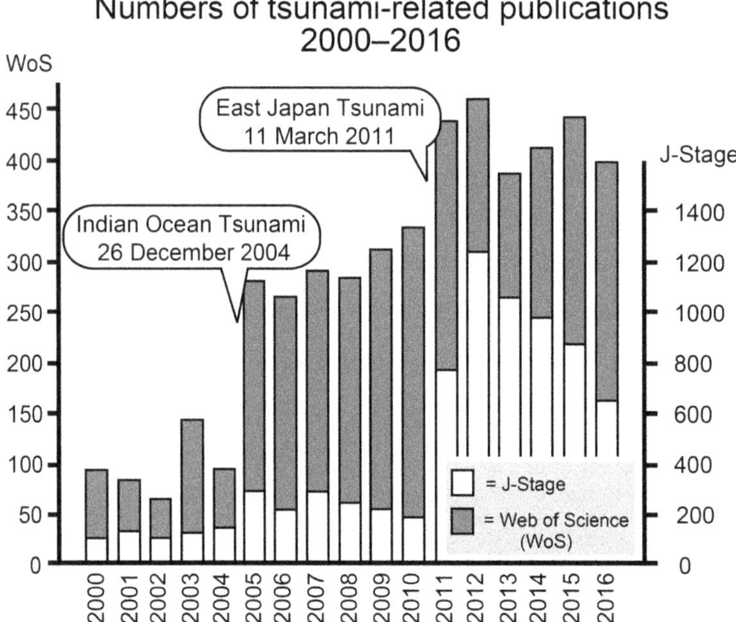

Fig. 1. Numbers of publications for the period 2000–16 related to tsunamis recorded in (i) Web of Science in the field of geosciences and (ii) J-Stage which closely reflects the state of research in science and technology within Japan. There are clear jumps in the numbers after both of the main destructive tsunami events of this century.

Mayuyama tsunami (or 島原肥後迷惑 in Japanese) in which about 15 000 people died (e.g. Aida 1975). Although such tsunamis are also a major potential hazard, here we restrict our discussion to earthquake-generated tsunamis with our main focus on the Tohoku tsunami of 2011. A fuller review of the state of tsunami research and tsunami-related disaster mitigation measures before and after the 2011 event is presented in Goto *et al.* (2014).

The effectiveness of tsunami countermeasures

Amongst all the destruction there were clearly cases where planning and the wisdom left to posterity by past generations helped lessen the effects of the tsunami. Pine forests began to be planted around the Japanese coast in the seventeenth century as a means of protecting against shifting sand. They also have the potential to help against storm surges and tsunamis. For 3.11 their contribution was mixed. In the Ishinomaki area, where the tsunami height was about 6 m, the forest significantly reduced the power and height of the tsunami. The trees also acted as a filter for some of the floating debris, reducing the impact on populated areas (Suppasri *et al.* 2013*a*, *b*). However, in other areas such as Rikuzentakata (Fig. 2), where the tsunami was about 11 m, the pine forests were completely destroyed, and the uprooted trees contributed to the destruction by forming floating debris (Suppasri

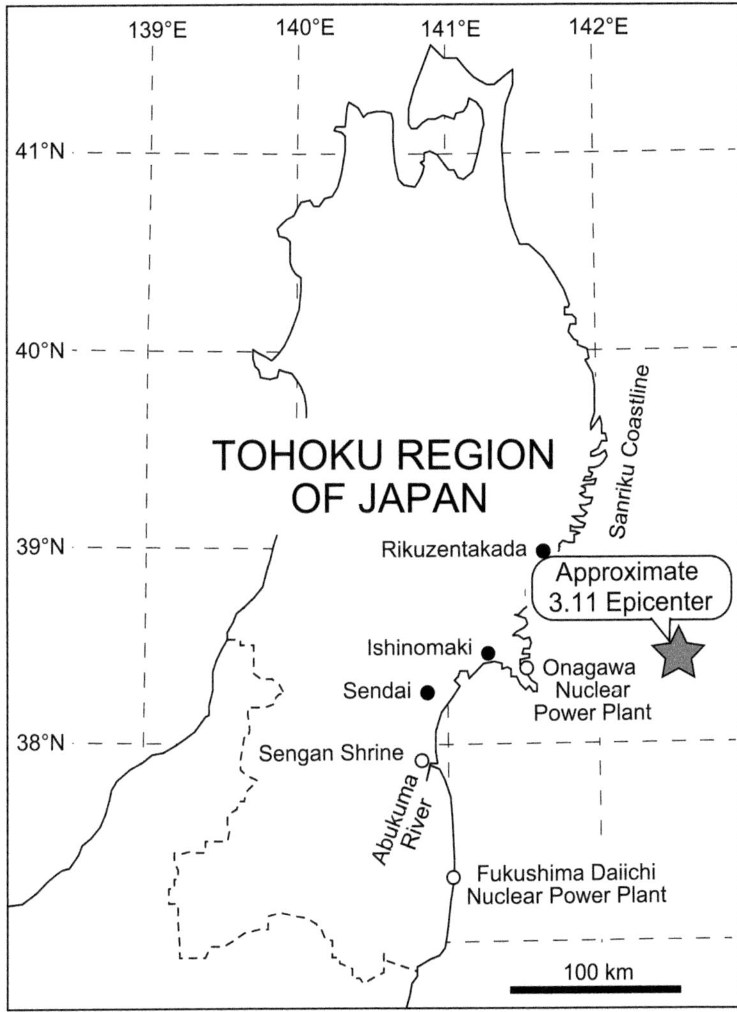

Fig. 2. Map showing main locations mentioned in the text concerning the Tohoku (or NE) area of Japan.

et al. 2013*b*). In some places breakwaters and sea walls also played their part in reducing the effect of the tsunami, but very few of the defensive measures offered much resistance in areas where the tsunami reached depths of 10 m or more (Suppasri *et al.* 2013*b*) and the overall effectiveness of sea walls as tsunami countermeasures has been brought into question (Aldrich & Sawada 2015). One noteworthy exception is the high sea defenses that protected the Onagawa nuclear power plant (Fig. 2). The Onagawa plant is a similar design to the ill-fated Fukushima Daiichi power plant and roughly 60 km closer to the earthquake epicentre, but suffered little damage from the tsunami. Yanosuke Hirai was a senior engineer in the electrical power business and a member of the Coastal Engineering Committee responsible for planning construction of the Onagawa plant. Hirai was keenly aware of the historical documents suggesting that NE Japan had suffered from large tsunamis in the past. In particular, he highlighted the Jogan tsunami of 869 (Ogata 2012). Records from several hundred years ago are imprecise, but some suggest that on one or more occasions tsunamis carried ships above the Sengan Shrine close to the Abukuma river (e.g. Hatori 1975; Fig. 2); this shrine is about 10 m above sea level. Hirai was successful in arguing that the Onagawa plant should be built 14.8 m above sea level specifically to reduce the risk of tsunami damage. This compares to the 10 m of defense for the Fukushima Daiichi plant. The actually tsunami height in Onagawa was close to 13 m and there was about 1 m of subsidence (IAEA 2012). It was a very close call, but thanks to the influence of Hirai, Japan was saved the ordeal of having to tackle multiple disasters at nuclear power plants.

The impact of the Indian Ocean Tsunami on tsunami studies

The 3.11 event generated the second large-scale tsunami of the twenty-first century and some of the recent trends in tsunami studies began as a result of the earlier 2004 Indian Ocean Tsunami. The widespread damage and large loss of life in Indian Ocean coastal regions raised awareness of the need to develop better responses to large earthquakes and in particular the associated tsunamis. This event also highlighted an important problem: the historical documentation of the area gathered over the last few hundred years was insufficient to accurately assess the earthquake and tsunami potential of the area. The only practical ways of accessing longer timescale information is by studying geological and archaeological records (Satake & Atwater 2007). Since the 2004 earthquake, studies of tsunami deposits in Indian Ocean coastal areas and geodynamic modelling have been used to show these areas have repeatedly been the sites of similar magnitude earthquakes and tsunamis in the geological past (e.g. Rajendran *et al.* 2006; Monecke *et al.* 2008; Malik *et al.* 2015). After the 2004 Indian Ocean event there was a clear increase in the attention being paid to palaeoseismicity and palaeotsunami studies based on geological observations by researchers – the following year saw a quantum leap in geoscience-related academic publications concerning tsunamis (Fig. 1). Such historical studies can help not only to reveal the past but also to indicate what the future holds. These studies are therefore an essential part in developing a clearer understanding of tsunamis in the scientific community as well as beyond it. Long-term records can also help guide efforts to develop effective tsunami countermeasures.

The unexpected magnitude of the East Japan Earthquake

Prior to 3.11 the Japanese government Headquarters for Earthquake Research Promotion had estimated that there was an 80–90 per cent chance of a magnitude 7.7–8.0 subduction-type earthquake associated with slip over lengths up to around 100 km occurring in the Sendai region of NE Japan (Fig. 3) sometime in the following 30 years (HERP 2002). But there was no mention of the possibility of a rupture over a length of 400–500 km or a magnitude 9 earthquake. There were suggestions from historical and geological studies that large earthquakes and associated tsunamis had affected the Pacific coastal region of the Tohoku region (e.g. Abe *et al.* 1990; Minoura & Nakaya 1991; Watanabe 2001; Namegaya *et al.* 2010) and there were limited examples of a very large earthquake and tsunami being taken into account in planning, such as the Onagawa nuclear power plant. In addition, the awareness of tsunami risk had been heightened as a result of the 2004 Indian Ocean event that occurred on a subduction zone in some ways comparable to NE Japan. However, this information was not reflected in changes in government forecasts. It is important to consider what signs were missed and what could have been done better. There are surely many reasons why the possibility of events such as 3.11 occurring in NE Japan were not more widely recognized (e.g. Sagiya 2011). We consider the following points to be particularly important: the limits of our ability to generalize about subduction processes; the limits of earthquake mechanism models; and insufficient use of geological data. Each of these points is discussed below.

(1) The limits of our ability to generalize about subduction processes

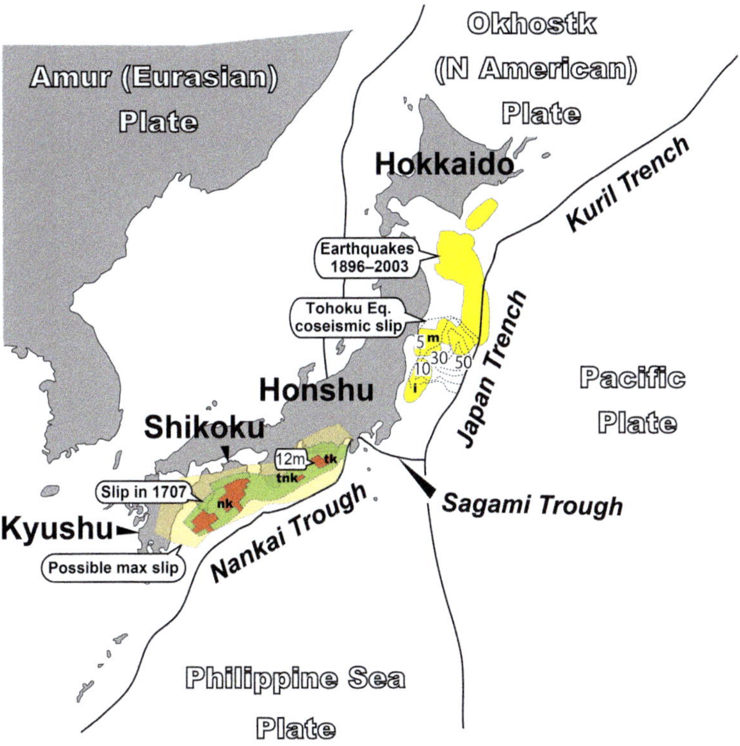

Fig. 3. Outline map of Japan and surrounding areas showing plate boundaries. The northern part of the map shows the estimated coseismic slip based on GPS data for the 3.11 Tohoku Earthquake (contours in metres) and domains of slip in the same region associated with earthquakes from 1896 to 2003. The previously documented slip domains wrap around the area of maximum estimated slip for the 2011 event. The southern part of the map shows the estimated slip region for the Hoei earthquake in 1707 (NIED 2015), which is the largest documented earthquake in this area, and the maximum slip region now considered in risk assessments in yellow. The domains with the largest estimated slip ($\geq$12 m) are shown with a separate contour. The slip domain follows the estimates presented in a report by the Disaster Management, Cabinet Office (2015). General areas of repeated large earthquakes are shown by the following symbols: m, Miyagi-Oki; i, Ibaragi-Oki; tk, Tokai; tnk, Tonankai; and nk, Nankai.

In the early 1980s, it was proposed that the magnitude of the largest earthquake that can occur at one particular subduction zone is closely related to the speed and age of the subducting oceanic plate (Ruff & Kanamori 1980). This idea has been highly influential in the field and is commonly generalized to signify that very large subduction-type earthquakes are characteristic of places where young oceanic slabs are subducting rapidly. Subsequent research showed that known Mw 9-class earthquakes occurred where the subducting plate is younger than 80 Ma and has a convergence rate of 30–70 mm a^{-1} (Stein & Okal 2007), offering support for the Ruff & Kanamori (1980) model. However, the Tohoku earthquake was caused by subduction of 130 Ma oceanic plate, some of the oldest oceanic lithosphere in the world, with a subduction rate of 80 mm a^{-1} (Muller *et al.* 1997). As more data become available on the rare mega earthquakes, it becomes clear that factors other than plate age and subduction speed need to be taken into account when looking for the controls on these very large subduction earthquakes.

(2) The limits of earthquake mechanism models

There are many different approaches to modelling earthquakes. Here we will briefly touch on the asperity and the characteristic models for earthquake mechanisms. These models have had a strong influence on the way that earthquake forecasting is considered in Japan and their limitations have been identified by some workers as a significant problem in failing to recognize the possibility of a Mw 9) earthquake in NE Japan (e.g. Kagan & Jackson 2013).

Many seismologists consider that the asperity model has been successful in accounting for a significant group of large earthquakes, in particular the repeating magnitude 7.5-class Miyagi-Oki earthquake that has a recurrence interval of around 40 years. In their long-term forecast, the Headquarters for Earthquake Research Promotion combined this model with historical earthquake records to warn that the next Miyagi-Oki earthquake was imminent. However, even when combined with observational data, long-term forecasting of large subduction earthquakes based on this type of modelling has its limitations. For the M 7.5-class Miyagi-Oki earthquakes the associated slip is about 2 m. However, this amount of slip is easily exceeded by the 3.2 m of movement expected in 40 years at the known plate convergence rate of about 80 mm a^{-1}. This means that the amount of movement recorded by the M 7.5 earthquakes did not sufficiently account for the total accumulated strain and that the amount of recorded slip during the known earthquakes was insufficient to account for the full amount of plate motion. Observations from the nationwide GPS network (GIAJ 2011) played a major role in demonstrating both a large slip deficit (Suwa *et al.* 2006) and a strong coupling between the plates. These features could be taken as warning signs that strain energy had not been fully released by the known earthquakes. We can now see that this information should have been used to reassess the earthquake forecast for the Tohoku region before 3.11. The amount of slip that occurred along the plate boundary in the Tohoku earthquake is thought to have reached a maximum that exceeded 50 m near the oceanic trough (e.g. Fujii *et al.* 2011) (Fig. 3). This amount of slip is equivalent to 600 years of accumulated plate movement. We now know that the amount of strain released every few tens of years in a series of well-documented large earthquakes was only part of the amount stored in the rocks, and only now are we developing an understanding that there is also a longer timescale series of major events on timescales of hundreds to thousands of years, possibly as a super cycle (Ikeda *et al.* 2012). With the benefit of hindsight, we can see that we should have been more aware of this possibility.

The characteristic earthquake model proposes that earthquake faulting is not random but tends to occur in the same area with similar time intervals and seismic magnitudes and is due to the breaking of similar patches, or asperities, along the fault surface. Before 2011, there was a general consensus amongst seismologists that this type of model could be applied to ocean-trench earthquakes. In terms of earthquake forecasting this model has far-reaching implications. When considering the long-term forecasting of earthquakes expected in the Pacific coastal region of Tohoku – which had been based on previous experience – it was expected that from offshore Aomori to Ibaraki the largest earthquake would be Mw 8.0–8.1, and that earthquakes would occur separately in these two areas. However, in 2011 four or five of these offshore asperity patches stretching from Iwate to Ibaraki ruptured in a chain of events that resulted in the very large Mw 9.0 earthquake. This can also be thought of as the combined effect of a series of domains, each of which has a potential for a Mw 7–8-class earthquake. It was known that the time interval between the Jogan-type events (*c.* Mw 9) in this area was 500–800 years (Sawai *et al.* 2012, 2015), and these domains may themselves represent one or multiple asperities.

The lack of an appropriate model to consider subduction-type earthquakes is part of the explanation for the forecasting failures, but a lack of sensitivity or openness to data that did not fit the available models is also important. It was already known before 2011 that in subduction zones where Mw 7–8-class earthquakes repeatedly occurred, there were also rare examples of these earthquake slip zones joining in one event to form a mega-thrust earthquake. Such behavior had been documented for subduction zones in the circum-Pacific and Indian Ocean areas for events since the sixteenth century (Satake & Atwater 2007), and this pointed to a weakness of the characteristic earthquake model. Two other indications that this model was not an adequate description of subduction-type events were: (i) the differences in the seismic waveforms of the five *c.* Mw 7 Miyagi-Oki earthquakes in the 1930s, 1978 and 2005, implying they were not repetitions of a characteristic event, and (ii) the recognition the seismic slip rate in large parts of the subduction zone along the Japan trench was about a quarter of the known plate convergence rate (Kanamori *et al.* 2006).

We do not have a good understanding of the conditions required for moderate-sized slip areas and earthquakes to link up and form a mega earthquake. But this is clearly one of the most important issues to be tackled in earthquake forecasting. This issue is particularly important for areas such as the Nankai

Trough coast where there is a known history of mega earthquakes and which has long been highlighted as a high-risk area.

(3) Insufficient use of geological data

A combination of historical and geological studies has revealed evidence for an ancient event in the Tohoku area that appears to be closely comparable to the 2011 earthquake. In July 869, a large tsunami reshaped the coast of NE Japan and the results are preserved in a thick layer of tsunami deposits. The 869 Jogan earthquake and tsunami can now be seen as a harbinger of 3.11 (e.g. Goto *et al.* 2011; Sawai *et al.* 2012; Namegaya & Satake 2014) and a contemporary description of the destruction sounds prophetically familiar:

> ... the sea roared like thunder and the tide surged, reversing the currents of rivers. Waves crested in long succession, pounding the shoreline. The waters reached far inland and it was not long before the centre of the town was awash. Fields and roads sank into the sea and a thousand people drowned, because they failed to escape by ship or uphill from the waves. There was almost nothing left of many people's properties and crops (Translation based on an annotated Japanese version of the original text – Takeda & Sato, 2009, p. 454).

Since the late 1980s there have been several studies of the Jogan tsunami and its effects. From studies of the distribution of sandy tsunami deposits, we now estimate that the minimum inundation area was 3–4 km inland from the contemporary shoreline (Minoura *et al.* 2001; Sawai *et al.* 2008, 2012) and is closely comparable with the effects of the 3.11 tsunami. The Jogan tsunami deposits can also be closely compared to those developed by the 3.11 tsunami (Fig. 4). Many of the features of the Jogan event were already known in 2011, and there was active discussion on how to publicize the latest scientific findings at the time of 3.11 (Okamura 2012). One of the difficulties of using such information is that the interpretations are associated with large uncertainties and there was no good consensus on how to incorporate the information in risk assessment for the region. It is important to point out that 3.11 has shown the need not only to more carefully incorporate geological and archaeological data into

Fig. 4. Section through the uppermost *c.* 1.5 m of the deposits of the Sendai coastal plain showing three major sand layers thought to be related to the 869 Jogan, 1611 Keicho, and 2011 Tohoku tsunamis, and the 915 Towada-a tephra layer (Kawamata 2015). Identification of the older deposits is provisional and verification will require more complete documentation of their ages and origins.

long-term risk assessment but also to revise interpretations of the geological data. In the light of information gathered from the 2011 event, the size of the earthquake estimated to be associated with the Jogan deposits has been revised from an original estimate of Mw >8.3–8.4 (Satake *et al.* 2008; Namegaya *et al.* 2010; Sugawara *et al.* 2011) to a new estimate of Mw >8.6 (Namegaya & Satake 2014). Estimating earthquake magnitude from tsunami deposits relies on using knowledge of the extent of such deposits to estimate the extent of tsunami inundation. The Jogan inundation extent was based on tsunami deposits identifiable in the field – in practice these were deposits with thicknesses of 5 mm or more (e.g. Goto *et al.* 2011). This was clearly a minimum estimate of the actual inundation and the estimates of the associated magnitude were also given as minimum. Not surprisingly, 3.11 provided a concrete and well-documented example of the relationships between tsunami deposits identifiable in the field and tsunami inundation. This showed that at the limit of tsunami deposits identifiable in the field the tsunami depth was still greater than 1 m. Applying this new insight to the Jogan tsunami resulted in an increase in the estimated minimum magnitude (Namegaya & Satake 2014). The current estimates for the recurrence interval of a Jogan-type earthquake based on the geological record over the last 3000 years is about 500–800 years (Sawai *et al.* 2012, 2015).

Using geological studies to build on lessons learned from 3.11

Prior to 3.11 there was no general appreciation in the Japanese seismological community that a Mw 9-class earthquake could occur in the Tohoku region. It was recognized that the amount of elastic energy stored by subduction of the Pacific Plate was considerably less than the amount of energy released by seismic events. However, the cause of this discrepancy was variously blamed on uncertainties in the measurements or the presence of non-seismic slip that could not be detected from land-based observations. The failure to consider the significance of these discrepancies and highlight the possibility of a mega earthquake in this region is one of the great regrets and failings of the earthquake-science community in Japan. It is also clear that more use can be made of geological and archaeological records of tsunamis. There are many lessons to take from the 2011 tsunami and earthquake in Tohoku. Below, we highlight three aspects of palaeoseismology with particular focus on sedimentary evidence for palaeotsunamis: first we consider the diversity of trench-type earthquakes and tsunamis; second we discuss the possible existence of seismic supercycles; and third we consider that mega earthquakes might not be limited to a few subduction zones.

The diversity of trench-type earthquakes and tsunamis

One of the main unforeseen characteristics of the 2011 earthquake was the simultaneous linking that occurred between several seismically active domains. But in terms of the tsunami formation, equally important and unexpected was the way that the slip region along the plate interface reached both shallower and deeper regions than normal. Rapid slip at shallow levels made a major contribution to the large size of the resulting tsunami. The complex composite slip over a wide region shown by the 2011 event, including movement at shallower and deeper levels than normal, shows both the sort of diversity that needs to be taken into account when considering forecasting of oceanic trench-type earthquakes, and some of the limitations of the characteristic earthquake model (Lay & Kanamori 2011).

Before 3.11, the area considered most likely as the location of the next mega earthquake was thought to be the Nankai Trough region. Depending on one's point of view, 3.11 has highlighted or confirmed that with our present understanding of earthquakes it is not possible to forecast the location or the amount and distribution of slip associated with subduction zone earthquakes. As a result, there is a need to recast the long-term evaluation of earthquakes and seismic risk in SW Japan that has for many years been based on the characteristic earthquake model and included separate consideration of the Tokai, Tonankai and Nankai earthquakes. The new approach is to treat the whole of the Nankai Trough seismogenic zone as a single large area with the potential for nucleation of large earthquakes throughout. This pattern is reflected in the record of Mw 8–9 earthquakes that have occurred with timescales of 100–200 years in this area (HERP 2013) (Fig. 3).

The reason for the diversity in earthquakes is still unclear and an important subject for research. In addition to seismic monitoring using modern geophysical techniques and mechanical modelling, it is also important to gather accurate information on as many past events as possible. Clarifying the diversity in seismic activity including rare events requires study over long timescales and is an activity where documenting geological records of past mega earthquakes plays a central role.

The existence of seismic supercycles

Based on work on several different plate convergence domains, it has been proposed that in order to understand the way in which energy is first accumulated and then released during subduction, we may have to look beyond the range of well-documented large earthquakes, with repeat times of tens to hundreds of years, and consider the existence of supercycles with much longer repeat times. Examples where the existence of such supercycles has been proposed are: Chile, based on the variation of sizes in crustal movements and tsunamis among the great earthquakes (Cisternas *et al.* 2005); Sunda Trough, based on the patterns of surface deformation (Sieh *et al.* 2008); and the North American Pacific region, based on the scale of turbidite deposits and their recurrent intervals (Goldfinger *et al.* 2013). A brief survey of these examples shows that even if such supercycles can in general be defined, there must exist a great variety. The geological information about timing and magnitude of earthquakes that is provided by tsunami deposits is key not only as a basis for examining the variety but also for testing predictions of this hypothesis.

If the proposed supercycles do exist, it is clear that the geophysical records and historical documents do not adequately cover the timescales needed to characterize them. In some cases, historical records may stretch back several hundred years, but even in the most complete cases, the information on the amount of shaking, the height of the tsunami and area of inundation is only semi-quantitative, and the coverage is limited both in the number of events and affected areas. Although historical records can incorporate the experiences of many generations, they still represent a short period in terms of geological processes. The result of these limitations is that the magnitudes of the recorded earthquakes and tsunamis are likely to be underestimated. A prime motivation for tsunami research is to assist in disaster mitigation. But if the estimates provided by academic studies turn out to be smaller than the long-term averages, this will surely lead to more failures like that of the earthquake forecasts before 3.11 (e.g. Sugimoto 2014). The need for information that covers timescales appropriate to assessing long-term earthquake and tsunami potential can only be addressed through geological studies.

Mega earthquakes are not limited to a few subduction zones

As outlined above, there is no generally accepted model to account for the formation of Mw 9-class mega earthquakes. The global rate of Mw 9 earthquakes in subduction zones predicted from the statistical analyses of seismicity and moment conservation is about five per hundred years, and since the start of the twentieth century there have been precisely five such events (Kagan & Jackson 2013). Because these events are rare it is likely that a long time will pass before one occurs in any given area. With our improved knowledge of tsunami deposits and the geological significance of their records, we can now see that even before the 2004 and 2011 tsunamis, studies of sediments in many subduction zones showed clear signs that many areas had suffered much larger earthquakes and tsunamis than any recorded in historical documents. Examples where this is the case are found not only in Japan but also along the North American Pacific coast (e.g. Atwater 1987), and in the Aleutians (Witter *et al.* 2016) and the Indian Ocean coastal regions (e.g. Monecke *et al.* 2008).

Post-2011 there is a greater appreciation of the long timescales that need to be considered in risk assessment. Geology and archaeology are the only effective approaches to access such long-timescale information, and the increased attention given to events that happened thousands of years ago means that there is now a greater focus on the recognition and interpretation of ancient tsunami deposits. One of the important issues is to determine the cause of unusual marine incursions. Distinguishing between storms and tsunami events in the geological record is a challenge (Morton *et al.* 2007; Goff *et al.* 2012; Fujiwara & Tanigawa 2014), but a necessary step in order to properly relate sedimentary information to sudden changes in the seafloor topography. There is also a growing awareness of some of the potential complexities involved in tsunami generation. Movement on the main plate boundary represented by the frontal thrust of a subduction system is clearly of prime importance, but locally smaller scale movements on splay faults and submarine sliding may also play crucial roles (e.g. Plafker 1965; Clarke & Carver 1992; Park *et al.* 2000; Tappin *et al.* 2001, 2014; Kawamura *et al.* 2014). Our understanding of the 2011 tsunami is improving as research continues in areas such as: on-land and submarine geodesy and geodynamic modelling of post-tsunami ground movements (e.g. Iinuma *et al.* 2011; Ozawa *et al.* 2011); physical examination of the seismic fault zone in bore cores collected from the frontal thrust (Chester *et al.* 2013; Fulton *et al.* 2013; Ujie *et al.* 2013); and documenting the changes in sedimentation on the submarine floor through seismic surveys and ocean-floor sampling (Strasser *et al.* 2013; Tamura *et al.* 2015; Yoshikawa *et al.* 2015; Ikehara *et al.* 2016).

Developing a new direction in tsunami forecasting

The scientific forecasting of earthquakes and tsunami needs to incorporate information from a range of different sources some of which may be only semi-quantitative such as historical documents. Predictions based on this information, including the appropriate uncertainties, can be made using physical modelling. It must also be borne in mind that the results should be of practical use to society. The statistical analysis of Kagan & Jackson (2013) suggests all subduction zones are liable to produce very large earthquakes up to Mw 10. However, such large earthquakes have never been recorded and if they do exist they would only be as very rare events. Geological data such as tsunami sediments can provide a long-term record of events that have actually occurred in a particular area (e.g. Garrett *et al.* 2016).

Furthermore, there are limits both to the resolution possible with age dating of sediment and the preservation potential of tsunami-related deposits. Offshore deposits may preserve a better record than on-land deposits (Ikehara *et al.* 2016). Another important issue is that with our present level of understanding it is not possible to determine whether a large palaeo event was the result of full-length ruptures or a swift series of shorter ruptures (Goldilnger *et al.* 2003; Satake & Atwater 2007).

One of the priorities for the next stage in tsunami studies is developing better quantitative estimates of the timing and size of palaeotsunamis. Planning for rare but very large events is a priority for society, and a recent trend is to consider what would be appropriate as the maximum magnitude or largest tsunami that should be taken into account. There is a need to consider the most appropriate definition of the terms 'maximum' and 'largest' when applied to earthquakes and tsunamis (the largest historical event, the largest theoretically possible event, or perhaps the largest event in the Holocene?), and consider which of these is most appropriate for the timescale of the future we wish to plan for. To move forward with these studies more work such as that already carried out in Hokkaido and the Sendai Plain and offshore is needed to document and quantify past earthquakes and tsunamis. In addition to an increased database, there is also a need to be more aware of the limitations of any model proposed and our own imperfections as interpreters of the information – the importance of data that do not fit the prevailing paradigm can easily be overlooked. A Japanese proverb taken from the Chinese *The Art of War* by Sun Tzu seems apt here:

彼を知り己を知れば百戦殆うからず.

彼を知らずして己を知れば，一勝一負す.

彼を知らず己を知らざれば，戦う毎に必ず殆し.

which translates as 'if you know the enemy and know yourself, you need not fear the result of a hundred battles; if you know yourself but not the enemy, for every victory gained you will also suffer a defeat; if you know neither the enemy nor yourself, you will succumb in every battle'.

Government policy adapting to new developments

Following 3.11 a major review was undertaken of the Japanese government's policies for tsunami disaster management and mitigation measures. The results were published in a series of reports by the Central Disaster Management Council (CDMC). The first and most important change was a recommendation to incorporate estimates for the maximum possible earthquake and tsunami that could affect the coastline of Japan. Such estimates are fraught with difficulties but one of the most valuable data sources to inform these decisions is the geological record of palaeotsunami deposits. This led to the second major recommendation, which is that risk assessments should be based on information from a wide variety of different sources including geology and archaeology. This second recommendation reflects the realization that long-term records are required that extend beyond the available seismic data. Consequently, the profile of tsunami geology has increased, and the results of such studies are sought after for inclusion in disaster mitigation planning (e.g. CDMC 2011). The third notable recommendation was made to assist local authorities in their disaster mitigation planning (CDMC 2011) and proposes that tsunamis be divided into two separate levels. A level 1 tsunami would be one that occurs relatively frequently on timescales of a few tens to hundreds of years. The estimated tsunami height of such tsunamis should form the basis for construction of shore protection facilities. A level 2 tsunami would be much rarer, occurring on timescales of 1000 years or more – similar to the type that affected NE Japan in 2011. Normal shore protection facilities would probably be insufficient to protect populated areas against the effects of tsunamis of this size; they would probably also be associated with severe damage to property and loss of life. However, natural tsunamis are unlikely to divide easily into two separate groups in this way, and the implementation of these ideas requires careful consideration of their societal implications.

The large fault slip of up to 69 m (Satake *et al.* 2013) and major changes in the stress field – many areas changing from horizontal compression to extension (Yoshida *et al.* 2012) – suggest the forces that caused the 2011 Tohoku earthquake have now been released and that it will be many generations before a similar-sized earthquake occurs in the same area. However, there are domains along the same plate boundary beneath Japan where large slip did not take place in 2011 (e.g. Simons *et al.* 2011). Areas to the south (Simons *et al.* 2011) and north of the 2011 slip zone are now more liable to large displacement slip associated with a large magnitude earthquake (Tanigawa *et al.* 2014). The Tsushima–Kamchatka Trench off Hokkaido (Fig. 3) is also a place of concern. Previous assessments of the earthquake hazard in eastern Hokkaido have pointed out the potential for very large earthquakes occurring on timescales as long as 500 years (e.g. Nanayama et al. 2003; CDMC 2006). This is a rare, if not unique, example of geological information being incorporated into policy-making related to disaster prevention countermeasures prior to 2011 event. But following the revisions in the CDMC recommendations, a reassessment of the risk was undertaken by the Hokkaido local government, and contingency planning now incorporates consideration of tsunamis several times larger than previously (see http://www.pref.hokkaido.lg.jp/sm/ktk/bsb/tunami/index.htm) and an earthquake sufficiently great to account for the recognized tsunami effects. This is an example of how studies of tsunami deposits have had a tangible effect on government policy and paves the way for similar future developments elsewhere (Hokkaido Local Government 2012).

Before 3.11, SW Japan was highlighted as the area most likely to be affected by a major earthquake; the area is underlain by the Philippine Sea plate, and the risks associated with the subduction that begins in the Nankai Trough (Fig. 3) remain high. Large centres of population such as Osaka, Kobe and Nagoya cities are all located in SW Japan and there is a potential for unprecedented loss of life and property after a mega earthquake in this area. The plate boundary beneath SW Japan has been the location of numerous historically documented Mw 8-class earthquakes with relatively short recurrence intervals of 90–250 years. Source regions of these destructive events have been semi-empirically divided into five distinct segments associated with three rupture zones along the plate boundary. Large earthquakes caused by movement on these rupture zones are referred to as the Tokai, Tonankai and Nankai earthquakes (Fig. 3). In some cases, the fault segments move separately, but the largest earthquakes develop when this slip propagates from one segment to another resulting in a larger total slip area. Since 3.11 there is a renewed urgency in the efforts to assess the potential for a massive Nankai Trough earthquake where the Tokai, Tonankai and Nankai earthquakes occur together. Following the policy changes mentioned above, the CDMC issued a report examining the effects of a magnitude 9.1 earthquake in the Nankai Trough. This report suggests a possible maximum tsunami run-up height of about 30 m, estimated fatalities of up to 320 000 people and an economic loss of 220 trillion yen (*c.* 2 trillion USD) (e.g. Disaster Management, Cabinet Office 2015). As mentioned in the report, it is important to bear in mind that these estimates are not based on historical or geological evidence and the damage far exceeds the known effects of the largest historical event: the 1707 Hoei earthquake of magnitude 8.6. The Hoei earthquake is thought to have formed when all the segments along the Nankai Trough slipped together.

In addition to reassessment of areas of well-documented high seismic risk, the CPPDM (2012) also proposes greater study of areas that hitherto have been given a relatively low priority. One example is the Sea of Japan coastline. Seismicity shows this area is under compression, and major seismic events have occurred. However, the maximum long-term potential for tsunami and earthquake damage has not been closely studied. The CPPDM (2012) report also proposes that risk assessment for Tokyo inland earthquakes should include possible major earthquakes along the Sagami Trough – in 2004 the maximum magnitude considered was Mw 8.1 but the Headquarters for Earthquake Research Promotion (2014) has increased this estimate to Mw 7.9–8.6.

Disseminating information of tsunami geological records

Earth science is a broad field that encompasses experts with a wide range of different skill sets and expertise. The Tohoku event has shown there is a greater need to bring these different groups of workers together so that future modelling of tsunamis could include information derived from seismology, studies of ground movement and geology. Interdisciplinary studies – including studies of archaeological sites and historical documents – can also provide valuable information to help reveal the frequency and size of past tsunamis. In addition, we have been reminded of the need to focus on gaps and inconsistencies in our understanding that may be concealing important unknown phenomena.

Better understanding of tsunamis and how they form can contribute to saving human life. Scientific research is an essential part of this process, but there is also a need to make the most of our current knowledge, even in its present incomplete state. An important issue is communication of the information that tsunami studies can usefully provide. The expectations placed upon researchers by people in local government and the general public at large can easily exceed that which is feasible with our current state of understanding. This requires efforts to develop an awareness campaign that includes scientific communication as a part of disaster education.

For tsunami geologists there is a need to improve the clarity of explanations of what is known and the associated degrees of uncertainty. This information should both inform and underpin development of new disaster mitigation policy.

A central issue in developing effective tools for communication is to consider the appropriate methods by which the results of scientific research are presented to the public. One of the most readily understood ways is to provide maps of inundation areas. In their pioneering work, Satake & Nanayama (2004) provide a sequence of maps for both historical and prehistoric tsunamis and how they affected the area of eastern Hokkaido. This work explained the estimated tsunami inundation area during the last '500 year-interval earthquake' that occurred in the early seventeenth century: its heights were in excess of 15 m in parts of Hokkaido. These geological data have been incorporated in the risk assessment for the areas given by the CDMC (2006). A recent development of this approach is provided by the Japan Tsunami Trace Database (http://tsunami-db.irides.tohoku.ac.jp/?LANG=-2), which includes in excess of 30 000 sets of information on the heights of tsunamis based on historical records. This database also includes estimates of the uncertainties associated with tsunami height values and is freely available on the web. The great strength of this type of approach is not only the ease of access to information by the general public, but also the possibility that it presents for incorporating new data as they become available. A similar idea lies behind the database systems for tsunami deposits that have been developed in some other countries (e.g. Goff *et al.* 2010; Goff & Chagué-Goff 2014). There is ongoing work to develop a comparable database for Japan fronted by the National Institute of Advanced Industrial Science and Technology (AIST) (https://gbank.gsj.jp/tsunami_deposit_db/) and other institutions. It is hoped that providing a user-friendly system with sufficient information will help local government and the general public gain a better understanding of their local tsunami histories and the type of information that can be extracted from historical and geological records.

Concluding remarks

This short review attempts to highlight key issues for the earth science community – in particular geologists – that have arisen from 3.11. Improved earthquake forecasting and understanding of the effects of tsunamis are required to assist redevelopment of the areas affected by 3.11, to identify areas liable to large tsunami events, and to take suitable and effective measures against future tsunamis. Failure to adequately plan for the 3.11 event in NE Japan show the need both to revise models of seismicity at convergent margins and to develop our knowledge of long-timescale records of tsunamis. This clearly requires us to develop our understanding of geological records and how to interpret them in terms of tsunamis.

We are grateful to Ellie Scourse for her patience in handling this manuscript and the two reviewers, David Tappin and Robert Geller, whose critical but thoughtful reviews helped improve the manuscript. Dylan McGee of Nagoya University is thanked for his advice concerning translation of the Nihonsandai Jitsuroku. This research was partially funded by the JSPS grant 16H06476.

References

Abe, H., Sugeno, Y. & Chigama, A. 1990. Estimation of the height of the Sanriku Jogan 11 earthquake-tsunami (A.D. 869) in the Sendai plain. *Jishin (Journal of the Seismological Society of Japan, 2nd ser.)*, **43**, 513–525 [in Japanese with English abstract].

Aida, I. 1975. Numerical experiments of the tsunami associated with the collapse of Mt. Mayuyama in 1792. *Jishin (Journal of the Seismological Society of Japan, 2nd ser.)*, **28**, 449–460 [in Japanese with English abstract].

Aldrich, D.P. & Sawada, Y. 2015. The physical and social determinants of mortality in the 3.11 tsunami. *Social Science and Medicine*, **124**, 66–75.

Asahi Newspaper 2011. Earthquake evacuation centers record 27 deaths: problems with cold and lack of medical resources [in Japanese], http://www.asahi.com/special/10005/TKY201103170134.html

Atwater, B.F. 1987. Evidence for great Holocene earthquakes along the outer coast of Washington State. *Science*, **336**, 942–944.

Brunt, K.M., Okal, E.A. & MacAyeal, D.R. 2011. Antarctic ice-shelf calving triggered by Honshu earthquake and tsunami, March 2011. *Journal of Glaciology*, **57**, 785–788, https://doi.org/10.3189/002214311798043681

CDMC (Central Disaster Management Council) 2006. *Report of damage estimation by the subduction zone earthquake along Japan and Kuril trenches*, http://www.bousai.go.jp/kaigirep/chousakai/tohokukyokun/pdf/Report.pdf

CDMC (Central Disaster Management Council) 2011. *Report of the Committee for Technical Investigation on Countermeasures for Earthquakes and Tsunamis based on the lessons learned from the 2011 off the Pacific coast of Tohoku earthquake*, http://www.bousai.go.jp/kaigirep/chousakai/tohokukyokun/pdf/Report.pdf

Chester, F.M., Rowe, C.R. et al. 2013. Structure and composition of the plate-boundary slip zone for the 2011 Tohoku-Oki Earthquake. *Science*, **342**, 1208–1211, https://doi.org/10.1126/science.1243719

Cisternas, M., Atwater, B.F. et al. 2005. Predecessors of the giant 1960 Chile earthquake. *Nature*, **437**, 404–407.

CLARKE, S.H., JR. & CARVER, G.A. 1992. Late Holocene tectonics and paleoseismicity, Southern Cascadia subduction zone. *Science*, **255**, 188–192, https://doi.org/10.1126/science.255.5041.188

CPPDM (COMMITTEE FOR POLICY PLANNING ON DISASTER MANAGEMENT) 2012. *Toward the reconstruction for sound and unwavering Japan*, http://www.bousai.go.jp/kaigirep/chuobou/suishinkaigi/english/ [last accessed 22 June 2017].

DISASTER MANAGEMENT, CABINET OFFICE 2015. *Regarding the expected damage after a massive Nankai earthquake: second report*, [in Japanese], http://www.bousai.go.jp/jishin/nankai/nankaitrough_info.html

FUJII, Y., SATAKE, K., SAKAI, S., SHINOHARA, M. & KANAZAWA, T. 2011. Tsunami source of the 2011 off the Pacific coast of Tohoku Earthquake. *Earth, Planets and Space*, **63**, 815–820, https://doi.org/10.5047/eps.2011.06.010

FUJIWARA, O. & TANIGAWA, K. 2014. Bedforms record the flow conditions of the 2011 Tohoku-oki tsunami on the Sendai Plain, northeast Japan. *Marine Geology*, **358**, 79–88.

FULTON, P.M., BRODSKY, E.E. *ET AL*. 2013. Low coseismic friction on the Tohoku-Oki fault determined from temperature measurements. *Science*, **342**, 1214–1217, https://doi.org/10.1126/science.1243641

GARRETT, E., FUJIWARA, O. *ET AL*. 2016. A systematic review of geological evidence for Holocene earthquakes and tsunamis along the Nankai-Suruga Trough, Japan. *Earth-Science Reviews*, **159**, 337–357.

GIAJ (GEOSPATIAL INFORMATION AUTHORITY OF JAPAN) 2011. *Crustal movements in the Tohoku district*. Report of the Coordinating Committee for Earthquake Prediction, Japan, **86**, 184–272 [in Japanese].

GOFF, J. & CHAGUÉ-GOFF, C. 2014. The Australian tsunami database – a review. *Progress in Physical Geography*, **38**, 218–240.

GOFF, J., NICHOL, S.L. & KENNEDY, D. 2010. Development of a palaeotsunami database for New Zealand. *Natural Hazards*, **54**, 193–208.

GOFF, J., CHAGUÉ-GOFF, C., NICHOL, S.L., JAFFE, B. & DOMINEY-HOWES, D. 2012. Progress in palaeotsunami research. *Sedimentary Geology*, **243–244**, 70–88.

GOLDILNGER, C., NELSON, C.H. & JOHNSON, J.E. 2003. Holocene earthquake records from the Cascadia subduction zone and northern San Andreas fault based on precise dating of offshore turbidites. *Annual Review of Earth and Planetary Sciences*, **31**, 555–577.

GOLDFINGER, C., IKEDA, Y., YEATS, R.S. & REN, J. 2013. Superquakes and supercycles. *Seismological Research Letters*, **84**, 24–32.

GOTO, K., CHAGUÉ-GOFF, C. *ET AL*. 2011. New insights of tsunami hazard from the 2011 Tohoku-oki event. *Marine Geology*, **290**, 46–50.

GOTO, K., CHAGUÉ-GOFF, C., GOFF, J. & JAFFE, B. 2012. The future of tsunami research following the 2011 Tohoku-oki event. *Sedimentary Geology*, **282**, 1–13.

GOTO, K., FUJINO, S., SUGAWARA, D. & NISHIMURA, Y. 2014. The current situation of tsunami geology under new policies for disaster countermeasures in Japan. *Episodes*, **37**, 258–264.

HATORI, T. 1975. Tsunami magnitude and wave source regions of historical Sanriku tsunamis in northeast Japan. *Bulletin of the Earthquake Research Institute*, **70**, 397–414 [in Japanese with English abstract].

HERP (HEADQUARTERS FOR EARTHQUAKE RESEARCH PROMOTION) 2002. *Evaluation of Long-Term Earthquake Activity from the Sanriku to Boso Offshore Regions*, HERP, Japan, [in Japanese], http://www.jishin.go.jp/main/chousa/kaikou_pdf/sanriku_boso.pdf [last accessed 22 June 2017].

HERP (HEADQUARTERS FOR EARTHQUAKE RESEARCH PROMOTION) 2013. *Evaluation of Long-Term Earthquake Activity along the Nankai Trough*. 2nd edn, HERP, Japan, [in Japanese], http://www.jishin.go.jp/main/chousa/13may_nankai/index.htm [last accessed 22 June 2017].

HERP (HEADQUARTERS FOR EARTHQUAKE RESEARCH PROMOTION) 2014. *Evaluation of Long-Term Earthquake Activity along the Sagami Trough*. 2nd edn, HERP, Japan, [in Japanese], http://www.jishin.go.jp/main/chousa/14apr_sagami/ [last accessed 22 June 2017].

HOKKAIDO LOCAL GOVERNMENT 2012. *Reassessment of the potential tsunami damage in Hokkaido for the Pacific Coast Region*. Working Group Report [in Japanese], http://www.pref.hokkaido.lg.jp/sm/ktk/bsb/tsunami/h240629/houkokusyo.pdf [last accessed 22 June 2017].

IAEA (INTERNATIONAL ATOMIC ENERGY AGENCY) 2012. *IAEA mission to Onagawa Nuclear power station to examine the performance of systems, structures and components following the Great East Japanese Earthquake and Tsunami*. Report to the Japanese government. IAEA, Vienna.

IINUMA, T., OHZONO, M., OHTA, Y. & MIURA, S. 2011. Coseismic slip distribution of the 2011 off the Pacific coast of Tohoku Earthquake (M 9.0) estimated based on GPS data–Was the asperity in Miyagi-oki ruptured? *Earth, Planets and Space*, **63**, 643–648.

IKEDA, Y., OKADA, S. & TAJIKARA, M. 2012. Long-term strain buildup in the Northeast Japan arc-trench system and its implications for gigantic strain-release events. *Journal of the Geological Society of Japan*, **118**, 294–312 [in Japanese with English abstract].

IKEHARA, K., KANAMATSU, T. *ET AL*. 2016. Documenting large earthquakes similar to the 2011 Tohoku-oki earthquake from sediments deposited in the Japan Trench over the past 1500 years. *Earth and Planetary Science Letters*, **445**, 48–56.

ISLAND ARC 2016. Geological records of storms, tusnamis, and other extreme events (Sept. 2016).

KAGAN, Y.Y. & JACKSON, D.D. 2013. Tohoku earthquake: a surprise? *Bulletin of the Seismological Society of America*, **103**, 1181–1194.

KANAMORI, H., MIYAZAKI, M. & MORI, J. 2006. Investigation of the earthquake sequence off Miyagi prefecture with historical seismograms. *Earth, Planets and Space*, **58**, 1533–1541.

KAWAMATA, T. 2015. Second report concerning the Takaose site in Iwanuma town. *Miyagi Archeology*, **17**, 29–32 [in Japanese].

KAWAMURA, K., LABERG, J.S. & KANAMATSU, T. 2014. Potential tsunamigenic submarine landslides in active margins. *Marine Geology*, **356**, 44–49.

LAY, T. & KANAMORI, H. 2011. Insights from the great 2011 Japan earthquake. *Physics Today*, **64**, 33–39.

MAHUL, O. & WHITE, E. 2012. *Earthquake Risk Insurance*. The World Bank, p. 19. https://openknowledge.

worldbank.org/bitstream/handle/10986/16149/7939 30BRI0drm000Box377374B00Public0.pdf?sequence=1

MALIK, J.N., BANERJEE, C., KHAN, A., JOHNSON, F.C., SHISHIKURA, M. & SATAKE, K. 2015. Stratigraphic evidence for earthquakes and tsunamis on the west coast of South Andaman Island, India during the past 1000 years. *Tectonophysics*, **661**, 49–65.

MIAC (MINISTRY FOR INTERNAL AFFAIRS AND COMMUNICATIONS) 2011. *Concerning the role of public statistics in the Great East Japan earthquake*, [in Japanese], http://www.stat.go.jp/training/2kenkyu/pdf [last accessed 22 June 2017].

MINOURA, K. & NAKAYA, S. 1991. Traces of tsunami preserved in inter-tidal lacustrine and marsh deposits: some examples from northeast Japan. *Journal of Geology*, **99**, 265–287.

MINOURA, K., IMAMURA, F., SUGAWARA, D., KONO, Y. & IWASHITA, T. 2001. The Jogan tsunami deposit and recurrence interval of large-scale tsunami on the Pacific coast of northeast Japan. *Journal of Natural Disaster Science*, **23**, 83–88.

MONECKE, K., FINGER, W., KLARER, D., KONGKO, W., MCADOO, B.G., MOORE, A.L. & SUDRAJAT, S.U. 2008. A 1000-year sediment record of tsunami recurrence in northern Sumatra. *Nature*, **455**, 1232–1234.

MORTON, R.A., GELFENBAUM, G. & JAFFE, B.E. 2007. Physical criteria for distinguishing sandy tsunami and storm deposits using modern examples. *Sedimentary Geology*, **200**, 184–207.

MULLER, R.D., ROEST, W.R., ROYER, J.Y., GAHAGAN, L.M. & SCLATER, J.G. 1997. Digital isochrones of the world's ocean floor. *Journal of Geophysical Research*, **102**, 3211–3214.

NAKAHARA, S. & ICHIKAWA, M. 2013. Mortality in the 2011 Tsunami in Japan. *Journal of Epidemiology*, **23**, 70–73.

NAMEGAYA, Y. & SATAKE, K. 2014. Reexamination of the A.D. 869 Jogan earthquake size from tsunami deposit distribution, simulated flow depth, and velocity. *Geophysical Research Letters*, **41**, 2297–2303.

NAMEGAYA, Y., SATAKE, K. & YAMAKI, S. 2010. Numerical simulation of the AD 869 Jogan tsunami in Ishinomaki and Sendai plains and Ukedo river-mouth lowland. *Annual Report on Active Fault and Paleoearthquake Researches*, **10**, 1–21 [in Japanese with English abstract].

NANAYAMA, F., SATAKE, K., FURUKAWA, R., SHIMOKAWA, K., ATWATER, B.F., SHIGENO, K. & YAMAKI, S. 2003. Unusually large earthquakes inferred from tsunami deposits along the Kuril trench. *Nature*, **424**, 660–663.

NPA (NATIONAL POLICE AUTHORITY) 2016. *Damage Situation and Police Countermeasures*, http://www.npa.go.jp/archive/keibi/biki/higaijokyo_e.pdf [last accessed 22 June 2017].

NATIONAL RESEARCH INSTITUTE FOR EARTH SCIENCE AND DISASTER RESILIENCE 2015. *Models for large seismic faults and tsunami faults associated with paleo seismicity in the Nankai trough region*, 32 [in Japanese], http://www.bousai.go.jp/jishin/nankai/pdf/jishinnankai20151217_03.pdf [last accessed 22 June 2017].

OGATA, T. 2012. *The Company Culture that Saved Onagawa Nuclear Power Plant.* Report for NTT Facilities Research Institute, NTT Facilities Research Institute Report, Tokyo [in Japanese].

OKAMURA, Y. 2012. Reconstruction of the 869 Jogan tsunami and lessons of the 2011 Tohoku earthquake – significance of ancient earthquake studies and problems in announcing study results to society. *Synthesiology*, **5**, 234–242.

OZAWA, S., NISHIMURA, T., SUITO, H., KOBAYASHI, T., TOBITA, M. & IMAKIIRE, T. 2011. Coseismic and postseismic slip of the 2011 magnitude-9 Tohoku-Oki earthquake. *Nature*, **475**, 373–376 (21 July 2011), https://doi.org/10.1038/nature10227

PARK, J.-O., TSURU, T., KODAIRA, S., NAKANISHI, A., MIURA, S., KANEDA, Y. & KONO, Y. 2000. Out-of-sequence thrust faults developed in the coseismic slip zone of the 1946 Nankai earthquake (Mw = 8.2) off Shikoku, southwest Japan. *Geophysical Research Letters*, **27**, 1033–1036.

PLAFKER, G. 1965. Tectonic deformation associated with the 1964 Alaska earthquake. *Science*, **148**, 1675–1687.

RAJENDRAN, C.P., RAJENDRAN, K., MACHADO, T., SATYAMURTHY, T., ARAVAZHI, P. & JAISWAL, M. 2006. Evidence of ancient sea surges at the Mamallapuram coast of India and implications for previous Indian Ocean tsunami events. *Current Science*, **91**, 1242–1247.

RUFF, L. & KANAMORI, H. 1980. Seismicity and the subduction process. *Physics of the Earth and Planetary Interiors*, **23**, 240–252.

SAGIYA, T. 2011. Rebuilding seismology. *Nature*, **473**, 146–148, https://doi.org/10.1038/473146a

SATAKE, K. & ATWATER, B.F. 2007. Long-term perspectives on giant earthquakes and tsunamis at subduction zones. *Annual Review of Earth and Planetary Sciences*, **35**, 49–374.

SATAKE, K. & NANAYAMA, F. 2004. *Tsunami inundation maps of the Pacific coast of Hokkaido.* Digital Geologic Map Series: EQ-1, Geological Survey of Japan, Tsukuba.

SATAKE, K., NAMEGAYA, Y. & YAMAKI, S. 2008. Numerical simulation of the AD869 Jogan Tsunami in Ishinomaki and Sendai plains. *Annual Reports of Active Fault Paleoearthquake Research*, **8**, 71–89, [in Japanese with English abstract].

SATAKE, K., FUJII, Y., HARADA, T. & NAMEGAYA, Y. 2013. Time and space distribution of coseismic slip of the 2011 Tohoku earthquake as inferred from tsunami waveform data. *Bulletin of the Seismological Society of America*, **103-2B**, 1473–1492, https://doi.org/10.1785/0120120122

SAWAI, Y., SHISHIKURA, M. & KOMATSUBARA, J. 2008. A study of paleotsunami using hand corer in Sendai plain (Sendai City, Natori City, Iwanuma City, Watari Town, Yamamoto Town), Miyagi, Japan. *Annual Reports of Active Fault Paleo-Earthquake Research*, **8**, 17–70 [in Japanese with English Abstract].

SAWAI, Y., NAMEGAYA, Y., OKAMURA, Y., SATAKE, K. & SHISHIKURA, M. 2012. Challenges of anticipating the 2011 Tohoku earthquake and tsunami using coastal geology. *Geophysical Research Letters*, **39**, L21309, https://doi.org/10.1029/2012GL053692

SAWAI, Y., NAMEGAYA, Y., TAMURA, T., NAKASHIMA, R. & TANIGAWA, K. 2015. Shorter intervals between great earthquakes near Sendai: scour ponds and a sand layer attributable to A.D. 1454 overwash. *Geophysical Research Letters*, **42**, 4795–4800.

SHUTO, N., KOSHIMURA, S., SATAKE, K., IMAMURA, F. & MATSUTOMI, H. 2007. *An Encyclopedia of Tsunami*. Asakura Publishing Co. Ltd, Tokyo [in Japanese].

SIEH, K., NATAWIDJAJA, D.H. ET AL. 2008. Earthquake supercycles inferred from sea-level changes recorded in the corals of West Sumatra. *Science*, **322**, 1674–1678.

SIMONS, M., MINSON, S.E. ET AL. 2011. The 2011 magnitude 9.0 Tohoku-oki earthquake: mosaicking the megathrust from seconds to centuries. *Science*, **332**, 1421–1425, https://doi.org/10.1126/science.1206731

STEIN, S. & OKAL, E. 2007. Ultralong period seismic study of the December 2004 Indian Ocean earthquake and implications for regional tectonics and the subduction process. *Bulletin of the Seismological Society of America*, **97**, 279–295.

STRASSER, M., KÖLLING, M. ET AL. 2013. A slump in the trench: tracking the impact of the 2011 Tohoku-Oki earthquake. *Geology*, **41**, 935–938.

SUGAWARA, D., IMAMURA, F., MATSUMOTO, H., GOTO, K. & MINOURA, K. 2011. Reconstruction of the AD869 Jogan earthquake-induced tsunami by using the geological data. *Journal of the Japanese Society of Natural Disaster Science*, **29**, 501–516 [in Japanese with English abstract].

SUGIMOTO, M. 2014. Chapter 26: Geoethics and risk-communication issues in Japan's disaster management system revealed by the 2011 Tohoku Earthquake and Tsunami. *In*: WYSS, M. & PEPPOLONI, S. (eds) *Geoethics, Ethical Challenges and Case Studies in Earth Sciences*. Elsevier, Amsterdam, 323–334.

SUPPASRI, A., FUKUTANI, Y., ABE, Y. & IMAMURA, F. 2013*a*. Relationship between earthquake magnitude and tsunami height along the Tohoku coast based on historical tsunami trace database and the 2011 Great East Japan Tsunami. *Report of Tsunami Engineering*, **30**, 37–49.

SUPPASRI, A., SHUTO, N., IMAMURA, F., KOSHIMURA, S., MAS, E. & YALCINER, A.C. 2013*b*. Lessons learned from the 2011 Great East Japan Tsunami: performance of tsunami countermeasures, coastal buildings, and tsunami evacuation in Japan. *Pure and Applied Geophysics*, **170**, 993–1018.

SUWA, Y., MIURA, S., HASEGAWA, A., SATO, T. & TACHIBANA, K. 2006. Interplate coupling beneath NE Japan inferred from three-dimensional displacement field. *Journal of Geophysical Research*, **111**, B04402, https://doi.org/10.1029/2004JB003203

TAKEDA, Y. & SATO, K. 2009. *Yomi kudashi Nihon sandai jitsuroku (jyokan)*. Ebisukosyo Publishing, Tokyo (in Japanese), p. 716.

TAMURA, T., SAWAI, Y., IKEHARA, K., NAKASHIMA, R., HARA, J. & KANAI, Y. 2015. Shallow-marine deposits associated with the 2011 Tohoku-oki tsunami in Sendai Bay, Japan. *Journal of Quaternary Science*, **30**, 293–297, https://doi.org/10.1002/jqs.2786

TANIGAWA, K., SAWAI, Y., SHISHIKURA, M., NAMEGAYA, Y. & MATSUMOTO, D. 2014. Geological evidence for an unusually large tsunami on the Pacific coast of Aomori, northern Japan. *Journal of Quaternary Science*, **29**, 200–208, https://doi.org/10.1002/jqs.2690

TAPPIN, D., WATTS, P., MCMURTY, G., LAFOY, Y. & MATSUMOTO, T. 2001. The Sissano, Papua New Guinea tsunami of July 1998: offshore evidence on the source mechanism. *Marine Geology*, **175**, 1–23.

TAPPIN, D.R., GRILLI, S.T. ET AL. 2014. Did a submarine landslide contribute to the 2011 Tohoku tsunami? *Marine Geology*, **357**, 344–361.

TOHOKU UNIVERSITY 2005. *Field survey reports for the 2004 Indian Ocean Tsunami*. Ch. 7, 178–186, http://www.tsunami.civil.tohoku.ac.jp/sumatra2004/report.html [last accessed 22 June 2017].

UJIE, K., TANAKA, H. ET AL. 2013. Low Coseismic Shear Stress on the Tohoku-Oki Megathrust Determined from Laboratory Experiments. *Science*, **342**, 1211–1214, https://doi.org/10.1126/science.1243485

WATANABE, H. 2001. Is it possible to clarify the real state of part earthquakes and tsunamis on the basis of legends? As an example of the 869 Jogan earthquake and tsunami. *Rekishi-Jishin*, **17**, 130–146 [in Japanese].

WITTER, R.C., CARVER, G.A. ET AL. 2016. Unusually large tsunamis frequent a currently creeping part of the Aleutian megathrust. *Geophysical Research Letters*, **43**, 76–84, https://doi.org/10.1002/2015GL066083

YOSHIDA, K., HASEGAWA, A., OKADA, T., IINUMA, T., ITO, Y. & ASANO, Y. 2012. Stress before and after the 2011 great Tohoku-oki earthquake and induced earthquakes in inland areas of eastern Japan. *Geophysical Research Letters*, **39**, L03302, https://doi.org/10.1029/2011GL049729

YOSHIKAWA, S., KANAMATSU, T. ET AL. 2015. Evidence for erosion and deposition by the 2011 Tohoku-oki tsunami on the nearshore shelf of Sendai Bay, Japan. *Geo-Marine Letters*, **35**, 315–328.

Tsunami simulations of mega-thrust earthquakes in the Nankai–Tonankai Trough (Japan) based on stochastic rupture scenarios

KATSUICHIRO GODA[1]*, TOMOHIRO YASUDA[2], P. MARTIN MAI[3], TAKUMA MARUYAMA[4] & NOBUHITO MORI[4]

[1]*Department of Civil Engineering, University of Bristol, Queen's Building, University Walk, Bristol BS8 1TR, UK*

[2]*Faculty of Environmental and Urban Engineering, Kansai University, Osaka 564-8680, Japan*

[3]*Earth Science & Engineering, King Abdullah University of Science & Technology, Thuwal 23955-6900, Saudi Arabia*

[4]*Disaster Prevention Research Institute, Kyoto University, Kyoto 611-0011, Japan*

**Correspondence: katsu.goda@bristol.ac.uk*

Abstract: In this study, earthquake rupture models for future mega-thrust earthquakes in the Nankai–Tonankai subduction zone are developed by incorporating the main characteristics of inverted source models of the 2011 Tohoku earthquake. These scenario ruptures also account for key features of the national tsunami source model for the Nankai–Tonankai earthquake by the Central Disaster Management Council of the Japanese Government. The source models capture a wide range of realistic slip distributions and kinematic rupture processes, reflecting the current best understanding of what may happen due to a future mega-earthquake in the Nankai–Tonankai Trough, and therefore are useful for conducting probabilistic tsunami hazard and risk analysis. A large suite of scenario rupture models is then used to investigate the variability of tsunami effects in coastal areas, such as offshore tsunami wave heights and onshore inundation depths, due to realistic variations in source characteristics. Such investigations are particularly valuable for tsunami hazard mapping and evacuation planning in municipalities along the Nankai–Tonankai coast.

Mega-thrust subduction earthquakes may trigger massive tsunamis that can lead to catastrophic consequences for coastal cities and towns. Recent devastating tsunamis include the 2004 Indian Ocean tsunami (Murata *et al.* 2010) and the 2011 Great East Japan (Tohoku) tsunami (Fraser *et al.* 2013). Globally, tsunami exposure is not negligible and more accurate assessment of tsunami hazard due to future mega-thrust subduction earthquakes is necessary to mitigate the potential tsunami risk and to enhance the tsunami resilience of coastal communities (Løvholt *et al.* 2014). One of the most challenging problems in tsunami preparedness and risk mitigation is to determine the range of possible tsunami scenarios for a given geographical area. Methods, such as probabilistic tsunami hazard analysis (Geist & Parsons 2006; Gonzalez *et al.* 2009; Horspool *et al.* 2014), are useful for identifying tsunami source regions and corresponding scenarios that have major impact to a site of interest. In developing such scenarios, many experts with different backgrounds are involved and the scientific views that are expressed by these experts may be diverse. Such expert judgements rarely lead to a consensus model, but rather generate a set of disparate scenarios that need to be weighted in a logic-tree approach. Nevertheless, in face of imminent tsunami risk in major subduction zones, scientists and engineers need to evaluate the risk quantitatively and effectively, and produce reliable and testable hazard/risk maps at national/regional/local levels. In the ideal (future) case, such hazard and risk maps do not depend on difficult-to-quantify expert solicitations, but on algorithmically developed and objective selection criteria. Typically, such testable hazard quantification will require a probabilistic/ensemble simulation-based approach.

Recent advances in tsunami hazard analysis aim at incorporating the 'anticipated variability' associated with source characteristics of future tsunamigenic earthquakes, such as location, magnitude, geometry and slip distribution (Geist 2002; McCloskey *et al.* 2008; Løvholt *et al.* 2012; Goda *et al.* 2014; Fukutani *et al.* 2015; Mueller *et al.* 2015).

From: Scourse, E. M., Chapman, N. A., Tappin, D. R. & Wallis, S. R. (eds) 2018. *Tsunamis: Geology, Hazards and Risks*. Geological Society, London, Special Publications, **456**, 55–74.
First published online February 22, 2017, https://doi.org/10.1144/SP456.1

Methods for developing stochastic earthquake source models, based on spectral analysis of slip heterogeneity of inverted source models and subsequent spectral synthesis of constrained-random slip fields (Mai & Beroza 2002; Lavallée *et al.* 2006), facilitate the generation of possible scenarios with different earthquake slips and fault geometry. The approaches also allow for incorporating epistemic uncertainty in the stochastic source characterization by considering one or more reference source models, if available (Goda *et al.* 2014). Through Monte Carlo tsunami simulation, probabilistic inundation depth maps can be evaluated using numerous synthesized source models (Mueller *et al.* 2015). Furthermore, by integrating tsunami hazard estimates with tsunami fragility models (Tarbotton *et al.* 2015), probabilistic tsunami risk maps, as well as tsunami loss curves, at local and regional scales can be developed (Goda & Song 2016). An advantage of such hazard and risk assessments is that the currently known sources of uncertainty related to the tsunami source characteristics are taken into account, whereby such simulation-based approaches also provide a basis to develop testable and easily updatable hazard maps.

Historically, the Nankai–Tonankai Trough hosted many M_w 8+ mega-thrust subduction earthquakes that triggered intense ground shaking and massive tsunami across the central to SW part of Japan (Ando 1975). In this region, the Philippines Sea Plate subducts underneath the Eurasian Plate with an average slip rate of 44 mm a^{-1} (Loveless & Meade 2010). The source areas span over offshore regions of Nankai (Kyushu–Shikoku–Kinki region) and Tonankai (Kinki–Tokai region), as shown in Figure 1. The magnitude of a future Nankai–Tonankai earthquake can be as large as M_w 9.0 (if the entire trough region ruptures in a single event), which is similar in size to the 2011 Tohoku earthquake. The Central Disaster Management Council (2012), under the auspices of the Cabinet Office of the Japanese Government, has developed a new tsunami source model for the Nankai–Tonankai earthquake based on expert judgement, and has produced updated estimates of tsunami hazard, risk and economic impact due to such a catastrophic event. Notable features of the Nankai–Tonankai tsunami source model by the Central Disaster Management Council (hereafter, the 2012 CDMC model) include: (i) consideration of 'large-slip areas' and 'very-large-slip areas' (so-called asperities); and (ii) consideration of multiple scenarios (in total, 11 cases) to represent different slip characteristics in terms of the number and locations of very large asperities. The large-slip areas are placed in the shallow part of the source region (depths of less than 20 km) and consist of about 20% of the entire fault plane, whereas the very-large-slip areas are located within the large-slip areas, consisting of about 5% of the entire fault plane. For the Nankai–Tonankai scenarios, assigned average slip values are about 10 m, whereas the large-slip areas and very-large-slip areas typically show 20 and 40 m fault displacement, respectively. These slip parameters reflect the characteristics of inverted source models for the 2011 Tohoku earthquake (e.g. Yamazaki *et al.* 2011; Gusman *et al.* 2012; Satake *et al.* 2013). Although the locations of asperities are varied among the 11 cases, it is unlikely that the considered slip distributions comprehensively cover possible future scenarios. Consequently, they are probably insufficient to

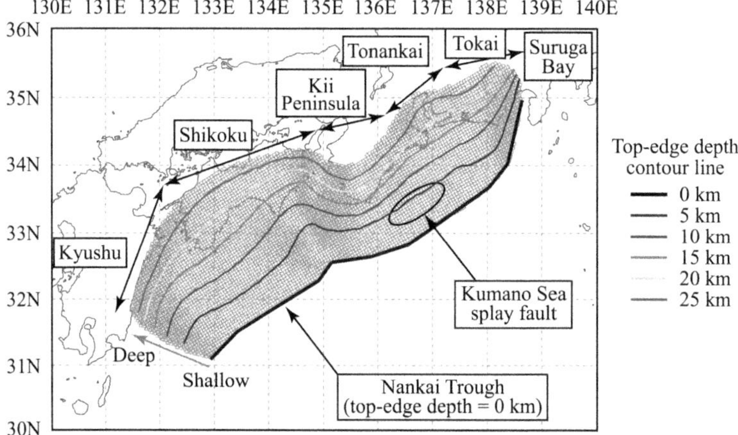

Fig. 1. Subduction region of the Nankai–Tonankai Trough showing geographical areas and colour-coded contour lines of depth to the fault plane. Grey boxes mark the sub-faults of the 2012 CDMC model.

fully quantify the uncertainties associated with tsunami hazard and risk predictions for the Nankai–Tonankai event (note: the 2012 CDMC model was not intended for modelling/evaluating such uncertainties). Nonetheless, such assessments are crucial for improving disaster preparedness by understanding the consequences of different scenarios, and by communicating uncertainties associated with hazard predictions among stakeholders (Goda & Song 2016).

This study develops constrained-stochastic earthquake slip models for the Nankai–Tonankai event. The models are based on the approach of generating stochastic source models for the 2011 Tohoku earthquake by Goda *et al.* (2014), but in addition reflecting key features of the 2012 CDMC tsunami source model. One of the innovative features of the newly developed stochastic tsunami source models is the incorporation of the kinematic fault rupture evolution: that is, specifying the point of rupture nucleation, the rupture propagation speed and the slip duration (rise time) on the fault. These models can then be used to carry out detailed analyses on tsunami sensitivity due to varying source characteristics for the Nankai–Tonankai event, as well as for the resulting hazard and risk predictions. Such investigations, albeit challenging, are particularly valuable for tsunami hazard mapping and evacuation planning in municipalities along the Nankai–Tonankai coast.

This paper is organized as follows. First, we briefly summarize geological aspects of the Nankai–Tonankai subduction region and the 2012 CDMC tsunami source model. Subsequently, details of the stochastic Nankai–Tonankai tsunami source model are described, focusing on the slip distribution generation and kinematic rupture modelling. Using the constrained-stochastic source models for the Nankai–Tonankai earthquake, we conduct Monte Carlo tsunami simulation. Variability of tsunami wave profiles and tsunami inundation depths due to varying source characteristics is evaluated to discuss the extent of uncertainty associated with tsunami hazard predictions for the Nankai–Tonankai tsunami. Finally, we discuss the main conclusions of this study, as well as future investigations that need to be conducted using our newly developed tool.

Nankai–Tonankai mega-thrust subduction earthquake

Nankai–Tonankai subduction region

The Nankai–Tonankai Trough accommodates plate movements between the Philippines Sea Plate and the Eurasian Plate. The convergence rate varies along the trough. Recent geodetic measurements in Japan indicate that the slip rate increases from about 36 mm a^{-1} at the eastern end near Suruga Bay to about 63 mm s^{-1} at the western end near Kyushu (Loveless & Meade 2010) (Fig. 1). The accumulated strain is released periodically during major subduction earthquakes. Since 1700, five M8+ earthquakes occurred (Ando 1975). The most recent ones were the 1944 M_w 8.1 Tonankai earthquake and the 1946 M_w 8.4 Nankai earthquake. In 1854, large dual earthquakes (Ansei earthquakes, M_w 8.4 and M_w 8.5) occurred in sequence, separated by 32 h (note: the eastern Tonankai segment ruptured prior to the western Nankai segment). Further dating back to 1707, the M_w 8.7+ Hoei earthquake ruptured the entire Nankai–Tonankai region (where it is possible that the actual event occurred as an earthquake doublet with a short time gap: Ando 1975). By examining the earthquake rupture patterns of past Nankai–Tonankai earthquakes since 684, Ando (1975) suggested two predominant rupture modes: segmented ruptures, in which only a single segment fails; and synchronized (or coupled) rupture modes in which two segments fail in rapid succession. The recurrence interval of large subduction events in the Nankai–Tonankai Trough varies from 90 to 264 years (Ando 1975).

Issues related to fault zone segmentation are crucial for earthquake and tsunami hazard assessments in general, but it appears that this is particularly important for the Nankai–Tonankai events. Seismic imaging surveys conducted by Kodaira *et al.* (2006) indicated that a physical segmentation boundary exists (just off the Kii Peninsula: Fig. 1) that separates the Nankai and Tonankai source regions, and thus controls the occurrence of segmented or synchronized rupture. The shallow portion of the crust offshore the Kii Peninsula is fractured, while the deeper portion of the crust is strongly coupled. When the rupture starts near the strongly coupled part, the synchronized rupture may be triggered (Kodaira *et al.* 2006), affecting a wider geographical region simultaneously. Considering the large-scale synchronized ruptures of the 2004 Sumatra earthquake and the 2011 Tohoku earthquake, it is difficult to predict whether the next mega-thrust earthquake along the Nankai–Tonankai Trough will rupture in segments or simultaneously, as this region has shown events occurring in both rupture modes in the historic past.

Central Disaster Management Council tsunami source model

The 2012 CDMC tsunami source model is developed for the synchronized rupture case. The model is intended to represent one of the more extreme scenarios that may need to be considered from an earthquake disaster mitigation viewpoint. Therefore, it corresponds to a M_w 9-class event and its rupture area spans over a vast region (Fig. 1). The key features of the 2012 CDMC model include:

- The entire fault plane is represented by a set of 5773 sub-faults (including the Kumano splay fault: Fig. 1); the size of each sub-fault is 5 × 5 km. The main fault plane consists of 5669 sub-faults, while the splay fault plane is composed of 104 sub-faults. The total fault-plane area is about 1.4×10^5 km^2. The top-edge depth for sub-faults along the Nankai–Tonankai Trough is set to 0 km or very shallow depths (i.e. the rupture plane reaches or is very close to the ocean bottom).
- The average slip (D) over the fault plane is determined based on a scaling law: $D = (16/7\pi^{2/3}) \Delta\sigma S^{0.5}/\mu$, where $\Delta\sigma$ is the average stress drop, S is the fault-plane area and μ is the rock rigidity (=40.9 GPa). The representative average stress drop is determined by analysing stress drop parameters estimated for the past M_w 8–9-class subduction-type events (i.e. 1944 Tonankai, 1946 Nankai, 2003 Tokachi-oki, 2004 Sumatra, 2010 Chile and 2011 Tohoku). The mean plus 1 SD (standard deviation) of the stress drop parameters from the considered inversion studies is 2.2 MPa (Central Disaster Management Council 2012): whereas, for smaller subduction events (M_w 7–8), a representative stress drop value is about 3.0 MPa. In the 2012 CDMC model, $\Delta\sigma$ is set to 3.0 MPa.
- Two types of asperities are defined. The large-slip areas take up about 20% of the entire fault plane, with an average slip-value twice as large as the average slip over the fault plane. The large-slip areas are positioned at depths shallower than 20 km. The very-large-slip areas take up about 5% of the entire fault plane,with an average slip-value four times as large as the average slip over the fault plane. Note that the very-large-slip areas are located within the large-slip areas and are constrained to be at depths less than 10 km.
- There are 11 scenarios (see Fig. 2). Cases (1)–(5) consider a single asperity region (i.e. large and very large-slip areas): the asperity is positioned off (1) Suruga Bay to Kii Peninsula, (2) Kii Peninsula, (3) Kii Peninsula to Shikoku, (4) Shikoku, or (5) Shikoku to Kyushu. Cases (6) and (7) take into account local splay fault sources in the Kumano Sea (see Fig. 1); for these cases, asperities are located off the Kii Peninsula (either on the western or eastern side of the Kumano Sea). Cases (8)–(11) consider two asperities; their locations are off: (8) Suruga Bay to Kii Peninsula and Kii Peninsula to Shikoku, (9) Suruga Bay to Kii Peninsula and Shikoku, (10) Kii Peninsula to Shikoku and Shikoku to Kyushu, or (11) Shikoku and Shikoku to Kyushu. Note that rupture nucleation is set to occur near the centre of asperity regions at a depth of about 20 km, and varies for individual cases (therefore, the total rupture duration for the 11 cases also differs).
- The slip values for individual sub-faults are determined based on spatially varying convergence rates at the sub-faults given that the total slip amounts (or seismic moments) within the large-slip and very-large-slip areas are conserved according to the above-mentioned average slip values.
- The average slip values for the 11 cases range between 8.8 and 11.2 m, whereas the maximum slip values for the 11 cases range between 39.8 and 52.0 m, increasing from east to west (Fig. 2).
- The moment magnitude is about 9.1, and the corresponding seismic moment ranges between 5.3×10^{22} and 6.7×10^{22} N m (for the 11 cases).
- The geometrical parameters (i.e. strike, dip and rake) are variable over the curved fault plane (approximated by the sub-faults).
- The kinematic slip distributions are generated at an interval of 10 s. The assumed rupture propagation velocity is 2.5 km s^{-1} and the rise time of individual sub-faults is set to 60 s.

Stochastic Nankai–Tonankai ssunami source model

Methodology

The stochastic source models for the Nankai–Tonankai tsunami combine the 2012 CDMC model and the spectral synthesis method for megathrust subduction earthquakes (Mai & Beroza 2002; Goda *et al.* 2014). Importantly, our rupture models are not intended for verifying or replacing the 2012 CDMC model, but to constrain our modelling. It is noteworthy that the fault-plane parameters of the 2012 CDMC model (as a whole) are heterogeneous in terms of fault geometry and slip-vector orientation. Both strike and dip angles of individual sub-faults vary (Fig. 1). Because the fault geometry and slip features of the 2012 CDMC model are based on current seismotectonic settings in the Nankai–Tonankai Trough region (e.g. Hirose *et al.* 2008), the fault-plane information of the 2012 CDMC model is incorporated into our stochastic slip modelling. Specifically, the sub-faults of the 2012 CDMC model are mapped onto a 2D (rectangular) matrix (note: sub-faults for the Kumano splay fault are excluded). The size of the 2D matrix is 54 (downdip) × 153 (along-strike), and its origin is set to the SW corner of the fault plane (i.e. offshore Kyushu: Fig. 1). In our stochastic simulations,

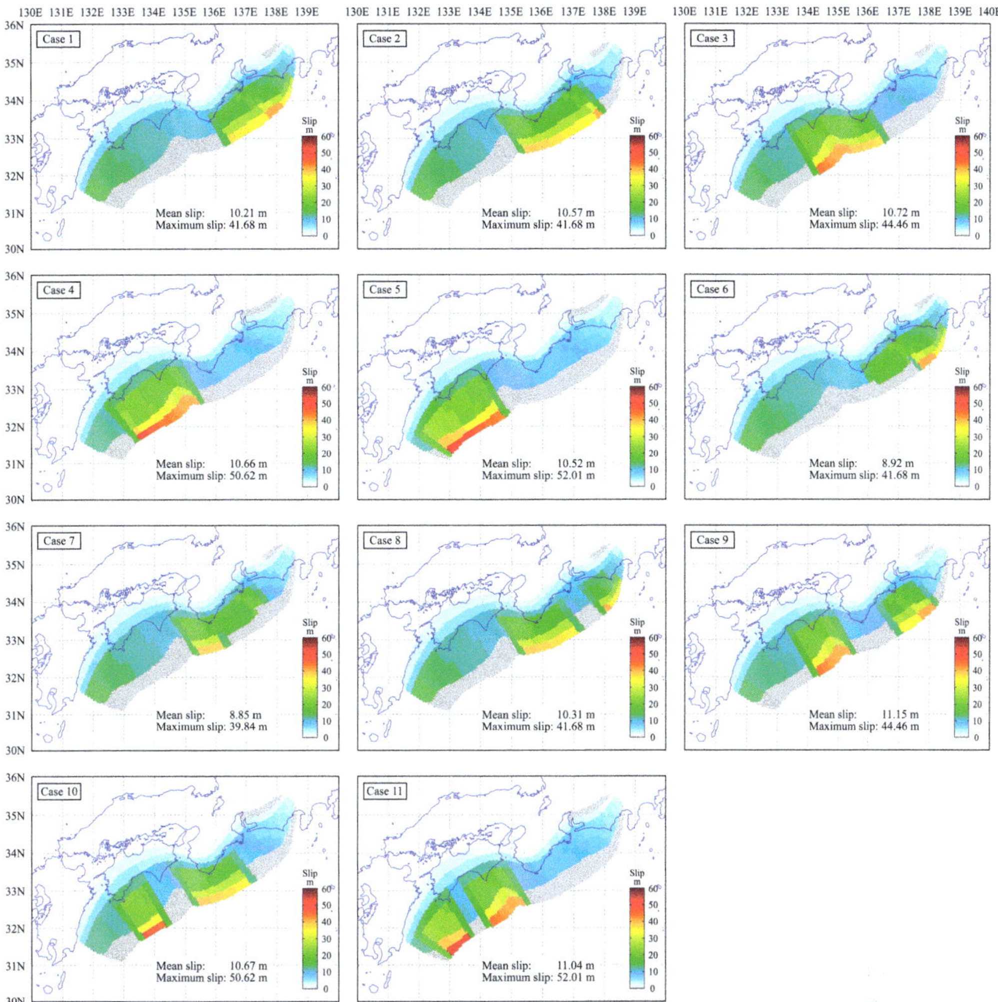

Fig. 2. Rupture cases (1)–(11) for the 2012 CDMC model. Note the variations in location and values of maximum fault slip, as well as the variable mean slip (hence, variable total seismic moment).

slip values for the 54 × 153 matrix are generated, and the synthesized slip values that correspond to the 5669 sub-faults (main fault plane only) are mapped back to the CDMC rupture surface by ignoring slip values outside the main fault plane.

The procedure for generating stochastic earthquake slip models for the Nankai–Tonankai tsunami and for simulating tsunami waves is illustrated in Figure 3. The method consists of stochastic synthesis of slip distributions and subsequent temporal rupture specification by considering uncertain rupture start locations. The model parameters for generating stochastic slip distributions include: stochastic synthesis parameters (i.e. correlation lengths in the downdip and along-strike directions, Hurst number for the von Kármán auto-correlation model, and Box–Cox power parameter), target slip parameters (i.e. mean and maximum slips) and kinematic rupture parameters (i.e. rupture nucleation point, rupture propagation velocity and rise time). For future Nankai–Tonankai scenarios, these parameters are uncertain or unknown, and thus are treated as random variables in stochastic slip simulations. In the developed computer code for Monte Carlo tsunami simulation, the probabilistic information of the above-mentioned parameters can be represented by: (i) multiple discrete values (i.e. a set of parameter values and corresponding weights); (ii) uniform random variable (i.e. lower and upper values); or (iii) truncated normal variable (i.e. mean, standard deviation, lower limit and upper limit). In this paper, we adopted the truncated

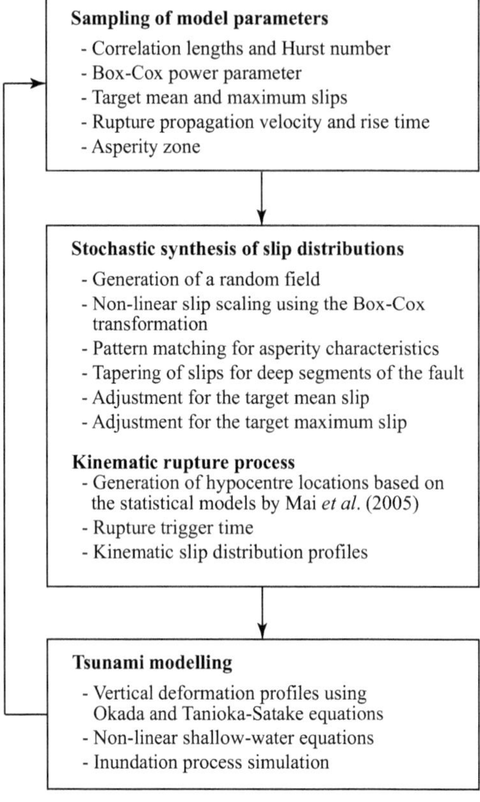

Fig. 3. Flowchart for generating stochastic earthquake slip realizations and kinematic rupture processes.

normal distribution to characterize the variability of the source parameters.

Slip distribution modelling

The key slip characteristics for stochastic source modelling are specified in terms of slip statistics, slip distribution parameters and asperity areas. The slip statistics, such as mean and maximum, are used to control the overall features of the resulting slip distribution. The spatial heterogeneity of earthquake slip is modelled in the wavenumber domain by adopting a von Kármán auto-correlation function (Mai & Beroza 2002), but other spectral models could be used. The relevant parameters of the von Kármán function are the correlation lengths (for the downdip and along-strike directions) and the Hurst number. The correlation length determines the absolute level of the power spectrum in the low wavenumber range, and captures the anisotropic spectral features of the slip distribution (when different correlation lengths are specified for downdip and along-strike directions). The Hurst number controls the slope of the power spectral decay in the high wavenumber range, and is theoretically constrained to range between 0 and 1. In addition, realistic tail characteristics of the slip distribution (i.e. slip values often have right-skewed distributions) are modelled via Box–Cox transformation to represent non-normal distribution of earthquake slip (Goda *et al.* 2014). The asperity areas, which are used for pattern matching for trial synthesized slip distributions, can be specified either based on the 2012 CDMC model (11 rupture cases: Fig. 2) or based on depth limits. For the former, multiple CDMC models can also be considered with different weights. For the latter, the asperity areas are determined by the lower and upper depth limits of the sub-faults (e.g. depth range between 1 and 10 km).

For a given set of stochastic synthesis parameters, slip distributions are generated using a Fourier integral method (Pardo-Iguzquiza & Chica-Olmo 1993). To ensure that synthesized slip distributions have some desirable asperity features (e.g. large asperities are located at shallow depths), multiple trial slip distributions are generated and evaluated by a pattern-matching algorithm that implements selected criteria. In the adopted procedure, two criteria are considered: (i) slip concentrations around the maximum slip are sufficient to form asperity areas; and (ii) the maximum value of the synthesized slip distribution is located within asperity areas of the fault plane. For instance, slip near the main asperity areas (about 10% of the total fault plane) should have more than 30% of the total slip over the fault plane (Mai *et al.* 2005; Goda *et al.* 2014).

Subsequently, additional (optional) adjustment of slip values may be needed. A taper function is applied to deeper segments of the fault plane; slip values for sub-faults with top-edge depths of 25–27, 27–29, 29–31, 31–33 and 33–35 km are multiplied by the reduction factors of 0.9, 0.7, 0.5, 0.3 and 0.1, respectively (note: the maximum top-edge depth of the 2012 CDMC model is <35 km, and a diminishing taper is applied in the 2012 CDMC model). The mean slip is adjusted to match the target mean slip (treated as a random variable in our approach). Then unrealistically large slip values, exceeding the target maximum slip (treated as a random variable), are resampled. For resampling of very large slip values, a histogram is constructed based on the synthesized slip distribution with lower and upper limits, for instance, by adopting slip values between two-thirds of the target maximum slip and the target maximum slip (i.e. slip values corresponding to the very large asperities). The simulated slip distributions, after the adjustments, are then used as stochastic slip models for the Nankai–Tonankai tsunami (as instantaneous rupture).

Kinematic rupture modelling

The kinematic rupture process is taken into account following Mai *et al.* (2005). The probability density functions (PDFs) for hypocentre locations are defined based on the statistical models developed by Mai *et al.* (2005). The preliminary PDFs for hypocentre locations are specified based on the fault dimensions and the mean/maximum slip ratios. Subsequently, further constraints are placed to exclude unlikely hypocentre locations for a given slip distribution, using empirical findings by Mai *et al.* (2005). By combining the preliminary PDF and the constraints, the final PDF (or strictly weighting function) for hypocentre locations is obtained. This is used to sample the location of hypocentres. Using the randomly generated rupture propagation velocity and rise time (assumed constant for all sub-faults), the kinematic rupture process of the synthesized slip distributions can be simulated.

Illustration of constrained-stochastic source rupture simulation

The above-mentioned method for stochastic source-rupture modelling for the Nankai–Tonankai tsunami is illustrated in the following. Consider that all model parameters are sampled from assumed probability distributions. For a particular illustration case, the correlation lengths for the downdip and along-strike directions are 60.95 and 104.70 km, respectively; the Hurst number is 0.84; the Box–Cox parameter is 0.24; the target mean and maximum slips are 11.74 and 57.67 m, respectively; the seismic moment and moment magnitude are 6.8×10^{22} N m and 9.19, respectively (using the constant rock rigidity of 40.9 GPa); the asperity areas are defined based on the depth limits of 1.0 and 12.5 km; the rupture propagation velocity is 2.83 km s^{-1}; and the rise time is 60.58 s.

Figure 4a shows a synthesized slip distribution prior to the non-linear scaling of slip values based on the Box–Cox transformation. The generated slip values on the fault plane are approximately standard normally distributed (i.e. mean equal to 0 and standard deviation equal to 1). This slip distribution already satisfies the criteria set for pattern matching (thus, it is adopted for further adjustments). The black grid marks the sub-faults, with top-edge depths between 1.0 and 12.5 km (as specified for the asperity areas). The grey square represents the maximum slip over the fault plane, and the grey rectangle is the asperity zone defined for pattern matching (fractional dimensions of the asperity zone are 25 and 40% of the fault width and length, respectively). The threshold for slip concentration is set to 30%. Figure 4b shows the final (instantaneous) slip distribution after non-linear scaling of slip and subsequent adjustments. Owing to the non-linear transformation, the asperity areas, slip values of which range around 40–55 m, become more prominent. The effects of tapering slip values for deep sub-faults can be seen by comparing slip values near

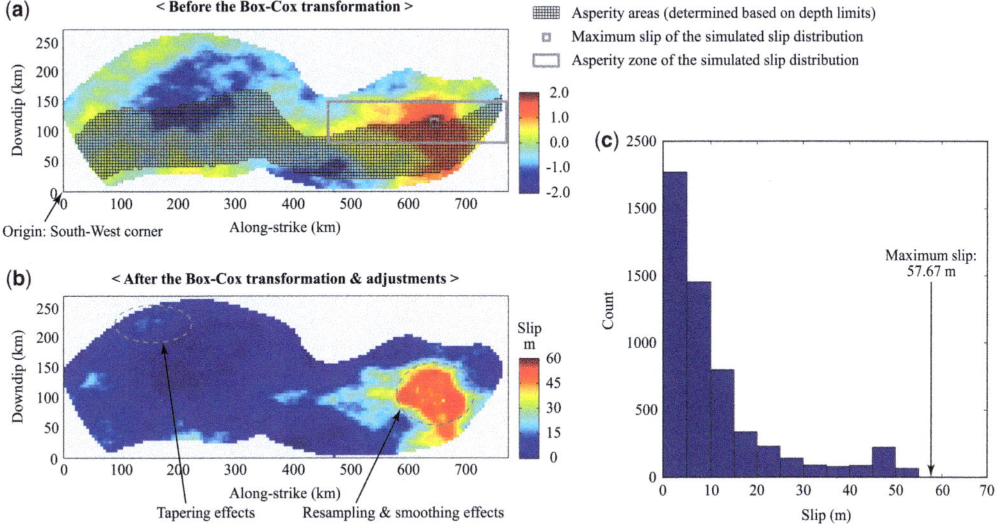

Fig. 4. Slip distribution modelling: (**a**) simulated slip distribution prior to non-linear scaling of slips and adjustments; (**b**) final simulated slip distribution after non-linear scaling of slips and adjustments; and (**c**) histogram of the final simulated slip distribution.

the downdip edges of the fault plane (Fig. 4a, b). The mean-slip value is adjusted to achieve the target mean-slip value of 11.74 m. The standard deviation of the slip values is 12.22 m; the coefficient of variation of the slip values is comparable to those from stochastic models for the 2011 Tohoku earthquake (Goda *et al.* 2014). The target maximum slip for this case is 57.67 m, whereas several slip values (after the Box–Cox transformation and the mean-slip adjustment) exceed this value. To limit the maximum slip value to within a reasonable range, resampling of the slip values for sub-faults with very large slips exceeding the target maximum value is carried out. The effects of resampling slip values for the asperity areas can be seen in Figure 4, which may change somewhat the location of maximum slip. Figure 4c shows the histogram of slip values after the adjustments, exhibiting heavy right-tail features but less than the target maximum value.

Subsequently, a hypocentre location (i.e. rupture starting point) is determined based on the method proposed by Mai *et al.* (2005). Figure 5 shows: (a) the preliminary PDF for hypocentre locations based on the fault dimensions and the mean/maximum slip ratios; (b) constraints based on large, as well as very large, asperity areas; and (c) the final PDF for hypocentre locations. All values shown in Figure 5 are normalized to 1. The grey square represents the maximum slip of the final simulated slip distribution. The preliminary PDF function indicates that the likely location of hypocentres is near the centre of the fault plane and around sub-faults with relatively large slip. The contour maps for the constraints on the hypocentre locations reflect empirical findings from the statistical analysis of source-inversion models (Mai *et al.* 2005). The resulting constraint function indicates that the hypocentre is more likely to be close to very large asperity areas (but not too close) and that it is very unlikely to have hypocentres far away from the very large asperity areas. The final PDF combines the preliminary PDF function and the constraint function, and can be used to sample the location of hypocentres. The sampled hypocentre location for the illustrated case is indicated as a grey circle in Figure 5c.

Figure 6 displays snapshots of the overall temporal rupture evolution for the simulated slip distribution (between 0 and 180 s), in which the hypocentre location (rupture nucleation) is marked by a grey circle. The slip distributions are generated at an interval of 10 s by considering the rupture propagation velocity of 2.83 km s^{-1} and the rise time of 60.58 s. The entire rupture process takes 239.10 s (i.e. 178.52 s to reach the farthest sub-fault from the hypocentre plus 60.58 s to complete the sub-fault rupture process). This information can be used in tsunami simulation to take into account the kinematic process of earthquake rupture.

Monte Carlo tsunami simulation for Nankai–Tonankai earthquake

In this section, we discuss our Monte Carlo tsunami simulation for the Nankai–Tonankai regions. The main purpose of the investigations is to demonstrate that the newly developed method can produce useful results for tsunami hazard mapping and preparedness. In the following, a computational set-up of the tsunami simulation for the Nankai–Tonankai earthquake is described, followed by tsunami simulation results for both offshore and inland areas. A rigorous assessment of uncertainty, as well as a detailed sensitivity analysis of tsunami hazard predictions to model parameters, is beyond the scope of this study, but will be subject of a subsequent investigation.

Tsunami modelling and input data

The tsunami modelling is carried out using a well-tested numerical code (Goto *et al.* 1997) that solves the non-linear shallow-water equations using a leap-frog staggered-grid finite-difference scheme and is capable of generating offshore tsunami propagation and inundation/run-up. The run-up calculation is based on a conventional moving boundary approach, where a dry/wet condition of a computational cell is determined based on total water depth relative to its elevation. The numerical tsunami calculation is performed for a 3 h duration, which is sufficient to model the most critical phases of tsunami waves for the Nankai–Tonankai scenarios. The multi-domain nesting from coarse to fine resolution is conducted to consider large- to small-scale tsunami waves, depending on water depth.

A complete dataset of bathymetry/elevation, coastal/riverside structures (e.g. breakwater and levees) and surface roughness is obtained from the Cabinet Office of the Japanese Government. The data are provided in the form of nested grids (2430 – 810 – 270 – 90 – 30 – 10 m), covering the geographical regions of the Nankai–Tonankai Tough and western–central Japan (Fig. 1). The low-lying land areas along the coast are covered by 10 m grids for accurate inundation modelling. Because this study focuses on regional-scale effects of variable slip distributions on simulated tsunami waves, the minimum considered grid size for the Monte Carlo tsunami simulation is set to 90 m.

The ocean-floor topography data are based on the 1:50 000 bathymetric charts and JTOPO30 database developed by the Japan Hydrographic Association, and the nautical charts developed by the Japan Coastal Guard. The raw data are gridded using a triangulated irregular network. The land elevation data are based on the 5 m grid digital elevation model (DEM) developed by the

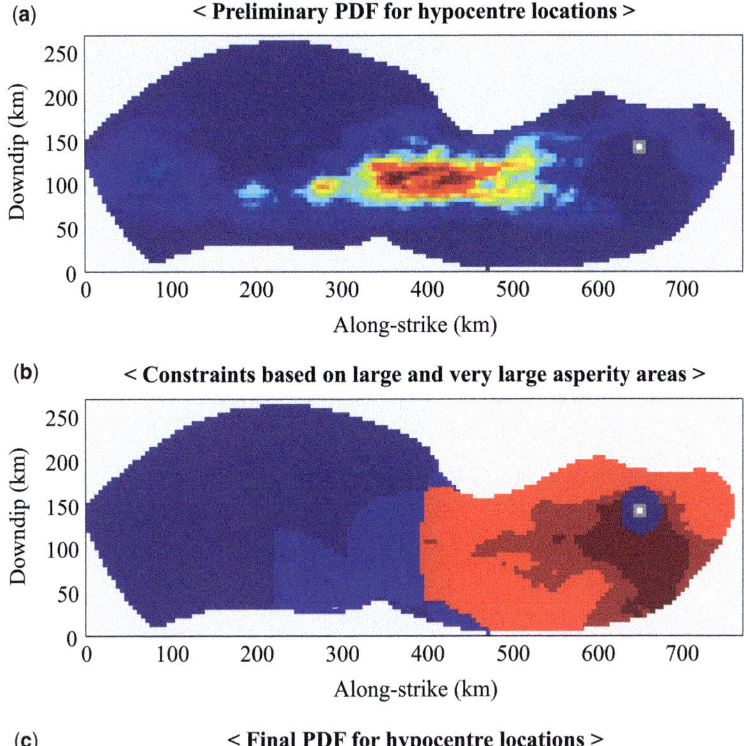

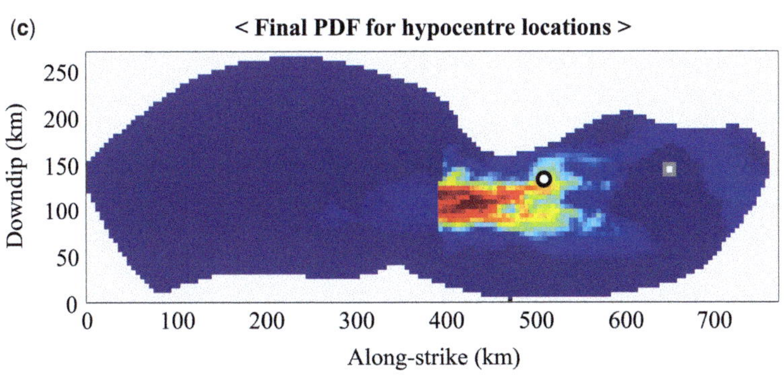

Fig. 5. Generation of hypocentre locations (the colour range is between 0 and 1): (**a**) preliminary probability density function for hypocentre locations; (**b**) constraints based on large and very large asperity areas; and (**c**) final probability density function for hypocentre locations.

Geospatial Information Authority of Japan. The raw data are based on airborne laser surveys and aerial photographic surveys. These data have measurement errors (in terms of standard deviation) of less than 1.0 m horizontally and of 0.3–0.7 m vertically. The reference elevation of the bathymetry and elevation data is the standard altitude in Japan (Tokyo Peil), and no variation of sea levels is taken into account.

The elevation data of the coastal/riverside structures are primarily provided by municipalities, supplemented also by the national coastline database. In the coastal/riverside structural dataset, only structures with dimensions less than 10 m are represented, noting that those with dimensions greater than 10 m are included in the DEM data. In the tsunami simulation, the coastal/riverside structures are represented by a vertical wall at one or two sides of

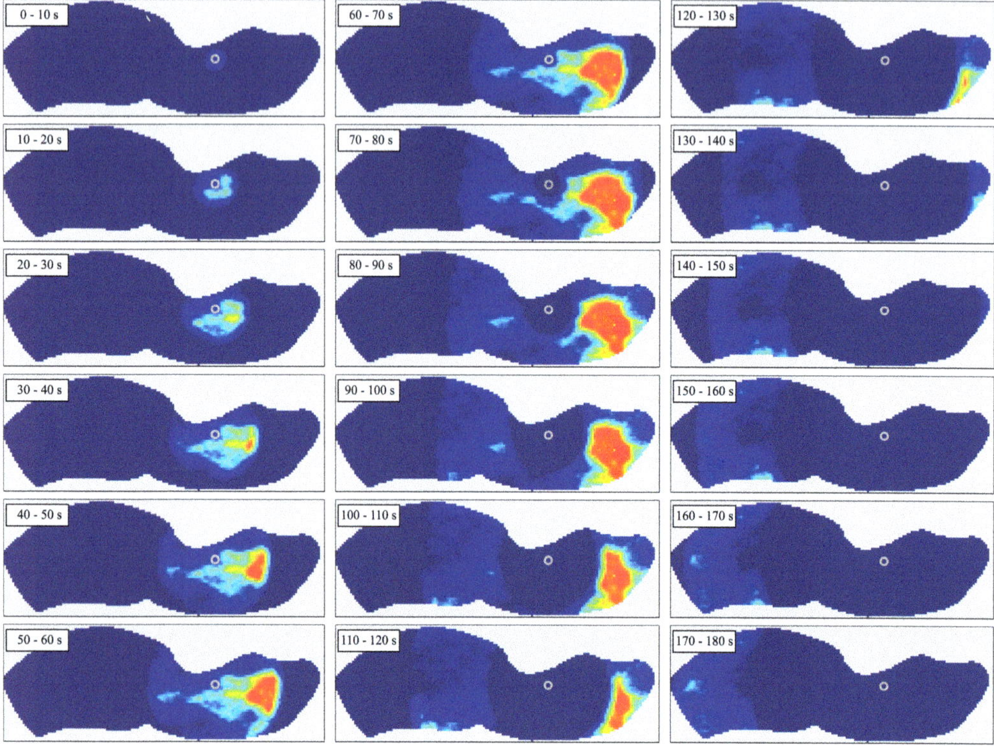

Fig. 6. Kinematic rupture process from 0 to 180 s (the colour range is between 0 and 10 m).

the computational cells. To evaluate the volume of water that overpasses these walls, the overflowing formulae of Honma (1940) are employed for coastal breakwater modelling at sub-grid scale (Japan Society of Civil Engineers 2002).

In the tsunami simulation, the bottom friction is evaluated using Manning's formula, following the Japan Society of Civil Engineers (2002) standard. Manning's coefficients are assigned to computational cells based on national land use data in Japan: 0.02 $m^{-1/3}$ s for agricultural land, 0.025 $m^{-1/3}$ s for ocean/water, 0.03 $m^{-1/3}$ s for forest vegetation, 0.04 $m^{-1/3}$ s for low-density residential areas, 0.06 $m^{-1/3}$ s for moderate-density residential areas and 0.08 $m^{-1/3}$ s for high-density residential areas. In the dataset provided by the Cabinet Office, no information on the roughness coefficient is provided for grids greater or equal to 90 m. In such cases, a uniform value of 0.025 $m^{-1/3}$ s (i.e. ocean/water) is adopted, and thus the extent of the inland inundation may be overestimated.

Simulated earthquake slips and slip patterns

The constrained-stochastic source modelling for the Nankai–Tonankai earthquake is applied in generating 100 slip distributions by taking into account variability in the stochastic synthesis parameters, target slip parameters and kinematic rupture parameters. All these parameters are modelled using the truncated normal distribution. The considered values of the parameters are summarized in Table 1. They are determined based on the stochastic source models for the Tohoku earthquake (Goda *et al.* 2014) and the 2012 CDMC model. The moment magnitudes of the source models are controlled by the mean-slip parameter, which can vary from 6 to 12 m (the average is 9 m). The adopted range of the mean-slip parameter is determined based on the observed variation of this parameter for the 11 inverted source models for the 2011 Tohoku earthquake (Goda *et al.* 2014). We propagate this variability in our tsunami simulations; therefore, the tsunami simulation results discussed in the following need to be interpreted with this consideration. Note that detailed assessments of the assumed model parameters – and their influence on the tsunami effects – need to be performed in future studies, in which alternative sets of model parameters may also need to be considered.

The 100 simulated source models have various slip (asperity) characteristics. To illustrate the

Table 1. *Summary of stochastic earthquake slip simulation parameters*

Model parameter	Distribution type	Mean	Standard deviation	[Lower, Upper]
Correlation length in downdip direction (km)	Truncated normal distribution with lower and upper bounds	50	20	[30, 70]
Correlation length in along-strike direction (km)		100	50	[50, 150]
Hurst number		0.8	0.2	[0.6, 1.0]
Box–Cox power		0.2	0.1	[0.1, 0.3]
Mean slip (m)		9	3	[6, 12]
Maximum slip (m)		50	10	[40, 60]
Rise time (s)		60	10	[50, 70]
Rupture velocity (km s^{-1})		2.5	0.5	[2.0, 3.0]
Depth limit for the asperity areas (km)	–	–	–	[1, 10]

variations of the synthesized slip distributions, six source models are shown in Figure 7. They are selected based on three slip patterns: west, middle and east. The classifications are determined by calculating the ratios of slip amount in the west, middle and east regions to the total slip over the entire fault plane. The boundaries of the west–middle and middle–east regions are set to the western end and eastern end of the Kii Peninsula (Fig. 1), dividing the region of the Nankai–Nonankai Trough into 40, 20 and 40% segments. The synthesized source models are categorized as west/middle/east slip pattern, when slip concentrations in the three regions exceed 60, 30 and 50% of the total slip; otherwise, the source models are left with no classification. This leads to 17 west slip patterns, 35 middle slip patterns, 25 east slip patterns and 23 unclassified patterns that have multiple asperities over two or three regions. Figure 7 presents two source models for the west/middle/east slip pattern. It is noted that these classifications are considered for the sake of presentations of the results only.

Figure 7 shows that the location and size of asperities vary significantly for different slip patterns (e.g. Fig. 7b v. Fig. 7d v. Fig. 7f). The maximum slip values also differ significantly (e.g. Fig. 7d v. Fig. 7e). These features are controlled by the stochastic synthesis parameters and the target slip parameters, and have a major influence on the tsunami simulation results (Goda *et al.* 2014). The developed stochastic source models can generate numerous realistic earthquake scenarios, and thus are useful for exploring the effects of different scenarios on the triggered tsunamis and their impacts on coastal communities. These are investigated in the following.

Offshore tsunami results

For each of the 100 synthesized slip distributions, initial boundary conditions for tsunami modelling (i.e. water surface elevation) are computed using the analytical formulae for elastic dislocation by Okada (1985) together with the equation by Tanioka & Satake (1996). The latter is to take into account the effects of horizontal movements of steep seafloor topography on the vertical water dislocation. Subsequently, the calculated water surface elevations are smoothed using a 9-cell × 9-cell moving average function (same as for the 2012 CDMC model) to avoid very steep initial water surface profiles. The water surface elevation profiles are calculated with an interval of 10 s to account for the temporal evolution of the kinematic earthquake rupture process. In the above-mentioned procedure for calculating the water surface elevation, the hydrodynamic response of seawater is not accounted for. Caution needs to be exercised when modelling tsunamigenic earthquakes with major asperities on a gently dipping fault plane at a shallow depth (applicable to the Nankai–Tonankai earthquake), because large peak vertical deformation is generated along the top edge of the dipping fault (Goda 2015) as a result of the assumption of elastic behaviour of the rock encoded in the Okada (1985) solutions. In reality, strongly non-linear (plastic) deformation occurs, which is typically not accounted for in tsunami simulations: instead, the seabed deformation calculated from the Okada (1985) equations is used directly as an input sea surface condition in the tsunami simulation.

First, offshore tsunami wave characteristics due to the stochastic source models are examined by focusing on the Shikoku and Tokai regions (see Fig. 1). Figures 8 and 9 show the maximum tsunami wave-height (maximum surface elevation – this convention is used throughout) contours for Shikoku and Tokai, respectively, by considering the six source models presented in Figure 7. As expected, slip patterns have a significant impact on the maximum wave heights: if major asperities are near to the locations of interest, larger tsunami

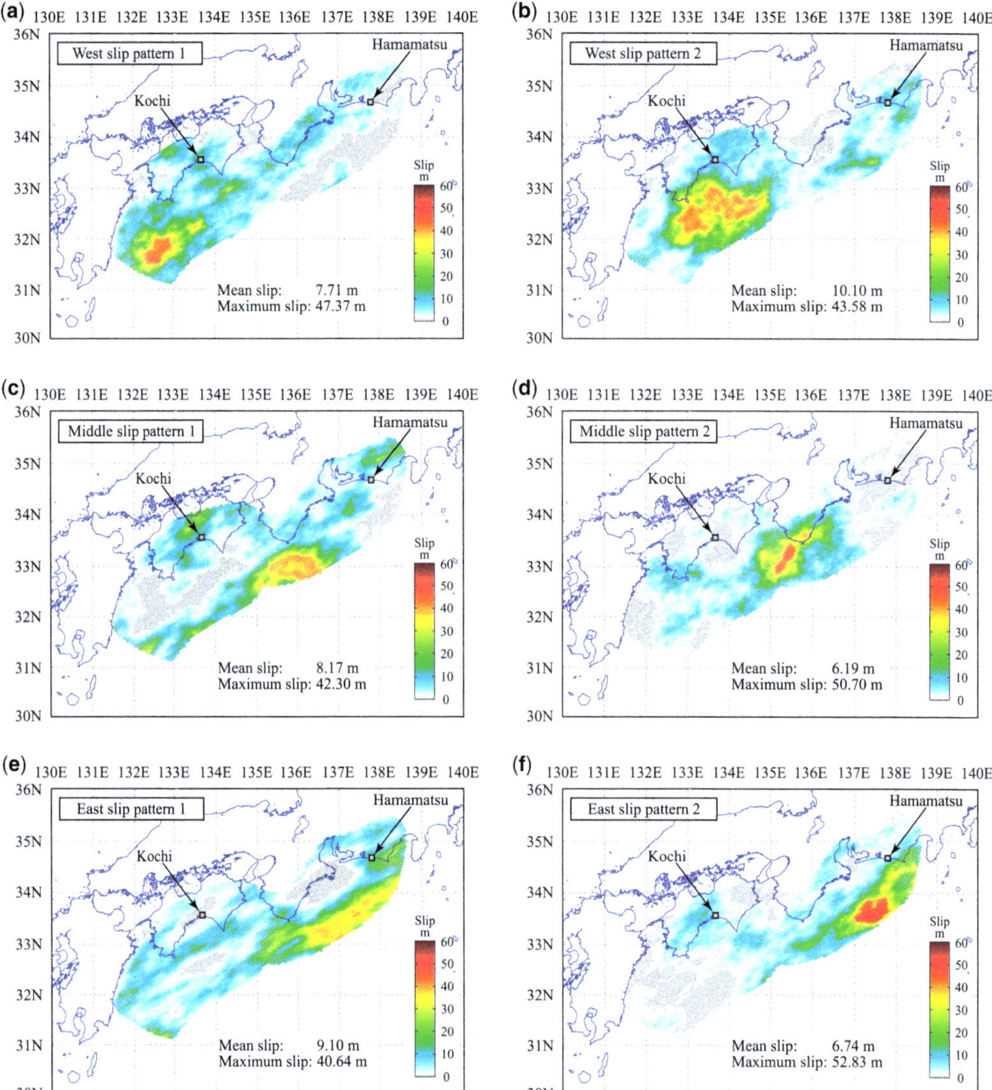

Fig. 7. Stochastic slip distributions: (**a**) west slip pattern 1; (**b**) west slip pattern 2; (**c**) middle slip pattern 1; (**d**) middle slip pattern 2; (**e**) east slip pattern 1; and (**f**) east slip pattern 2.

wave-amplitudes are generated. By inspecting the source model and the corresponding wave heights, the causal relationship between the source and the impact can be easily understood. The comparison of the results for Figure 8a, b (both are for west slip patterns) also suggests that the local features of the slip patterns have a major influence on the tsunami waves along the coastal line due to the near-shore bathymetry. The complex spatial distribution of tsunami wave heights can be observed in the regions between offshore and coastline (Figs 8a–c & 9b, e, f). The tsunami wave profiles are significantly affected by refraction due to bathymetry and by overlapping of wave propagation due to complex slip patterns, giving non-uniform incoming tsunami profiles to the land. It highlights the importance of the spatial slip distribution, in conjunction with the local bathymetry and coastline, on the tsunami hazard assessment.

To further illustrate the variability of temporal tsunami wave profiles, two specific sites are selected. The first location is off Kochi in the Shikoku region, and the second one is off Hamamatsu in the Tokai region (see Fig. 7 for the locations on

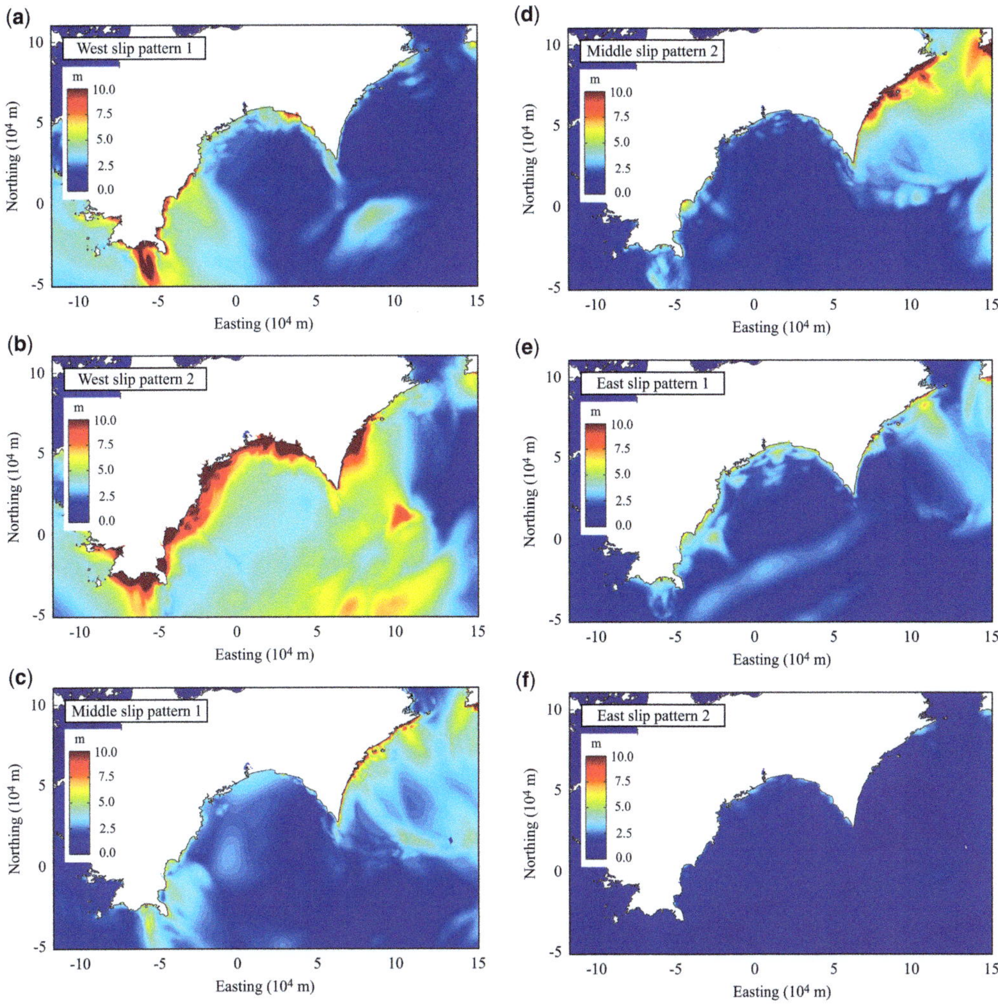

Fig. 8. Maximum wave-height contours in Shikoku: (**a**) west slip pattern 1; (**b**) west slip pattern 2; (**c**) middle slip pattern 1. (**d**) middle slip pattern 2; (**e**) east slip pattern 1; and (**f**) east slip pattern 2.

the maps). The water depths at the two locations are about 45 and 60 m for Kochi and Hamamatsu, respectively. The time series of water surface elevation from Monte Carlo tsunami simulation results based on the 100 stochastic source models are shown in Figure 10a, b for Kochi and Hamamatsu, respectively. In the figure panel, individual tsunami wave profiles are shown with light blue lines, whereas the statistics (i.e. median and 10th/90th percentiles) of the tsunami simulation results are shown by thick black (solid/broken) lines to represent major trends of the tsunami wave profiles. It can be observed that the tsunami wave profiles at these locations are highly variable. Although the average wave profiles are not so large, rare tsunami waves (e.g. exceeding 90th percentile waves) can be several times greater than the average tsunami wave levels. For example, at both locations, the maximum tsunami wave heights attained by the severest scenarios are more than twice as large as the 90th percentile wave levels. To further investigate the effects of regional slip patterns on the simulated tsunami profiles, average tsunami wave profiles for the west, middle and east slip patterns are compared with the statistics of the 100 tsunami simulations, and the results are presented in Figure 10c, d for Kochi and Hamamatsu, respectively. As one intuitively expects, tsunami wave profiles are strongly affected by the proximity to the major asperities (i.e. for Kochi, the west slip patterns represent more critical scenarios: while, for Hamamatsu, the east slip patterns have more influence),

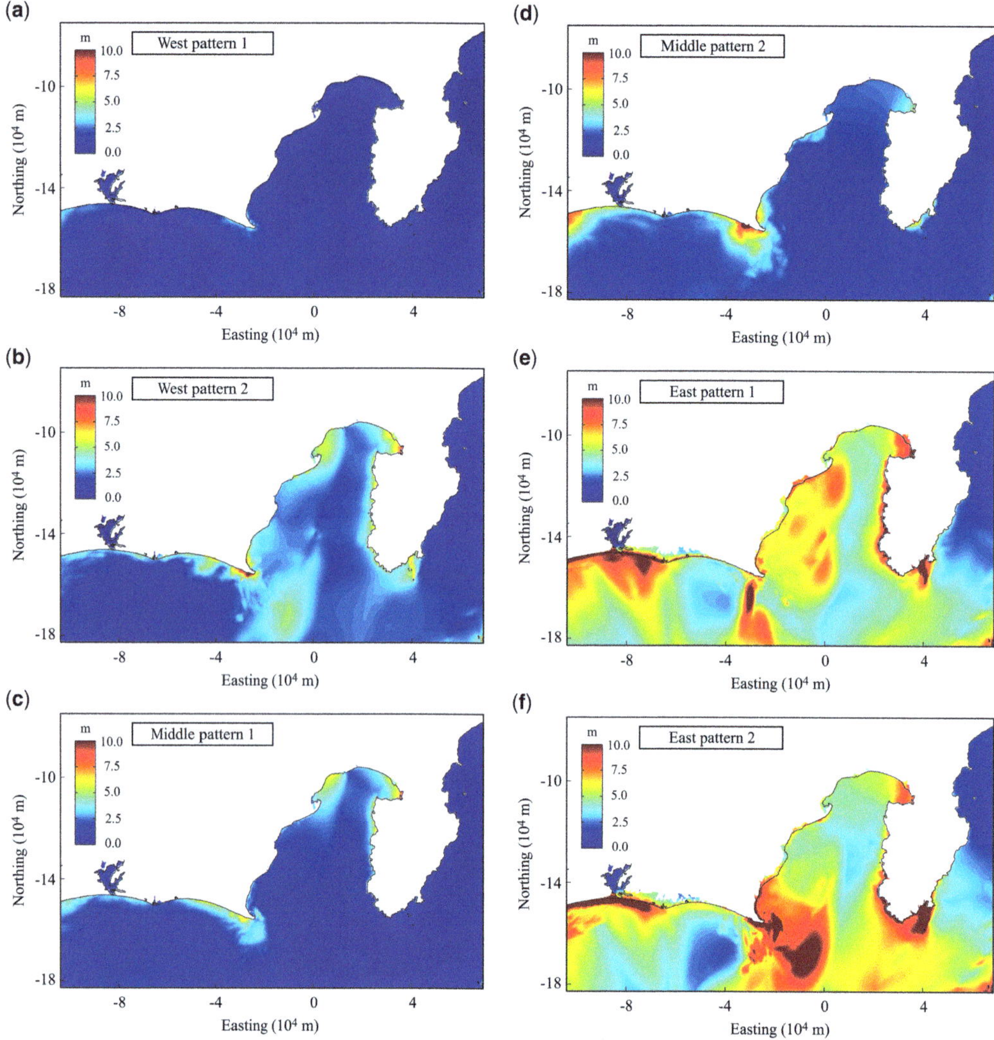

Fig. 9. Maximum wave-height contours in Tokai: (**a**) west slip pattern 1; (**b**) west slip pattern 2; (**c**) middle slip pattern 1. (**d**) middle slip pattern 2; (**e**) east slip pattern 1; and (**f**) east slip pattern 2.

corroborating the importance of the spatial slip distribution.

Lastly, the overlay of tsunami wave profiles provides useful information on the arrival times of major tsunami waves. For instance, large tsunami waves exceeding 5 m are expected to arrive within 20–30 min at Kochi, whereas such significant waves may arrive earlier (15–20 min) at Hamamatsu (the differences are caused by the proximity to the asperities and the seafloor topography). The difference of 5–10 min has important implications in successfully completing horizontal/vertical evacuations in coastal cities and towns (Federal Emergency Management Agency 2008). Hence, the stochastic tsunami simulations facilitate the quantification and visualization of the tsunami-hazard prediction uncertainty, and are particularly useful for emergency managers who wish to be informed of the range of possible consequences they need to deal with.

Onshore tsunami results

The tsunami wave heights from the source to nearshore regions are significant, as shown in Figures 8 and 9. The spatial variability of the maximum tsunami inundation height along the coastal line in the Shikoku and Tokai regions is shown in Figure 11.

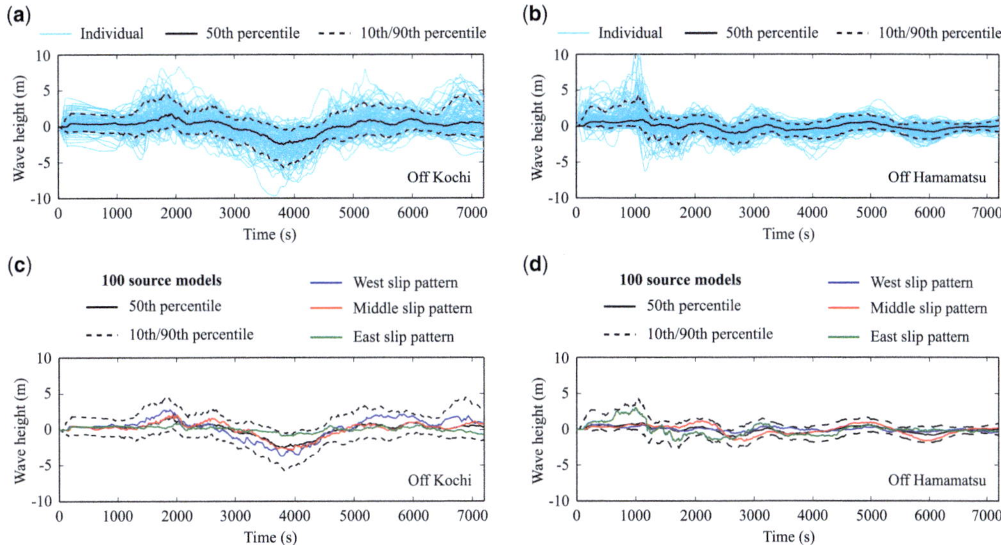

Fig. 10. (a) 100 tsunami wave profiles off Kochi, (**b**) 100 tsunami wave profiles off Hamamatsu, (**c**) statistics of 100 tsunami wave profiles and average wave profiles for west/middle/east slip patterns off Kochi, and (**d**) statistics of 100 tsunami wave profiles and average wave profiles for west/middle/east slip patterns off Hamamatsu.

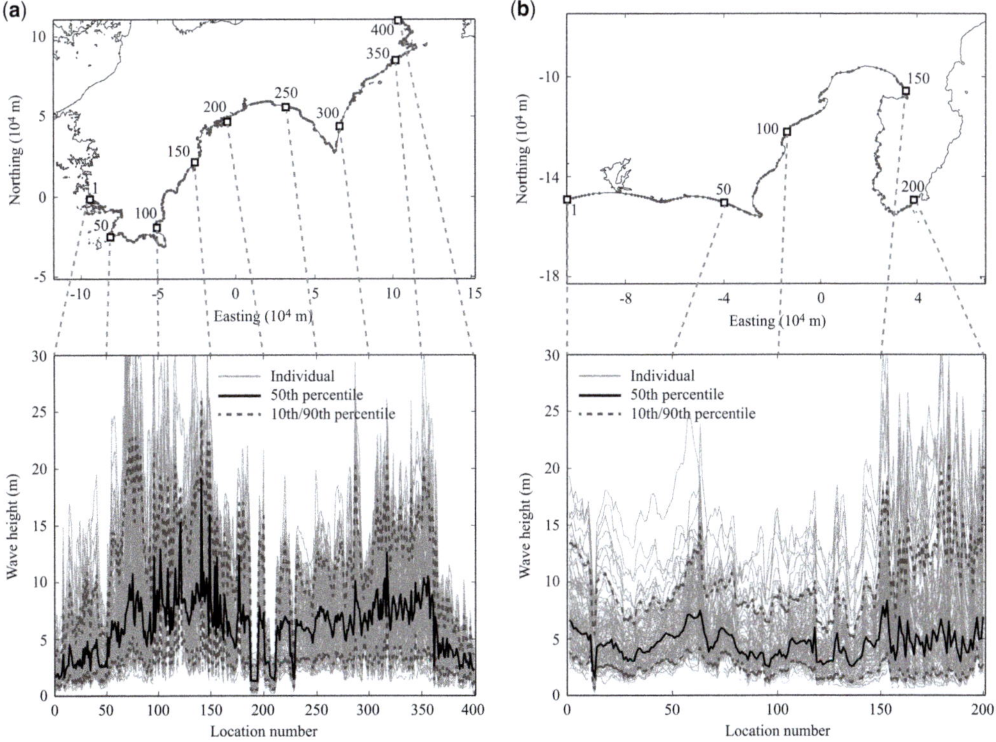

Fig. 11. Maximum wave-height predictions along the coastal line: (**a**) Shikoku and (**b**) Tokai.

For this analysis, computational cells that are close to zero (i.e. depth between −1.0 and 1.0 m) are defined based on DEM data at a horizontal spacing of approximately 1.0 km. In Figure 11, individual simulation results as well as 10th, 50th, and 90th percentiles of the simulation results at individual locations are shown. Individual results from the simulation exhibit significant variability of the inundation height due to stochastic tsunami scenarios. Tsunami wave heights for average and rare cases are spatially variable, influenced by source, path and topographical effects. Although the mean and extreme tsunami wave heights vary along the coast, the ratios between the 50th and 90th percentiles do not change significantly except for particular areas. This is due to local tsunami amplification effects near-shore (e.g. shape and characteristic water depth of a bay). It is important to know both the average tsunami wave height and range of variations for tsunami disaster reduction.

The stochastic tsunami simulations are also useful for assessing the extent of onshore tsunami

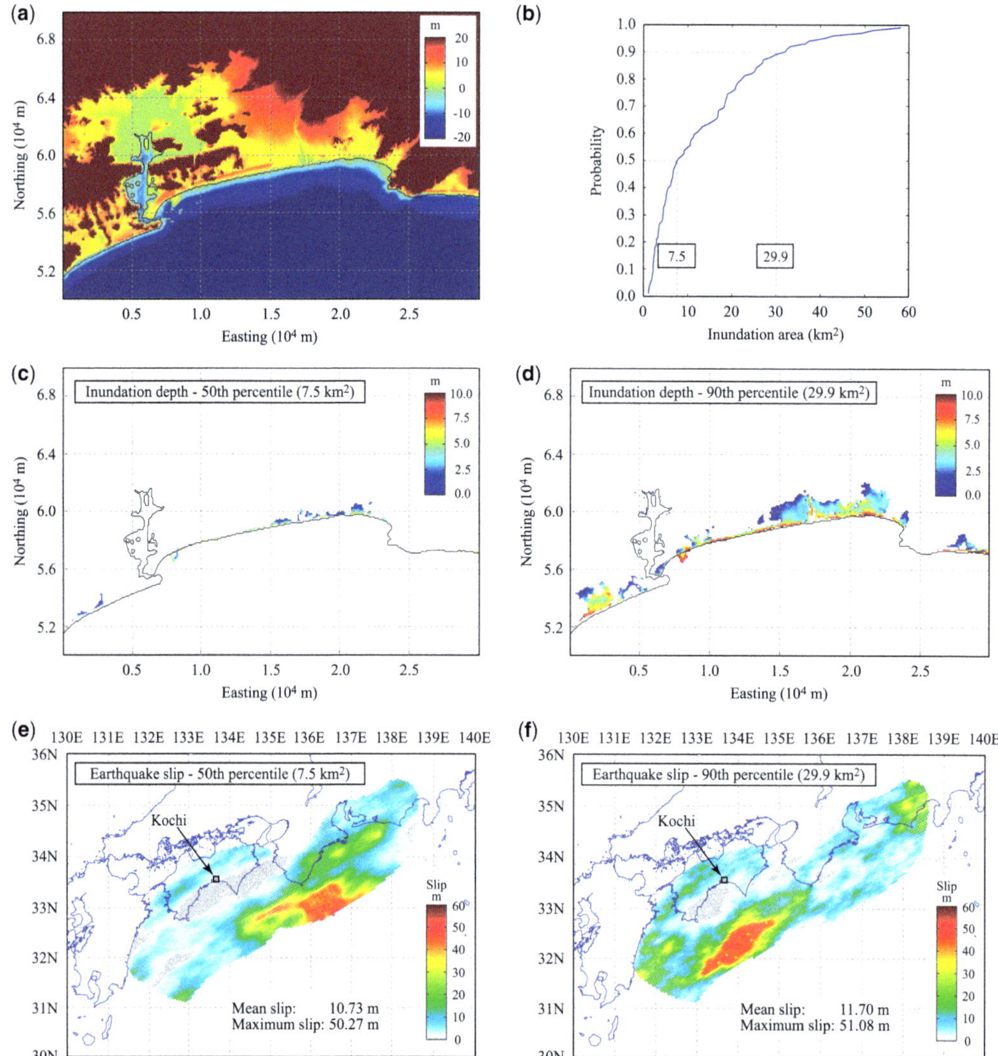

Fig. 12. Tsunami inundation results in Kochi: (**a**) elevation data; (**b**) cumulative probability distribution of inundation areas above 1 m depth; (**c**) inundation depth distribution at 50th percentile; (**d**) inundation depth distribution at 90th percentile; (**e**) slip distribution corresponding to the 50th percentile inundation map; and (**f**) slip distribution corresponding to the 90th percentile inundation map.

inundation by developing stochastic inundation maps. The probabilistic hazard maps can be further utilized in tsunami risk assessment and loss estimation (Goda & Song 2016). To demonstrate the procedure for developing stochastic inundation maps for the Nankai–Tonankai earthquake, results are shown for Kochi and Hamamatsu in Figures 12 and 13, respectively. The land elevation data are displayed in Figures 12a and 13a, respectively. The western section of the low-lying areas in Kochi is protected by hills/high grounds. On the other hand, the topography in Hamamatsu is flat along the coastal line. Although both locations have flat bathymetry offshore, the different land-side topographical features affect the predicted inundation maps, especially under extreme situations.

To develop stochastic inundation maps, a tsunami hazard parameter that is suitable for characterizing the city-level inundation extent needs to be chosen. In this demonstration, inundated areas with depths of more than 1 m are considered. It is noteworthy that the chosen parameter is relevant

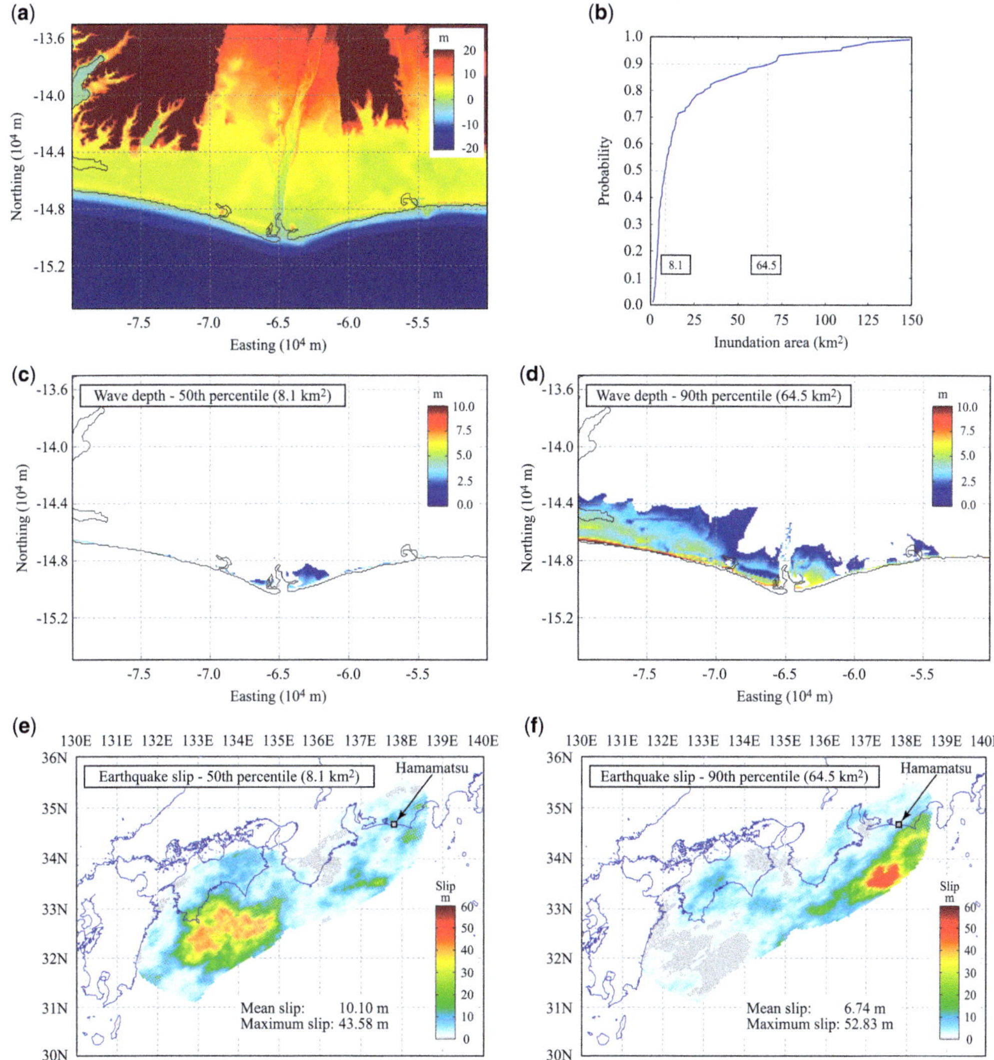

Fig. 13. Tsunami inundation results in Hamamatsu: (**a**) elevation data; (**b**) cumulative probability distribution of inundation areas above 1 m depth; (**c**) inundation depth distribution at 50th percentile; (**d**) inundation depth distribution at 90th percentile; (**e**) slip distribution corresponding to the 50th percentile inundation map; and (**f**) slip distribution corresponding to the 90th percentile inundation map.

for tsunami inundation damage to wooden houses; different parameters may be adopted depending on the main features of the building portfolio of interest. Using the tsunami inundation results from stochastic tsunami simulations, city-level inundation values can be evaluated for individual source models and the cumulative distribution function (CDF) of the inundation parameter can be derived (Figs 12b & 13b). The developed CDF of the inundation area above 1 m depth is then used to define probabilistic hazard scenarios. For instance, risk managers may be interested in knowing typical (expected) scenarios and worst-case scenarios, and relevant probability levels may be chosen. In this study, two probability levels are selected for illustration: the 50th percentile and the 90th percentile. The inundation areas at the 50th and 90th percentiles are 7.5 and 29.9 km^2, respectively, for Kochi (Fig. 12b), and are 8.1 and 64.5 km^2, respectively, for Hamamatsu (Fig. 13b). It is noteworthy that the inundation areas in Hamamatsu can be significantly greater than those in Kochi (Figs 12b v. Fig. 13b) because the total low-lying areas susceptible to tsunami inundation in Hamamatsu are larger than those in Kochi (no major natural barriers; Fig. 12a v. Fig. 13a).

Figure 12c, d show the 50th and 90th percentile tsunami inundation maps for Kochi, respectively. The counterparts for Hamamatsu are shown in Figure 13c, d. The comparison of the 50th and 90th percentile maps clearly indicate spatial expansion of the inundated areas and increased inundation depth along the shoreline. Consequently, the more extreme scenarios cause much significant damage and consequences to coastal cities and towns. The visual inspection of the changes of the inundation profiles from the 50th to the 90th percentile scenarios in Kochi and Hamamatsu clearly indicate that topographical features of the low-lying land have paramount influence on the tsunami inundation result. It is important to emphasize that such dramatic changes of the inundation results may be overlooked when a limited number of tsunami source models are used in tsunami hazard analysis. Using a wide range of stochastic source models, such cases can be captured and appropriate risk mitigation measures can be implemented.

Furthermore, the generated hazard maps can be related to the causative source models, helping define critical hazard scenarios. Figure 12e, f show slip distributions that correspond to the 50th and 90th percentile inundation maps for Kochi; Figure 13e, f are for Hamamatsu. The source models shown in Figures 12 and 13 indicate that the locations of the major asperities move closer to the site of interest in case more critical scenarios are considered (this is expected and similar observations were made for Figs 8 & 9). This information is useful for managing tsunami risk proactively, as local municipalities may be interested in preparing a set of tsunami hazard maps for evacuation purposes. Therefore, such stochastic tsunami scenarios and corresponding inundation maps for local communities are useful planning tools. On the other hand, from a wider perspective, emergency managers should prepare several tsunami scenarios and corresponding inundation maps for the region to assess spatial dependency of tsunami hazard at different locations for a range of critical scenarios.

Conclusions

Tsunami hazard assessment for future mega-thrust subduction earthquakes is challenging, because it involves various kinds of uncertainties and assumptions. The stochastic earthquake source models can be used in Monte Carlo tsunami simulation to quantify the uncertainty of tsunami hazard parameters at regional/local scales. This study developed stochastic source models for the M_w 9-class Nankai–Tonankai earthquake based on the inversion-based source models for the 2011 Tohoku earthquake and the 2012 CDMC tsunami source model. The modelling procedure took into account uncertainties of the location of rupture nucleation point, the rupture propagation speed and the rise time to capture the kinematic fault rupture process. The applicability of the developed models was demonstrated by generating stochastic source models with different slip patterns, and by comparing the maximum tsunami wave-height contours and temporal wave-height profiles in the Nankai–Tonankai regions. Moreover, the models were employed to develop stochastic inundation maps in Kochi and Hamamatsu. The numerical examples highlighted that the developed tool is useful for exploring the effects of different scenarios on the triggered tsunamis and their impacts on coastal communities, and produces useful results for tsunami hazard mapping and preparedness. In particular, the stochastic tsunami simulations facilitate the uncertainty quantification and visualization of the tsunami hazard predictions.

The developed stochastic source models for the Nankai–Tonankai earthquake can be improved in several ways. Firstly, a critical review of the assumed model parameters should be conducted as the current models are primarily based on the source characteristics for the 2011 Tohoku-type earthquake, while other subduction earthquakes may also be applicable. For such purposes, a broad range of earthquake source models (e.g. those from the 2004 Indian Ocean event and the 2010 Chile event) should be explored to obtain the suitable sets of stochastic synthesis parameters, target slip parameters and kinematic rupture parameters.

Updating of the scaling relationships for the stochastic synthesis parameters is necessary (Mai & Beroza 2002; Murotani *et al.* 2013). For such a purpose, new scaling relationships for large tsunamigenic earthquakes by Goda *et al.* (2016) can be adopted; the models have been developed based on an extensive set of 226 inverted rupture models from the SRCMOD database (Mai & Thingbaijam 2014). Subsequently, rigorous sensitivity analysis needs to be carried out by varying key model parameters and by evaluating the effects of these parameters on tsunami hazard predictions. In particular, the effects of changing source rupture parameters (e.g. rupture speed and rise time) should be investigated more thoroughly in future studies. Another important issue is to develop a formal method to incorporate the geological evidence of the anticipated earthquake fault rupture behaviour, such as segmentation, rupture limit along dip and asperity locations based on past events, in a subduction region of interest. This will enable the construction of more realistic stochastic source models. Such extended models are also useful for testing the effects of adopted assumptions on the tsunami hazard predictions.

The bathymetry and elevation data for the Nankai–Tonankai regions were provided by the Cabinet Office of the Japanese Government. This work was supported by the Engineering and Physical Sciences Research Council (EP/M001067/1), as well as by the King Abdullah University of Science and Technology (BAS/1339-01-01).

References

ANDO, M. 1975. Source mechanisms and tectonic significance of historical earthquakes along the Nankai Trough, Japan. *Tectonophysics*, **27**, 119–140.

CENTRAL DISASTER MANAGEMENT COUNCIL 2012. *Working Group Report on Mega-Thrust Earthquake Models for the Nankai Trough, Japan.* Cabinet Office of the Japanese Government, Tokyo, http://www.bousai.go.jp/jishin/nankai/nankaitrough_info.html

FEDERAL EMERGENCY MANAGEMENT AGENCY 2008. *Guidelines for Design of Structures for Vertical Evacuation from Tsunamis.* FEMA P646. Federal Emergency Management Agency, Washington, DC.

FRASER, S., POMONIS, A. *ET AL.* 2013. Tsunami damage to coastal defences and buildings in the March 11th 2011 M_w 9.0 Great East Japan earthquake and tsunami. *Bulletin of Earthquake Engineering*, **11**, 205–239.

FUKUTANI, Y., SUPPASRI, A. & IMAMURA, F. 2015. Stochastic analysis and uncertainty assessment of tsunami wave height using a random source parameter model that targets a Tohoku-type earthquake fault. *Stochastic Environmental Research and Risk Assessment*, **29**, 1763–1779.

GEIST, E.L. 2002. Complex earthquake rupture and local tsunamis. *Journal of Geophysical Research: Solid Earth*, **107**, https://doi.org/10.1029/2000JB000139

GEIST, E.L. & PARSONS, T. 2006. Probabilistic analysis of tsunami hazards. *Natural Hazards*, **37**, 277–314.

GODA, K. 2015. Effects of seabed surface rupture v. buried rupture on tsunami wave modeling: a case study for the 2011 Tohoku, Japan earthquake. *Bulletin of the Seismological Society of America*, **105**, 2563–2571.

GODA, K. & SONG, J. 2016. Uncertainty modeling and visualization for tsunami hazard and risk mapping: a case study for the 2011 Tohoku earthquake. *Stochastic Environmental Research and Risk Assessment*, **30**, https://doi.org/10.1007/s00477-015-1146-x

GODA, K., MAI, P.M., YASUDA, T. & MORI, N. 2014. Sensitivity of tsunami wave profiles and inundation simulations to earthquake slip and fault geometry for the 2011 Tohoku earthquake. *Earth, Planets and Space*, **66**, https://doi.org/10.1186/1880-5981-66-105

GODA, K., YASUDA, T., MORI, N. & MARUYAMA, T. 2016. New scaling relationships of earthquake source parameters for stochastic tsunami simulation. *Coastal Engineering Journal*, **58**, https://doi.org/10.1142/S0578563416500108

GONZALEZ, F.I., GEIST, E.L. *ET AL.* 2009. Probabilistic tsunami hazard assessment at Seaside, Oregon, for near- and far-field seismic sources. *Journal of Geophysical Research: Oceans*, **114**, https://doi.org/10.1029/2008JC005132

GOTO, C., OGAWA, Y., SHUTO, N. & IMAMURA, F. 1997. *Numerical Method of Tsunami Simulation with the Leap-Frog Scheme.* IOC Manuals and Guides, **35**. UNESCO, Paris, France.

GUSMAN, A.R., TANIOKA, Y., SAKAI, S. & TSUSHIMA, H. 2012. Source model of the great 2011 Tohoku earthquake estimated from tsunami waveforms and crustal deformation data. *Earth and Planetary Science Letters*, **341–344**, 234–242.

HIROSE, F., NAKAJIMA, J. & HASEGAWA, A. 2008. Three-dimensional seismic velocity structure and configuration of the Philippine Sea slab in southwestern Japan estimated by double-difference tomography. *Journal of Geophysical Research: Solid Earth*, **113**, https://doi.org/10.1029/2007JB005274

HONMA, J. 1940. Discharge coefficient for trapezoidal weir. *Journal of Japan Society of Civil Engineers*, **26**, 635–645.

HORSPOOL, N., PRANANTYO, I. *ET AL.* 2014. A probabilistic tsunami hazard assessment for Indonesia. *Natural Hazards and Earth System Sciences*, **14**, 3105–3122.

JAPAN SOCIETY OF CIVIL ENGINEERS 2002. *Tsunami Assessment Method for Nuclear Power Plants in Japan.* Japan Society of Civil Engineers, Tokyo, https://www.jsce.or.jp/committee/ceofnp/Tsunami/eng/JSCE_Tsunami_060519.pdf

KODAIRA, S., HORI, T. *ET AL.* 2006. A cause of rupture segmentation and synchronization in the Nankai trough revealed by seismic imaging and numerical simulation. *Journal of Geophysical Research: Solid Earth*, **111**, https://doi.org/10.1029/2005JB004030

LAVALLÉE, D., LIU, P. & ARCHULETA, R.J. 2006. Stochastic model of heterogeneity in earthquake slip spatial distributions. *Geophysical Journal International*, **165**, 622–640.

LOVELESS, J.P. & MEADE, B.J. 2010. Geodetic imaging of plate motions, slip rates, and partitioning of deformation in Japan. *Journal of Geophysical Research: Solid Earth*, **115**, https://doi.org/10.1029/2008JB006248

LØVHOLT, F., PEDERSEN, G., BAZIN, S., KUHN, D., BREDESEN, R.E. & HARBITZ, C. 2012. Stochastic analysis of tsunami runup due to heterogeneous coseismic slip and dispersion. *Journal of Geophysical Research: Oceans*, **117**, https://doi.org/10.1029/2011JC007616

LØVHOLT, F., GLIMSDAL, S., HARBITZ, C.B., HORSPOOL, N., SMEBYE, H., DE BONO, A. & NADIMA, F. 2014. Global tsunami hazard and exposure due to large co-seismic slip. *International Journal of Disaster Risk Reduction*, **10**, 406–418.

MAI, P.M. & BEROZA, G.C. 2002. A spatial random field model to characterize complexity in earthquake slip. *Journal of Geophysical Research: Solid Earth*, **107**, https://doi.org/10.1029/2001JB000588

MAI, P.M. & THINGBAIJAM, K.K.S. 2014. SRCMOD: An online database of finite-fault rupture model. *Seismological Research Letters*, **85**, 1348–1357.

MAI, P.M., SPUDICH, P. & BOATWRIGHT, J. 2005. Hypocenter locations in finite-source rupture models. *Bulletin of the Seismological Society of America*, **95**, 965–980.

MCCLOSKEY, J., ANTONIOLI, A. ET AL. 2008. Tsunami threat in the Indian Ocean from a future megathrust earthquake west of Sumatra. *Earth and Planetary Science Letters*, **265**, 61–81.

MUELLER, C., POWER, W.L., FRASER, S. & WANG, X. 2015. Effects of rupture complexity on local tsunami inundation: implications for probabilistic tsunami hazard assessment by example. *Journal of Geophysical Research: Solid Earth*, **120**, https://doi.org/10.1002/2014JB011301

MURATA, S., IMAMURA, F., KATOH, K., KAWATA, Y., TAKAHASHI, S. & TAKAYAMA, T. 2010. *Tsunami: To Survive from Tsunami*. World Scientific Publishing, Hackensack, NJ.

MUROTANI, S., SATAKE, K. & FUJII, Y. 2013. Scaling relations of seismic moment, rupture area, average slip, and asperity size for $M \sim 9$ subduction-zone earthquakes. *Geophysical Research Letters*, **40**, 5070–5074.

OKADA, Y. 1985. Surface deformation due to shear and tensile faults in a half-space. *Bulletin of the Seismological Society of America*, **75**, 1135–1154.

PARDO-IGUZQUIZA, E. & CHICA-OLMO, M. 1993. The Fourier integral method: an efficient spectral method for simulation of random fields. *Mathematical Geology*, **25**, 177–217.

SATAKE, K., FUJII, Y., HARADA, T. & NAMEGAYA, Y. 2013. Time and space distribution of coseismic slip of the 2011 Tohoku earthquake as inferred from tsunami waveform data. *Bulletin of the Seismological Society of America*, **103**, 1473–1492.

TANIOKA, Y. & SATAKE, K. 1996. Tsunami generation by horizontal displacement of ocean bottom. *Geophysical Research Letters*, **23**, 861–864.

TARBOTTON, C., DALL'OSSO, F., DOMINEY-HOWES, D. & GOFF, J. 2015. The use of empirical vulnerability functions to assess the response of buildings to tsunami impact: comparative review and summary of best practice. *Earth-Science Review*, **142**, 120–134.

YAMAZAKI, Y., LAY, T., CHEUNG, K.F., YUE, H. & KANAMORI, H. 2011. Modeling near-field tsunami observations to improve finite-fault slip models for the 11 March 2011 Tohoku earthquake. *Geophysical Research Letters*, **38**, https://doi.org/10.1029/2011GL049130

Spatial variability in sediment lithology and sedimentary processes along the Japan Trench: use of deep-sea turbidite records to reconstruct past large earthquakes

KEN IKEHARA[1]*, KAZUKO USAMI[1], TOSHIYA KANAMATSU[2], KAZUNO ARAI[2,3], ASUKA YAMAGUCHI[4] & RINA FUKUCHI[4]

[1]*Institute of Geology and Geoinformation, Geological Survey of Japan, National Institute of Advanced Industrial Science and Technology (AIST), Tsukuba Central 7, 1-1-1 Higashi, Tsukuba, Ibaraki 305–8567, Japan*

[2]*R&D Centre for Earthquake and Tsunami, Japan Agency of Marine Science and Technology (JAMSTEC), 2-15 Natsushima-cho, Yokosuka, Kanagawa 237-0061, Japan*

[3]*Present address: Center for Advanced Marine Core Research, Kochi University, B200 Monobe, Nankoku, Kochi 783-8502, Japan*

[4]*Atmosphere and Ocean Research Institute, University of Tokyo, 5-1-5 Kashiwanoha, Kashiwa, Chiba 277-8564, Japan*

**Correspondence: k-ikehara@aist.go.jp*

Abstract: Large earthquakes and related tsunamis serve as triggering mechanisms that generate turbidity currents which form turbidites. The event deposits from the recent 2011 Tohoku-oki earthquake and tsunami are observed throughout a wide area along the Pacific coast of Tohoku, northern Japan, extending from the coast through the shelf and slope, to the trench floor. Spatio-temporal correlation of turbidites and other tsunamigenic deposits, such as those generated in the 2011 event, can be used to reconstruct the recurrence history of large earthquakes and tsunamis. Here we use sediment cores and sub-bottom profiles to analyse the depositional setting along the Japan Trench, and show that the environment is ideal for preserving turbidites. The subducting Pacific Plate creates graben or basins along the trench floor that accommodate the episodic deposition of fine-grained turbidites; and interseismic hemipelagic deposits that form with high sedimentation rates along the Japan Trench effectively cover earthquake-induced turbidites and preserve the deposits as a geological record of large earthquakes. Therefore, small deep-sea basins with high sedimentation rates, such as in and around the Japan Trench floor, are favourable environments for studies of turbidite palaeoseismology.

The 2011 earthquake off the Pacific coast of Tohoku (the 2011 Tohoku-oki earthquake: M_w 9.0; epicentre location, 38° 06.2′ N, 142° 51.6′ E; hypocentre depth, 24 km) (Hirose *et al.* 2011) (Fig. 1) occurred on 11 March 2011 and triggered a devastating tsunami with a run-up height of more than 40 m in northern Honshu, Japan (Mori *et al.* 2011). Large slip was generated by the earthquake near the Japan Trench (e.g. Fujii *et al.* 2011; Suzuki *et al.* 2011; Yokota *et al.* 2011). Repeated observations of bathymetry and seismic profiles show large (>50 m) displacements of the trench floor (Fujiwara *et al.* 2011; Kodaira *et al.* 2012). A major rotational slump on the lowermost trench slope is also interpreted from sediment cores (Strasser *et al.* 2013).

Onshore tsunami deposits serve as evidence that similar significant earthquakes and tsunamis have impacted the Pacific coast of Tohoku in the past, including the 869 Jogan tsunami (Sawai *et al.* 2007, 2012; Shishikura *et al.* 2007; Ishimura & Miyauchi 2015). Numerical simulation of the 869 Jogan tsunami suggests that the source was an earthquake similar to the 2011 Tohoku-oki earthquake (Namegaya & Satake 2014). Furthermore, Sawai *et al.* (2012) showed the repeated occurrence of large tsunamis based on onshore tsunami deposits recorded in sediment sequences on the Sendai and Ishinomaki plains. Sawai *et al.* (2015) suggested the large 1454 Kyotoku historical tsunami was also generated by an earthquake along the Japan Trench, based on geological evidence on the Sendai Plain. The spatio-temporal distribution of onshore tsunami deposits is consistent with the characteristics of interplate earthquakes along the northern

From: SCOURSE, E. M., CHAPMAN, N. A., TAPPIN, D. R. & WALLIS, S. R. (eds) 2018. *Tsunamis: Geology, Hazards and Risks*. Geological Society, London, Special Publications, **456**, 75–89.
First published online March 3, 2017, https://doi.org/10.1144/SP456.9

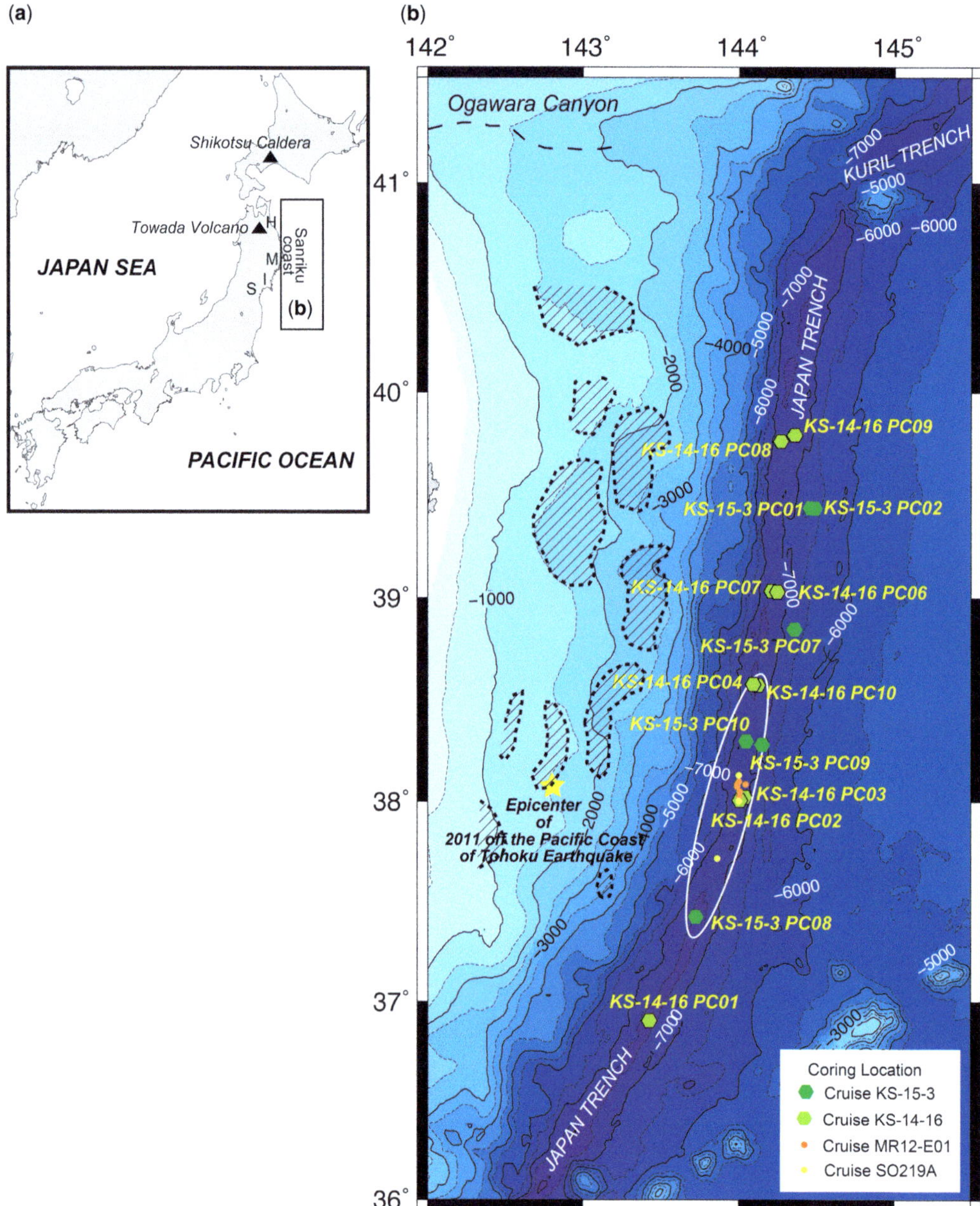

Fig. 1. (**a**) Location of the study area: H, Hachinohe; M, Miyako; I, Ishinomaki; S, Sendai; (**b**) bathymetry of the central and northern Japan Trench, showing core sites. The white circle indicates area where three thick turbidite units were observed in the upper part of the sediment cores. The distribution of isolated basins is after Arai *et al.* (2014).

Japan Trench, although the tsunami deposits might also reflect large far-field earthquakes that occurred in other areas of the northern Japan Trench, such as that which produced the 1961 Chilean tsunami (Kon'no *et al.* 1961).

Large earthquakes and tsunamis also trigger submarine slope failure, surface sediment resuspension and turbidity current generation. Therefore, turbidites deposited by turbidity currents are a potential tool in the study of subaqueous palaeoseismology (e.g. Adams 1990; Goldfinger *et al.* 2003, 2012; Noda *et al.* 2008; Goldfinger 2011; Patton *et al.* 2013, 2015; Pouderoux *et al.* 2014; Ikehara *et al.* 2016). Indeed, the earthquake- and/or tsunami-induced seafloor disturbance and turbidite depositions resulting from the 2011 Tohoku-oki earthquake and tsunami are documented throughout an extensive region, from the coastal sea to the Japan Trench floor offshore from Tohoku (Ikehara *et al.* 2011, 2014*a*, 2016; Noguchi *et al.* 2012; Arai *et al.* 2013; Oguri *et al.* 2013; Miura *et al.* 2014; Toyofuku *et al.* 2014; Tamura *et al.* 2015; Nomaki *et al.* 2016; Usami *et al.* In press). However, the deposition and preservation of turbidites varies by location and event, making spatio-temporal correlation of deposits and events challenging when reconstructing the palaeoseismic record (Sumner *et al.* 2013; Atwater *et al.* 2014; Ikehara *et al.* 2014*b*). Detailed characterization of the depositional history of a site can reveal the causes of apparent gaps in the palaeoseismic record, and identify sites ideal for preserving turbidites and other tsunamigenic deposits; therefore, site characterization is essential for turbidite palaeoseismology (Goldfinger *et al.* 2012, 2014; Atwater *et al.* 2014).

Here, we examine the deposition and preservation of fine-grained turbidites that were possibly generated by past earthquakes near the central and northern Japan Trench. The results show that small, tectonically subsiding basins along the Japan Trench floor are ideal sites for studies of turbidite palaeoseismology in this region.

Physiographical setting

The Pacific Plate is subducting beneath the Okhotsk Plate along the Japan Trench, moving to the NW at a rate of 8.0–8.6 cm a^{-1} (DeMets *et al.* 2010). The trench strikes north–south to NNE–SSW, originating at the triple junction of the Pacific, Philippine Sea and Okhotsk plates in the south, and intersecting the Kuril Trench to the north. Trench floor depths are of the order of 6800 m below sea level in the northern part of the trench, and deepen to more than 7500 m in the central and southern parts (Fig. 1). The plate interface is erosional, with subduction erosion producing tectonic subsidence (von Huene & Lallemand 1990) that forms a low-gradient upper-slope terrace (the average gradient of 1°–2° calculated from bathymetric maps). Although there is no clear forearc basin along the northern Japan Trench off Sanriku, isolated basins occur on the upper-slope terrace (Arai *et al.* 2014) (Fig. 1b). The lower slope is steeper than the upper slope, with an average gradient of approximately 5°. Knick-points separate the upper and lower slopes at water depths of 2500–3000 m. The narrow mid-slope terrace at water depths of 4000–6000 m is formed by active faulting along the subduction margin (Tsuru *et al.* 2002). Horst-and-graben structures trending north–south to NNW–SSE on the subducting Pacific Plate are evident in the bathymetry, and are generally more pronounced along the central and northern parts of the trench. Vertical relief within the graben is typically of the order of a few hundreds of metres.

Surface waters around the northern Japan Trench are mainly within the path of the Oyashio cold-water current. The Kuroshio warm-water current flows south to north, mixing with the Oyashio cold-water current. Further mixing of the water mass occurs as the Tsugaru warm-water current flows from the Japan Sea through the Tsugaru Strait. The mixed waters form eddies that induce the upwelling of nutrient-rich subsurface water masses. Supported by a large nutrient supply to the surface water, diatoms flourish and primary productivity is high in this area (Saino *et al.* 1998); diatomaceous (biosiliceous) sediments continuously deposit on the seafloor.

Sandy sediments, composed of medium to coarse sand, distribute on the shelf off the Sanriku coast. Very fine sand and muds are deposited on the upper slope in water depths of 150–500 m, and finer muddy sediments are found deeper than 500 m (Arita & Kinoshita 1984). Upper-slope sediments consist of calcareous and diatomaceous clay with abundant clastic and pumiceous grains (Mann & Müller 1980; Arita & Kinoshita 1984; Ikehara *et al.* 2013), whereas diatomaceous silt to silty clay and diatom ooze are distributed on the lower slope and trench floor (Mann & Müller 1980; Ikehara *et al.* 2016). Holocene and late Pleistocene sedimentation rates range from 6 to 100 cm ka^{-1} on the upper slope (von Huene *et al.* 1980; Yamane & Oba 1999; Minoshima *et al.* 2007), approximately 23 cm ka^{-1} on the mid-slope terrace (Shipboard Scientific Party 1980), and from 100 to 440 cm ka^{-1} on the trench floor (Ikehara *et al.* 2013, 2016).

Materials and methods

We obtained 19 sediment cores from the Japan Trench floor during cruises KS-14-19 and KS-15-3 aboard the R/V *Shinsei-Maru*. Fifteen of the cores were collected from the trench floor or graben depressions, and five were from horsts. Herein, we utilize these 15 trench and graben cores (Fig. 1; Table 1) to examine sediment lithology, sedimentation rates and sediment variability along the central and northern Japan Trench. The sediment cores were analysed at Kochi University, Japan, with an X-ray computed tomography (CT) scanner (GE Light Speed Ultra 16) and multi-sensor core logger (GEOTEK multi-sensor core logger (MSCL)) to

Table 1. *Sediment core locations*

Cruise	Core	Latitude (N)	Longitude (E)	Water depth (m)	Total core length (cm)	Location
KS-14-16	KS-14-16 PC01	36° 53.50′	143° 24.89′	7619	751.5	Trench
	KS-14-16 PC02	38° 00.47′	144° 00.28′	7484	818.5	Trench
	KS-14-16 PC03	38° 01.32′	144° 02.20′	7110	944.5	Graben
	KS-14-16 PC04	38° 34.78′	144° 05.46′	7591	884.5	Trench
	KS-14-16 PC06	39° 01.89′	144° 14.84′	7394	883.5	Graben
	KS-14-16 PC07	39° 02.21′	144° 12.64′	7282	874.0	Trench
	KS-14-16 PC08	39° 45.94′	144° 16.23′	7368	843.7	Trench
	KS-14-16 PC09	39° 47.63′	144° 21.42′	7210	832.3	Graben
	KS-14-16 PC10	38° 34.56′	144° 07.37′	7590	797.0	Graben
KS-15-3	KS-15-3 PC01	39° 25.66′	144° 28.35′	6795	1114.8	Graben
	KS-15-3 PC02	39° 25.57′	144° 29.67′	6694	652.0	Graben (edge)
	KS-15-3 PC07	38° 49.93′	144° 22.24′	6919	219.1	Graben (edge)
	KS-15-3 PC08	37° 25.03′	143° 44.04′	7790	697.0	Trench
	KS-15-3 PC09	38° 16.64′	144° 09.71′	7160	736.0	Graben
	KS-15-3 PC10	38° 17.57′	144° 03.43′	7433	835.8	Trench

obtain vertical X-ray CT images and profiles of sediment bulk density and magnetic susceptibility, respectively. Vertical X-ray CT images were reconstructed from axial images of 0.625 mm intervals. The voxel size of the axial image was around 0.2 mm. MSCL measurement was conducted every 2 cm. The cores were measured, split longitudinally and logged based on visual inspection. Soft X-radiographs were collected for detailed observations of sedimentary structures using an X-ray generator (Sofron SR0-i503-2) and digital radiography charge-coupled device imaging sensor (RF Systems NX-04SN). The morphology and refractive index of volcanic glass shards and the mineral composition of tephra beds were assessed by Kyoto Fission Track, Kyoto, Japan, for correlation with known onshore tephra deposits. Sub-bottom profiles were also obtained using the ship-mounted CHIRP sub-bottom profiler system (Kongsberg TOPAZ) at five coring sites during the KS-15-3 cruise. Ikehara *et al.* (2016) reported the sediment lithologies and sedimentation rates of five piston cores and three gravity cores collected during the R/V *Mirai* MR12-E01 and R/V *Sonne* SO219A cruises, respectively, on the trench floor near the epicentre of the 2011 Tohoku-oki earthquake (Fig. 1b). We compare our data and interpretations with their data.

Results

Lithological characteristics and sedimentation rates

Sediment cores obtained for this study contain mostly diatomaceous clay (Fig. 2) that contains alternating black and greyish olive/olive black beds and is moderately to highly bioturbated with local white sponge spicules. Three types of coarse-grained beds were observed in the diatomaceous clay. The first is a fining-upwards unit a few tens of centimetres to a few metres in thickness deposited above a sharp erosional surface (Fig. 3a, b). This unit is generally composed of multiple amalgamated fining-upwards subunits. Sandy layers in each subunit range from coarse silt to fine sand with parallel- or cross-laminations, and a distinct erosional basal contact. Relatively thick muddy layers that thicken upwards overlie the sandy layers, producing a fining-upwards structure. Sediment bulk density and magnetic susceptibility both decrease with decreasing grain size (Fig. 4). The muddy layers are generally diatomaceous, greyish olive in colour, homogeneous, lack evidence of bioturbation and occasionally contain calcareous nannoplankton. Layers with calcareous nannofossils are shown in Figure 2; generally, these deposits are lighter in colour (grey/greyish olive) than the diatomaceous mud. The characteristics of these sedimentary structures are consistent with those of fine-grained turbidites (subdivision T0–T8) described by Stow & Shanmugam (1980).

The second type of coarse-grained bed observed in the sediment cores is a single coarse-grained bed without mud (Fig. 3c). Typical examples were encountered in cores KS-14-16 PC03, KS-15-3 PC02 and KS-15-3 PC09. These deposits are less than a few centimetres in thickness, are usually massive and contain medium coarse silt to very fine sand; however, some layers show parallel- or cross-laminations. An upwards-fining grading structure is also found in some layers. The basal contact of the

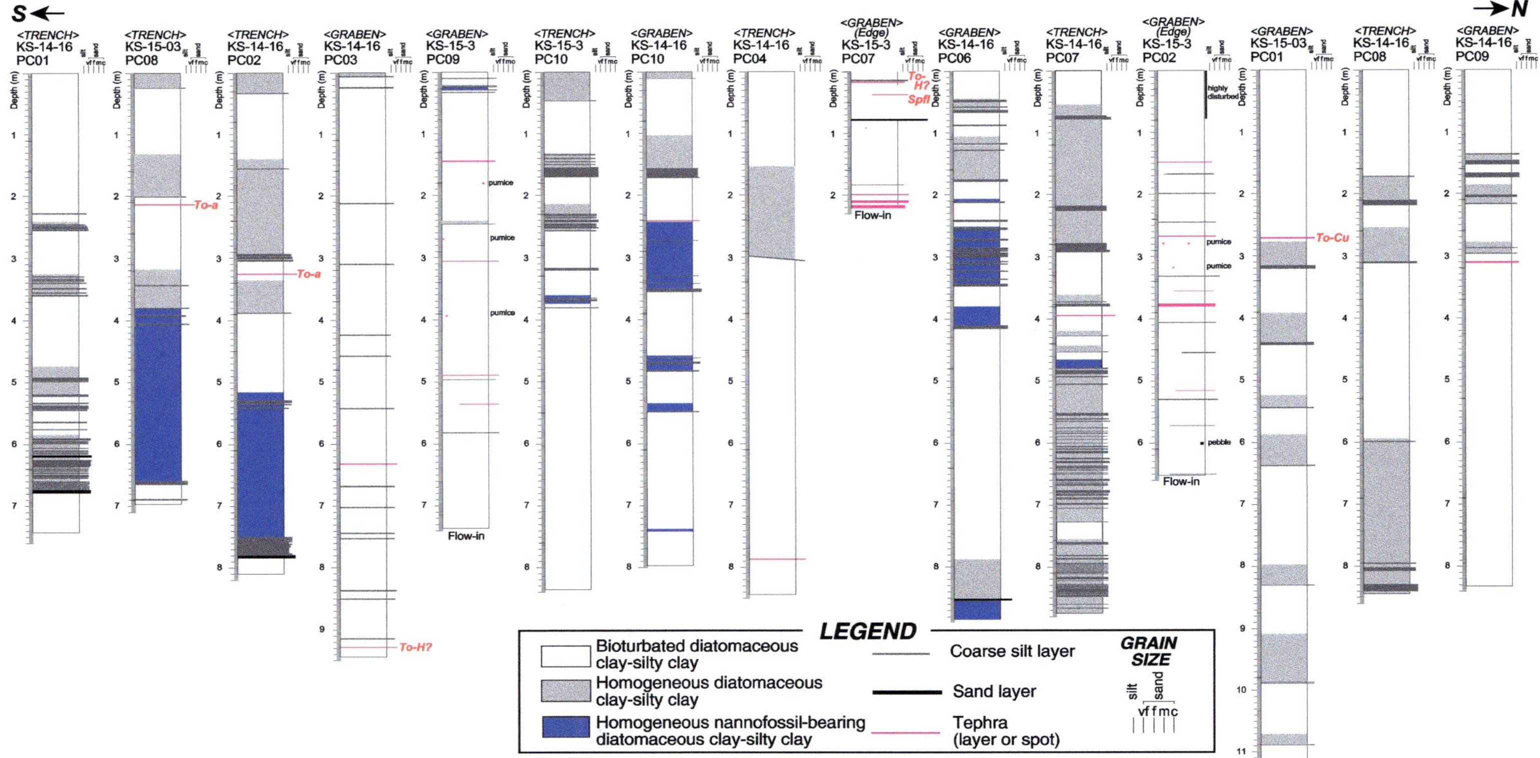

Fig. 2. Columnar sections of the obtained cores.

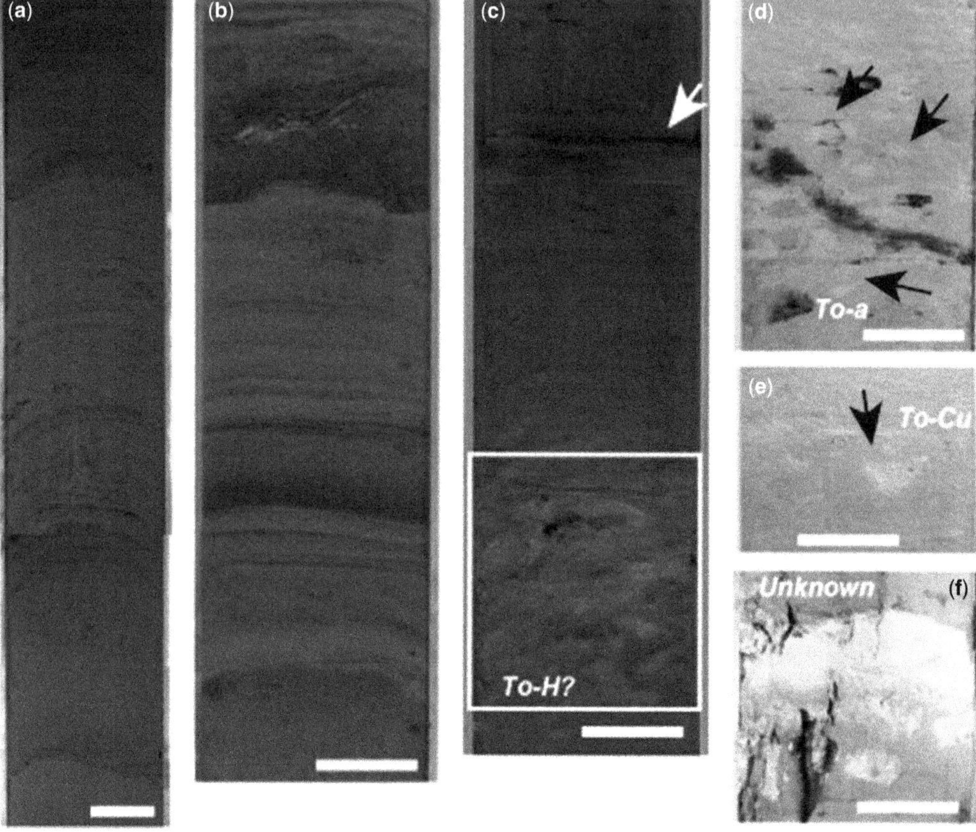

Fig. 3. Typical examples (core photograph) of coarse-grained beds observed in the collected sediment cores. (**a**) Portion of thick, fine-grained turbidite with numerous thin coarse layers (KS-14-16 PC07). (**b**) Portion of thick, fine-grained turbidite mud containing calcareous nannofossils (KS-14-16 PC10). (**c**) Single-bed turbidite without turbidite mud (white arrow), containing a bioturbated tephra layer (KS-14-16 PC03). (**d**) Bioturbated tephra (To-a tephra) layer (black arrows) (KS-15-3 PC08). (**e**) Bioturbated tephra (To-Cu tephra) layer (black arrow) (KS-15-3 PC01). (**f**) Thick tephra layer with no known correlative tephra unit (KS-15-3 PC07).

bed is sharp and erosional. The structure of this bed type corresponds to the basal sand layer of fine-grained turbidites (subdivision T1) of Stow & Shanmugam (1980), although the samples we observed lacked the muddy layer (subdivision T2–T8).

In some cores, distinct tephra beds were recognized as the third type of coarse-grained bed (Fig. 3c–f). Tephra grains range from coarse silt to fine sand in size. Isolated rounded pumice grains were observed in some cores, and some tephra beds were bioturbated (Fig. 3c–e). The upper and lower boundaries of the tephra beds are occasionally sharp (Fig. 3f), but usually irregular and gradual. Petrographical characteristics of the tephra beds are listed in Table 2. Four late Quaternary tephras are identified based on petrographical characteristics: the Towada-a tephra (To-a, AD 915: Machida & Arai 2003), Towada-Chuseri (To-Cu, 6 ka: Machida & Arai 2003) and Towada-Hachinohe (To-H, 15 ka: Ikehara *et al.* 2013) tephras from the Towada volcano in NE Japan, and the Shikotsu dai-ichi tephra (Spfl, 40–45 ka: Machida & Arai 2003) from the Shikotsu Caldera in Hokkaido. Based on the depths and ages of the known tephra deposits, the average late Quaternary sedimentation rates in the collected cores range from 0.9 to 295 cm ka^{-1}.

Acoustic facies and patterns of sediment accumulation on the Japan Trench floor

Sub-bottom profile surveys conducted across small depressions and graben along the Japan Trench reveal well-stratified acoustic facies (Fig. 5). The reflectors are nearly horizontal, and are slightly oblique to the general trend of the bathymetry of the subducting Pacific Plate. Relatively thick

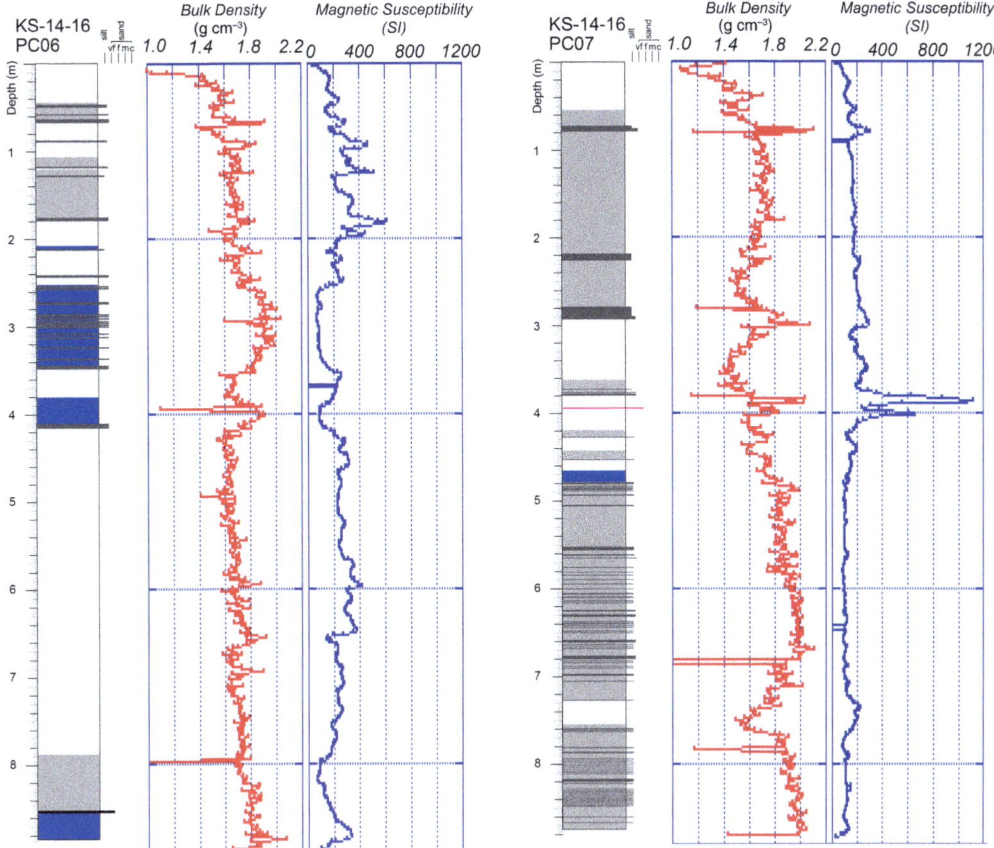

Fig. 4. Bulk density and magnetic susceptibility profiles from a graben depression (KS-14-16 PC06) and from a trench depression (KS-14-16 PC07). The legend of the sediment columns is the same as in Figure 2.

acoustically transparent layers occur in the profile (Fig. 5a). Trench-fill and graben-fill deposits thicken to the west in a basin (Fig. 5b, c). Well-stratified deposits onlap underlying sediments at the edge of the graben (Fig. 5b).

Discussion

Lithological variability of sediments along the Japan Trench

Most of sediment cores recently collected from the Japan Trench small basins contains several thick muddy turbidites (Fig. 2). In some cores, however, single and thin coarse-grained layers without turbidite mud are intercalated in bioturbated diatomaceous clay. Thus, lithology of the Japan Trench sediments has large variability. In contrast, turbidite stratigraphy is similar among the cores collected near the epicentre of the 2011 Tohoku-oki earthquake.

While some of the sediments examined in the cores of the present study appear to correlate with sediments described previously (e.g. Ikehara *et al.* 2016), we note the lithology in our cores varies significantly along the trench axis. For example, cores KS-14-16 PC10 and KS-14-16 PC06 contain at least three or four nannoplankton-bearing turbidite mud layers, suggesting a more frequent supply of sediments from the upper slope than is documented in the cores collected from sites near the epicentre of the 2011 Tohoku-oki earthquake (cores KS-15-3 PC08 and KS-14-16 PC02). Furthermore, when comparing cores of the trench-fill and nearby graben-fill (e.g. KS-14-16 PC02 v. KS-14-16 PC03, KS15-3 PC10 v. KS-15-3 PC09, KS-14-16 PC07 v. KS-14-16 PC06 and KS-14-16 PC08 v. KS-14-16 PC09), the sediment lithology is quite different (Figs 2 & 4). Such dissimilarities may be due to differences in the source materials in the turbidity currents, the pathway between the trench-fill and the graben-fill, and/or core sample locations relative to the turbidity current path. Differences

Table 2. *Results of tephra analyses and correlation of tephra beds*

Sample ID	Cruise	Core	Depth (cm)		Thickness (cm)	Occurrence	Glass type*	Glass morphology†	Light minerals‡
			Top	Bottom					
KS1416P2-1	KS-14-16	PC02	325	330	5	Spots	pm > bw	T, H, C	pl, Qz
KS14P0301	KS-14-16	PC03	927	929	2	Spots	pm > bw	T > C, H	pl (Qz)
KS15P0101	KS-15-3	PC01	270	272	2	Spots	pm > It > bw	T ≫ C, It > H	pl, Qz
KS15P0701	KS-15-3	PC07	16	16.5	0.5	Spots	pm > bw, It	T ≫ H, C, It	pl, Qz
KS15P0702			37	39	2	Spots	pm > bw	T, H > C	pl, Qz, beta-Qz
KS15P0703			199.1	200.1	1	Layer	pm	T ≫ C	pl, Qz, Kf
KS15P0704			208.1	212.1	4	Layer	pm > bw, It	T ≫ C, H	pl, Qz
KS15P0705			216.1	219.1	3	Layer	pm = bw	T > H ≫ C	pl, Qz
KS15P0706			734.2	736.2	2	Block	pm	T ≫ C	pl, Qz
KS15P0801	KS-15-3	PC08	219.5	219.5	0	Spot	pm > bw, It	T ≫ H, C	pl, Qz

*Glass type: pm, pumiceous; bw, bubble wall; It, intermediate.
†Glass morphology (after Yoshikawa 1976): H, large broken bubble wall; T, fibrous and pumice type; C, intermediate form between H-type and T-type; It, irregular shaped.

in the source materials or materials incorporated in the mobilized turbidity current also affect the velocity of the current, which, in turn, impacts the composition and areal distribution of the turbidite. While sedimentation rates appear to differ among the basins, there are not enough data (i.e. insufficient age control) to discuss the spatio-temporal variability of sedimentation rates in the sediments cores.

A preliminary correlation of the sediments in the cores of the present study suggests a turbidite stratigraphy (occurrence of three thick turbidite units in the upper sequence) in three small basins (from south to north): core KS-15-3 PC08, core KS-14-16 PC02, core KS-15-3 PC10 and core KS-14-16 PC10 (which is near the epicentre of the 2011 Tohoku-oki earthquake: Fig. 2). The Towada-a tephra (To-a) is recognized between the second and third thick turbidites in cores KS-15-3 PC08 and KS-14-16 PC02. Ikehara *et al.* (2016) characterized the turbidite stratigraphy of the uppermost sequence along the central Japan Trench, near the epicentre of the 2011 Tohoku-oki earthquake. They recognized three thick fine-grained turbidite units (TT1, TT2 and TT3 in ascending stratigraphic order) deposited within the past 1500 years. The TT1 turbidite unit is most likely formed by the 2011 earthquake. A historical tephra bed, the To-a tephra, was identified between the TT2 and TT3 turbidite units, and was used to correlate the deep-sea turbidite stratigraphy with onshore tsunami deposits on the Sendai and Ishinomaki plains (Sawai *et al.* 2007, 2012, 2015; Shishikura *et al.* 2007). Ikehara *et al.* (2016) also postulated a correlation of their TT2 and TT3 units with onshore tsunami deposits from the 1454 Kyotoku and 869 Jogan tsunamis, respectively. If the correlation is correct, the source of the two historical large tsunamis might also be within the subduction zone, similar to the 2011 Tohoku-oki earthquake. Sawai *et al.* (2012, 2015), Namegaya & Satake (2014) and Namegaya & Yata (2014) also suggested the subduction zone as the source of both the 1454 and 869 tsunamis, based on onshore tsunami records and historical documents. Thus, a comparison and correlation of onshore records with deep-sea records might provide new insights into past tsunamigenic earthquakes, and the spatial distribution of the TT2 and TT3 turbidites throughout the Japan Trench is important for understanding the source area of the 1454 and 869 tsunamis, respectively. If our turbidites in cores KS-15-3 PC08, KS-14-16 PC02, KS-15-3 PC10 and KS-14-16 PC 10 are correlative with the units documented by Ikehara *et al.* (2016), then the lateral extent of these turbidites is of the order of 120 km along the Japan Trench (Fig. 1b). This area is similar to the rupture area of the 2011 Tohoku-oki earthquake (>30–40 m slip: Iinuma *et al.* 2011; Suzuki *et al.* 2011; Yagi & Fukahata 2011; Yokota *et al.* 2011) and also consistent with the spatial distribution of observed seismo-turbidites (TT1 unit) most likely to have been generated by the 2011 Tohoku-oki earthquake.

Sedimentary processes of the central and northern Japan Trench

Turbidites near the central Japan Trench, generated by the 2011 Tohoku-oki earthquake, were likely

Heavy minerals§	Glass Index						Correlation	Age (ka)	Sedimentation rate (cm ka^{-1})
	Min.	Max.	Average	No.	Mode	Morphology			
Opx, Cpx, Opq, GHb > Ap, Bt	1.5061	1.5099	1.5080	45/60	1.508	T, H, C	To-a	AD 915	295.5
Opq, Opx > Cpx > GHb, Ap, Bt	1.5000	1.5179	1.5087	55/60	1.504, 1.508	T > C, H	To-H?	15.7	59.0
Opq, Opx ≫ Cpx, GHb (Ap, BHb)	1.4958	1.5038	1.5002	13/86	1.502	C, H, T	To-Cu	6	45.0
	1.5100	1.5156	1.5133	60/86	1.513–1.514	T > It, C			
Opx > Cpx, Opq, GHb > Bt (Ap, BHb)	1.4970	1.5037	1.5002	12/79		H, C, T, It	To-H?	15.7	1.0
	1.5053	1.5081	1.5072	60/79	1.507	T > H, C			
Opq, Opx > GHb, Cpx > Bt (BHb, Cum, Zr)	1.5010	1.5045	1.5031	60/82	1.503	T, H, C	Spfl	40–45	0.9
Bt ≫ Opx, Opq (Ap, GHb, Cpx, Zr)	1.4986	1.5009	1.4997	60/70	1.500	T > C, H, It	Unknown		
GHb ≫ Ap, Opq, Bt	1.4979	1.5007	1.4993	60/72	1.499	T > C, H, It	Unknown		
Bt ≫ GHb > Opq, Cpx, Opx, Ap	1.4995	1.5022	1.5012	60/67	1.501	T, H > C > It	Unknown		
Opq > Opx, Cpx, GHb > Ap (BHb, Bt)	1.4985	1.5050	1.5022	60/73	1.503	T > C > H, It	Unknown		
	1.5089	1.5435	1.5248	10/73		T, It			
Opq > Opx ≫ Cpx > GHb, Bt, Ap	1.5045	1.5106	1.5075	60/69	1.507	T > H, C, It	To-a	AD 915	199.5

‡Light minerals: pl, plagioclase; Qz, quartz; Kf, K-feldspar; beta-Qz, beta-quartz.
§Heavy minerals: Opx, orthopyroxene; Cpx, clinopyroxene; GHb, green hornblende; BHb, blue hornblende; Opq, opaque minerals; Ap, apatite; Bt, biotite; Zr, zircon.

to have been produced by seafloor sediment remobilization along the lower trench slope, as the turbidites consist of diatomaceous clay that contains radiocesium (^{137}Cs) originating from atmospheric nuclear bomb tests and high excess ^{210}Pb concentrations (Ikehara *et al.* 2016). However, the occurrence of nannoplankton-bearing turbidite muds at >7500 m depth on the seafloor (Ikehara *et al.* 2016, this study) is below the carbonate compensation depth (CCD; *c.* 4000–4500 m depth in the area of the Japan Trench: Berger *et al.* 1976) and suggests that at least some turbidity currents were generated in slope environments located at water depths shallower than 4000 m. Because surface sediments on the upper and lower slopes of the Japan Trench are muddy (Mann & Müller 1980; Arita & Kinoshita 1984), turbidity currents generated in these zones might consist of a mix of seawater and muddy sediments with fewer sandy particles, resulting in fine-grained turbidites being deposited on the Japan Trench floor.

As the water depth along the Japan Trench axis deepens to the south, the turbidity currents initiated upon the upper or lower slope will flow down the slope under gravity and then southwards along the trench floor, following the trench axis. The small elongate basins along the trench serve as loci of deposition for the turbidity currents. Although some currents might flow over features with minor relief, the turbidity currents are trapped in the deeper terminal basins. That is, some turbidity currents may flow across trench-connected graben depressions and deposit fine-grained turbidites in the deeper graben basins. Such depressions along the trench axis are filled with well-stratified horizontal acoustic units that onlap lower units (Fig. 5). Similar horizontally stratified deposits have been observed in seismic reflection profiles (Tsuru *et al.* 2000, 2002; Nakamura *et al.* 2013), although the resolution and/or distribution of the fill is insufficient to elucidate details of the internal structure. Our recently collected sub-bottom profiles indicate the occurrence of thick acoustically transparent layers in the trench-fill (Fig. 5a). These transparent layers might correspond to thick muddy turbidite beds, which we suggest formed from episodic turbidity currents along the Japan Trench slope.

In some deep troughs and trenches, the deep-sea currents were observed (e.g. Matsuyama *et al.* 1993; Okada & Ohta 1993; Mitsuzawa & Halloway 1998). These are tidally affected, and the total amplitude of the current reached 50 cm s^{-1} at the axis of the Suruga Trough (Matsuyama *et al.* 1993). Therefore, the deep-sea current might influence to the sedimentation along the trench. The southwards Deep Western Boundary Current was observed with its mean speed of 3–7 cm s^{-1} and maximum speed of 15–20 cm s^{-1} along the northern (Mitsuzawa & Halloway 1998) and with its mean speed of 1–2 cm s^{-1} along the southern (Hallock & Teague 1996) Japan Trench landwards slopes. Furthermore, a similar SW current with a mean speed of 4.28 cm s^{-1} was reported along the Kuril Trench landwards slope off SE Hokkaido (Owens & Warren 2001). Although these observations suggest the occurrence of an anticlockwise deep western boundary current along the landwards slope of the Japan and Kuril trenches, a northwards current

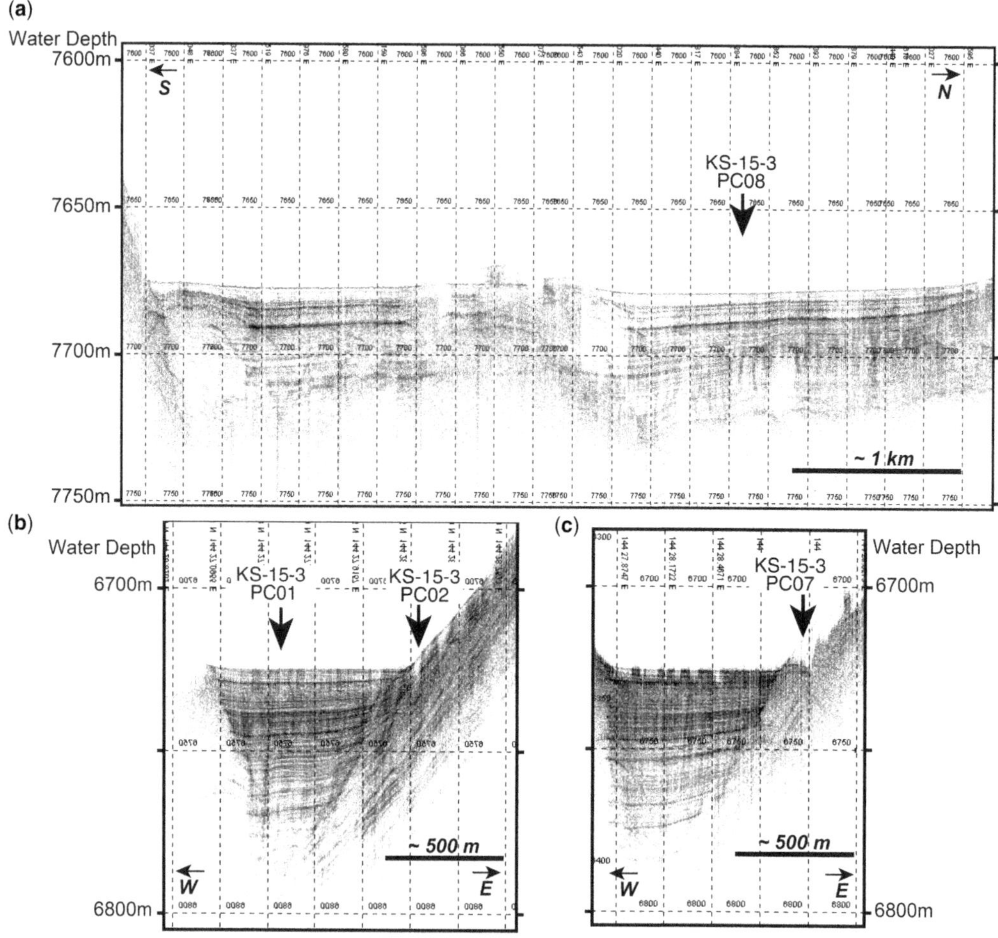

Fig. 5. Sub-bottom profile records of the trench basin traversing site KS-15-3 PC08 (**a**), and of graben sites KS-15-3 PC01 and PC02 (**b**) and KS-15-3 PC07 (**c**). A sound velocity of 1500 m s^{-1} is used to convert the two-way travel time to water depth.

was illustrated in the schematic diagram of the abyssal circulation in the NW Pacific (Warren & Owens 1988), and a northwards deep flow of approximately 4 cm s^{-1} and northeastwards flow of $>$8 cm s^{-1} were observed on the oceanwards side of the southern Japan (Hallock & Teague 1996) and Kuril (Owens & Warren 2001) trenches. Mitsuzawa & Halloway (1998) also inferred a similar northwards current along the seawards slope of the northern Japan Trench from the preferred orientation of the benthic fauna. Hirano *et al.* (1998) conducted a repeated submersible observation at a crack on the oceanwards slope of northern Japan Trench, and suggested an influence of the deep bottom currents on the local sedimentation. However, there is no clear evidence of sediment characteristics collected during this study and sediment accumulation (stacking) patterns on the seismic reflection profiles (Tsuru *et al.* 2000, 2002; Nakamura *et al.* 2013; Arai *et al.* 2014). For example, no typical contourite sequence (Faugeres *et al.* 1984; Gonthier *et al.* 1984) and sortable silt (McCave *et al.* 1995; McCave 2008) were observed in the obtained cores from the trench and graben depressions. Sediment particles near the trench axis, however, have been moved by the deep bottom currents, as suggested by Hirano *et al.* (1998). Thus, the reworked particles might contribute a high sedimentation rate of hemipelagic mud in the trench and graben depressions. Because our surveys have been concentrated near the trench axis, turbidite sedimentation might be more dominant than bottom-current-controlled sedimentation.

The influence of the bottom currents on the trench floor is a future question for trench sedimentology.

Potential for deposition and preservation of earthquake-induced turbidites in basins of the Japan Trench

Interpretation of the core lithologies and acoustic facies of the seismic reflection profiles suggests that the small basins along the central and northern Japan Trench have been filled by turbidites. Several other studies have also identified seismically induced turbidity currents along the Japan Trench and its lower slope. For example, Itou *et al.* (2000) identified a significant increase in terrigenous material following the 1994 M_w 7.7 Sanriku-haruka-oki earthquake (located near 40° 27′ N, 143° 43′ E, with a hypocentre depth of <10 km) in sediment traps deployed on the northern Japan Trench floor and the lower slope, about 100 km from the earthquake epicentre. Shirasaki *et al.* (2012) reported submarine cable breaks on the slopes along the southern and central Japan Trench due to the 2011 Tohoku-oki earthquake, and interpreted that the earthquake-generated turbidity currents cut these cables. Oguri *et al.* (2013) and Ikehara *et al.* (2016) found that fine-grained turbidites were likely to have formed in response to the 2011 Tohoku-oki earthquake on the Japan Trench floor, near the epicentre of the earthquake. Similar fine-grained turbidites were observed at the core top in several cores collected for this study (KS-15-3 PC08, KS-14-16 PC02, KS-14-16 PC03, KS-15-3 PC10 and KS-14-16 PC10: Fig. 2) and were most probably formed by the 2011 Tohoku-oki earthquake.

Earthquake strong ground motions and sediment resuspension in large tsunami waves are both known mechanisms for triggering turbidity currents and submarine slope failure. Several alternative mechanisms have also been proposed, including storm surge, hyperpycnal flows from rivers (floods), large storm waves, sediment loading, submarine groundwater discharge, volcanic explosions and bolide impacts (Nakajima 2000; Goldfinger *et al.* 2012; Pickering & Hiscott 2016). Along the central and northern Japan Trench, there are only a few submarine canyons that incise both the upper and lower slope. Some isolated basins, related to tectonic subsidence, are located on a wide terrace on the upper slope (Arai *et al.* 2014) (Fig. 1b). These isolated basins might serve as natural sediment traps receiving particulate matter from the shelf. With the exception of the Ogawara submarine canyon, the Japan Trench slope contains no submarine canyon systems directly connecting the shore or shelf to the deep-water trench floor. The Ogawara submarine canyon extends from the Shimokita shelf edge to the Hidaka Trough floor, near the Kuril Trench (Fig. 1). The lack of submarine canyons along the Japan Trench limits the formation and mobility of any meteorologically induced turbidity currents such as by storm surges, hyperpycnal flows from rivers (floods) and large storm waves: therefore, turbidites observed in this area are not likely to be generated by storms. While there are many active and explosive volcanoes on the NE Japan arc (e.g. Nakada *et al.* 2016), all of these locate along the back-bone range of the arc. Therefore, volcanic explosions are not the main mechanism involved in forming the turbidites. The occurrence of bivalve Calyptogena and bacteria mats along the landwards slope of Japan Trench is believed to be the result of cold-water discharge from the seafloor (Fujioka *et al.* 1993; Ogawa *et al.* 1996). The contribution of fluid migration in the lower-slope sediments to the generation of turbidity currents, however, is unknown. Observation, historical documents and onshore tsunami deposits (geological records) indicate that large earthquakes and tsunamis have occurred repeatedly along the Japan Trench (e.g. Earthquake Research Committee 1998; Sawai *et al.* 2012; Usami *et al.* 2012; Ishimura & Miyauchi 2015; Toda 2016). Therefore, large earthquakes and related tsunamis are most likely to be the origin of the turbidites in this area.

Small elongate basins along the Japan Trench floor and the trench-connecting graben are subjected to ongoing tectonic subsidence, creating and maintaining accommodation space for the deposition of sediments, including the episodic deposition of fine-grained turbidites.

Oguri *et al.* (2016) reported surface sediment resuspension within the upper-slope terrace off the Sanriku coast at a water depth of approximately 1000 m generated by a M_w 7.3 aftershock of the 2011 Tohoku-oki earthquake on 7 December 2012. Therefore, earthquakes with a magnitude of around 7 have the potential to resuspend unconsolidated seafloor sediments. This implies that during moderate-sized earthquakes, sediment resuspension and remobilization in regions with high productivity may supply or redistribute a large amount of sediment to the trench floor, and small basins may act as natural sediment traps. Sediment focusing may contribute to the accumulation of material in the basins, thereby maintaining a high sedimentation rate (Lehman 1975). High sedimentation rates also prevent the destruction of thick fine-grained turbidites by benthos activity. We note that intense macrobenthos activity has been reported from submersible observations on the flat floor of the seawards slope of the Japan Trench (Fujioka *et al.* 1993), and our core data indicate relatively high benthos activity (moderate to strong bioturbation), even at water depths exceeding 7000 m. High rates of hemipelagic sediment deposition in the Japan Trench, in excess of 1 m ka^{-1} (Ikehara *et al.* 2016), may effectively cover and preserve the fine-grained turbidites.

Conclusions

Recently collected sediment cores from small basins in the Japan Trench floor contain several thick muddy turbidite beds. Examination of the acoustic facies of the trench- and graben-fill deposits also suggests that the deposits are turbiditic sediments. Large earthquakes and related tsunamis are known to trigger turbidity currents that form turbidites, and we suggest that this is the most probable origin of the turbidites we observed because meteorologically induced turbidity currents may not reach to the trench floor due to few shelf-connected submarine canyons, which act as pathways of meteorologically induced turbidity currents formed in shallow-water areas, and the occurrence of isolated basins, which act as the natural sediment traps receiving particles from the shelf, on the upper and lower slope of the Japan Trench. Resuspension and remobilization of muddy deep-sea sediments might form fine-grained turbidites in deep basins. The bathymetric and sedimentological setting of the central and northern Japan Trench suggest that these fine-grained turbidites were most likely to have been seismically generated. Subduction of the Pacific Plate results in subsidence and basin formation on the trench floor, providing loci of deposition for fine-grained turbidites. High sedimentation rates of interseismic hemipelagic deposits effectively bury the earthquake-induced turbidites, preserving a geological record of large earthquakes. Therefore, these basins are favourable environments for studies of turbidite palaeoseismology.

Analysis of the new sediment cores reported in this study indicates that shallow turbidites are observed along at least 120 km of the Japan Trench near the epicentre of the 2011 Tohoku-oki earthquake. This area overlaps with earthquake-induced turbidites reported previously, and the turbidites are tentatively attributed to the 2011 earthquake. Comparison of the deep-sea turbidite record with known onshore tsunami deposits, constrained by robust To-a tephra correlations, suggests that the subduction zone along the central Japan Trench is a possible source area of the 869 Jogan and 1454 Kyotoku tsunamis. Such comparisons of deep-sea records with onshore tsunami records might provide new insights into past tsunamigenic earthquakes.

We thank the captains, officers, crews, marine technicians and onboard scientists of the R/V *Shinsei-maru* KS-14-16 and KS15-3 cruises for help in taking sediment samples and sub-bottom profiles. Discussions with Michael Strasser (Innsbruck University), Hiske Fink (MARUM, University of Bremen), Cecillia McHugh (Queens College, New York City University), Yoshitaka Nagahashi (Fukushima University), Yasuyuki Nakamura, Shuichi Kodaira, Yasuhiro Yamada and Kazumasa Oguri (JAMSTEC), Kotaro Ujiie (University of Tsukuba), Hajime Naruse (Kyoto University), and Tomohisa Irino (Hokkaido University) were useful in interpreting depositional processes. We are grateful to two anonymous reviewers for their constructive comments. This work was financially supported in part by the Ministry of Education, Culture, Sports, Science and Technology (MEXT) of Japan via the research projects 'Geophysical and Geological Studies of Earthquakes and Tsunamis for off-Tohoku District, Japan' and 'Regional Study on Hazard Mitigation Along the Nankai Trough', the AIST research project 'Study on Coastal Sediments', the supporting funds for important research by the Research Institute of Geology and Geoinformation, GSJ, AIST 'Tephra Study in the Japan Trench Cores', and Kakenhi 'Japan Trench Deep-Sea Research Project for Assessing Shallow Seismic Slips and Their History (No. 26000002)'. Part of the work was conducted under the cooperative research programme of the Centre for Advanced Marine Core Research, Kochi University 'Sedimentological Research on the Development of Plate Boundary Faults in the Japan Trench' (No. 15A041 and 15B036).

References

ADAMS, J. 1990. Paleoseimicity of the Cascadia subduction zone: evidence from turbidites off the Oregon–Washington margin. *Tectonics*, **9**, 569–583.

ARAI, K., NARUSE, H. *ET AL.* 2013. Tsunami-generated turbidity current of the 2011 Tohoku-Oki earthquake. *Geology*, **41**, 1195–1198.

ARAI, K., INOUE, T., IKEHARA, K. & SASAKI, T. 2014. Episodic subsidence and active deformation of the forearc slope along the Japan Trench near the epicenter of the 2011 Tohoku Earthquake. *Earth and Planetary Science Letters*, **408**, 9–15.

ARITA, M. & KINOSHITA, Y. 1984. *Sedimentological Map of off Kamaishi.* Marine Geology Map Series, **25**. Geological Survey of Japan, AIST, Tsukuba.

ATWATER, B.F., CARSON, B., GRIGGS, G.B., JOHNSON, H.P. & SALMI, M.S. 2014. Rethinking turbidite paleoseismology along the Cascadia subduction zone. *Geology*, **42**, 827–830.

BERGER, W.H., ADELSECK, C.G., JR. & MAYER, L.A. 1976. Distribution of carbonate in surface sediments of the Pacific Ocean. *Journal of Geophysical Research*, **81**, 2617–2627.

DEMETS, C., GORDON, R.G. & ARGUS, D.F. 2010. Geologically current plate motions. *Geophysical Journal International*, **181**, 1–80.

EARTHQUAKE RESEARCH COMMITTEE 1998. *Seismic Activity in Japan – Regional Perspectives on the Characteristics of Destructive Earthquakes.* Headquaters for Earthquake Research Promotion, Tokyo.

FAUGERES, J.-C., STOW, D.A.V. & GONTHIER, E. 1984. Contourite drift moulded by deep Mediterranean outflow. *Geology*, **12**, 296–300.

FUJII, Y., SATAKE, K., SAKAI, S., SHINOHARA, M. & KANAZAWA, T. 2011. Tsunami source of the 2011 off the Pacific coast of Tohoku Earthquake. *Earth, Planets and Space*, **63**, 815–820.

FUJIOKA, K., TAKEUCHI, A., HORIUCHI, K., OKANO, H., MURAYAMA, M. & HORII, Y. 1993. Constrated nature between landward and seaward slopes of the Japan Trench off Miyako, Northeast Japan. *Proceedings of JAMSTEC Symposium on Deep Sea Research*, **9**, 1–26 [in Japanese with English abstract].

FUJIWARA, T., KODAIRA, S., NO, T., KAIHO, Y., TAKAHASHI, N. & KANEDA, Y. 2011. The 2011 Tohoku-Oki earthquake: displacement reaching the trench axis. *Science*, **334**, 1240, https://doi.org/10.1126/science.1211554

GOLDFINGER, C. 2011. Submarine paleoseismology based on turbidite records. *Annual Review of Marine Science*, **3**, 35–66, https://doi.org/10.1146/annurev-marine-120709-142852

GOLDFINGER, C., HANS NELSON, C. & JOHNSON, J.E. & THE SHIPBOARD SCIENTIFIC PARTY 2003. Deep-water turbidites as Holocene earthquake proxies: The Cascadia subduction zone and Northern San Andreas Fault system. *Annals of Geophysics*, **46**, 1169–1194, https://doi.org/104401/ag-3452

GOLDFINGER, C., HANS NELSON, C. ET AL. 2012. *Turbidite Event History: Methods and Implications for Holocene Paleoseismicity of the Cascadia Subduction Zone*. United States Geological Survey, Professional Papers, **1661-F**.

GOLDFINGER, C., PATTON, J., VAN DAELE, M., MOERNAUT, J., HANS NELSON, C., DE BATIST, M. & MOREY, A. 2014. Can turbidites be used to reconstruct a paleoearthquake record for the central Sumatra margin?: Comment. *Geology*, **42**, e344. https://doi.org/10.1130/G35558C1

GONTHIER, E.G., FAUGERES, J.-C. & STOW, D.A.V. 1984. Contourite facies of the Faro drift, Gulf of Cadiz. *In*: STOW, D.A.V. & PIPER, D.J.W. (eds) *Fine-Grained Sediments: Deep-Water Processes and Facies*. Geological Society, London, Special Publications, **15**, 275–292, https://doi.org/10.1144/GSL.SP.1984.015.01.18

HALLOCK, Z.R. & TEAGUE, W.J. 1996. Evidence for a North Pacific deep western boundary current. *Journal of Geophysical Research*, **101**, 6617–6624.

HIRANO, S., OGAWA, Y., FUJIOKA, K. & KAWAMURA, K. 1998. Temporal changes of cracks in the oceanward slope of northern Japan Trench off Sanriku: Six-year observation by submersibles. *JAMSTEC Journal of Deep Sea Research*, **14**, 445–454.

HIROSE, F., MIYAOKA, K., HAYASHIMOTO, N., YAMAZAKI, T. & NAKAMURA, M. 2011. Outline of the 2011 off the Pacific coast of Tohoku Earthquake (M_w 9.0) – Seismicity: foreshocks, mainshock, aftershocks, and induced activity. *Earth, Planets and Space*, **63**, 513–518.

IINUMA, T., OHZONO, M., OHTA, Y. & MIURA, S. 2011. Coseismic slip distribution of the 2011 off the Pacific coast of Tohoku Earthquake (M9.0) estimated based on GPS data- Was the asperity in Miyagi-oki ruptured? *Earth, Planets and Space*, **63**, 643–648.

IKEHARA, K., USAMI, K., JENKINS, R. & ASHI, J. 2011. Occurrence and lithology of seismo-turbidites by the 2011 off the Pacific coast of Tohoku earthquake. *In*: *Abstracts of IGCP the Fifth International Symposium on Submarine Mass movements and Their Consequences*, ISSMMTC-5 Organizing Committee, Kyoto, 74.

IKEHARA, K., OHKUSHI, K., NODA, A., DANHARA, T. & YAMASHITA, T. 2013. A new local reservoir correction for the last deglacial period in the Sanriku region, northwestern North Pacific, based on radiocarbon dates from the Towada-Hachinohe (To-H) tephra. *The Quaternary Research (Daiyonki-kenkyu)*, **52**, 127–137.

IKEHARA, K., IRINO, T., USAMI, K., JENKINS, R., OMURA, A. & ASHI, J. 2014*a*. Possible submarine tsunami deposits on the outer shelf of Sendai Bay, Japan resulting from the 2011 earthquake and tsunami off the Pacific coast of Tohoku. *Marine Geology*, **358**, 120–127.

IKEHARA, K., ITAKI, T., TUZINO, T. & HOYANAGI, K. 2014*b*. Deep-sea turbidite evidence on the recurrence of large earthquakes off Shakotan Peninsula, northeastern Japan Sea. *In*: KRASTEL, S., BEHRMANN, J.-H. ET AL. (eds) *Submarine Mass Movements and their Consequences*. Springer, Dordrecht, The Netherlands, 639–647.

IKEHARA, K., KANAMATSU, T. ET AL. 2016. Documenting large earthquakes similar to the 2011 Tohoku-oki earthquake from sediments in the Japan Trench over the past 1,500 years. *Earth and Planetary Science Letters*, **445**, 48–56.

ISHIMURA, D. & MIYAUCHI, T. 2015. Historical and paleotsunami deposits during the last 4000 years and their correlations with historical tsunami events in Koyadori on the Sanriku Coast, northeastern Japan. *Progress in Earth and Planetary Science*, **2**, 16, https://doi.org/10.1186/s40645-015-0047-4

ITOU, M., MATSUMURA, I. & NORIKI, S. 2000. A large flux of particulate matter in the deep Japan Trench observed just after the 1994 Sanriku-Oki earthquake. *Deep-Sea Research Part I: Oceanographic Research Papers*, **47**, 1987–1998.

KODAIRA, S., NO, T. ET AL. 2012. Coseismic fault rupture at the trench axis during the 2011 Tohoku-oki earthquake. *Nature Geoscience*, **5**, 646–650, https://doi.org/10.1038/ngeo1547

KON'NO, E., IWAI, J. ET AL. 1961. Geological observations of the Sanriku coastal region damaged by the tsunami due to the Chile earthquake in 1960. Contributions from the Institute of Geology and Paleontology, Tohoku University, **52**, 1–45.

LEHMAN, J.T. 1975. Reconstructing the rate of accumulation of lake sediment: the effect of sediment focusing. *Quaternary Research*, **5**, 541–550.

MACHIDA, H. & ARAI, F. 2003. *Atlas of Tephras in and Around Japan*. 2nd edn. University of Tokyo Press, Tokyo [in Japanese].

MANN, U. & MÜLLER, G. 1980. Composition of sediments of the Japan Trench transect, Legs 56 and 57, Deep Sea Drilling Project. *In*: *Scientific Party, Initial Reports of the Deep Sea Drilling Project, 56, 57, Pt. 2*. United States Government Printing Office, Washington, DC, 939–977.

MATSUYAMA, M., OHTA, S., HIBIYA, T. & YAMADA, H. 1993. Strong tidal currents observed near the bottom in the Suruga Trough, central Japan. *Journal of Oceanography*, **49**, 683–696.

McCave, I.N. 2008. Size sorting during transport and deposition of fine sediments: sortable silt and flow speed. *In*: Robesco, M. & Camerlenghi, A. (eds) *Contourites*. Elsevier, Amsterdam, 121–142.

McCave, I.N., Manighetti, B. & Robinson, S.G. 1995. Sortable silt and fine sediment size/composition slicing: parameter for paleocurrent speed and paleoceanography. *Paleoceanography*, **10**, 593–610.

Minoshima, K., Kawahata, H. & Ikehara, K. 2007. Changes in biological production in the mixed water region (MWR) of the northwestern North Pacific during the last 27 kyr. *Palaeogeography, Palaeoclimatology, Palaeoecology*, **254**, 430–447.

Mitsuzawa, K. & Halloway, G. 1998. Characteristics of deep currents along trenches in the northwest Pacific. *Journal of Geophysical Research*, **103**, 13 085–13 092.

Miura, R., Hino, R., Kawamura, K., Kanamatsu, T. & Kaiho, Y. 2014. Accidental sediments trapped in ocean bottom seismometers during the 2011 Tohoku-Oki earthquake. *Island Arc*, **23**, 365–367.

Mori, N., Takahashi, T., Yasuda, T. & Yanagisawa, H. 2011. Survey of 2011 Tohoku earthquake tsunami inundation and run-up. *Geophysical Research Letters*, **38**, L00G14, https://doi.org/10.1029/2011GL049210

Nakada, S., Yamamoto, T. & Maeno, F. 2016. Miocene–Holocene volcanism. *In*: Moreno, T., Wallis, S., Kojima, T. & Gibbons, W. (eds) *The Geology of Japan*. Geological Society, London, 273–308.

Nakajima, T. 2000. Initiation processes of turbidity currents: implications for assessment of recurrence intervals of offshore earthquakes using turbidites. *Bulletin of Geological Survey of Japan*, **51**, 79–87 [in Japanese with English abstract].

Nakamura, Y., Kodaira, S., Miura, S., Regalla, C. & Takahashi, N. 2013. High-resolution seismic imaging in the Japan Trench axis area off Miyagi, northeastern Japan. *Geophysical Research Letters*, **40**, 1–6, https://doi.org/10.1002/grl.50364

Namegaya, Y. & Satake, K. 2014. Reexamination of the AD 869 Jogan earthquake size from tsunami deposit distribution, simulated flow depth, and velocity. *Geophysical Research Letters*, **41**, 2297–2303.

Namegaya, Y. & Yata, T. 2014. Tsunamis which affected the Pacific coast of eastern Japan in medieval times inferred from historical documents. *Zisin*, **66**, 73–81, https://doi.org/10.4294/zisin.66.73

Noda, A., TuZino, T., Kanai, Y., Furukawa, R. & Uchida, J. 2008. Paleoseismology along the southern Kuril Trench deduced from submarine fan turbidites. *Marine Geology*, **254**, 73–90.

Noguchi, T., Tanikawa, W. *et al.* 2012. Dynamic process of turbidity generation triggered by the 2011 Tohoku-Oki earthquake. *Geochemistry, Geophysics, Geosystems*, **13**, Q11003, https://doi.org/10.1029/2012GC004360

Nomaki, H., Arai, K. *et al.* 2016. Sedimentary organic matter contents and porewater chemistry at upper bathyal depths influenced by the 2011 off the Pacific coast of Tohoku Earthquake and tsunami. *Journal of Oceanography*, **72**, 99–111.

Ogawa, Y., Fujioka, K., Fujikura, K. & Iwabuchi, Y. 1996. En echelon patterns of *Caplyptogena* colonies in the Japan Trench. *Geology*, **24**, 807–810.

Oguri, K., Kawamura, K. *et al.* 2013. Hadal disturbance in the Japan Trench induced by the 2011 Tohoku-Oki earthquake. *Scientific Reports*, **3**, 1–6, https://doi.org/10.1038/srep01915

Oguri, K., Furushima, Y. *et al.* 2016. Long-term monitoring of bottom environments of the continental slope off Otsuchi Bay, northeastern Japan. *Journal of Oceanography*, **72**, 151–166.

Okada, H. & Ohta, S. 1993. Photographic evidence of variable bottom-current activity in the Suruga and Sagami Bays, central Japan. *Sedimentary Geology*, **82**, 221–237.

Owens, W.B. & Warren, B.A. 2001. Deep circulation in the northwest corner of the Pacific Ocean. *Deep-Sea Research Part I: Oceanographic Research Papers*, **48**, 959–993.

Patton, J.R., Goldfinger, C., Morey, A.E., Romos, C., Black, B., Djadjadhardia, Y. & Udrekh, , 2013. Seismoturbidite record at the Cascadia and Sumatra–Andaman subduction zones. *Natural Hazards and Earth System Sciences*, **13**, 833–867.

Patton, J.R., Goldfinger, C. *et al.* 2015. A 6600 year earthquake history in the region of the 2004 Sumatra–Andaman subduction zone earthquake. *Geosphere*, **11**, 1–62, https://doi.org/10.1130/GES0 1066.1

Pickering, K.T. & Hiscott, R.N. 2016. *Deep Marine Systems: Processes, Deposits, Environments, Tectonics and Sedimentation*. Wiley, Chichester, UK.

Pouderoux, H., Proust, J.-N. & Lamarche, G. 2014. Submarine paleoseismology of the northern Hikurangi subduction margin of New Zealand as deduced from turbidite record since 16ka. *Quaternary Science Review*, **84**, 116–131.

Saino, T., Shang, S. *et al.* 1998. Short term variability of particle fluxes and its relation to variability in sea surface temperature and Chlorophyll a field detected by ocean color and temperature scanner (OCTS) off Sanriku, Northwestern North Pacific in the spring of 1997. *Journal of Oceanography*, **54**, 583–592.

Sawai, Y., Shishikura, M. *et al.* 2007. A study on paleotsunami using handy geoslicer in Sendai Plain (Sendai, Natori, Iwanuma, Watari, and Yamamoto), Miyagi, Japan. *Annual Report on Active Fault and Paleoearthquake Researches*, **7**, 47–80.

Sawai, Y., Namegaya, Y., Okamura, Y., Satake, K. & Shishikura, M. 2012. Challenges of anticipating the 2011 Tohoku earthquake and tsunami using coastal geology. *Geophysical Research Letters*, **39**, L21309, https://doi.org/10.1029/2012GL053692

Sawai, Y., Namegaya, Y., Tamura, T., Nakashima, R. & Tanigawa, K. 2015. Shorter intervals between great earthquakes near Sendai: Scour ponds and a sand layer attributable to A.D.1454 overwash. *Geophysical Research Letters*, **42**, 4785–4800, https://doi.org/10.1002/2015GL064167

Shipboard Scientific Party 1980. Site 440: Japan Trench midslope terrace, Leg 57. *In*: *Scientific Party, Initial Reports of the Deep Sea Drilling Project, 56, 57, Pt. 1*. United States Government Printing Office, Washington, DC, 225–317.

Shirasaki, Y., Ito, K., Kuwazuru, M. & Shimizu, K. 2012. Submarine landslides as cause of submarine cable break. *Journal of Japan Society of Marine Survey and Technology*, **24**, 17–20 [In Japanese].

SHISHIKURA, M., SAWAI, Y. *ET AL.* 2007. Age and distribution of tsunami deposit in the Ishinomaki Plain, Northeastern Japan. *Annual Report on Active Fault and Paleoearthquake Researches*, **7**, 31–46.

STOW, D.A.V. & SHANMUGAM, G. 1980. Sequence of structures in fine-grained turbidites: Comparison of recent deep-sea and ancient flysch sediments. *Sedimentary Geology*, **25**, 23–42.

STRASSER, M., KOLLING, M. *ET AL.* 2013. A slump in the trench: Tracking the impact of the 2011 Tohoku-Oki earthquake. *Geology*, **41**, 935–938, https://doi.org/10.1130/G34477.1

SUMNER, E.J., SITI, M.I. *ET AL.* 2013. Can turbidites be used to reconstruct a paleoearthquake record for the central Sumatran margin? *Geology*, **41**, 763–766, https://doi.org/10.1130/G34298.1

SUZUKI, W., AOI, S., SEKIGUCHI, H. & KUNUGI, T. 2011. Rupture process of the 2011 Tohoku-Oki mega-thrust earthquake (M9.0) inverted from strong-motion data. *Geophysical Research Letters*, **38**, L00G16, https://doi.org/10.1029/2011GL049136

TAMURA, T., SAWAI, Y., IKEHARA, K., NAKASHIMA, R., HARA, J. & KANAI, Y. 2015. Shallow-marine deposits associated with the 2011 Tohoku-oki tsunami in Sendai Bay, Japan. *Journal of Quaternary Science*, **30**, 293–297.

TODA, S. 2016. Crustal earthquakes. *In*: MORENO, T., WALLIS, S., KOJIMA, T. & GIBBONS, W. (eds) *The Geology of Japan*. Geological Society, London, 371–408.

TOYOFUKU, T., DUROS, P. *ET AL.* 2014. Unexpected biotic resilience on the Japanese seafloor caused by the 2011 Tohoku-Oki tsunami. *Scientific Reports*, **4**, 7517, https://doi.org/10.1038/srep07517

TSURU, T., PARK, J.-O., TAKAHASHI, N., KODAIRA, S., KIDO, Y., KANEDA, Y. & KONO, Y. 2000. Tectonic features of the Japan Trench convergent margin off Sanriku, northeastern Japan, revealed by multichannel seismic reflection data. *Journal of Geophysical Research*, **105**, 16 403–16 413.

TSURU, T., PARK, J.-O., MIURA, S., KODAIRA, S., KIDO, Y. & HAYASHI, T. 2002. Along-arc structural variation of the plate boundary at the Japan Trench margin: Implication of interplate coupling. *Journal of Geophysical Research*, **107**, 2357.

USAMI, K., IKEHARA, K., JENKINS, R.G. & ASHI, J. In press. Benthic foraminiferal evidence of deep-sea sediment transport by the 2011 Tohoku-oki earthquake and tsunami. *Marine Geology*, https://doi.org/10.1016/j.margeo.2016.04.001

USAMI, T., ISHII, H., IMAMURA, T., TAKEMURA, M. & MATSUURA, R. 2012. *Materials for Comprehensive list of Destructive Earthquakes in Japan, 599-2012*. University of Tokyo Press, Tokyo.

VON HUENE, R., LANGSETH, M., NASU, N. & OKADA, H. 1980. Summary, Japan Trench transect. *In*: *Scientific Party, Initial Reports of the Deep Sea Drilling Project, 56, 57, Pt. 1*. United States Government Printing Office, Washington, DC, 473–488.

VON HUENE, R. & LALLEMAND, S. 1990. Tectonic erosion along the Japan and Peru convergent margins. *Geological Society of America Bulletin*, **102**, 704–720.

WARREN, B.A. & OWENS, W.B. 1988. Deep currents in the central sub-arctic Pacific Ocean. *Journal of Physical Oceanography*, **18**, 529–551.

YAGI, Y. & FUKAHATA, Y. 2011. Rupture process of the 2011 Tohoku-oki earthquake and absolute elastic strain release. *Geophysical Research Letters*, **38**, L19307, https://doi.org/10.1029/2011GL048701

YAMANE, M. & OBA, T. 1999. Paleoceanographic change of the Sanriku area during the last 90,000 years based on the analysis of a sediment core (KH94-3, LM-8). *The Quaternary Research (Daiyonki-kenkyu)*, **38**, 1–16 [in Japanese with English abstract].

YOKOTA, Y., KOKETSU, K., FUJII, Y., SATAKE, K., SAKAI, S., SHINOHARA, M. & KANAZAWA, T. 2011. Joint inversion of strong motion, teleseismic, geodetic, and tsunami datasets for the rupture process of the 2011 Tohoku earthquake. *Geophysical Research Letters*, **38**, L00G21, https://doi.org/10.1029/2011GL050098

YOSHIKAWA, S. 1976. The volcanic ash layers of the Osaka Group. *Journal of Geological Society of Japan*, **82**, 497–515 [in Japanese with English abstract].

Tsunami hazard in Central America: history and future

CONRAD LINDHOLM[1]*, WILFRIED STRAUCH[2] & MARIO FERNÁNDEZ[3]

[1]*NORSAR, PO Box 53, 2027 Kjeller, Norway*

[2]*Instituto Nicaragüense de Estudios Territoriales (INETER), Frente a Hospital Solidaridad, Apartado Postal 2110, Managua, Nicaragua*

[3]*Escuela de Geografía, Universidad de Costa Rica, San Jose, Costa Rica*

**Correspondence: conrad@norsar.no*

Abstract: Central America is a small and culturally homogeneous region that, since the 1990s, has experienced economic and political integration of its six countries, which share the same threats of volcanic eruptions, disastrous earthquakes and tsunamis. The Pacific coastline of 1700 km is common for Guatemala, El Salvador, Honduras, Nicaragua, Costa Rica and Panama, and the Pacific subduction zone has the potential for creating huge tsunamis that threaten this coast. In addition to the natural hazard, the growing tourist industry is expanding its infrastructure along the Pacific beaches, which again enhances the exposure and tsunami risk. Even though the 1992 tsunami disaster in Nicaragua did not severely hit the tourist beaches, it raised the risk awareness, and special attention is now given to 'slow' earthquakes that may be modest in shaking while still having a large tsunami potential. The tsunami hazard mapping is well advanced in Nicaragua, Costa Rica and El Salvador, and initiatives are ongoing to improve the mapping in all countries. National systems for early warning were established in Nicaragua and El Salvador, while the other four countries rely on rapid information from the Pacific Tsunami Warning Center. Mitigation measures and information campaigns are presently conducted on a national basis in all countries, but a regional centre for early tsunami warning and coordinated information campaigns (CATAC) is expected to become operational in the near future.

The Central American countries of Guatemala, El Salvador, Honduras, Nicaragua, Costa Rica and (partly) Panama are all bordered by the Cocos Plate to the SW, which subducts under the Caribbean Plate with a plate convergence velocity ranging around 75 mm a^{-1} (DeMets 2001; see also http://earthquake.usgs.gov/earthquakes/eqarchives/poster/regions/caribbean.php). The high convergence velocity is the cause for the volcanic activity, as well as the high earthquake hazard, in Central America, and the subduction has created a 1700 km-long trench with ocean depths down to 6700 m. While Central America is accustomed to volcanic activity and earthquakes in the subduction zone, the tsunami risk has long been underestimated.

Tsunamis are now recognized as a real threat along the Pacific and Caribbean coasts of Central America. The sources of tsunamis are mainly underwater earthquakes: however, subsea slides (sometimes triggered by earthquakes) cannot be ignored. The earthquakes are generated in four tectonic environments: the Caribbean and North American plates (PMCHFS and Swan Fault); the Middle American Trench (MAT); the Panama Fracture Zone (PFZ); and the North Panama Deformed Belt (NPDB) (Fig. 1). The Polochic–Motagua–Chamalecón Fault System (PMCHFS), the Swan Fault and the PFZ are strike-slip boundaries, while the MAT and the NPDB are convergent margins. The MAT is the source region that generates most of the seismicity of Central America (see Fig. 1) and the boundary with capacity to cause tsunamis that may reach all the Central American countries.

The coast of Central America can naturally also be impacted by distant tsunamis originating in the far Pacific or the Caribbean, but the main concerns are local and regional tsunamis, and destructive local events have hit along both coasts. Historically, the most destructive tsunamis of the region have impacted the coastlines of Panama, Costa Rica, Nicaragua, El Salvador and Honduras, while neither Guatemala nor Belize have been similarly severely impacted by historical tsunamis (Fig. 2). Fernández *et al.* (2000) carried out the first and, until now, most complete review of the tsunamis history in Central America, where it is estimated that tsunamis historically have claimed about 500 lives in the region.

Tsunami occurrences

According to the tsunami catalogue for Central America (Molina 1997; overview in Fernández

From: Scourse, E. M., Chapman, N. A., Tappin, D. R. & Wallis, S. R. (eds) 2018. *Tsunamis: Geology, Hazards and Risks*. Geological Society, London, Special Publications, **456**, 91–104.
First published online February 23, 2017, https://doi.org/10.1144/SP456.2

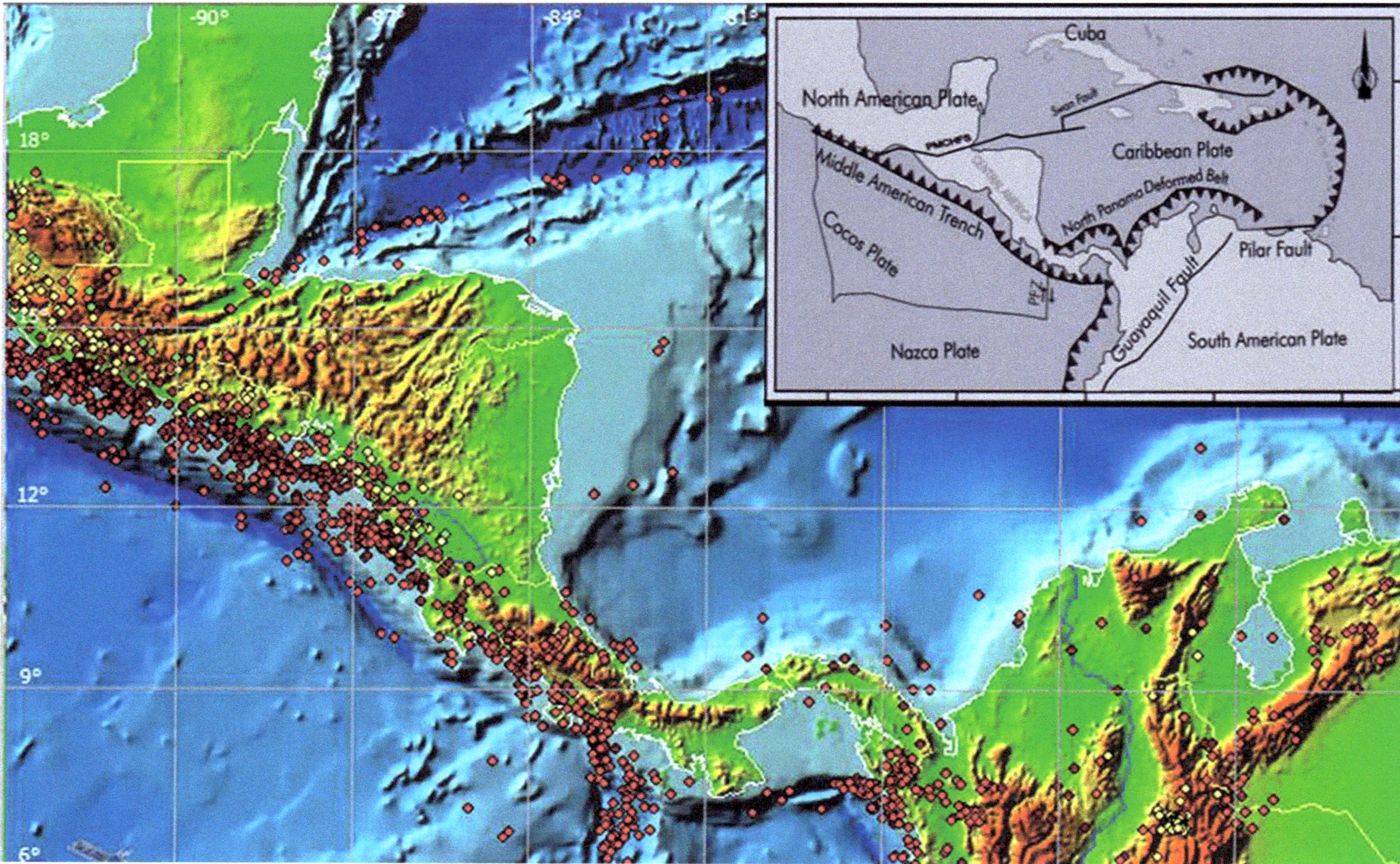

Fig. 1. Seismicity and tectonic setting of Central America. Dots indicate earthquakes larger than magnitude 5.0 at shallow (red), intermediate (yellow) and deep (green) depths. PMCHFS, Polochic–Motagua–Chamalecon Fault System; PFZ, Panama Fracture Zone. Inset from Fernández (2013).

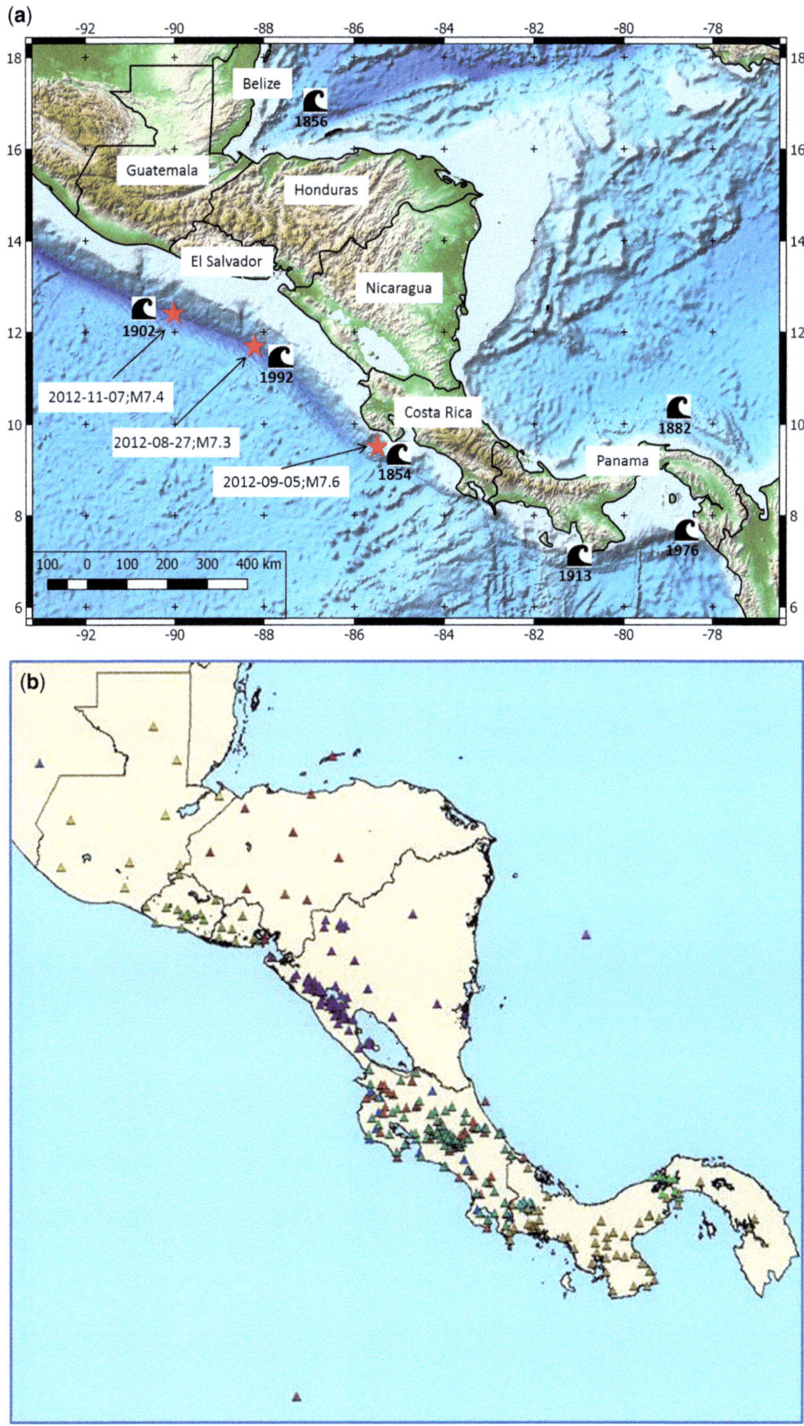

Fig. 2. Upper: Major historical and destructive tsunamis of Central America, which were all generated by earthquakes. The 2012 extraordinary sequence of tsunami generating earthquakes are indicated with dates and magnitudes. For each of these earthquakes a tsunami warning was issued. Lower: Seismic stations in Central America operated by the various national earthquake observatories.

et al. 2000) and the NOAA (2015) world catalogue, there are reports of 50 significant tsunamis detected in Central America from 1539 to present, with only one significant tsunami reaching Nicaragua and El Salvador in recent years (August 2012). See also Figure 2 and Table 1 for the locations of tsunami waves from three magnitude 7+ earthquakes in 2012. (It should be noted that the NOAA catalogue for Central American tsunamis may be less reliable for some reports.)

Distant tsunamis originating from Alaska, South America and Asia have hit the Pacific coast. For example, the 2011 Tohoku earthquake tsunami arrived in Central America with relatively low amplitude but was observed on the Cocos Island, Costa Rica, with two waves running up a river channel. In comparison, the tsunami risk and frequency is lower on the Caribbean side than on the Pacific side. This is because the earthquake sources on the Caribbean side are generally smaller and less frequent, and also because of the shallower water depths in the Caribbean. After 1539, only 12 tsunamis on the Caribbean side were documented to have had significantly damaging impacts, and then mainly in Honduras, Costa Rica and Panama (Fernández *et al.* 2000).

A key issue in the mapping of historical tsunami earthquakes has been that it has been difficult to get access to original and detailed descriptions of the effects, and seminal research on this problem was conducted by Fernández *et al.* (2000). Of the 49 events reported by Molina (1997) and Fernández *et al.* (2000), seven have been destructive: two on the Caribbean coast and five on the Pacific coast. The Caribbean tsunami damage was from Honduras and Panama, and the earthquakes causing the damaging Pacific tsunamis were distributed as follows: two off Panama, one off Costa Rica, one off Nicaragua and one off El Salvador. To the best of our knowledge today, the largest event was the 1992 Nicaragua tsunami, with a maximum wave height close to 10 m (Abe *et al.* 1993).

From the nineteenth century, six tsunami-events are known to have affected the Caribbean coast, of which four (1822, 1825, 1855 and 1873) had a more uncertain impact and size. One of the main tsunami events was most likely to have been caused by a large earthquake on the offshore section of PMCHFS in 1856: Its impact was so great that it destroyed and left several coastal communities in Honduras flooded and in ruins. Most probably, it also claimed lives, but we have not been able to identify reports to support this. The other major nineteenth century tsunami occurred in 1882 on the Caribbean coast of Panama, not far from the Panama Canal entrance. Its source was an estimated 7.9 magnitude earthquake (the largest known in this region) on the North Panama Deformed Belt and it caused a tsunami that also severely affected the San Blas Islands, with 75–100 reported deaths.

On the Pacific coast, four damaging tsunami events were recorded in the nineteenth century: 1844, 1854 and two events in 1859. The 1844 event particularly affected Lake Nicaragua, with great waves (seiches?) and was associated with a magnitude 7.4 earthquake. The 1854 tsunami hit the Golfo Dulce region of southern Costa Rica where it was claimed to have wiped out the small Villa Golfo Dulce community. Five years later, two wave-generating earthquakes hit the El Salvador coasts in 1859, one south in the Gulf of Fonseca and another further north near Acajutla. Both earthquakes were reported as causing significant waves.

According to historical records, the deadliest tsunami in Central America took place in 1902 on the coast of El Salvador. The coastline was inundated along a 120 km stretch, resulting in 185 casualties. Eleven years later, in 1913, the water level of the Pacific Ocean on the Panama coast rose suddenly after a great earthquake, and the Villa Pedassi community on the southern tip of the Azuero peninsula were reported to have been inundated. In 1957, a tsunami from Alaska caused waves several metres in height in Port Acajutla in El Salvador, which resulted in casualties. In 1976, Pacific coastal communities in Panama were moderately affected by a tsunami following an earthquake, and in 1991 the Limon earthquake caused significant tsunami waves on the Caribbean coast.

Finally, in 1992, a magnitude 7.7 earthquake generated a tsunami nearly 10 m high at the Nicaraguan Pacific coast: 170 people were killed or

Table 1. *Details of the three tsunamis recorded in 2012 in Central America and depicted in Figure 2*

Date	Depth (Km)	Magnitude	Maximum tsunami height	Location
7 November 2012	24.0	7.4	6 m	El Salvador/Nicaragua
5 September 2012	35.0	7.6	Centimetres	Costa Rica
27 August 2012	28.0	7.3	Centimetres	Guatemala

The Nicaragua earthquake on 27 August 2012 was a ‘slow’ earthquake (similar to the disastrous 1992 earthquake).

disappeared (e.g. see Abe *et al.* 1993; Baptista *et al.* 1993 for more details), and this tsunami significantly increased the consciousness about the inherent tsunami danger. It was caused by a 'slow' earthquake with significantly less ground shaking than from other subduction zone earthquakes of similar magnitudes, for which reason the coastal population took less notice of the earthquake and, consequently, did not prepare for the tsunami that hit the coast some 45 minutes later. It remains unknown if some of the major historical tsunamis shown in Figure 2 were caused by similar slow earthquakes, but it has been claimed that the August 2012 earthquake (see Fig. 2 and below) that hit El Salvador and northern Nicaragua was also a 'slow' earthquake (C. Synolakis pers. comm.).

In more recent times, a number of tsunami events have impacted the region (although less dramatic than in 1992):

- A minor tsunami of possibly 50 cm height was observed after the destructive magnitude 7.7 earthquake of 2001 in El Salvador (Koshimura 2001).
- In 2009, a tsunami affected the coast of northern Guatemala after a magnitude 7.3 earthquake in the Gulf of Honduras (Molina & Rosales 2009).
- On 27 August 2012, an important tsunami hit parts of the Pacific coasts of El Salvador and Nicaragua with wave heights of around 6 m (see Acosta 2012; Borrero 2012; Tenorio & Strauch 2012; Borrero *et al.* 2014). The waves did not cause major harm, as the population density in the affected areas is very low.
- In 2012, two other local tsunamis in Central America were recorded by tide gauges after the large earthquake (M7.6) off the coast of Costa Rica on 5 September 2012 (Han *et al.* 2013; Linkimer *et al.* 2013), and the M7.4 earthquake off the coast of Guatemala on 7 November 2012 (INSIVUMEH 2012).

These events did not cause severe destruction but should be used as reminders of the constant threat tsunamis pose to the coastal areas of Central America, and in particular north of Panama.

The Pacific subduction trench has steep slopes on the Central America side, and oceanographical investigations suggest the occurrence of large submarine slides in the steep deep-sea trenches off the Pacific coast of Nicaragua and Costa Rica. These submarine slides might have been triggered by earthquakes and may have generated significant tsunamis (von Huene *et al.* 2004). However, a mapping of the slides in time and space relative to the historical tsunami record has not been possible.

Seiches and waves in lakes and bays caused by earthquakes, volcanic processes and landslides

Central America is a region with active volcanoes, and also the location of some major lakes. The largest lakes are Lake Nicaragua and Lake Managua in Nicaragua, where a major part of the population live close to the shorelines. Seiches were reported in Lake Nicaragua after an earthquake in 1884 (Molina 1997), and several volcanoes on the shores of the lakes have very steep flanks with slide scars evidencing that large slides have hit the lakes: slides that could have caused significant waves even if the water is not very deep. Evidence has been found for the occurrence of tsunami-like waves in Lake Managua that have been associated with volcanic eruptions of the Apoyeque volcano (Freundt *et al.* 2007). Furthermore, 'the isletas' (hundreds of small islands in Lake Nicaragua) near Granada were formed by a giant landslide from the Mombacho volcano (Shea *et al.* 2008), which certainly caused huge waves along the shorelines of Lake Nicaragua. Similar observations were made in Guatemala in 2005: a wave of about 4 m in height was observed in Lake Atitlan, caused by a landslide from the San Pedro volcano during heavy rain from Hurricane Stan (Girón & Matias 2005). Volcanic eruptions and related landslides may, in particular, generate tsunami-like waves in the Gulf of Fonseca, a large bay on the Pacific coast shared by El Salvador, Honduras and Nicaragua, which is bordered by the Cosiguina volcano in Nicaragua and by the Conchagua volcano in El Salvador. Cosiguina erupted in 1835 (Scott *et al.* 2006), while Conchagua at present appears dormant: however, the coastline around the Gulf of Fonseca has important port entrances for three countries, and is very vulnerable to huge waves generated from slides or even partial collapses on the steep flanks of these two volcanoes.

The tsunami-like waves generated from local sources represent a significant threat to people living on the shorelines of the larger lakes: however, this site specific tsunami hazard needs to be investigated more thoroughly. Possible wave heights and travel times to the endangered population need to be calculated/modelled to design new early warning systems, even when recognizing that the warning time for waves from local sources are significantly shorter than for the tsunami waves generated in the subduction zone.

Travel times from tsunami source areas

The travel times for tsunamis generated by subduction earthquakes along the Pacific coast are generally in the range of 25–45 min in Guatemala, El

Salvador and Nicaragua. In Costa Rica and Panama, the minimal travel times are for some situations less than 10 min for the Nicoya, Osa and Burica peninsulas, which are situated very near the subduction trench. On the Caribbean coast, the minimum travel time of local tsunamis might also be less than 10 min for the islands in the Bay of Honduras, for the border area between Costa Rica and Panama, as well as for the Island of El Porvenir, northern Panama and other islands in the Gulf of San Blas, north of Panama. Tsunamis from earthquakes in the Western Colombia subduction zone can also affect Panama and reach the Darien Province in Panama in less than 10 min and the Azuero peninsula in 30 min.

The challenge of 'slow' earthquakes

In traditional subduction zones, strong shaking from large earthquakes will warn the population located near the shoreline that a tsunami may be associated with the earthquake. The threatened population, if adequately informed and prepared, can then take protective measures and possibly save lives and valuables from potential inundation. However, in the Pacific subduction zone of Central America, destructive tsunamis have also been generated by 'slow earthquakes' of large magnitude, as recently seen in 1992 and 2012 in Nicaragua and El Salvador (Satake 1994; Borrero 2012; Tenorio & Strauch 2012; Borrero *et al.* 2014). These earthquakes are characterized as having less energy on frequencies felt by people and may therefore (when felt) be misinterpreted as not being very strong, and consequently the need for immediate evacuation may not be recognized by the population. For these cases, the warning from the tsunami warning centres is indispensable and the centres must be able to determine rapidly the magnitude of slow earthquakes using special procedures (e.g. see Lomax & Michelini 2012). The CATAC regional centre (see below) recognizes this challenge.

Regional and national hazard mapping

Tsunamis remain relatively rare events and people tend to forget about the impacts. In the current setting, where large parts of the Central American coastline is developed and continues to develop as tourist destinations, this 'deficient memory' increases the risk significantly (as also tragically evidenced in the 2004 Sumatra tsunami). To improve this situation, tsunami hazard mapping in all the Central American countries has started and a first step is the mapping of low-lying coastal areas. It is important to recognize that detailed elevation mapping of the coastline and bathymetry is critical for tsunami hazard mapping, and, generally, the existing maps for both (bathymetry and coast zone elevation) are not available with the optimal precision level. The second step is to improve on the bathymetric data and thereby provide the bathymetric basis for the tsunami modelling, and the third step is the modelling itself leading to more realistic hazard estimates. However, since hazard mapping is a national responsibility, the tsunami hazard maps remain quite different in terms of resolution and dissemination, and more so in terms of how they are used as a basis for national disaster mitigation actions. In most of the countries, the tsunami hazard mapping is more focused on the Pacific than on the Caribbean coast.

Owing to the tragic 1992 tsunami, hazard mapping is now relatively complete for the Pacific coast of Nicaragua (Strauch 2005; Strauch *et al.* 2014). Digital elevation models (DEMs) with 2 m spacing were used for hazard zoning, and the maps are completed with evacuation routes, concentration points and vulnerability information, together with the local population, as shown for one area in Figure 3. Such maps are now used to train the coastal population and raise awareness of the risks for tsunami waves hitting the coast. For the four largest population centres near the coast, numerical tsunami modelling was also used to estimate the impact of local tsunamis (Yamazaki *et al.* 2007). In a similar manner, tsunami hazard maps were elaborated for 20 villages and towns in the Gulf of Fonseca (Acosta *et al.* 2009), which is shared by El Salvador, Honduras and Nicaragua.

In El Salvador, a catalogue of tsunami hazard and risk maps were developed that cover the whole Pacific coast of the country (González-Riancho *et al.* 2014). The hazard zonation is here largely based on DEMs with 10 m spacing (i.e. lower resolution than in Nicaragua) and numerical tsunami simulation.

In Costa Rica, simulations of tsunamis and estimates of flood-prone areas began in 2001 and, since then, the analysis has been focused on the city of Puntarenas (Fig. 4). That city is located on a sandbank that grew from the coast into the ocean and the region is consequently extremely vulnerable to a possible tsunami; the city area is flat with more than 10 000 permanent residents, and Puntarenas is also a popular tourist destination. The first simulation of a tsunami in Costa Rica was made to investigate the influence of local and distant tsunamis (Ortiz *et al.* 2001). The findings indicated that both local and distant tsunamis can reach the city with wave heights of up to 3.5 m, representing a serious threat to the city.

The University of Costa Rica also executed a project studying the possible tsunami inundation of Puntarenas and Playa del Coco, both tourist

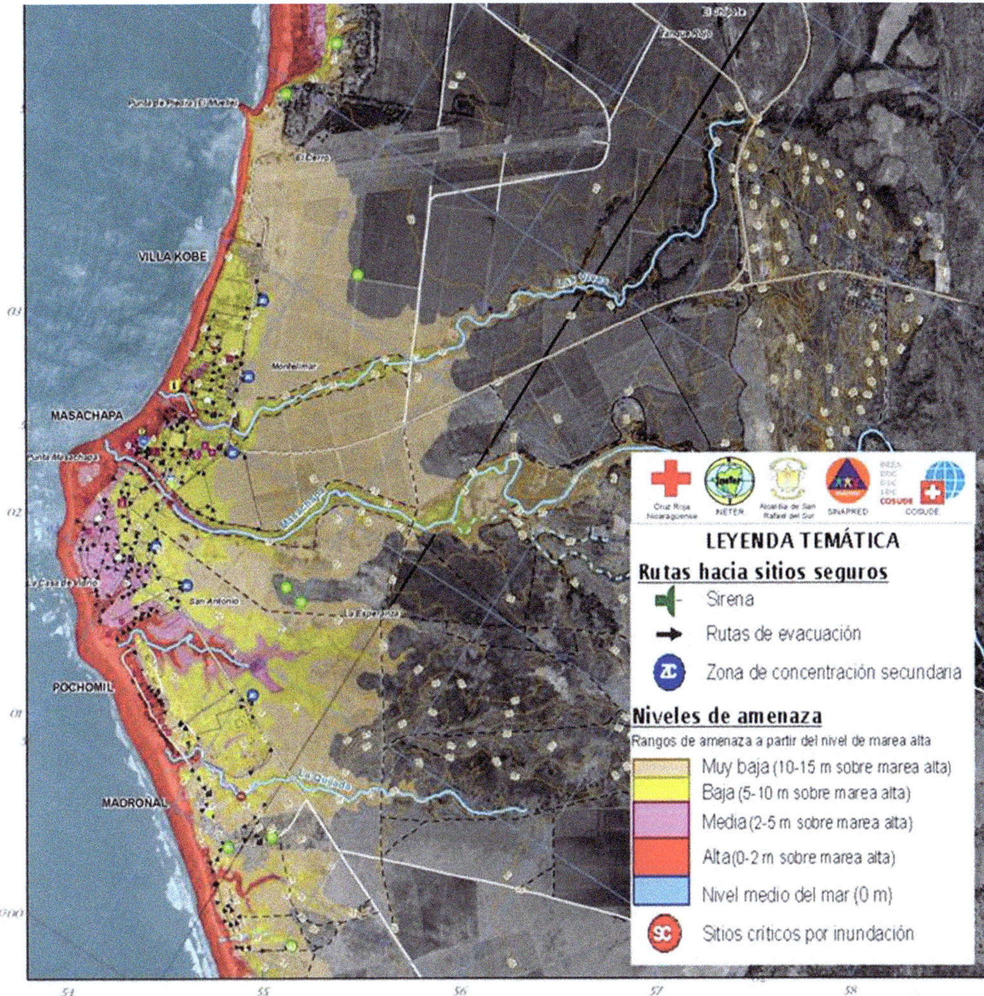

Fig. 3. Example of a tsunami risk map for the Masachapa region in Nicaragua showing the varying degree of hazard along the coast, as well as evacuation routes and assembly locations.

magnets on the Pacific coast, in which the HyFlux2 code (http://publications.jrc.ec.europa.eu/repository/handle/JRC36005); was applied to estimate the wave propagation and inundation potential (Zamora *et al.* 2012). The team made the most accurate analysis allowed by the existing data and concluded that a hypothetical earthquake with an epicentre near the Gulf of Nicoya could generate a tsunami capable of completely inundating Puntarenas. Similarly, Chacón-Barrantes & Protti (2011) simulated a possible tsunami associated with a future earthquake on the subduction zone near Puntarenas and assessed its impact on the city, and they reached the same conclusion: the city is very vulnerable to tsunami inundation. The magnitude 7.6 earthquake on 5 September 2012 only produced a minor rise of the ocean (centimetres); perhaps partly because the epicentre was close to the coast in shallow waters, which reduced the possibility of a large and destructive tsunami. This should, however, not lead to the belief that a major tsunami cannot take place in a future scenario. Finally, Gutiérrez (2016) made an analysis for the Tamarindo beach resort located on the Pacific north coast of Costa Rica aimed at developing a tsunami emergency response plan for this important tourist beach.

The other countries in Central America (Guatemala, Honduras and Panama) have done less to map the tsunami hazard and prepare for emergency plans: however, studies have been conducted and now are on the way. In Panama, overview hazard maps were elaborated for both the Pacific and

Fig. 4. The 'sand bar' of Puntarenas city located in the Gulf of Nicoya on the Pacific coast of Costa Rica. Photography by Jean Mercier.

Caribbean coasts using 30 m DEMs, and a detailed hazard map was elaborated for Puerto Armuelles in the Gulf of Chiriquí in Panama (Strauch 2015).

Seismic and geophysical monitoring

Initial tsunami warnings today are mainly based on seismological information, and, for more than two decades, the seismic observatories of the six Central American countries have closely collaborated in data exchange and the use of compatible recording and processing systems (e.g. see Alvarenga *et al.* 1998). Figure 2 demonstrates the density of the seismic stations. Automatic seismic data-processing systems based on SeisComP3 software (Weber *et al.* 2007) were installed in 2011–13 in all Central American countries in cooperation with the Seismic Observatory of Western Panama (OSOP 2015), and the national observatories routinely use data from national networks augmented with data from the global seismic network (IRIS and GFZ international data centres). Following a proposal by INETER, Nicaragua, the seismic network operators in the region have agreed to join in a consortium and to integrate their stations in a virtual regional seismic network which now includes several hundred seismic stations in Central America (CEPREDENAC 2015*a*). This network will initially be used for a regional tsunami warning and later for other cooperation programmes. The cooperation will also be extended to global positioning systems (GPS/GNSS), which are another important tool for tsunami warnings. In cooperation with UNAVCO, a regional mirror of the COCONet regional GPS network (http://coconet.unavco.org/science/center-awards.html) was established at INETER and, besides other applications, it is planned to use this network capacity for the regional tsunami warnings.

Tsunami warning systems

An effective, real-time tsunami warning system is an important goal for the Central American countries. To this side, there are two main challenges: (a) to quickly and reliably identify potential tsunamis arriving at a specified coastline; and (b) an effective warning system for the exposed population. The time and space reliability are critical factors: a warning must not arrive too late, and many false warnings undermine the credibility of the system.

Real-time tests with seismic data from the University of Panama and from other Central American and global stations are carried out by OSOP (http://www.osop.com.pa); to acquire practical knowledge about the software tools for moment tensor inversion and TOAST (Tsunami Observation And Simulation Terminal) for real-time tsunami forecasting (http://www.gempa.de). This experience is expected to be applied and further tested at the Central American Tsunami Advisory Center (CATAC, see below) and at the national tsunami warning centres in the region.

The seismic network is the main system for early detection of tsunamis, and currently some 350 seismic stations are installed in Central America, all operated by the national agencies (Fig. 2). The rather unique situation in Central America in comparison with other regions is that data are shared between the countries, and two of the countries have 24/7 manning of the observatories, which enables quick manual location and magnitude determination when an earthquake occurs. Honduras has recently also improved its seismic network, and now comprises 10 telemetric stations along the northern coast and in the Gulf of Fonseca (ICG/PTWS-XXVI 2015*a*). Tide gauges are additionally used for the confirmation of the tsunami occurrence and the estimation of its impact on the coast, and the countries have recently extended their network of tide gauges to more reliably monitor the impact of tsunami waves. Tide gauge measurements are yet to be shared between neighbouring countries.

After the tsunami disaster in 1992, Nicaragua joined the Pacific Tsunami Warning System (PTWC) and has since then participated actively in the system. Over the last years, all countries in the region have joined the international tsunami warning system or increased their activity in the system, which is developed under the guidance of the Intergovernmental Oceanographic Commission (IOC) of UNESCO. At present, the responsible international tsunami centre for both the Pacific and Atlantic coasts of the Central American countries is therefore the PTWC, which provides the location and magnitude of earthquakes, estimates of tsunami arrival times, and wave amplitudes in both alphanumeric and graphic formats. The decision about the issuance of warnings to the population remains the responsibility of the national authorities in the respective countries, in most cases handled by the national agencies for civil protection.

The messages sent out by the PTWC sometimes come too late and are not sufficiently detailed for local tsunamis, and therefore the IOC promotes the development of national and regional tsunami warning systems. Nicaragua established its national tsunami warning system in 1996 (Strauch 1998; Strauch *et al.* 2014) based on the national/regional seismic monitoring, and defined a national tsunami warning centre. At the time of writing, the Nicaraguan system is considered the most advanced system in the region (JICA 2015), and remarkable features of the system are: the very short response time due to the automatic data processing within the 24/7 watch system; and the 24/7 Civil Defence office with its alarm system that includes 60 automatic sirens along the Pacific coast and a robust voice radio communications system that reaches the coastal communities (ICG/PTWS-XXVI 2015*b*). El Salvador likewise developed a national tsunami warning system following the large and damaging 2001 earthquake (which only caused minor waves).

In 2014, the National University (UNA) and the University of Costa Rica (UCR) joined efforts to create the National System of Monitoring of Tsunamis (SINAMOT: http://www.facebook.com/sinamot.cr). The aim of this institution is to monitor all effects associated with the occurrence of tsunamis that may influence the Costa Rica coastlines. When a distant tsunami occurs, SINAMOT models the potential tsunami impact and provides the simulation results to the National System for Risk Management as a basis for evacuation and alert decisions. This national initiative is important because both the press and local authorities sometimes make inappropriate use of the PTWC tsunami warnings and add 'sensationalism' promoting unnecessary and uncoordinated evacuations. An example of this happened on 15 August 2007 when the Pisco Peru earthquake occurred. As usual, the PTWC estimated the tsunami potential from that event and issued an alert that included Costa Rica as one of the countries that could be affected. The National System for Risk Management immediately issued statements in the national press, instructing the coastal residents or visitors to the beaches to move away from the shoreline. This generated panic in Puntarenas. Such was the stress that many thought they were going to perish. A tsunami never arrived at Costa Rica and the mobilization was recognized as unnecessary. SINAMOT was created to provide more accurate national information and to avoid false or erroneous press releases leading to panic.

Present state; a newly established Central American Tsunami Center

The idea of a regional tsunami warning system in Central America dates from 1998 (INETER 1998). After years of detailed discussions in the coordination groups of the regional tsunami warning systems for the Pacific and the Caribbean (ICG/PTWS-CA 2014; ICG/CARIBE EWS-X 2015; ICG/PTWS-XXVI 2015*a*), it was decided in 2015 to accept an offer from Nicaragua to establish a regional Central American Tsunami Advisory Center (CATAC) at INETER in Managua. The geographical location and the fact that INETER for several years has had the capacity to integrate data from seismic stations all over Central America were important for establishing the CATAC at INETER. CATAC will provide tsunami services for the region, which includes the seismic monitoring, near real-time determination of the hypocentre and magnitude of large earthquakes, evaluation of focal parameters, estimation of the probability of tsunami occurrence, determination of the arrival times, and maximum amplitudes of the possible tsunami waves on both coasts of Central America. The results will be available within minutes after a large earthquake and provided to the designated institutions of the Central American countries.

CATAC will be affiliated with the global tsunami warning system directed by the IOC of UNESCO (IOC 2015), and will closely cooperate within the regional tsunami warning systems of the Pacific (PTWS) and the Caribbean (CARIBE EWS). CATAC will also support the exchange of experiences on tsunami risk reduction in Central America, and promote the development of the seismic and geophysical monitoring capacities. The monitoring base of CATAC will be the real-time data of the seismic networks in Central America and the data of the mareographic networks (tide gauges placed and designed for tsunami detection). CEPREDENAC (Centro de Coordinación para la Prevención de los Desastres Naturales en América Central) and the national civil protection agencies have expressed firm support for CATAC (CEPREDENAC 2015*b*), and CATAC is considered to become a part within the System of the Central American Integration (SICA).

Tsunami risk prevention and mitigation measures

Besides early tsunami warnings, the countries have also developed other aspects of prevention and mitigation measures. A first 'invisible' measure is the development of national protocols that regulate the responsibilities for prevention and mitigation of tsunami disasters among the national institutions. In this respect, El Salvador and Nicaragua may be in the forefront in the region. In addition, the first activities for increasing population awareness (supported by INETER/CATAC) consist of warning signs and drills, as exemplified in Figure 5.

The level of information and education in the population about tsunami hazard and adequate response measures is high in Nicaragua, largely due to the experience with the 1992 tsunami and the persistent development of tsunami mitigation measures since then. In Nicaragua, several projects were carried out to establish local tsunami warning systems integrating local stakeholders (local authorities, local police, health facilities, Red Cross and

Fig. 5. (**a**) Placement of evacuation route signs in areas of tsunami hazard (Corinto beach in Nicaragua). (**b**) Evacuation drill at Masachapa, Nicaragua.

private enterprises). Information and education campaigns are frequently carried out with the local authorities, and schools and the general coastal population. During Easter week, the main holiday season in Nicaragua, TV and radio programmes inform the coastal population and tourists about the tsunami hazard and appropriate behaviour in case of a tsunami warning. Tens of thousands of leaflets are distributed to the population and tourists (ICG/PTWS-XXVI 2015*b*).

In Costa Rica, the UCR, through the seismological network and Preventec, is working on public education to prevent and mitigate disasters generated by tsunamis (Fernández & Alvarado 2005). The main ideas behind this initiative are: (a) to change attitudes and influence behaviours with respect to the threat of tsunamis; and (b) to enable people to realize that they can protect themselves from tsunamis. Under a joint umbrella, the managers on disaster risk cooperate with members of local groups to disseminate basic information about tsunamis and tsunami protection. Another important project of the Costa Rican group is to promote the knowledge of tsunamis in coastal elementary schools (Fernández *et al.* 2014), and under this initiative a toolkit on tsunamis information has been designed (Porras *et al.* 2014) aimed at training and preparing those who are studying and working at educational centres to adequately address the tsunami threat. The toolkit is designed for use throughout Central America, and contains a brochure, a board game and an animation video about tsunamis.

An example of the challenges faced in Costa Rica is exemplified at Tamarindo beach, which was originally a small fishing village that has now swelled to an international hotspot of thousands of foreigners that enjoy the sunny beaches. Its location is attractive for tourists, but also vulnerable to tsunamis generated from large subduction-type earthquakes. The multi-ethnical and multi-lingual population during the daytime creates a significant challenge for tsunami preparation and mitigation activities.

Finally, all countries in the region participate in the yearly tsunami warning exercises promoted by the IOC, PTWS and CARIBE EWS, and, additionally, several countries organize exercises on national and local levels. For example, the latest national exercise focusing on the earthquake and tsunami impact in Nicaragua was organized on 30 June 2015, and integrated 1.5 million participants on the national, municipal and local levels (El 19 2015*a*, *b*).

Summary

The present article has attempted to review the tsunami risk situation in Central America as it has developed through the last decades. Since politically turbulent years in the 1980s and early 1990s, this small and culturally homogeneous region has experienced a nearly continuous process towards economic growth and political integration, and this has also involved coordination and cooperation between the six countries within the natural disaster coordination: Central America shares the threats of volcanic eruptions, disastrous earthquakes and tsunamis (mainly, but not exclusively, along the 1700 km-long Pacific coastline). Indeed, tsunamis have affected all countries of the region and have claimed close to 500 victims. A particular threat in Central America relates to slide-induced waves washing the shorelines of the great lakes. This threat is still only tentatively mapped, but has been documented, in particular, for lakes in Nicaragua and Guatemala, and also applies to many of the larger lakes situated near the volcanoes. In addition to the natural hazard, the growing tourist industry focused largely on the Pacific beaches is now greatly enhancing the tsunami risk.

The 1992 tsunami disaster in Nicaragua raised awareness and tsunami hazard mapping, and this is now well advanced in Nicaragua, Costa Rica and El Salvador, with continuing initiatives to improve mapping in all countries. National systems for early warning and risk reduction have been established in Nicaragua, Costa Rica and El Salvador, while the other countries mainly rely on the Pacific Tsunami Warning Center information. However, recently Guatemala, Honduras and Panama have also embarked on national tsunami risk reduction programmes. The hope is that the planned regional tsunami centre (CATAC, with its physical location in Nicaragua) will significantly improve both early warning and coordinated information campaigns. The education activities and emergency plans that are presently carried out by the individual countries can, through CATAC, be enhanced in quality and outreach through the regional coordination.

Six small countries now work together and move forward in their efforts to mitigate the natural disaster threats: their efforts are applauded and can be taken as examples for other earthquake- and tsunami-affected regions.

The present manuscript was prepared after two decades of collaboration between countries in Central America and Norway. We are indebted to the Norwegian government for the grants that allowed the collaboration. We also direct our sincere appreciation to two anonymous reviewers who significantly helped to improve the first manuscript, both by patient language correction, and by technical hints and comments. We finally thank Ellie Scourse and Ian Terry for improving the language.

References

Abe, K., Abe, K. & Tsuji, Y. 1993. Field survey of the Nicaragua earthquake and tsunami of September 2, 1992. *Bulletin of the Earthquake Research Institute, University of Tokyo*, **68**, 23–70.

Acosta, N. 2012. *Tsunami en Nicaragua después del terremoto del 26 de agosto de 2012* [The tsunami in Nicaragua after the earthquake on 26 August 2012]. Unpublished Report INETER, Managua, Nicaragua.

Acosta, N., Strauch, W., Castellon, A., Larreynaga, A. & Funes, G. 2009. GIS based tsunami hazard mapping in Nicaragua, El Salvador and Honduras and application to disaster prevention measures. *In*: Wörner, G. & Möller-McNett, S. (eds) *International Lateinamerika-Kolloquium 2009, Göttingen, April 7–9, 2012. Abstracts and Programs*. Universitätsverlag Göttingen, Göttingen, Germany, 17–19.

Alvarenga, E., Barquero, R. *et al.* 1998. Central American Seismic Center (CASC). *Seismological Research Letters*, **69**, 394–399, https://doi.org/10.1785/gssrl.69.5.394

Baptista, A.M., Priest, G.R. & Murty, T.S. 1993. Field survey of the 1992 Nicaragua tsunami. *Marine Geodesy*, **16**, 2.

Borrero, J.C. 2012. *Field Survey Report of Tsunami Effects Caused by the August, 2012 Offshore El Salvador Earthquake*. UNESCO/IOC Report.

Borrero, J.C., Kalligeris, N., Lynett, P.J., Fritz, H.M., Newman, A.V. & Convers, J.A. 2014. Observations and Modeling of the August 27, 2012 Earthquake and Tsunami affecting El Salvador and Nicaragua. *Pure and Applied Geophysics*, **171**, 3421–3435.

CEPREDENAC 2015*a*. Acta de Acuerdos número 1, 2015. *Primera Sesión Ordinaria del Consejo de Representantes de CEPREDENAC*, 06 de febrero 2015, Ciudad de Guatemala, Guatemala [Minutes of Agreements. First Session of Ordinary Board Meeting, 6. February 2015, Guatemala City].

CEPREDENAC 2015*b*. *2ª Sesión Técnica Regional de Discusión sobre la Integración de Capacidades en*

Sismología de América Central. 4–5 Junio, 2015, Managua, Nicaragua [Minutes. Second Technical Meeting on Integration of Technical Capacities within Seismology in Central America. Session of Ordinary Board Meeting. 4–5 June 2015, Managua].

CHACÓN-BARRANTES, S.E. & PROTTI, M. 2011. Modeling a tsunami from the Nicoya, Costa Rica, seismic gap and its potential impact in Puntarenas. *Journal of South America Earth Science*, **1**, 372–382.

DEMETS, C. 2001. A new estimate for present-day Cocos-Caribbean Plate motion: Implications for slip along the Central American Volcanic Arc. *Geophysical Research Letters*, **28**, https://doi.org/10.1029/2001GL013518

EL 19. 2015*a*. http://www.el19digital.com/articulos/ver/titulo:30986-nicaragua-desarrolla-segundo-simulacro-nacional-ante-sismo

EL 19. 2015*b*. http://www.el19digital.com/articulos/ver/titulo:30988-masachapa-desarrolla-con-exito-ejercicio-de-simulacro-nacional-ante-terremoto

FERNÁNDEZ, M. 2013. Seismotectonic and the hypothetical Tectonic Boundary of Central Costa Rica. *In*: D'AMICO, S. (ed.) *Earthquake Research and Analysis, Book 2*. In-Tech, Rijeka, Croatia.

FERNÁNDEZ, M. & ALVARADO, G. 2005. Tsunamis and tsunami preparedness in Costa Rica, Central America. *ISET Journal of Earthquake Technology*, Paper 466, **42**, 203–212.

FERNÁNDEZ, M., MOLINA, E., HAVSKOV, J. & ATAKAN, K. 2000. Tsunamis and tsunami hazard in Central America. *Natural Hazards*, **22**, 91–116.

FERNÁNDEZ, M., SOLÍS, D., PORRAS, J. & GONZÁLEZ, G. 2014. Proposal to disseminate the knowledge of tsunamis in the Caribbean Coast of Costa Rica, Central America. *World Journal of Engineering and Technology*, **2**, 85–90, https://doi.org/10.4236/wjet.2014.23B013

FREUNDT, A., STRAUCH, W., KUTTEROLF, ST. & SCHMINCKE, H.U. 2007. Volcanogenic tsunamis in lakes: examples from Nicaragua and general implications. *Pure and Applied Geophysics*, **164**, 527–545.

GIRÓN, J.R. & MATIAS, O. 2005. *Pequeño Tsunami (Seiche) en la Bahía de Santiago, Lago de Atitlán, Sololá, Guatemala* [Small tsunami (Seiche) in the Santiago Bay, Lake Atitlan, Solola, Guatemala]. Unpublished report Instituto Nacional de Sismología, Vulcanología, Meteorología e Hidrología (INSIVUMEH), Guatemala City, Guatemala.

GONZÁLEZ-RIANCHO, P., AGUIRRE-AYERBE, I. ET AL. 2014. Integrated tsunami vulnerability and risk assessment: application to the coastal area of El Salvador. *Natural Hazards and Earth System Sciences*, **14**, 1223–1244, https://doi.org/10.5194/nhess-14-1223-2014

GUTIÉRREZ, H. 2016. *Analysis of spatial data for the implementation of a contingency plan to tsunamis in the Tamarindo and Langosta beaches, Santa Cruz, Guanacaste, Costa Rica*. Bachelor thesis, School of Geography, University of Costa Rica.

HAN, Y., LAY, T. ET AL. 2013. The 5 September 2012 Nicoya, Costa Rica Mw 7.6 earthquake rupture process from joint inversion of high-rate GPS, strong-motion and teleseismic P wave data and its relationship to adjacent plate boundary interface properties. *Journal of Geophysical Research: Solid Earth*, **118**, 5453–5466, https://doi.org/10.1002/jgrb.50379

ICG/CARIBE EWS-X 2015. *Tenth Session of the Intergovernmental Coordination Group for the Tsunami and Other Coastal Hazards Warning System for the Caribbean and Adjacent Regions*, 19–21 May 2015, Philipsburg, Sint Maarten, http://www.ioc-unesco.org/index.php?option=com_oe&task=viewEventRecord&eventID=1634

ICG/PTWS-CA 2014. *Grupo de Trabajo Regional para América Central del Grupo Intergubernamental de Coordinación del Sistema de Alerta contra los Tsunamis y Atenuación de sus Efectos en el Pacífico (ICG/PTWS), Tercera Reunión*, Del 29 al 30 de septiembre de 2014, Managua, Nicaragua, http://ioc-unesco.org/index.php?option=com_oe&task=viewEventDocs&eventID=1571

ICG/PTWS-XXVI 2015*a*. *Twenty-Sixth Session of the Intergovernmental Coordination Group for the Pacific Tsunami Warning and Mitigation System (ICG/PTWS-XXVI)*, 22–24 April 2015, Honolulu, USA, http://ioc-unesco.org/index.php?option=com_oe&task=viewEventRecord&eventID=1598

ICG/PTWS-XXVI 2015*b*. Country report for Nicaragua. *Twenty-Sixth Session of the Intergovernmental Coordination Group for the Pacific Tsunami Warning and Mitigation System (ICG/PTWS-XXVI)*, 22–24 April 2015, Honolulu.

INETER 1998. *Letter to CEPREDENAC about the establishment of a Regional Tsunami Warning System in Central America*. INETER, Managua, Nicaragua.

INSIVUMEH 2012. *Sismo de Magnitud 7,2 del 7 de Noviembre de 2012* [Earthquake of M7.2 on 7th November 2012. Technical report]. Informe técnico 2012 Instituto Nacional de Sismología, Vulcanología, Meteorología e Hidrología (INSIVUMEH), Guatemala City, Guatemala.

IOC 2015. *28th Session of the IOC Assembly*, 18–25 June 2015, Paris, France, http://ioc-unesco.org/index.php?option=com_oe&task=viewEventRecord&eventID=1576

JICA 2015. *Data Collection Survey on the Observation Capacity of Earthquakes and Tsunami in Central America*. Final Report *Japan International Cooperation Agency (JICA)/Oriental Consultants Global Co. Ltd/Japan Meteorological Business Support Centre*, March, 2015.

KOSHIMURA, S. 2001. *Modeling a Tsunami Generated by an Earthquake of Mw 7.7 in Central America [2001-01-13 17:33:31 UTC]. Tsunami Research Program*, Pacific Marine Environmental Laboratory, NOAA.

LINKIMER, L., ARROYO, I. ET AL. 2013. El terremoto de Sámara (Costa Rica) del 5 de Setiembre del 2012 (Mw 7, 6). [The Samara (Costa Rica) earthquake of 5 September 2012 (Mw 7.6).] *Revista Geológica de América Central*, **49**, 73–82.

LOMAX, A. & MICHELINI, A. 2012. Tsunami early warning within 5 minutes. *Pure and Applied Geophysics*, **170**, 1385–1395, https://doi.org/10.1007/s00024-012-0512-6

MOLINA, E. 1997. *Tsunami Catalogue for Central America, 1539–1996*. Technical Report No. II 1-04: Project on Reduction of Natural Disasters in Central America, Earthquake Preparedness and Hazard Mitigation Phase

II 1996–2000. Institute of Solid Earth Physics, University of Bergen & Instituto Nacional de Sismología, Vulcanología, Meteorología e Hidrología (INSIVUMEH), Guatemala City, Guatemala.

Molina, E. & Rosales, M. 2009. *Tsunamis en Guatemala* [Tsunamis in Guatemala]. Unpublished report Instituto Nacional de Sismología, Vulcanología, Meteorología e Hidrología (INSIVUMEH), Guatemala City, Guatemala.

NOAA 2015. *National Geophysical Data Center/World Data Service (NGDC/WDS): Global Historical Tsunami Database*. National Geophysical Data Center, NOAA, Boulder, CO, USA, https://doi.org/10.7289/V5PN93H7, http://www.ngdc.noaa.gov/hazard/tsu_db.shtml [last accessed August 2015]

Ortiz, M., Fernández-Arce, M. & Rojas, W. 2001. Análisis de Riesgo de Inundación por Tsunamis en Puntarenas, Costa Rica [Analysis of tsunami inundation risk in Puntarenas, Costa Rica]. *GEOS*, **21**, 108–113.

OSOP 2015. *Short-period Seismometers. Remastered.* OSOP, Volcán, Panama, http://www.osop.com.pa/

Porras, J., González, G. & Fernández, M. 2014. A tsunami kit for elementary schools of Costa Rica. The Science of Tsunami Hazards. *Abstract presented at the 6th International Tsunami Symposium of Tsunami Society International*, 2–5 September 2014, Nicoya, Costa Rica.

Satake, K. 1994. Mechanism of the 1992 Nicaragua Tsunami Earthquake. *Geophysical Research Letters*, **21**, 2519–2522, https://doi.org/10.1029/94GL02338

Scott, W., Gardner, C., Devoli, G. & Alvarez, A. 2006. The A.D. 1835 eruption of Volcán Cosigüina, Nicaragua: a guide for assessing local volcanic hazards. *In*: Rose, W.I., Bluth, G.J.S., Carr, M.J., Ewert, J.W., Patino, L.C. & Vallance, J.W. (eds) *Volcanic Hazards in Central America.* Geological Society of America, Special Papers, **412**, 167–187, https://doi.org/10.1130/2006.2412(09)

Shea, T., van Wyk de Vries, B. & Pilato, M. 2008. Emplacement mechanisms of contrasting debris avalanches at Volcán Mombacho (Nicaragua), provided by structural and facies analysis. *Bulletin of Volcanology*, **70**, 899–921.

Strauch, W. 1998. The tsunami warning system in Nicaragua. *Paper presented at the International Conference on Early Warning Systems for Natural Disaster Reduction EWC 98*, 7–11 September 1998, Potsdam, Germany.

Strauch, W. 2005. Tsunami warning and tsunami hazard mitigation efforts in Nicaragua. *Paper presented at the Tsunami Workshop, IST/PTWS XX*, 29–30 September 2005, Santiago, Chile.

Strauch, W. 2015. *Analysis and Modeling of Geophysical, Seismological and Hydrometeorological Data for the Implementation of Automated Risk Management Systems to Strengthen OSOP and the Disaster Prevention System of Panama.* Report, Phase II, SENACYT – Contract 4-CAP11-001. OSOP, Volcan, Panama.

Strauch, W., Talavera, E., Muñoz, A., Tenorio, V. & Izaguirre, A. 2014. El Desarrollo del Sistema de Alerta de Tsunami en Nicaragua [Development of a tsunami early warning system in Nicaragua]. *Paper presented at the First Regional Assembly of the Latin-American and Caribbean Seismological Commission (LACSC – IASPEI)*, 23–25 July 2014, Bogota, Colombia.

Tenorio, V. & Strauch, W. 2012. *Evaluation of the earthquake, on August 26, 2012, in the Pacific Ocean of El Salvador and Nicaragua.* Nicaraguan Institute of Territorial Studies (INETER), Directorate General of Geophysics, Monthly Bulletin 'Earthquakes and Volcanoes of Nicaragua', August 2012.

von Huene, R., Ranero, C.R. & Watts, P. 2004. Tsunamigenic slope failure along the Middle America Trench in two tectonic settings. *Marine Geology*, **203**, 303–317.

Weber, B., Becker, J. et al. 2007. SeisComP3 – automatic and interactive real time data processing. *Geophysical Research Abstracts*, **9**, 09219.

Yamazaki, Y., Katayama, I., Strauch, W., Palacios, L., Trana, M. & Cordonero, S. 2007. Advances in seismic microzoning in Latin American countries and their implications for public policies. *EOS Trans. AGU*, **88**(23), Jt. Assem. Suppl., S33B–03.

Zamora, N., Fernández, M., Bergoeing, J. & González, C. 2012. Posible Inundación por Tsunamis en Puntarenas, Costa Rica [Possible inundation of Puntarenas by tsunamis]. *Revista Geográfica del Instituto Panamericano de Historia y Geografía*, **151**, 106–112.

Block and boulder accumulations on the southern coast of Crete (Greece): evidence for the 365 CE tsunami in the Eastern Mediterranean

SARAH J. BOULTON[1]* & MICHAEL R. Z. WHITWORTH[1,2]

[1]*School of Geography, Earth and Environmental Sciences, University of Plymouth, Drake Circus, Plymouth, Devon PL4 8AA, UK*

[2]*AECOM, Plumer House, Tailyour Road, Plymouth, PL6 5DH, UK*

**Correspondence: sarah.boulton@plymouth.ac.uk*

Abstract: The Eastern Mediterranean is one of the most seismically active regions in Europe. Crete, located in the centre of the Eastern Mediterranean, should experience tsunamis resulting from large magnitude earthquakes or volcanic eruptions. At three locations, boulders were observed that may relate to tsunami or storm events. At Lakki, the size of the boulders slightly favours a tsunami origin for deposition. By contrast, at Kommos, boulder size and geomorphology are consistent with storm parameters in the Mediterranean. The most compelling evidence for tsunami transport is found at Diplomo Petris, where a lithologically varied grouping of large boulders ($\leq$690 t) is exposed at sea level. The calculated storm wave heights (15 m) required to transport the observed boulders significantly exceeds winter averages: therefore, these accumulations are interpreted as tsunami deposits. Radiocarbon dating of encrusting biological material was undertaken to constrain periods of boulder motion. Encrustations from Diplomo Petris and Lakki predate the 365 CE earthquake, suggesting that this event transported the largest boulders; the first time that boulder deposits have been identified on Crete from this tsunami. Therefore, these data are important for developing local and regional hazard assessments but also to inform numerical models of tsunami propagation in the Mediterranean.

In recent years there has been increased interest in coastal boulder accumulations worldwide owing to the recognition that these deposits are not only the result of high-frequency storm processes but also of low-frequency tsunami events (i.e. Dawson 1994; Dawson & Shi 2000; Bryant & Nott 2001; Nott 2003*a*, *b*; Kennedy *et al.* 2007). Interest in boulder deposits was sparked in the aftermath of the 2004 Boxing Day Indian Ocean tsunami as a result of the extensive boulder transport witnessed (e.g. Etienne *et al.* 2011). Although determining the transport history and origin of coastal boulders remains challenging, a range of criteria and characteristics that can be used to separate storm from tsunami deposits are beginning to emerge from the study of recent events (e.g. Scheffers 2008).

In Thailand, boulders derived from coastal defences and natural settings, such as reefs and beachrock, were observed to have been transported up to 900 m inland in fields of scattered boulders with local imbrication or lines of deposition, so called 'boulder trains' (Goto *et al.* 2007; Paris *et al.* 2009, 2010). Paris *et al.* (2009) reported the presence of landwards-imbricated tabular boulders in Indonesia, while Paris *et al.* (2010) also documented that there was no relationship between boulder size and transport distance.

Extensive boulder deposits were also reported from the 2009 South Pacific tsunami affecting American Samoa and other South Pacific islands (Etienne *et al.* 2011). There, Richmond *et al.* (2011*a*) described a well-developed boulder field with no trend in boulder-size distribution and with boulder long axes perpendicular to the flow direction. Additionally, Goto *et al.* (2012) documented tuffaceous and concrete boulders that were observed to have been reworked from sea defences and rockfalls after the 2011 Tohoku-oki tsunami, Japan, observing that platy concrete boulders exhibited a landwards fining in clast size, as well as landwards imbrication. By contrast, the blocky tuffaceous boulders showed either a landwards coarsening in clast size or no trend in clast-size distribution. These differences were interpreted to be the result of the concrete slabs being transported and deposited by the incoming flow and the light tuffaceous blocks during the return flow (Goto *et al.* 2012). This suggests that the lack of size trends observed elsewhere may be due to a combination of boulder transport and deposition during the incoming and

From: SCOURSE, E. M., CHAPMAN, N. A., TAPPIN, D. R. & WALLIS, S. R. (eds) 2018. *Tsunamis: Geology, Hazards and Risks*. Geological Society, London, Special Publications, **456**, 105–125.
First published online February 9, 2017, https://doi.org/10.1144/SP456.4

outgoing flow (Bahlburg & Spiske 2012). Nandasena *et al.* (2013) also measured boulders following the 2011 Japanese earthquake in northern Japan, demonstrating that the long axes of the boulders was broadly perpendicular to the transport direction. Similar relationships have been observed after the 2004 Indian Ocean tsunami (Goto *et al.* 2007; Etienne *et al.* 2011), the 2009 South Pacific tsunami (Richmond *et al.* 2011*a*) and on Hawaii (Richmond *et al.* 2011*b*), as well as in ancient deposits (i.e. Ramalho *et al.* 2015).

These characteristics are somewhat different to the boulder beaches, ridges and ramparts that appear to be characteristic of storm-derived boulder accumulations (e.g. Oak 1984; Kortekaas & Dawson 2007; Morton *et al.* 2007, 2008; Etienne & Paris 2010; Paris *et al.* 2011). In comparison, tsunamis form distributed debris fields not the stratified, shore-parallel ridges associated with storm waves (Morton *et al.* 2008). Furthermore, Etienne & Paris (2010) showed that storm boulder beaches generally show a landwards decrease in boulder size, imbrication may be present, and the shape and size of clasts reflects the lithology of the underlying bedrock.

Therefore, we can summarize that tsunami boulders could be recognized by their large size that cannot be explained easily by known storm parameters for the study areas (e.g. Goto *et al.* 2010), a lack of grain size trends along the flow direction, yet orientation of the boulder long axis perpendicular to flow direction. The imbrication of large boulders may also be significant and has been used as a tsunami indicator (Kennedy *et al.* 2007; Maouche *et al.* 2009) but equally has also been recognized in storm boulder deposits (Morton *et al.* 2008; Etienne & Paris 2010). Furthermore, evidence of a littoral or foreshore origin for boulders displaced far inland is also an important tsunami signature (Scheffers & Scheffers 2007) but not unique as storm-transported boulders can also originate in the near shore; features that can be used as diagnostic indicators of a marine origin include calcareous algal encrustations, bivalve borings, bioerosive notches and tilted rock pools (Scheffers 2008). Finally, these geometrical observations and interpretations can be tested by the use of hydrodynamic models, such as those of Nott (1997, 2003*a*, *b*), Nandasena *et al.* (2011) and Benner *et al.* (2010), to determine the wave heights and velocities needed to transport observed boulders.

Numerous studies have attempted to recognize palaeo-tsunami deposits using these characteristics (e.g. Nott 1997; Kelletat *et al.* 2004; Goff *et al.* 2006; Kennedy *et al.* 2007; Shah-hosseini *et al.* 2011; Hoffman *et al.* 2013; Ramalho *et al.* 2015) but are hampered in study areas where significant storm wave heights can be generated (i.e. May *et al.* 2015), complicating the distinction between tsunami and storm deposits through overlap in calculated wave heights needed for transport and reworking of boulder deposits. However, the Mediterranean Basin offers a good opportunity for the study of palaeo-tsunamis due to a micro-tidal regime, limited fetch reducing maximum possible storm wave heights, and many active faults and other tsunamigenic sources. As a result, tsunami boulder deposits have been described from across the Mediterranean region (Algeria: Maouche *et al.* 2009; France: Vella *et al.* 2011; Sicily: Scicchitano *et al.* 2007; Barbano *et al.* 2010; and the Adriatic: Mastronuzzi & Sansò 2004; Malta: Mottershead *et al.* 2014, 2015; Cyprus: Kelletat & Schellmann 2002; and Greece: Scheffers & Scheffers 2007; Scheffers *et al.* 2008; Vacchi *et al.* 2012). In addition, fine-grained to cobble deposits attributed to tsunamis are also common in the Mediterranean, and have been described mainly from Italy, Greece and Turkey (i.e. Dominey-Howes 1998; Vött *et al.* 2007, 2008, 2009, 2011; Bruins *et al.* 2008; Smedile *et al.* 2011, 2012; May *et al.* 2012).

Here we document locations from southern Crete where unusual boulder accumulations were observed. This study focuses on boulder accumulations for two reasons: firstly, the coarsest grain-size fraction is required to determine maximum flow characteristics, such as velocity and wave height, that can be used to test numerical models. The fine-grained fractions washed further inland cannot give information on these parameters. Secondly, the southern coast of Crete is dominated by steep mountainous topography with deeply incised drainage lines unsuitable for the preservation of fine-grained material. Qualitative and quantitative evidence is described and combined with recent wave data and radiocarbon dating to interpret these accumulations in terms of storm or tsunami wave processes. These data provide important data for future validation of tsunami models for the Mediterranean region.

Geological setting

The island of Crete, located at the centre of the Hellenic Arc, is the topographical expression of the accretionary wedge above the subducting African Plate below the Eurasian Plate. The Hellenic boundary accommodates approximately 40 mm a^{-1} of relative motion between the African and Eurasian plates (Reilinger *et al.* 2006; Nocquet 2012). This fast plate convergence rate corresponds to high levels of historical and recent seismicity, with $\leq$20% of the relative plate motion accommodated as earthquakes (Vernant *et al.* 2014) and the remainder of the plate motion occurring through aseismic slip (Reilinger *et al.* 2010).

Therefore, Crete should have been affected by the numerous recorded tsunamigenic earthquakes that have occurred in the region since antiquity (e.g. Kelletat & Schellmann 2002; Ambraseys & Synolakis 2010; England *et al.* 2015) and an unknown number in prehistory (Fig. 1). In addition, tsunamis caused by volcanic eruptions also have the potential to affect this coast (Dominey-Howes *et al.* 2000). However, many of these events were small in nature or of dubious provenance (Ambraseys & Synolakis 2010). Yet there is clear evidence for at least five large tsunamigenic earthquakes along the Hellenic plate boundary zone where earthquake magnitudes exceeded $M = 7$ (Fig. 1). These devastating tsunamis took place on 21 July 365 CE, Crete; 8 August 1303 CE, Crete; 3 May 1481 CE, Rhodes; 28 February 1629 CE, Kythera; and 22 January 1899 CE, Kyparissia (England *et al.* 2015).

However, up to now, only traces of the tsunami generated by the eruption of Santorini in approximately 1650 BCE have been observed in the far east of the island (Bruins *et al.* 2008), and a series of fine-grained tsunami deposits have been observed in and nearby the ancient harbour of Phalasarna in the far west of Crete, dating to tsunamis that occurred in 66 CE and 365 CE (Pirazzoli *et al.* 1992; Stiros & Papageorgiou 2001; Scheffers & Scheffers 2007). Boulder clusters have also been described by Scheffers & Scheffers (2007) in western Crete but limited dating suggests only that these deposits are probably younger than 365 CE. Furthermore, investigations by Dominey-Howes (2004) failed to document any tsunami traces along the north coast of Crete. This lack of geological evidence for extreme events seems rather incongruous given the evidence for large tsunamis in the region.

One of the largest earthquakes ever documented for the Mediterranean was the approximately $M_w = 8$ earthquake and tsunami in 365 CE (Pirazzoli 1986; Pirazzoli *et al.* 1996; Shaw *et al.* 2008), which caused the coast of Crete to be uplifted by as much as 9 m absl (above sea-level), as evidenced by a raised marine notch observed across western Crete (Pirazzoli *et al.* 1996). The event also generated a tsunami that caused devastation throughout the Eastern Mediterranean, especially along the north coast of Africa (e.g. Guidoboni *et al.* 1994; Ambraseys & Synolakis 2010). The location of the fault causing this rupture is still under debate but there are a number of numerical models using a shallow thrust ramp (*c.* 30° dip) within the upper Aegean Plate as the causative fault (i.e.

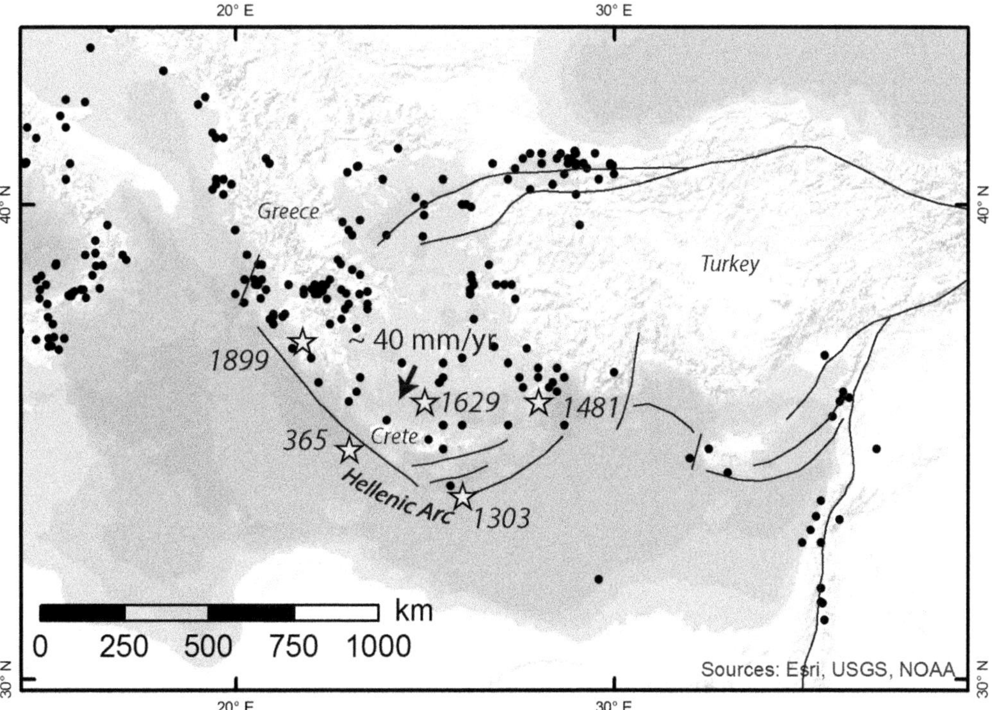

Fig. 1. Distribution of tsunami source events from 2000 BCE to 2004 CE in the Eastern Mediterranean (NGDC/WDS 2015); stars indicate the sources of the five largest tsunamigenic events according to England *et al.* (2015).

Papadimitriou & Karakostas 2008; Shaw *et al.* 2008; England *et al.* 2015), while others suggest that the plate interface ruptured (i.e. Ganas & Parsons 2009; Stiros 2010). Similar debates exist regarding the location and morphology of the causative faults for the other major plate boundary earthquakes (England *et al.* 2015). Therefore, new field data are required in order that models of tsunami propagation and fault structure can be tested and refined.

Methods

Sedimentology

In this study, >75 km of coastline along the southern edge of Crete (Fig. 2) was investigated for evidence of tsunami deposits. This area was targeted due to the small amount of vertical uplift (*c.* 0–2 m) felt in this region as a result of the 365 CE earthquake (Fig. 2). As this area did not experience significant uplift, it is more likely to record tsunami deposits from this event as well as possible tsunamis that predate and post-date this event. Furthermore, due to the generally steep and mountainous topography of the southern coast, there is less tourist development than elsewhere on the island. However, this rugged geomorphology also means there are few areas where fine-grained tsunami sediments could accumulate, and so the focus of this study was on potential tsunami boulder deposits and care had to be taken to rule out material from rock falls.

In particular, sediments from the littoral or foreshore that now occupy an onshore position were examined and measured (maximum length of the *a*-, *b*- and *c*-axes; dip of imbricated clasts). Here we use the extended Udden–Wentworth grain-size scale proposed by Blair & McPherson (1999) to describe the size of observed clasts, specifically the grades for 'boulders' and 'blocks' are most applicable to this study. Small boulders were measured using a tape measure, while larger boulders were measured using a TruPulse laser range finder. In addition, we used the TruPulse to survey the beach profiles and record the location of the boulders. Features characteristic of normal shallow-marine processes were recorded, if present (such as encrusting material, tilted wave-cut notches, tilted rockpools). Locations where boulders were clearly being transported down local slopes were not studied further.

Density of boulder lithologies were calculated in the laboratory using the water displacement method to calculate the volume, and an electronic microbalance was used to determine the mass of boulder samples collected from each location. These data were then combined with the measurement of the three major axes to determine the weight of the observed boulders. However, as this method approximates the boulders as cuboids, the method is likely to overestimate the volume and weight of many of the boulders measured (e.g. Spiske *et al.* 2008; Engel & May 2012).

Recent work by Hoffmeister *et al.* (2012) and Hoffman *et al.* (2013) have demonstrated (using a 3D terrestrial scanner) that this is, indeed, the case, and boulder volume can be overestimated by as much as 30–50%, resulting in an overestimate of calculated wave heights by 28–83%. In this study, at two locations, the boulders are composed of beachrock slabs that closely approximate cuboids due to their regular shape and, although some boulders possess more trapezoid and triangular shapes, we focused our measurements on the regular boulders to eliminate this bias. A similar approach was used by Lau *et al.* (2015), who studied beachrock boulder emplacement in Taiwan. Where the boulders are of variable shape and composition, we trialled a low-cost technique using a 'structure from motion' (SfM) approach, this required a photographic record of the boulder (normally at least 12 photographs are needed) and the AgiSoft Photoscan software to create detailed 3D models of several boulders from which volumes can be calculated and compared to the traditional calculations.

^{14}C dating

Encrusting marine material was found to be very rare at all locations studied: however, samples of material (Serpulidae, coralline algae) were collected from all sites for accelerator mass spectrometry (AMS) radiocarbon dating (Table 1). Samples were processed and analysed at the Centre for Climate, the Environment and Chronology at Queen's University Belfast. Conventional radiocarbon ages were calibrated using the MARINE13 curve in Calib 7.1 (Stuiver & Reimer 1993; Reimer *et al.* 2013) and a marine reservoir age of $\Delta R = 65 \pm 20$.

Extreme value analysis of wave data

The Mediterranean experiences a micro-tidal regime of approximately 0.1–0.3 m that can easily be exceeded by storm waves. However, the southern coast of Crete is sheltered from the worst of the winter storms from the north by the mass of the island. The offshore profile of this stretch of the coast is variable, with a steep offshore transition in the western part of the study area and a wide shallow offshore zone in the east (Fig. 3). Therefore, the western sites are more exposed to high-energy waves than the more sheltered eastern sites. Furthermore, all areas are additionally sheltered by the presence of the island of Gavdos to the SW. The difference in offshore topography would also result

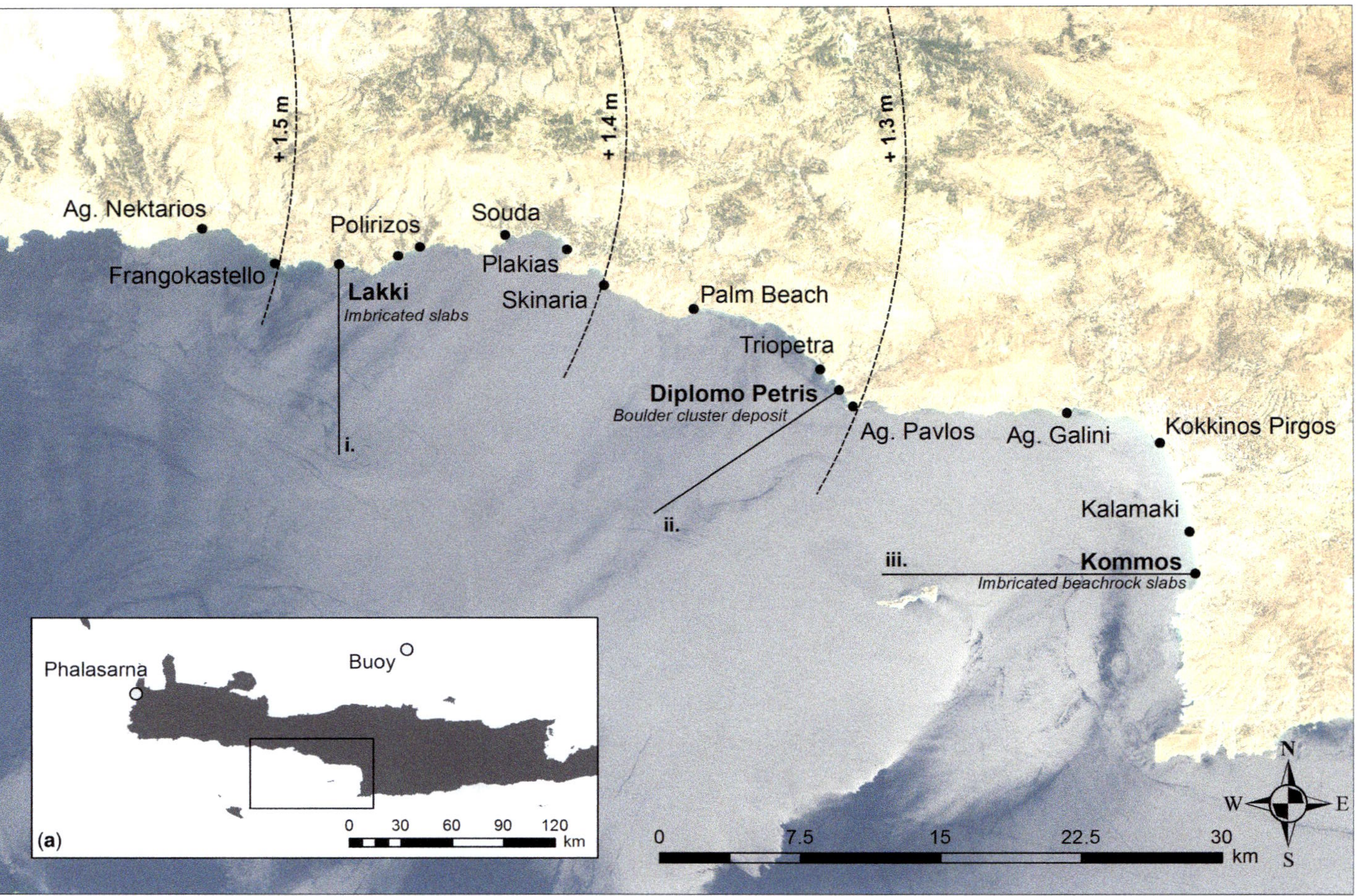

Fig. 2. Landsat image of the study area (NASA Landsat Program 2003) showing locations visited without tsunami deposits and locations discussed herein (bold), with a summary of deposits in italics below; note the abbreviation Ag. for Agios. The contours of co-seismic uplift of the 365 CE event as measured in the field during this work are also indicated by dashed lines. Extent of the image is shown in the inset (**a**), with the location of the buoy used to calculate modern storm wave heights indicated.

Table 1. *Radiocarbon analysis results*

Location	Sample ID	Laboratory ID	Material	^{14}C age	±	1σ	2σ	Median probability
Lakki	CR14-01	UBA-28172	Serpulid	2149	29	AD 243–350	AD 181–397	AD 293
Diplomo Petris	CR14-02	UBA-28173	Algal	2290	27	AD 73–168	AD 32–229	AD 122
Kommos	CR14-07	UBA-28174	Serpulid	Modern				
Kommos	CR14-09	UBA-28175	Serpulid	1282	36	AD 1152–1255	AD 1079–1278	AD 1192

in different amounts of shoaling as waves approach the coastline.

Published data suggest that waves >2.5 m rarely affect the coast, with the average height of northerly waves in winter being approximately 1 m (DHI 1971). However, in order to test these published data, we obtained annual wave data from the POSEIDON system (Table 2) from a buoy to the north of Crete (Fig. 2). Unfortunately, similar buoys are not present to the south of the island so this is the geographically closest system to the study area. To analyse the wave data return periods and to allow a comparison to be made with boulder kinematics requires the evaluation of wave height extremes: therefore, an extreme value analysis (EVA) was undertaken.

EVA is a statistical technique that is based on an understanding of the probability of a series of observed data, which can allow extrapolation to evaluate the unseen highs and lows (extremes) of the data. There are a variety of techniques available to undertake an EVA (see Coles 2001), with the

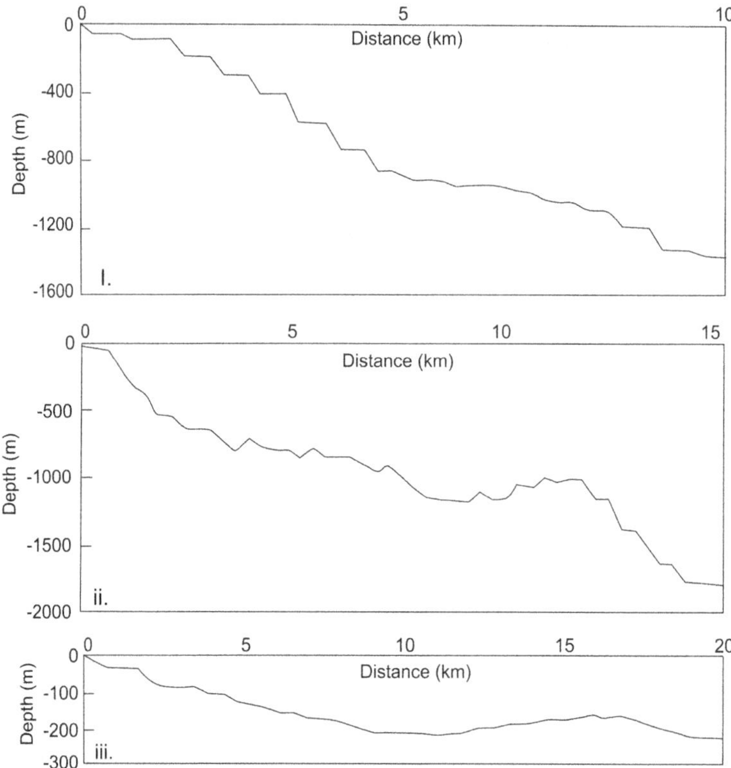

Fig. 3. Bathymetric profiles (i, Lakki; ii, Diplomo Petris; iii, Kommos) taken perpendicular to the coast for each significant boulder accumulation identified; the locations of these transects are shown in Figure 1. Data are from the EMODnet Portal for bathymetry (http://portal.emodnet-bathymetry.eu/mean-depth-full-coverage); resolution is nominally 0.125 arc-minute (*c.* 230 m)

Table 2. *Ten largest independent annual wave heights recorded for the period 2000–06*

Year	*R*1	*R*2	*R*3	*R*4	*R*5	*R*6	*R*7	*R*8	*R*9	*R*10
2000	3.87	2.99	2.99	2.98	2.89	2.87	2.86	2.85	2.83	2.80
2001	4.78	4.60	3.83	3.79	3.77	3.55	3.52	3.44	3.41	3.20
2002	4.65	4.61	4.43	4.17	4.15	4.07	4.04	3.83	3.53	3.05
2003	5.43	5.23	4.83	4.02	3.88	3.58	3.16	3.14	3.13	3.09
2004	4.97	4.44	4.18	4.09	4.04	3.96	3.47	3.47	3.33	3.31
2005	4.17	3.53	3.38	3.30	3.06	2.92	2.90	2.85	2.84	2.80
2006	3.90	3.65	3.62	3.43	3.17	3.07	3.01	2.92	2.91	2.78

These data are used as part of the *R* largest EVA of the data.

most common technique being the generalized extreme value (GEV) distribution. However, for a temporally short dataset, there are several drawbacks with the GEV in that it requires typically >20 years of data (Tawn & Vassie 1990; Coles 2001; Whitworth 2015). To combat the limitations of the GEV, the *r*-largest approach was developed by Smith (1986), so that temporally shorter datasets could be analysed. In an extension to the GEV distribution, the *r*-largest values for each year, $z = (z^{(1)}, \ldots, z^{(r)})$ where $r \geq 1$, are extracted from each year's data and ordered. The joint probability density function is then of the form:

$$f(z) = \exp\left[-\left\{1 + \xi\left(\frac{x^{(r)} - \mu}{\sigma}\right)\right\}^{-1/\xi}\right] \times \prod_{k=1}^{r} \frac{1}{\sigma}\left\{1 + \xi\left(\frac{x^{(k)} - \mu}{\sigma}\right)\right\}^{-(1/\xi)-1}. \quad (1)$$

The three parameters (μ, σ, ξ) correspond to the parameters of the GEV model (Butler *et al.* 2007).

To undertake the *r*-largest analysis, the 10 (*r*) largest observed maxima for each year of the dataset were compiled in an Excel spreadsheet and then converted into a text file. The text file was then imported into a software package called 'Extremes' (Gilleland & Katz 2006). The value of *r* (2–10) was then varied to identify the value of *r* that provided the best fit to the data by evaluating the various plots. The software was then interrogated to evaluate the precise return level and return period with the required confidence limits (Fig. 4).

Figure 4 shows that, for a 1 in 100 return period, the maximum wave height could be of the order of 5.5 m based on EVA (*R* largest) of 7 years' data (2000–06). We therefore consider that this is the maximum storm wave height that could affect the more sheltered southern coast.

Hydrodynamic equations

Hydrodynamic equations for storm and tsunamis were initially derived by Nott (1997, 2003*a*, *b*) to estimate the minimum wave height and velocity

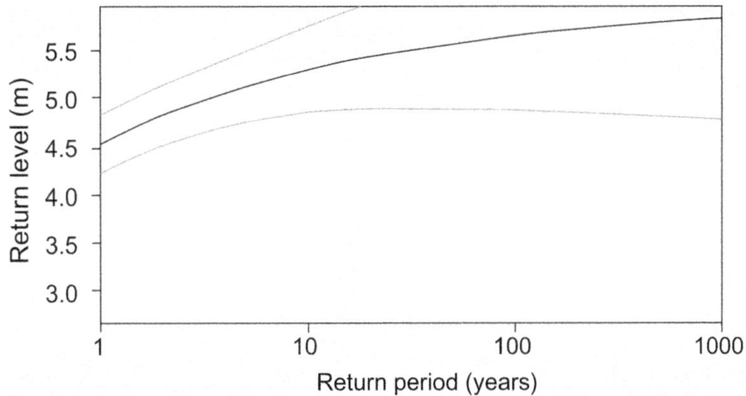

Fig. 4. Wave height return period plot based on a *R* largest extreme value analysis of the data presented in Table 1. The black line indicates the prediction and the grey lines show the 95% confidence limits.

needed to initiate boulder transport located in three environments; submerged, subaerial and joint-bounded blocks. These original equations have subsequently been improved to increase accuracy and include variables for the processes of dislodgement, emplacement and transport style (sliding, rolling, saltation), as well as the effect of slope at the pre-transport location (Noormets *et al.* 2004; Imamura *et al.* 2008; Pignatelli *et al.* 2009; Benner *et al.* 2010; Nandasena *et al.* 2011).

To calculate the flow velocity required to initiate clast transport, the equations of Nandasena *et al.* (2011) were used (Appendix A, equations A1–A4). As the transportation mechanisms and exact initial location (submerged, subaerial, joint bounded) of all boulders cannot be determined, a range of values were calculated for each boulder where appropriate (Table 3). In order to quantify the potential storm (H_s) or tsunami (H_t) wave height required to displace the boulders, the hydrodynamic approach of Benner *et al.* (2010) was used for a submerged pre-transport scenario and the equations of Pignatelli *et al.* (2009) were used for the joint-bounded boulder scenario (Appendix A, equations A5–A8): we also use the modified Nott equations proposed by Barbano *et al.* (2010) in comparison (Appendix A, equations A9–A14).

Observations

Field evidence from Lakki

Lakki (35.18° N, 24.27° E) is a small embayment approximately 350 m long near Frangokastello (Fig. 2). Here, the low cliffs are formed from Neogene cream-coloured bioclastic sandy limestone with unconformably overlying dark grey, coarse-grained litharenites from Quaternary alluvial fans (Pope *et al.* 2008). In the centre of the bay, a linear exposure of the bioclastic limestone extends approximately 20 m into the sea, with the upper surface 1.35 ± 0.5 m above present sea level dipping seawards at approximately 3° (Fig. 5). Wave-formed pools have developed in the upper surface of the rocky exposure. Several large tabular, coarse–very coarse boulders of litharenite are emplaced on top of one another on this exposure dipping seawards (174°–214°) with slight imbrication (Fig. 5). Other slabs are observed adjacent to the outcrop but it is unclear if the placement of these blocks has been altered by recent development of the area. Of note is boulder 'b', on the upper surface the boulder has a karstic pool, which shows evidence of reworking in the intertidal zone due to the base of the pool having been tilted from horizontal and then further erosion forming the present horizontal pool (Fig. 5). On the base of this boulder were a mass of encrusting serpulid worm tubes: sample CR14-01 was sampled from here giving a conventional radiocarbon age of 2149 ± 29 years BP, which equates to a 2σ calibrated date of 181–397 CE (Table 2). Boulder c is also interesting as it is broken into two parts, presumably upon placement in its current position, which may suggest transport through saltation rather than sliding or rolling.

The density of the litharenite was determined as 2.88 g cm^{-3}: thus, the largest boulder with a long axis >2.8 m (Table 4) has a mass of approximately 5.8 t. Hydrodynamic calculations (Benner *et al.* 2010) indicate that the minimum wave height needed to move this slab, assuming a subaerial origin (as indicated by the tilted pool), is 3.2 m for a storm and 0.15 m for a tsunami. By contrast, the equations of Pignatelli *et al.* (2009) for a joint-bounded block scenario suggest a minimum H_s = 8.4 m and H_t = 1.6 m. Flow velocity can also be calculated for submerged, subaerial block and joint-bounded transport (Nandasena *et al.* 2011), which suggests that minimum flow velocities of 0.1–4.5 m s^{-1} are required to initiate the transport of these blocks. Although this value is considerably higher if a joint-bounded block scenario is considered, in this case a flow velocity of >12 m s^{-1} would be required.

Field evidence from Diplomo Petris

South of the small village of Triopetra is a sheltered 1 km stretch of sandy beach (Diplomo Petris; 35.11° N, 24.555° E) where bedded beachrock is exposed along much of the beach, cropping out partly below and partly above present sea level, dipping shallowly seawards (≤10°) and is currently being eroded by wave action. To the north, the beach is backed by low cliffs of Upper Miocene marl and sandy limestones (Tortorici *et al.* 2012), with a discontinuous raised marine shoreline composed of cobbles at approximately 5 m absl. Heading southwards, the cliffs come closer to the beach and bedrock outcrops form a small headland. The bedrock is formed of steeply dipping and fractured dark grey to brown litharenite characteristic of the lower unit of the Cretan ophiolitic nappe, probably of Paleocene age (Tortorici *et al.* 2012). The boundary between the Palaeogene and Neogene rocks is poorly exposed but appears to be a thin sliver (<100 m thick) of tectonic melange. To the south of this bedrock exposure is the mouth of a river, which dominantly drains flysch and limestone bedrock.

At the headland (Table 4; Figs 2 & 6), a cluster of fine boulders to fine blocks is observed cemented into the beachrock at the shoreline. The clasts are of mixed lithologies, with limestone, quartzite and litharenite being the most common but boulders of sheared metamorphics, red sandstone and green

Table 3. *Location, size (maximum length of the three major axes) and weight of boulders measured along the Cretan coast, velocities required to initiate transport are derived from the equations of Nandasena* et al. *(2011), and the tsunami* (H_t) *and storm wave* (H_s) *heights required for their transport are found using the equations of Pignatelli* et al. *(2009), Barbano* et al. *(2010) and Benner* et al. *(2010). Note that boulders 8b, 9, 11, 16–19, 25, 26, 28–30 at Diplomo Petris were surveyed but long axes were not measured.*

Location	Boulder	Boulder axes (m)						Nandasena *et al.* (2011)				Pignatelli *et al.* (2009)		Benner *et al.* (2010)		Barbano *et al.* (2010)					
		A	B	C	Volume (m^3)	Mass (kg)	Mass (T)	SM Sliding ($m\ s^{-1}$)	SM Rolling ($m\ s^{-1}$)	SA ($m\ s^{-1}$)	JB ($m\ s^{-1}$)	H_t (m)	H_s (m)	SM – H_t (m)	SM – H_s (m)	SM H_t (m)	SM H_s (m)	SA H_t (m)	SA H_s (m)	JB H_t (m)	JB H_s (m)
Lakki	B	2.17	1.41	0.20	0.61	1761	1.8	1.94	0.45	0.68	20.41	0.81	4.22					0.91	3.66	1.05	4.22
	C	2.82	2.10	0.34	2.01	5793	5.8	2.53	0.79	0.71	23.27	1.38	7.18					1.50	5.99	1.79	7.17
	14a	0.79	0.80	0.40	0.25	727	0.7	2.69	2.26	1.96	12.32	1.62	8.45					0.74	2.97	2.11	8.44
Diplomo Petris	1	7.80	3.30	3.20	82.37	256988	257.0	8.42	9.97	9.49				2.20	8.79	2.79	9.14	2.35	9.42		
	2	3.00	2.10	1.80	11.34	35381	35.4	6.68	8.30	7.94				1.56	6.25	1.97	6.17	1.65	6.61		
	3	1.50	1.10	1.00	1.65	5148	5.1	4.85	5.89	5.62				0.78	3.11	0.98	3.15	0.83	3.31		
	4	2.40	1.65	1.50	5.94	18533	18.5	5.94	7.21	6.88				1.17	4.66	1.47	4.72	1.24	4.96		
	5	4.65	2.10	1.60	15.62	48747	48.7	6.63	8.61	8.28				1.72	6.88	2.16	6.45	1.80	7.19		
	6	5.00	5.00	4.20	105.00	327600	327.6	10.29	12.89	12.35				3.79	15.14	4.76	14.81	3.99	15.97		
	7	4.50	4.60	3.10	64.17	200210	200.2	9.74	13.17	12.73				4.12	16.49	5.18	14.59	4.25	17.00		
	8	2.70	4.30	2.80	32.51	101425	101.4	9.39	12.84	12.43				3.94	15.77	4.96	13.73	4.05	16.20		
	10	3.00	1.80	1.40	7.56	23587	23.6	6.15	7.92	7.61				1.45	5.81	1.82	5.49	1.52	6.08		
	12	2.10	2.60	2.10	11.47	35774	35.8	7.40	9.41	9.03				2.03	8.14	2.55	7.82	2.14	8.55		
	13	3.10	1.80	1.40	7.81	24373	24.4	6.15	7.92	7.61				1.45	5.81	1.82	5.49	1.52	6.08		
	14	2.80	1.80	1.20	6.05	18870	18.9	6.09	8.26	7.99				1.63	6.50	2.04	5.72	1.67	6.69		
	15	1.40	1.30	1.00	1.82	5678	5.7	5.22	6.75	6.49				1.06	4.23	1.33	3.98	1.11	4.42		
	20	6.00	4.50	3.50	94.50	294840	294.8	9.72	12.52	12.04				3.63	14.52	4.55	13.72	3.80	15.19		
	21	2.50	1.60	1.50	6.00	18720	18.7	5.85	7.03	6.70				1.10	4.40	1.39	4.51	1.17	4.69		
	22	2.50	1.50	1.10	4.13	12870	12.9	5.59	7.35	7.09				1.27	5.06	1.59	4.66	1.32	5.27		
	22a	2.20	1.50	1.40	4.62	14414	14.4	5.67	6.81	6.49				1.04	4.14	1.31	4.24	1.10	4.42		
	23	12.00	4.50	4.10	221.40	690768	690.8	9.80	11.90	11.36				3.17	12.70	4.00	12.86	3.38	13.51		
	24	1.20	1.60	1.00	1.92	5990	6.0	5.71	7.91	7.66				1.51	6.02	1.90	5.14	1.54	6.16		
	27	1.30	1.30	0.90	1.52	4746	4.7	5.19	6.95	6.71				1.14	4.57	1.44	4.10	1.18	4.73		
	31	3.60	3.40	3.40	41.62	129842	129.8		10.01	9.51				2.20	8.80	2.80	9.26	2.36	9.46		
	32	3.00	2.20	1.50	9.90	30888	30.9		9.08	8.77				1.96	7.82	2.46	6.96	2.02	8.08		
	33	1.70	1.30	1.20	2.65	8274	8.3		6.37	6.07				0.91	3.63	1.14	3.69	0.97	3.86		
Kommos	1	2.50	1.40	0.25	0.88	1893	1.9	3.00	4.66	4.57	5.06	0.35	2.84	0.63	2.09	1.14	1.47	0.65	2.60	0.80	3.19
	2	1.50	1.00	0.15	0.23	487	0.5	2.45	3.71	3.65	3.92	0.21	1.70	0.40	1.32	0.75	0.91	0.41	1.66	0.48	1.91
	3	1.80	0.90	0.18	0.29	631	0.6	2.45	3.87	3.78	4.29	0.26	2.04	0.43	1.44	0.76	1.04	0.45	1.79	0.57	2.30
	4	2.00	1.90	0.18	0.68	1480	1.5	3.03	4.24	4.20	4.29	0.26	2.04	0.52	1.72	1.05	1.14	0.55	2.18	0.57	2.30
	5	2.20	1.10	0.20	0.48	1047	1.0	2.67	4.16	4.07	4.52	0.28	2.27	0.50	1.66	0.90	1.18	0.52	2.07	0.64	2.55
	6	1.70	1.50	0.35	0.89	1931	1.9	3.25	5.19	5.06	5.98	0.50	3.97	0.77	2.60	1.32	1.93	0.80	3.21	1.12	4.47
	7	1.15	0.95	0.15	0.16	355	0.4	2.41	3.68	3.62	3.92	0.21	1.70	0.39	1.30	0.73	0.90	0.41	1.63	0.48	1.91
	8	1.40	1.20	0.20	0.34	727	0.7	2.74	4.22	4.14	4.52	0.28	2.27	0.51	1.71	0.95	1.19	0.53	2.14	0.64	2.55
	9	1.00	0.45	0.20	0.09	195	0.2	1.92	2.99	2.84	4.52	0.28	2.27	0.25	0.87	0.36	0.78	0.26	1.03	0.64	2.55
	10	2.15	1.50	0.20	0.65	1396	1.4	2.93	4.35	4.29	4.52	0.28	2.27	0.55	1.81	1.05	1.23	0.57	2.28	0.64	2.55
	11	3.15	0.90	0.42	1.19	2576	2.6	1.93	4.20	3.98	6.56	0.60	4.77	0.49	1.73	0.70	1.58	0.51	2.04	1.34	5.36
	12	1.40	1.05	0.19	0.28	604	0.6	2.60	4.05	3.97	4.41	0.27	2.16	0.47	1.58	0.86	1.12	0.49	1.97	0.61	2.43
	13	1.10	1.04	0.12	0.14	297	0.3	2.36	3.42	3.37	3.50	0.17	1.36	0.34	1.12	0.66	0.75	0.35	1.41	0.38	1.53
	14	1.54	0.95	0.18	0.26	570	0.6	4.20	3.91	3.83	4.29	0.26	2.04	0.44	1.47	0.79	1.05	0.46	1.83	0.57	2.30
	15	2.52	1.77	0.14	0.62	1351	1.4	2.79	3.78	3.75	3.79	0.20	1.59	0.42	1.37	0.84	0.90	0.43	1.73	0.45	1.79
	16	1.40	1.13	0.13	0.21	445	0.4	2.46	3.56	3.51	3.65	0.18	1.48	0.37	1.21	0.72	0.81	0.38	1.53	0.41	1.66
	17	1.66	1.50	0.23	0.57	1239	1.2	3.01	4.58	4.51	4.85	0.33	2.61	0.61	2.02	1.14	1.39	0.63	2.52	0.73	2.94
	18	2.05	1.62	0.20	0.66	1437	1.4	2.99	4.38	4.33	4.52	0.28	2.27	0.56	1.84	1.08	1.24	0.58	2.32	0.64	2.55

SM, submerged; SA, subaerial; JB, joint bounded.

Fig. 5. Photographs of the Lakki site: (**a**) sketch profile of the beach showing the location of boulders; (**b**) plan view and (**c**) side view of inset rock pools showing evidence of movement of the boulder 14 b (lens cap for scale); (**d**) photograph of the imbricated nature of the boulders 14 a and b (tape measure 1 m); and (**e**) general view of the location.

conglomerate are also present. While the litharenites, metamorphic rocks and other exotic boulders were probably derived from the adjacent cliffs, the limestone boulders are typical of the Pindos and Tripolitiza units that do not outcrop in this location nor were these lithologies observable in the melange, indicating that these large blocks are not derived from the adjacent cliff.

Table 4. *Comparison of boulder volumes derived by different techniques*

Boulder No.	Volume from 3D model (m^3)	Volume from field measurement (m^3)	Ratio of estimated volume to 3D volume (%)
4	2.48	5.94	239.5
10	2.95	7.56	256.3
32	10.72	9.9	92.4

The lithology of the boulders could be consistent with them having been transported downstream in the river system. However, several lines of evidence suggest that this is unlikely. First and, perhaps, most compellingly, is the presence of a wave cut or tidal notch on a number of the limestone boulders (Fig. 6c). Observation of the notch shows that it is remarkably similar to that preserved *in situ* elsewhere and uplifted by the 365 CE earthquake (i.e. Pirazzoli *et al.* 1992; Shaw *et al.* 2008), and is

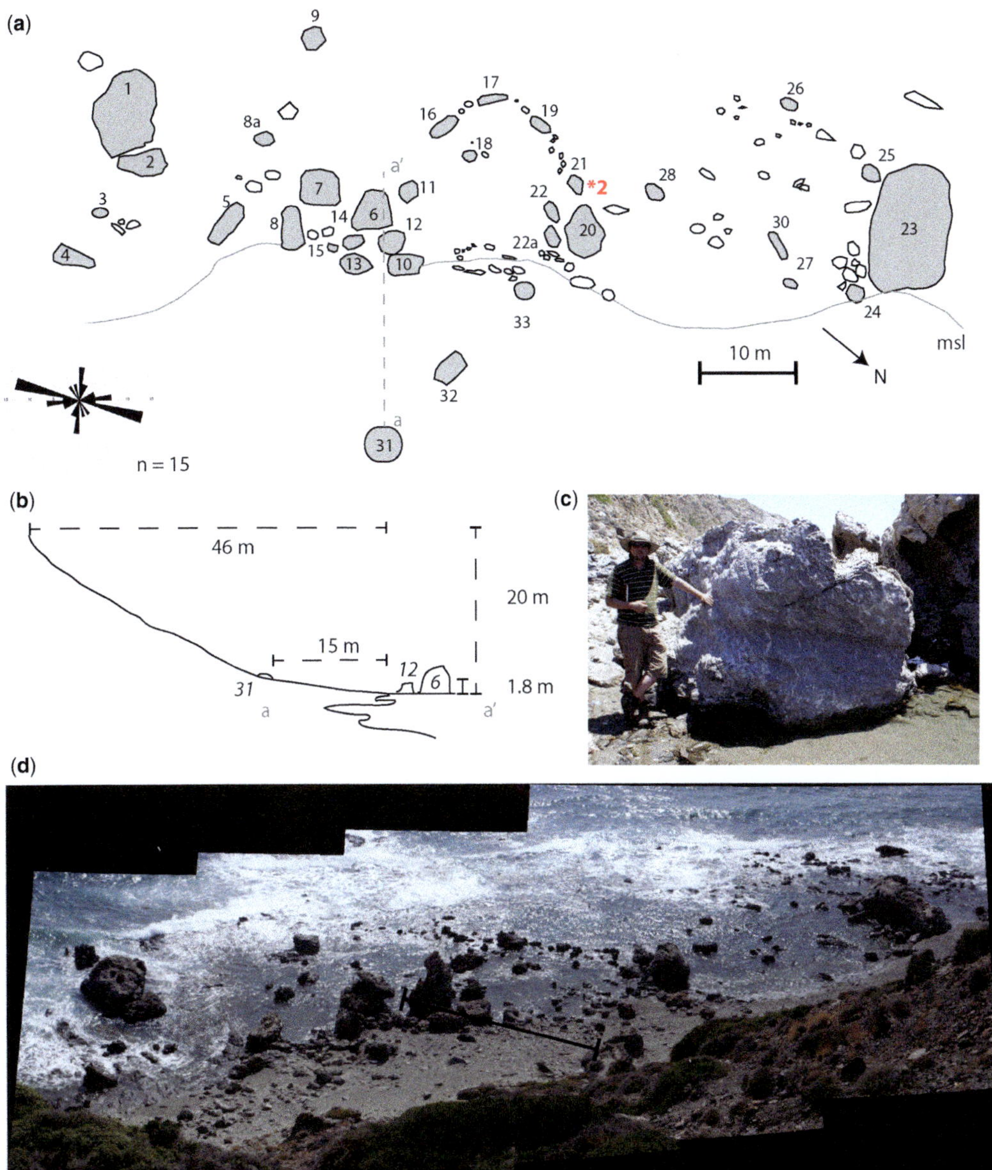

Fig. 6. Photographs and maps of Diplomo Petris: (**a**) map of the boulder deposit with the inset rose diagram showing the orientation of the long axes of the boulders; *2 indicates the location of radiocarbon sample CR14-02; (**b**) sketch profile of the beach section along the line a–a′; (**c**) photograph of boulder 12 showing a tilted wave-cut notch; and (**d**) panoramic view of the boulder deposits from the cliff top with the line of transect shown.

further evidence that the boulders are not the result of recent cliff collapse. Sample CR14-02 was taken from the bioencrustation associated with the notch observed on boulder 21 (Fig. 7b), this returned a conventional radiocarbon date of 2290 $\pm$ 27 years BP (calibrated 2σ age 23–229 CE: Table 2) consistent with this interpretation. The notch is observed to be orientated either horizontally or tilted at up to 12°, and it is observed to lie across a range of elevations from 0.4 to 0.9 m absl. By contrast, our measurements indicate that the 365 CE uplift in this location should result in a notch apex at approximately 1.3 m in elevation (Fig. 2). The presence of different notch heights, heights lower than

Fig. 7. (**a**) Photograph of Lithophagid-type borings observed at Diplomo Petris (pencil for scale); (**b**) algal bioencrustation sampled for ^{14}C dating at Diplomo Petris (tape measure for scale); (**c**) example of bivalve material embedded within the beachrock at Kommos indicating a marine origin for the litharenite (tip of hammer for scale); and (**d**) encrusting serpulid material on a beach rock from Kommos sampled for ^{14}C dating (lens cap for scale).

the predicted elevation for an uplifted notch and the different orientations of the notches strongly indicate that these boulders have been reworked within the intertidal zone after notch formation. Additionally, Lithophagid borings are visible on the side of many of the boulders and blocks above present sea level (Fig. 7a), further evidence of previous submergence. Secondly, the largest of the blocks weighs in excess of 690 t (longest axis 12 m, intermediate axis 4.5 m), with many blocks weighing >20 t (Table 4). Observations of the modern bedload of the river show that these boulders are significantly larger than the cobbles transported at the present time. Furthermore, the boulder cluster is not located at the mouth of the river but 100 m north, although some of the smaller clasts could and probably have been reworked locally by waves, longshore drift is expected to be directed to the SE as this is the direction of the prevailing winds (Cavaleri & Sclavo 2006). It is possible that the largest limestone boulders have been transported from Agios Pavlos, the closest coastal limestone outcrop with clear tidal notches, located approximately 1 km to the south (Fig. 2).

The limestone boulders are generally tabular to sub-spherical, sub-angular in shape and moderately weathered, with joint or fractures related to *in situ* deformation not later fracturing. Two samples of limestone were collected for density analysis and give an average density of 3.12 g cm^{-3}, consistent with the recrystallized nature of the limestone. Owing to the position in the intertidal zone, some boulders were inaccessible due to excessive water depth and/or the presence of a deep pool surrounding the boulders. As a result, the maximum axes of 23 boulders were measured but the location of 12 other boulders with an intermediate axis >0.5 m were also recorded (Fig. 6). Fine boulders and smaller were not measured as it is likely that these can be reworked during winter storms. In addition, the orientation of the long axis of the boulders was recorded.

The largest block (No. 23: Fig. 6) weighs approximately 690 t, there are another five boulders and fine blocks with a mass exceeding 100 t (Table 3). Block No. 23 would require a storm wave approximately 15 m high to initiate transport and a tsunami wave <4.5 m high. The presence of the raised tidal notch, however, strongly suggests a subaerial origin for the blocks and boulders, resulting in slightly lower predicted wave heights ($H_s = 12.7$ m; $H_t = 3.2$ m). Current velocity calculations indicate that, for a subaerial origin, a current of velocity of 12 m s^{-1} is required to initiate boulder transport. Boulder long axes are generally orientated approximately east–west but there is a subset of boulders that are orientated north–south (Fig. 6a).

Tsunami field evidence from Kommos

Kommos (35.01° N, 24.76° E) on the southern edge of the Mesaras Plain is at the end of a long sandy beach running for many kilometres (Fig. 8). Uplift from the 365 CE earthquake is negligible this far to the east (Pirazzoli *et al.* 1992) and the ruins of the Minoan-aged port of Kommos are located behind the recent dunes. At two locations along this stretch of beach, dislocated and imbricated medium- to coarse-grained beachrock boulders (Fig. 7c), with a slab morphology, are present approximately 20 m inland and 1.5–2 m absl. *In situ* beachrock is exposed, as at Diplomo Petris, above and below present sea level, dipping shallowly seawards.

At the first location, the beachrock slabs dipped to the west (seawards: Fig. 9a), have a tabular shape with the *a*- and *b*-axes typically 1–3 m long and the *c*-axis ≤0.2 m (Table 4), and were partly covered by dune sands. Unfortunately, when we returned to this location to undertake further field data collection, we discovered that the blocks had been bulldozed from their previous locations in order to develop the beach amenities – this prevented further meaningful data collection.

At the second location, 500 m to the north, the beachrock slabs are very similar in composition, size and morphology, although the dip is more variable (Fig. 9b) which is probably due to their location at the foot of a small cliff of Messinian cream-coloured marl (ten Veen & Kleinspehn 2003). The bedrock marls are lithologically distinct to the beachrock forming the displaced blocks, and although boulders of marl are present at the base of the cliff these were not subsequently measured.

Several beachrock slabs at Kommos were observed to have encrusting Serpulid worms tubes on the surface, which were sampled for radiocarbon dating (Fig. 7d). Two samples were submitted for dating: one returned a modern age, while the other gave a calibrated age of 1152–1255 CE (Table 1). Given the location of the boulders above high tide, this modern age implies that this boulder has been recently transported onshore.

The size and shape of the beachrock boulders is fairly tightly grouped, with long axes typically in the range of 2.5–1 m and a thickness of approximately 0.2 m. The beachrock is porous, so a wet density was calculated at 2.16 g cm^{-3} (dry density = 2.0 g cm^{-3}), resulting in the majority of the larger slabs having a mass in the order of 1–2.5 t. Given that beachrock can be observed *in situ* in the intertidal zone at the present day, it is probable that these blocks have been reworked from a joint-bounded location. To emplace these boulders onto the shore, a maximum H_s of 5 m and a H_t of 1.5 m is needed with a flow velocity of ≤6.5 m s^{-1} (Table 4).

Photographic analysis of blocks

Recent work by Hoffmeister *et al.* (2012) and Hoffman *et al.* (2013) demonstrated that approximating the irregularly-shaped boulders as cuboids often results in the overestimation of the boulder mass and, therefore, the results of the hydrodynamic equations. Hoffman *et al.* (2013) measured three boulders in Oman, and found that the high-precision LiDAR scans resulted in volumes approximately 30–40% lower than the volume derived by multiplying the three axes of the boulders. Using a similar method, Hoffmeister *et al.* (2012) investigated sites in western Greece, demonstrating that LiDAR scans resulted in volume differences of up to 70%. In each case, this would indicate that the results from the hydrodynamic equations would overestimate the tsunami wave height and velocities needed to move the tsunami blocks and boulders.

At Lakki and Kommos, the boulders investigated are tabular in shape and, therefore, are close to the ideal cuboid. Therefore, the simple mass approximation should be accurate for calculating the volume and resultant hydrological parameters. However, at Diplomo Petris, some of the boulders are highly irregular in shape. In order to assess the accuracy of our measurements, we constructed 3D models using a structure from motion approach (SfM) of three boulders that had 360° accessibility (Fig. 10). The process is straightforward in that a series of scaled photographs (>12) are taken around the boulder in question. We then used the Agisoft Photoscan software package to generate a 3D model of each boulder and calculate a volume to compare with the volume calculated from field measurements of the major axes (Table 4).

Our photogrammetric analysis indicates that the volumes of boulders 4 and 10 are significantly overestimated (Table 4) if the traditional method is applied, supporting the previous analyses of Hoffmeister *et al.* (2012) and Hoffman *et al.* (2013). However, the volume of boulder 32 is underestimated using the traditional method compared to the 3D photogrammetric analysis. This result reflects that boulders 4 and 10 diverge from an ideal cuboid by a significant degree, whereas boulder 32 is a more regular shape, as are the majority of the largest boulders surveyed. Therefore, although we

Fig. 8. Photographs and maps of Kommos: **(a)** sketch map of a boulder deposit located at the foot of a small cliff, with measured boulders indicated in grey with a number; **(b)** photograph of the boulders shown in (a); and **(c)** photograph of the boulders further to the south prior to redevelopment.

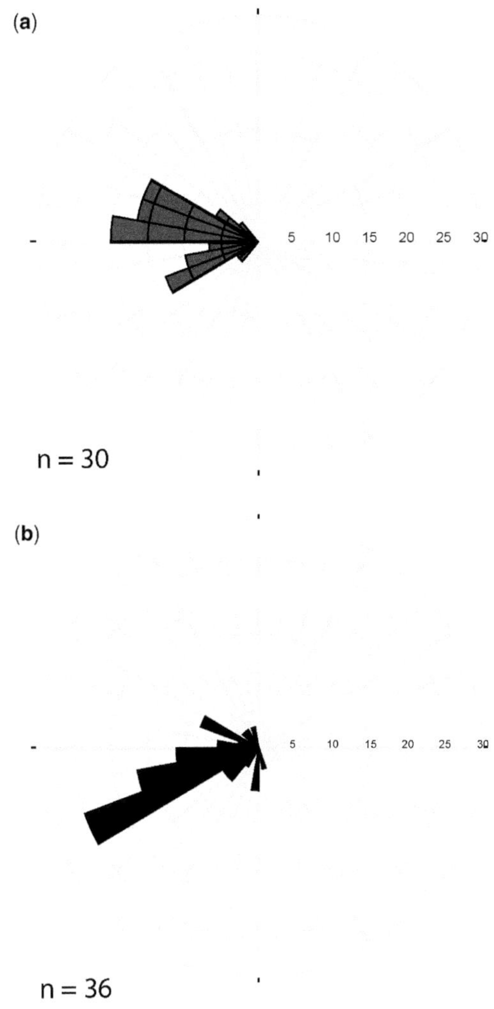

Fig. 9. Rose diagrams showing the dip directions of beachrock slabs measured at the two locations at Kommos (Fig. 2). (**a**) Measurements taken at UTM zone 35° S 0295620/3876465 (Fig. 8c). (**b**) Measurements taken at UTM zone 35° S 0295639/3876960 (Fig. 8b).

recognize that the calculated values for some of the small boulders will be significantly overestimated, this method indicates that, for the more cuboid boulders, measuring the major axes can result in reliable data. This photogrammetric method also is a low-cost option for the assessment of boulder volume, in comparison to LiDAR-based systems, and so offers a high potential for future studies.

Discussion: storm or tsunami deposit?

By considering the field observations and the wave heights estimated from the hydrodynamic models, it is now possible to make an assessment of which, if any, of the boulder clusters identified may have been deposited by tsunamis. The evidence is most ambiguous at Kommos. The imbrication direction combined with a near-shore origin for these boulders is strongly indicative of emplacement by extreme waves. Yet imbrication has been identified in both storm and tsunami boulder deposits, and therefore is not discriminatory (Paris *et al.* 2009; Etienne & Paris 2010; Goto *et al.* 2012). Given that beachrock is exposed at sea level, it is likely that an event not only transported the slabs to the current position but also plucked them from an *in situ* location. Therefore, the joint-bounded scenario was used for the calculation of extreme waves. Many of the boulders only weigh approximately 1 t, thus storm waves of $\leq$5.5 m would be needed to move the largest from a subaerial position when applying the equations of Barbano *et al.* (2010). By contrast, a tsunami would need to be only 1.3 m high to initiate transport. These values are reduced by approximately 10% when the equations of Pignatelli *et al.* (2009) are used. Therefore, transport and deposition by a large storm cannot be ruled out at this location as, although 5 m high storm waves are rare, they are possible in the Mediterranean-based upon EVA. In addition, velocities for the initiation of transportation fall in the range of 3–6 m s^{-1} depending on the equation used. These values are well within the tolerance for a large storm, where wave velocities of up to 8 m s^{-1} in coastal settings have been recorded (Dally 2005; EurOtop 2007). Furthermore, the boulders form a beach-parallel ridge rather than a distributed field, more characteristic of storms than tsunamis. This interpretation of a storm deposition is supported by the dating that returned a modern date for one of the samples, suggesting recent removal from the intertidal zone. Although, the other sample gives a much older date of 1192 CE, which would not be inconsistent with the 1303 CE tsunami (Guidoboni & Comastri 1997), the evidence is not compelling. Therefore, all things considered, the boulder deposit observed at Kommos cannot be confidently ascribed to a tsunami as it equally could be the result of storm deposition.

Further north at Lakki, a small number of blocks were observed lying on top of a raised platform. This part of the coast was uplifted by 1.7–2.0 m in 365 CE, greater than the height of the surface upon which the slabs rest, indicating that the platform was below sea level prior to that time. Again, these slabs show evidence for reworking in the intertidal zone due to the presence of a tilted tidal pool, suggesting reworking from an originally subaerial location or possibly a subaerial joint-bounded position. For the blocks to be transported, a storm wave 6.0–8.5 m high would be required.

Fig. 10. Photogrammetric 3D models of boulders 4, 10 and 32 from Diplomo Petris (Fig. 6a).

Current velocities for a subaerial origin are modest ($\leq 5\ \mathrm{m\ s^{-1}}$) but much higher if a joint-bounded origin is considered ($23\ \mathrm{m\ s^{-1}}$). These values are at the extreme for that expected within the Mediterranean, especially as the location is somewhat sheltered within a small bay and exceed values derived from the EVA. Therefore, on balance, it is slightly more likely that these blocks may have been transported by a tsunami wave approximately 2 m high but the sedimentological evidence is equivocal. However, the calibrated radiocarbon age of 293 CE is consistent with reworking by the 365 CE tsunami.

The final location of Diplomo Petris has the most convincing evidence for block transport by tsunamis. Wave height calculations indicate that storm waves of >14–16 m and current velocities of >13 m s^{-1} would be needed to move these boulders considering a submerged position. Furthermore, as indicated by bio-erosional notches part way up the boulders, as subaerial origin is more likely for the boulder origin this increases the required storm wave heights to 17 m. These values far exceed the possible storm wave height along this part of the Cretan coast.

However, a tsunami wave 4–5 m high could transport blocks of this size and morphology, and the calculated velocities are consistent with observations of recent tsunamis, which indicate tsunami wave velocities of 10–20 m s^{-1} (Titov & Synolakis 1997; Phuwieng *et al.* 2008). The presence of tidal notches (dated to 365 CE by Shaw *et al.* 2008 and our sample) also indicates that these boulders were present at sea level prior to the 365 CE uplift and were subsequently dislocated from their original position, most likely by this catastrophic event. A tsunami origin is also supported by the scattered and distributed debris field with no trend in clast-size distribution.

Measurements of other tsunami boulder deposits indicate that the long axis of the boulders is typically orientated perpendicular to the flow direction (i.e. Goto *et al.* 2007; Etienne *et al.* 2011; Richmond *et al.* 2011*a*). In this case that would suggest flow from the south, as boulders are orientated approximately east–west (Fig. 6a), consistent with the closest exposure of Mesozoic limestones (forming the largest of the boulders) at Agios Pavlos. This direction is similar to the NE direction of flow generally inferred from tsunami models of the event (i.e. Shaw *et al.* 2008; England *et al.* 2015). Furthermore, the calculated wave heights are comparable with predicted values from tsunami models of this event (Shaw *et al.* 2008; England *et al.* 2015). Therefore, on balance, we interpret these boulders to have been emplaced by a significant tsunami, probably the 365 CE tsunami.

Conclusions

At three locations (Lakki, Kommos and Diplomo Petris) on the coast of Crete, bolder deposits were investigated in order to determine the mechanism for emplacement. At Lakki and Kommos, the boulders are tabular in shape and formed of local alluvial fan material and beachrock, respectively. The size of the slabs at Kommos indicates that these could have been deposited by either storm or tsunami waves, whereas at Lakki it is probably more likely that a tsunami resulted in the emplacement of the observed blocks but a storm cannot be ruled out. The most compelling site for having been the result of tsunami deposition is those boulders seen at Diplomo Petris. The large size, geomorphic characteristics and orientation are all strongly indicative of tsunami transport, most likely by the 365 CE tsunami. Radiocarbon dating from Lakki is also consistent with the 365 CE tsunami. Although, fine-grained deposits have been identified for this event (i.e. Scheffers & Scheffers 2007), these are the first coarse-grained material identified on Crete resulting from this large event. Given that large-magnitude earthquakes and tsunamis are infrequent in the Mediterranean Basin, these observations provide important data to inform and test models of tsunami occurrence and risk in this populous region.

This project was supported by a fieldwork grant from the Geological Society of London. Thanks to the Institute of Oceanography and Y. Antoniou, Hellenic Centre for Marine Research, for providing the wave data from the POSEIDON Operational Oceanography System (http://www.poseidon.hcmr.gr). The Landsat data were obtained through the Global Land Cover Facility, University of Maryland (http://glcf.umd.edu/). We also thank Ellie Scourse and two anonymous reviewers for useful comments that have improved the manuscript

Appendix A

Boulder transport equations

For all equations the variables are as follows:

H is the wave height at breaking point; H_s denotes storm waves and H_t denotes tsunami waves;

ρ_s = density of the boulder (Table 1);
ρ_w = density of water (1.02 g cm^{-3});
a = A-axis of the boulder;
b = B-axis of the boulder;
c = C-axis of the boulder;
C_D = coefficient of drag (1.2) (cf. Benner *et al.* 2010);
C_L = coefficient of lift (0.178) (cf. Noormets *et al.* 2004);
C_m = coefficient of mass (2);
μ = coefficient of friction (0.65);
$\ddot{u}$ = instantaneous flow acceleration (1 m s^{-2});
u = velocity of a wave (m s^{-1});
g = acceleration due to gravity (9.81 m s^{-2});
Θ = angle of the beach.

Equations of Nandasena et al. *(2011)*

- Submerged boulder – sliding:

$$u^2 \geq \frac{2(\rho_s/\rho_w - 1)gc(\mu\cos\theta + \sin\theta)}{C_d(c/b) + \mu_s C_L}. \quad \text{(A1)}$$

- Submerged boulder – rolling:

$$u^2 \geq \frac{2(\rho_s/\rho_w - 1)gc(\cos\theta + (c/b)\sin\theta)}{C_d(c^2/b^2) + C_L}. \quad (A2)$$

- Subaerial boulder:

$$u^2 \geq \frac{2(\rho_b/\rho_w - 1)g(bc^2\sin\theta + b^2c\cos\theta) - 2C_m(bc^2)\dot{u}}{C_d c^2 + C_L b^2}. \quad (A3)$$

- Joint-bounded boulder:

$$u^2 \geq \frac{2(\rho_s/\rho_w - 1)gc(\cos\theta + \mu_s\sin\theta)}{C_L}. \quad (A4)$$

Equations of Benner et al. *(2010)*

- Submerged boulder:

$$H_t \geq \frac{0.5bc\,[b(\rho_s - \rho_w)/\rho_w - (\rho_s\,C_m\ddot{u}c)/(\rho_w g)]}{C_D c^2 + C_L b^2}. \quad (A5)$$

$$H_s \geq \frac{2bc\,[b(\rho_s - \rho_w)/\rho_w - (\rho_s C_m\ddot{u}c)/(\rho_w g)]}{C_D c^2 + C_L b^2}. \quad (A6)$$

Equations of Pignatelli et al. *(2009)*

- Joint-bounded boulder:

$$H_t \geq \frac{0.25c\,[(\rho_s - \rho_w)/\rho_w]}{C_L}. \quad (A7)$$

$$H_s \geq \frac{2c[(\rho_s - \rho_w)/\rho_w]}{C_L}. \quad (A8)$$

Equations of Barbano et al. *(2010)*

- Submerged boulder:

$$H_t \geq \frac{[(\rho_s - \rho_w/\rho_w)\,b^2c]}{2(C_D c^2 + C_L b^2)}. \quad (A9)$$

$$H_s \geq \frac{[(\rho_s - \rho_w/\rho_w)b^2c]}{0.5(C_D c^2 + C_L b^2)}. \quad (A10)$$

- Subaerial boulder:

$$H_t \geq \frac{[0.5(\rho_s - \rho_w/\rho_w)b^2cg - C_m bc^2\ddot{u}]}{g(C_D c^2 + C_L b^2)}. \quad (A11)$$

$$H_s \geq \frac{[2(\rho_s - \rho_w/\rho_w)b^2cg - 4C_m bc^2\ddot{u}]}{g(C_D c^2 + C_L b^2)}. \quad (A12)$$

- Joint-bounded boulder:

$$H_t \geq \frac{0.5c\,[(\rho_s - \rho_w)/\rho_w]}{C_L}. \quad (A13)$$

$$H_s \geq \frac{2c\,[(\rho_s - \rho_w)/\rho_w]}{C_L} \quad (A14)$$

References

Ambraseys, N. & Synolakis, C. 2010. Tsunami catalogs for the Eastern Mediterranean, revisited. *Journal of Earthquake Engineering*, **14**, 309–330.

Bahlburg, H. & Spiske, M. 2012. Sedimentology of tsunami inflow and backflow deposits: key differences revealed in a modern example. *Sedimentology*, **59**, 1063–1086.

Barbano, M.S., Pirrotta, C. & Gerardi, F. 2010. Large boulders along the south-eastern Ionian coast of Sicily: storm or tsunami deposits? *Marine Geology*, **275**, 140–154.

Benner, R., Browne, T., Brückner, H., Kelletat, D. & Scheffers, A. 2010. Boulder transport by waves: progress in physical modelling. *Zeitschrift für Geomorphologie*, **54**, 127–146.

Blair, T.C. & McPherson, J.G. 1999. Grain-size and textural classification of coarse sedimentary particles. *Journal of Sedimentary Research*, **69**, 6–19.

Bruins, H.J., MacGillivray, J.A. et al. 2008. Geoarchaeological tsunami deposits at Palaikastro (Crete) and the Late Minoan IA eruption of Santorini. *Journal of Archaeological Science*, **35**, 191–212.

Bryant, E.A. & Nott, J. 2001. Geological indicators of large tsunami in Australia. *Natural Hazards*, **24**, 231–249.

Butler, A., Heffernan, J.E., Tawn, J.A., Flather, R.A. & Horsburgh, K.J. 2007. Extreme value analysis of decadal variations in storm surge elevations. *Journal of Marine Systems*, **67**, 189–200.

Cavaleri, L. & Sclavo, M. 2006. The calibration of wind and wave model data in the Mediterranean Sea. *Coastal Engineering*, **53**, 613–627.

Coles, S. 2001. *An Introduction to Statistical Modelling of Extreme Values*. Springer, London.

Dally, W.R. 2005. Surf zone processes. *In*: Schwartz, M.L. (ed.) *Encyclopedia of Coastal Science*. Encyclopedia of Earth Sciences Series. Springer, Dordrecht, 929–935.

Dawson, A.G. 1994. Geomorphological effects of tsunami run-up and backwash. *Geomorphology*, **10**, 83–94.

Dawson, A.G. & Shi, S. 2000. Tsunami deposits. *Pure and Applied Geophysics*, **157**, 875–897.

DHI. 1971. *Mittelmeer-Handbuch. 4. Teil: Griechenland und Kreta*. 5. Auflage. Deutsches Hydrographisches Institut, Hamburg.

Dominey-Howes, D. 2004. A re-analysis of the Late Bronze Age eruption and tsunami of Santorini, Greece, and the implications for the volcano-tsunami hazard. *Journal of Volcanism and Geothermal Research*, **130**, 107–132.

Dominey-Howes, D.T.M. 1998. Assessment of tsunami magnitude and implications for urban hazard planning in Greece. *Disaster Prevention and Management*, **7**, 176–182.

Dominey-Howes, D.T.M., Papadopoulos, G.A. & Dawson, A.G. 2000. Geological and historical investigation of the 1650 Mt. Columbo (Thera Island) eruption and tsunami, Aegean Sea, Greece. *Natural Hazards*, **21**, 83–96.

Engel, M. & May, S.M. 2012. Bonaire's boulder fields revisited: evidence for Holocene tsunami impact on the Leeward Antilles. *Quaternary Science Reviews*, **54**, 126–141.

England, P., Howell, A., Jackson, J. & Synolakis, C. 2015. Palaeotsunamis and tsunami hazards in the Eastern Mediterranean. *Philosophical Transactions of the Royal Society A*, **373**, 20140374.

Etienne, S. & Paris, R. 2010. Boulder accumulations related to storms on the Reykjanes Peninsula, Iceland. *Geomorphology*, **114**, 55–70.

Etienne, S., Buckley, M. *et al.* 2011. The use of boulders for characterising past tsunamis: Lessons from the 2004 Indian Ocean and 2009 South Pacific tsunamis. *Earth-Science Reviews*, **107**, 76–90.

EurOtop. 2007. *Wave Overtopping of Sea Defences and Related Structures: Assessment Manual.* Die Küste, **73**, www.overtoppingmanual.com

Ganas, A. & Parsons, T. 2009. Three-dimensional model of Hellenic Arc deformation and origin of the Cretan uplift. *Journal of Geophysical. Research: Solid Earth*, **114**, B06404.

Gilleland, E. & Katz, R.W. 2006. Analyzing seasonal to interannual extreme weather and climate variability with the extremes toolkit (extRemes). *Paper presented at the 18th Conference on Climate Variability and Change, 86th American Meteorological Society (AMS) Annual Meeting*, 29 January–2 February 2006, Atlanta, Georgia, USA.

Goff, J.R., Dudley, W.C., de Maintenon, M.J., Cain, C. & Coney, J.P. 2006. The largest local tsunami in 20th century Hawaii. *Marine Geology*, **226**, 65–79.

Goto, K., Chavanich, S.A. *et al.* 2007. Distribution, origin and transport process of boulders deposited by the 2004 Indian Ocean tsunami at Pakarang Cape, Thailand. *Sedimentary Geology*, **202**, 821–837.

Goto, K., Shinozaki, T., Minoura, K., Okada, K., Sugawara, D. & Imamura, F. 2010. Distribution of boulders at Miyara Bay of Ishigaki Island, Japan: a flow characteristic indicator of tsunami and storm waves. *Island Arc*, **19**, 412–426.

Goto, K., Sugawara, D., Ikema, S. & Miyagi, T. 2012. Sedimentary processes associated with sand and boulder deposits formed by the 2011 Tohoku-oki tsunami at Sabusawa Island, Japan. *Sedimentary Geology*, **282**, 1–13.

Guidoboni, E. & Comastri, A. 1997. The large earthquake of 8 August 1303 in Crete: seismic scenario and tsunami in the Mediterranean area. *Journal of Seismology*, **1**, 55–72.

Guidoboni, E., Comastri, A. & Traina, G. 1994. *Catalogue of Ancient Earthquakes in the Mediterranean Area up to the 10th Century, Volume 1.* Instituo Nazionale di Geofisica, Rome.

Hoffman, G., Reicherter, K., Wiatr, T., Grützner, C. & Rausch, T. 2013. Block and boulder accumulations along the coastline between Fins and Sur (Sultanate of Oman): tsunamigenic remains? *Natural Hazards*, **65**, 851–873.

Hoffmeister, D., Tilly, N. *et al.* 2012. Terrestrial laser scanning for coastal geomorphologic research in Western Greece. *International Archives of the Photogrammetry, Remote Sensing and Spatial Information Sciences*, **39**, 511–516.

Imamura, F., Goto, K. & Ohkubo, S. 2008. A numerical model for the transport of a boulder by tsunami. *Journal of Geophysical Research*, **113**, C01008, https://doi.org/10.1029/2007JC004170

Kelletat, D. & Schellmann, G. 2002. Tsunamis on Cyprus – field evidences and ^{14}C dating results. *Zeitschrift für Geomorpholgie, NF*, **46**, 19–34.

Kelletat, D., Scheffers, A. & Scheffers, S. 2004. Holocene tsunami deposits on the Bahaman Islands of Long Island and Eleuthera. *Zeitschrift für Geomorpholgie*, **48**, 519–540.

Kennedy, D.M., Tannock, K.L., Crozier, M.J. & Rieser, U. 2007. Boulders of MIS 5 age deposited by a tsunami on the coast of Otago, New Zealand. *Sedimentary Geology*, **200**, 222–231.

Kortekaas, S. & Dawson, A.G. 2007. Distinguishing tsunami and storm deposits: an example from Martinhal, SW Portugal. *Sedimentary Geology*, **200**, 208–221.

Lau, A.A., Terry, J.P., Switzer, A.D. & Pile, J. 2015. Advantages of beachrock slabs for interpreting high-energy wave transport: Evidence from Ludao Island in south-eastern Taiwan. *Geomorphology*, **228**, 263–274.

Maouche, S., Morhange, C. & Meghraoui, M. 2009. Large boulder accumulation on the Algerian coast evidence tsunami events in the western Mediterranean. *Marine Geology*, **262**, 96–104.

Mastronuzzi, G. & Sansò, P. 2004. Large boulder accumulations by extreme waves along the Adriatic coast of southern Apulia (Italy). *Quaternary International*, **120**, 173–184.

May, S.M., Vött, A., Bruckner, H. & Smedile, A. 2012. The Gyra washover fan in the Lefkada Lagoon, NW Greece – possible evidence of the 365 AD Crete earthquake and tsunami. *Earth, Planets and Space*, **64**, 859–874.

May, S.M., Engel, M., Brill, D., Cuadra, C., Lagmay, A.M.F., Santiago, J., Suarez, J.K., Reyes, M. & Brückner, H. 2015. Block and boulder transport in Eastern Samar (Philippines) during Supertyphoon Haiyan. *Earth Surface Dynamics Discussions*, **3**, 739–771.

Morton, R.A., Gelfenbaum, G. & Jaffe, B.E. 2007. Physical criteria for distinguishing sandy tsunami and storm deposits using modern examples. *Sedimentary Geology*, **200**, 184–207.

Morton, R.A., Richmond, B.M., Jaffe, B.E. & Gelfenbaum, G. 2008. Coarse-clast ridge complexes of the Caribbean: a preliminary basis for distinguishing tsunami and storm-wave origins. *Journal of Sedimentary Research*, **78**, 624–637.

Mottershead, D., Bray, M., Soar, P. & Farres, P.J. 2014. Extreme wave events in the central Mediterranean: Geomorphic evidence of tsunami on the Maltese Islands. *Zeitschrift für Geomorphologie*, **58**, 385–411.

Mottershead, D.N., Bray, M.J., Soar, P.J. & Farres, P.J. 2015. Characterisation of erosional features associated with tsunami terrains on rocky coasts of the Maltese islands. *Earth Surface Processes and Landforms*, **40**, 2093–2111.

Nandasena, N.A.K., Paris, R. & Tanaka, N. 2011. Reassessment of hydrodynamic equations: Minimum flow velocity to initiate boulder transport by high energy events (storms, tsunamis). *Marine Geology*, **281**, 70–84.

Nandasena, N.A.K., Tanaka, N., Sasaki, Y. & Osada, M. 2013. Boulder transport by the 2011 Great East Japan tsunami: Comprehensive field observations and whither model predictions? *Marine Geology*, **346**, 292–309.

NASA Landsat Program. 2003. Landsat ETM+ scene p182r036_7dt20000630_z34, SLC-Off. *United States Geological Survey*, 26 October 2003.

National Geophysical Data Center/World Data Service (NGDC/WDS). 2015. *Global Historical Tsunami Database*. National Geophysical Data Center, NOAA, Boulder, CO, USA, https://doi.org/10.7289/V5PN93H7 [accessed on 23 February 2015].

Nocquet, J.-M. 2012. Present-day kinematics of the Mediterranean: a comprehensive overview of GPS results. *Tectonophysics*, **579**, 220–242.

Noormets, R., Crook, K.A.W. & Felton, E.A. 2004. Sedimentology of rocky shorelines: 3. Hydrodynamics of megaclast emplacement and transport on a shore platform, Oahu, Hawaii. *Sedimentary Geology*, **172**, 41–65.

Nott, J. 1997. Extremely high-energy wave deposits inside the Great Barrier Reef, Australia: determining the cause – tsunami or tropical cyclone. *Marine Geology*, **141**, 193–207.

Nott, J. 2003*a*. Tsunami or Storm waves? – Determining the origin of spectacular field of wave emplaced boulders using numerical storm surge and wave models and hydrodynamic transport equations. *Journal of Coastal Research*, **19**, 348–356.

Nott, J. 2003*b*. Waves, coastal boulder deposits and the importance of the pre-transport setting. *Earth and Planetary Science Letters*, **210**, 269–276.

Oak, H.L. 1984. The boulder beach, a fundamentally distinct sedimentary assemblage (New South Wales, Australia). *Annals of the Association of American Geographers*, **74**, 71–82.

Papadimitriou, E.E. & Karakostas, V.G. 2008. Rupture model of the great AD 365 Crete earthquake in the southwestern part of the Hellenic Arc. *Acta Geophysica*, **56**, 293–312.

Paris, R., Wassmer, P. *et al.* 2009. Tsunamis as geomorphic crises: lessons from the December 26, 2004 tsunami in Lhok Nga, West Banda Aceh (Sumatra, Indonesia). *Geomorphology*, **104**, 59–72.

Paris, R., Fournier, J., Poizot, E., Etienne, S., Morin, J., Lavigne, F. & Wassmer, P. 2010. Boulder and fine sediment transport and deposition by the 2004 tsunami in Lhok Nga (western Banda Aceh, Sumatra, Indonesia): A coupled offshore–onshore model. *Marine Geology*, **268**, 43–54.

Paris, R., Naylor, L.A. & Stephenson, W.J. 2011. Boulders as a signature of storms on rock coasts. *Marine Geology*, **283**, 1–11.

Phuwieng, P., Imamura, F. & Koshimura, S. 2008. The correlation between tsunami run-up height and coastal characteristics index of Phang Nga and Phukey Provinces, Thailand. *Bulletin of the International Institute of Seismology and Earthquake Engineering*, **42**, 145–150.

Pignatelli, C., Sansò, P. & Mastronuzzi, G. 2009. Evaluation of tsunami flooding using geomorphologic evidence. *Marine Geology*, **260**, 6–18.

Pirazzoli, P.A. 1986. The Early Byzantine tectonic paroxysm. *Zeitschrift für Geomorphologie Supplement*, **62**, 31–49.

Pirazzoli, P.A., Ausseil-Badie, J., Giresse, P., Hadjidaki, E. & Arnold, M. 1992. Historical environmental changes at Phalasarna harbor, West Crete. *Geoarchaeology*, **7**, 371–392.

Pirazzoli, P.A., Laborel, J. & Stiros, S.C. 1996. Earthquake clustering in the Eastern Mediterranean during historical times. *Journal of Geophysical Research: Solid Earth*, **101**, 6083–6097.

Pope, R., Wilkinson, K., Skourtsos, E., Triantaphyllou, M. & Ferrier, M. 2008. Clarifying stages of alluvial fan evolution along the Sfakian piedmont, southern Crete: New evidence from analysis of post-incisive soils and OSL dating. *Geomorphology*, **94**, 206–225.

Ramalho, R.S., Winckler, G. *et al.* 2015. Hazard potential of volcanic flank collapses raised by new megatsunami evidence. *Science Advances*, **1**, e1500456.

Reilinger, R., McClusky, S. *et al.* 2006. GPS constraints on continental deformation in the Africa–Arabia–Eurasia continental collision zone and implications for the dynamics of plate interactions. *Journal of Geophysical Research: Solid Earth*, **111**, B05411.

Reilinger, R., McClusky, S., Paradissis, D., Ergintav, S. & Vernant, P. 2010. Geodetic constraints on the tectonic evolution of the Aegean region and strain accumulation along the Hellenic subduction zone. *Tectonophysics*, **488**, 22–30.

Reimer, P.J., Bard, E., Bayliss, A., Beck, J.W., Blackwell, P.G., Bronk Ramsey, C. & van der Plicht, J. 2013. IntCal13 and Marine13 radiocarbon age calibration curves 0–50,000 years cal BP. *Radiocarbon*, **55**, 1869–1887.

Richmond, B.M., Buckley, M. *et al.* 2011*a*. Deposits, flow characteristics, and landscape change resulting from the September 2009 South Pacific tsunami in the Samoan islands. *Earth-Science Reviews*, **107**, 38–51.

Richmond, B.M., Watt, S., Buckley, M.L., Jaffe, B.E., Gelfenbaum, G. & Morton, R.A. 2011*b*. Recent storm and tsunami coarse-clast deposit characteristics, Southeast Hawaii. *Marine Geology*, **283**, 79–89.

Scheffers, A. 2008. Tsuanmi boulder deposits. *In*: Shiki, T., Tsuji, Y., Yamazaki, T. & Minoura, K. (eds) *Tsunamiites: Features and Implications*. Elsevier, Amsterdam, 299–317.

Scheffers, A. & Scheffers, S. 2007. Tsunami deposits on the coastline of west Crete (Greece). *Earth and Planetary Science Letters*, **259**, 613–624.

Scheffers, A., Kelletat, D., Vött, A., May, S.M. & Scheffers, S. 2008. Late Holocene tsunami traces on the western and southern coastlines of the Peloponnesus (Greece). *Earth and Planetary Science Letters*, **269**, 271–279.

SCICCHITANO, G., MONACO, C. & TORTORICI, L. 2007. Large boulder deposits by tsunami waves along the Ionian coast of south-eastern Sicily (Italy). *Marine Geology*, **238**, 75–91.

SHAH-HOSSEINI, M., MORHANGE, C., BENI, A.N., MARRINER, N., LAHIJANI, H., HAMZEH, M. & SABATIER, F. 2011. Coastal boulders as evidence for high-energy waves of the Iranian coast of Makran. *Marine Geology*, 17–28.

SHAW, B., AMBRASEYS, N.N. *ET AL.* 2008. Eastern Mediterranean tectonics and tsunami hazard inferred from the AD 365 earthquake. *Nature Geoscience*, **1**, 268–276.

SMEDILE, A., DE MARTINI, P.M. *ET AL.* 2011. Possible tsunami signatures from an integrated study in the Augusta Bay offshore (Eastern Sicily – Italy). *Marine Geology*, **281**, 1–13.

SMEDILE, A., DE MARTINI, P.M. & PANTOSTI, D. 2012. Combining inland and offshore paleotsunamis evidence: the Augusta Bay (eastern Sicily, Italy) case study. *Natural Hazards and Earth System Science*, **12**, 2557–2567.

SMITH, R.L. 1986. Extreme value theory based on the r largest annual events. *Journal of Hydrology*, **86**, 27–43.

SPISKE, M., BÖRÖCZ, Z. & BAHLBURG, H. 2008. The role of porosity in discriminating between tsunami and hurricane emplacement of boulders – a case study from the Lesser Antilles, southern Caribbean. *Earth and Planetary Science Letters*, **3–4**, 384–396.

STIROS, S.C. 2010. The 8.5+ magnitude, AD365 earthquake in Crete: Coastal uplift, topography changes, archaeological and historical signature. *Quaternary International*, **216**, 54–63.

STIROS, S.C. & PAPAGEORGIOU, S. 2001. Seismicity of western Crete and the destruction of the town of Kisamos at AD 365: archaeological evidence. *Journal of Seismology*, **5**, 381–397.

STUIVER, M. & REIMER, J. 1993. Extended ^{14}C data base and revised CALIB 3.014 C age calibration program. *Radiocarbon*, **35**, 215–230.

TAWN, J.A. & VASSIE, J.M. 1990. Spatial transfer of extreme sea-level data for use in revised joint probability method. *Proceedings of the Institute of Civil Engineering, Part 2*, **89**, 433–438.

TEN VEEN, J.H. & KLEINSPEHN, K.L. 2003. Incipient continental collision and plate-boundary curvature; late Pliocene–Holocene transtensional Hellenic forearc, Crete, Greece. *Journal of the Geological Society, London*, **160**, 161–181, https://doi.org/10.1144/0016-764902-067

TITOV, V.V. & SYNOLAKIS, C.E. 1997. Extreme inundation flows during the Hokkaido-Nansei-Oki Tsunami. *Geophysical Research Letters*, **24**, 1315–1318.

TORTORICI, L., CATALANO, S., CIRRINCIONE, R. & TORTORICI, G. 2012. The Cretan ophiolite-bearing mélange (Greece): a remnant of Alpine accretionary wedge. *Tectonophysics*, **568**, 320–334.

VACCHI, M., ROVERE, A., ZOUROS, N. & FIRPO, M. 2012. Assessing enigmatic boulder deposits in NE Aegean Sea: importance of historical sources as tool to support hydrodynamic equations. *Natural Hazards and Earth Systems Science*, **12**, 1109–1118.

VELLA, C., DEMORY, F., CANUT, V., DUSSOUILLEZ, P. & FLEURY, T.J. 2011. First evidence of accumulation of mega boulders on the Mediterranean rocky coast of Provence (southern France). *Natural Hazards and Earth Systems Science*, **11**, 905–914.

VERNANT, P., REILINGER, R. & MCCLUSKY, S. 2014. Geodetic evidence for low coupling on the Hellenic subduction plate interface. *Earth and Planetary Science Letters*, **385**, 122–129.

VÖTT, A., BARETH, G. *ET AL.* 2007. Beachrock-type calcarenitic tsunamites along the shores of the eastern Ionian Islands and the western Peloponnese. *Zeitschrift für Geomorphologie*, **54**, 1–50.

VÖTT, A., BRÜCKNER, H., MAY, M., LANG, F., HERD, R. & BROCKMÜLLER, S. 2008. Strong tsunami impact on the Bay of Aghios Nikolaos and its environs (NW Greece) during Classical–Hellenistic times. *Quaternary International*, **181**, 105–122.

VÖTT, A., BRÜCKNER, H. *ET AL.* 2009. Traces of Holocene tsunamis across the Sound of Lefkada, NW Greece. *Global and Planetary Change*, **66**, 112–128.

VÖTT, A., LANG, F. *ET AL.* 2011. Sedimentological and geoarchaeological evidence of multiple tsunamigenic imprint on the Bay of Palairos-Pogonia (Akarnania, NW Greece). *Quaternary International*, **242**, 213–239.

WHITWORTH, M.R.Z. 2015. *Utilising probabilistic techniques in the assessment of extreme coastal flooding frequency–magnitude relationships using a case study from south-west England.* PhD thesis, University of Plymouth, UK.

Tsunami landfalls in the Maltese archipelago: reconciling the historical record with geomorphological evidence

DEREK N. MOTTERSHEAD, MALCOLM J. BRAY* & PHILIP J. SOAR

Department of Geography, University of Portsmouth, Buckingham Building, Lion Terrace, Portsmouth, Hampshire PO1 3HE, UK

**Correspondence: malcolm.bray@port.ac.uk*

Abstract: The Maltese Islands lie in the middle of the tsunamigenic Mediterranean domain, around whose margins and islands evidence of historical tsunami landfall has been increasingly recognized in recent years. Critical review of historical evidence of events in 1693 and 1908 indicates extremely modest tsunami impacts. In marked contrast, though, recently discovered geomorphological evidence summarized herein suggests that Malta's coastlines have been overwashed up to elevations of >20 m above sea level by an exceptional event. A new perspective is provided by a review of the central Mediterranean context within which the Maltese evidence is located. Recent advances in understanding the Holocene sequence forming the floor of the Mediterranean Sea present a new stratigraphic and temporal framework within which to elucidate tsunami history. Within 100 km of Malta, terrestrial stratigraphy on Sicily also provides supporting evidence of tsunami impact. Review of these advances suggests that the exceptional event required to emplace the most extreme sedimentary and geomorphological signatures on and around Malta is likely to have had a far-field origin. The currently available circumstantial evidence points strongly towards a probability that the AD 365 earthquake and tsunami were responsible. This, in turn, enables critical reassessment of the exposure of Malta to tsunami hazard.

The Maltese archipelago sits on the Malta Plateau atop the Pelagius Platform, a northern promontory of the African Plate which extends towards Sicily and separates the deep Ionian Basin from the Western Mediterranean. Eastwards, at a distance of 120 km, the submerged Malta Escarpment forms the steep margin of the Ionian Basin, eastwards of which the seafloor falls to −4000 m (Galea 2007; Micallef *et al.* 2013). The islands are formed of Neogene sedimentary rocks tilted towards the NE, and affected by transverse faults creating horst and graben structures and topography.

There is a long and locally detailed history of recorded tsunami linked with tectonically active regions in the Mediterranean, summarized by Guidoboni *et al.* (1994) and Papadopoulos *et al.* (2014). Until quite recently, however, there was a reluctance to believe that tsunami pose a threat in the wider Mediterranean domain (Papadopoulos 2009). The Maltese Islands are located at a focal point in the central Mediterranean Sea and, thereby, exposed to marine influences from across the entire Mediterranean including both near-field and far-field sources (Fig. 1). Tinti *et al.* (2005) modelled tsunami originating from these zones, and showed how those originating from Algeria (*c.* 1000 km distant), the Western Hellenic Arc (600–800 km) and, especially, the Calabrian Arc off eastern Sicily (250–350 km) have the potential to create significant tsunami impacts on Malta. To the north, the near-field tsunamigenic zone of eastern Sicily and Messina Straits is rated as a threat of intermediate tsunamigenic potential, and to the east lies the Western Hellenic arc of high potential (Papadopoulos 2009). Even over these distances, a tsunami would have the potential to travel to Malta in approximately 70 min, and thereby to pose a significant threat to the archipelago and its inhabitants.

The tsunami history of Malta is recorded in both historical written records and by geomorphological evidence in the field. The former expresses the human experience of tsunami landfalls in Malta across a span of over 300 years and embraces two events, namely 1693 and 1908. The latter preserves tangible evidence of a tsunami landfall event in the form of recognizable tsunami signatures, of both erosional and depositional origin, which are available for scientific observation and analysis.

The objectives of this paper are therefore to:

- critically review and interpret available documentary evidence of historical tsunami in Malta;
- summarize recently acquired geomorphological and sedimentary evidence of tsunami landfall on Malta;
- highlight the apparent contrast between the historical and the geomorphological evidence;

From: Scourse, E. M., Chapman, N. A., Tappin, D. R. & Wallis, S. R. (eds) 2018. *Tsunamis: Geology, Hazards and Risks*. Geological Society, London, Special Publications, **456**, 127–141.
First published online February 23, 2017, updated March 3, 2017, https://doi.org/10.1144/SP456.8

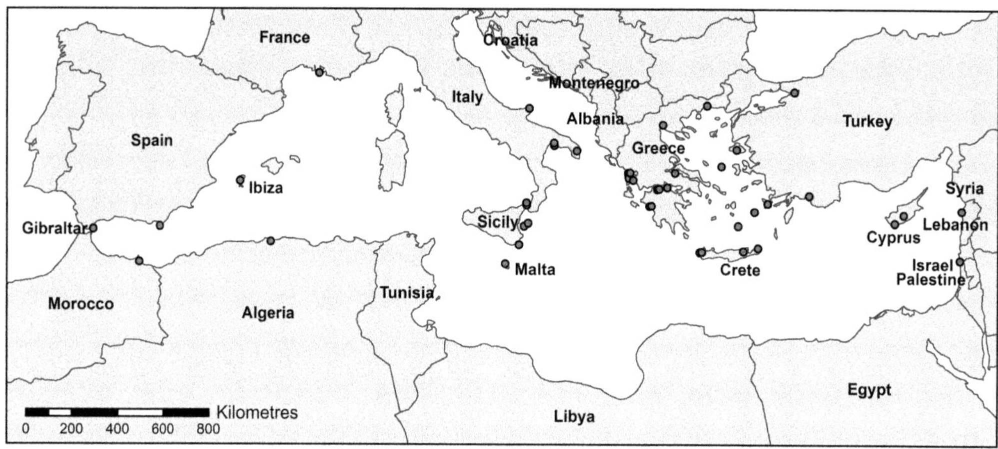

Fig. 1. Distribution of field sites around the Mediterranean with published evidence of extreme wave impacts linked to tsunami. Near-field sites are clustered around Sicily with the most relevant far-field sites located from western Greece to Crete comprising the Western Hellenic Arc.

- interpret the combined Maltese evidence in the context of current understanding of the Holocene and tsunami histories of the Mediterranean;
- re-evaluate the tsunami threat to Malta.

The historical experience of tsunami landfalls in Malta

There are known historical records for two significant tsunami landfalls in Malta: one in 1693, predating the era of scientific understanding of such phenomena, and a second one in 1908, within the modern era.

The tsunami of 10–11 January 1693

On 11 January 1693, an earthquake of magnitude (M_{aw}) 7.4 and scoring up to 10 on the Mercalli intensity scale occurred across most of SE Sicily, devastating many towns and villages and causing an estimated >60 000 deaths (Guidoboni *et al.* 1994; de la Torre 1995). Modern reconstructions of the earthquake event suggest five potential sources ranging from the Malta Escarpment offshore of east Sicily including other nearby faults, to the Sicilian Basal Thrust in east and central Sicily (Visini *et al.* 2009). It created a tsunami which impacted on a coastline length of 230 km, with a maximum reported run-up of up to 8 m above sea level (asl) at Augusta, and maximum inundation distance of 1.5 km at Mascali (Boccone 1697, in Gerardi *et al.* 2008). Gutscher *et al.* (2006) modelled associated tsunami wave heights of 1–3 m.

This earthquake is the most intense seismic event to have struck Malta since AD 1500 (Galea 2007). It caused severe damage to many large stone-built structures, including palaces and the Norman cathedral in Mdina (Shower 1693; Galea 2007; Borg *et al.* 2008). In Malta, its magnitude was calculated at K_0 V, and EMS-98 intensity at VII–VIII (Boschi *et al.* 2000). Historical accounts of the Maltese impact of both earthquake and tsunami are provided by Shower (1693) and de Soldanis (1746).

In contrast to the earthquake, information specific to the associated tsunami of 1693 is rather more limited. It is presented by Agius de Soldanis (1712–70) in his manuscript 'Il Gozo Antico Moderno e Sacro-Profano', first published in Italian (de Soldanis 1746). In 1936, Mgr Giuseppe Farrugia Gioioso published a Maltese translation (de Soldanis 1746 [1936]), which was itself the basis for a subsequent translation from Maltese into English by Rev. Fr A. Mercieca (de Soldanis 1746 [1999]). In the original text, de Soldanis provided two statements which are of potential assistance in gleaning pertinent information from the testimonies of Gozitan inhabitants regarding the tsunami event at Xlendi, Gozo (Fig. 2), as follows: 'il mare lascio il propio letto, e in fuomi si ritiro per un miglio circa', translated by the authors as 'the sea withdrew for about one mile exposing the sea bed', or by Camilleri (2006) as 'rolled out to about one mile'; 'ma poco dopo con grande impeto e mormorio cerco il suo luogo nel tempo stesso', translated by the authors as 'but shortly afterwards returned with great force and noise to its normal position', or by Camilleri (2006) as 'and swept back a little later with great impetus and murmur'.

Notably, these statements predate the modern era of quantitative scientific observations and therefore merit some caution in their interpretation, on several

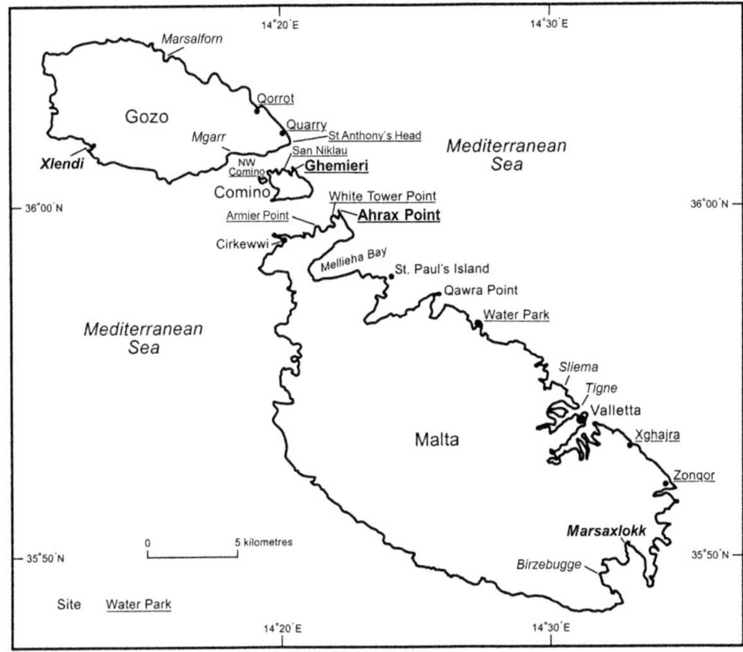

Fig. 2. Malta site locations covering locations of historical tsunami reports (italics) and geomorphological evidence (underlined). Modified after Mottershead *et al.* (2014), with permission from Schweizerbart Science Publishers (http://www.schweizerbart.de).

grounds. First, it is evident that Agius de Soldanis himself, who was not born until 1712, could not have been an eyewitness to this event. Furthermore, he does not reveal his source or sources. If we allow him 21 years to attain adulthood, then it is possible that he acquired his information of the event any time between 1733 and 1746. At this time, it would have been a memory of between 40 and 53 years, raising the question of how reliable the memory may have been. A sentient 10 year-old child witnessing the event would be 50–63 years of age by the time Agius de Soldanis obtained the information; a 30 year-old adult would, if still alive, have been 83 years of age. It can only be conjectured how many eyewitnesses actually survived to provide valid testimony or how much of that was hearsay from folk as yet unborn in 1693 who had received information from their own elders.

Second, there is the accuracy of the observed information: in this case, the 'mile'. Certainly, given the circumstances of the moment, the receding sea margin, the lack of landmarks and the sheer practicalities, it would not have been a measured mile. It is also the case that there were many definitions of the mile extant at the time: the international mile was not subject to a standardized definition until 1959. A more practical approach to this issue is to test the plausibility of a tsunami drawdown over a distance of 1 mile at this site by examining the local bathymetry. At a distance of 1 mile offshore of Xlendi, bathymetry reveals that the seafloor lies at an elevation of between -100 and -200 m. These values are far in excess of drawdown values reported and/or modelled for Mediterranean tsunami. Piatanese & Tinti (1998), for example, modelled mareograms for this event which generated four surges. Their model portrayed withdrawals of up to -4 m and -6 m at Augusta and Syracuse respectively, and run-ups to between $+1.75$ and $+2.0$ m. Although the spatial limits of the model did not extend as far as Malta, it is likely that the tsunami effects were not greater than these values on the more distant Maltese shores.

The observation concerning the mile could then well be a subjective interpretation, swelled by the memory (conceivably of an awestruck mediaeval peasant lacking the precision equipment to survey it accurately) and the passage of several decades. It is, however, possible to make a reasonable scientific interpretation based on information on hydrographic charts. From the shoreline at Xlendi, an enclosed, elongated (>0.5 km) bay of shallow depth extends seawards beyond the Ras il Badja Point at its entrance, where the depth is still a modest 4 m. It is reasonable, therefore, to suggest that a drawdown to -4 m may have taken place, implying a withdrawal of upwards of 0.5 km prior to the tsunami arrival. This interpretation would appear to place a reasonable constraint on the far less plausible claim of the 1 mile withdrawal prior to the tsunami advance.

Third, much depends on the quality of the translation. At the detailed level, translation of individual words may affect the sense of the text and there is no guarantee that a particular word has retained an unchanged meaning over a period of 300 years. Translation of the phrase describing the advance of the tsunami wave itself, 'con grande impeto e mormorio', is a case in point: 'Impeto' may reasonably be translated as force; 'mormorio' is more catholic in its implications, depending on context, and to avoid unnecessary and possibly subjective assumptions, we prefer the more general sense offered by the word 'noise'. The significance of this point lies in its influence on and potential value in assessing tsunami magnitude according to subjective scales such as those of Ambraseys (1962) and Iida *et al.* (1967).

Some reasonable inferences, however, can be made from these interpretations. There is no mention of damage and, probably of greater significance, no mention of fatalities. The latter would surely have been reported had any been caused by this event, especially in a scientifically primitive era when such catastrophes were likely to be interpreted as representing the wrath of God. On a very gently shelving inshore seabed, a withdrawal of 0.5 km (startling though it may have appeared to a population of very limited education and experience) is not particularly abnormal as meteorologically induced seiches occur in this area. In terms of current scales of tsunami magnitude (Ambraseys 1962; Iida *et al.* 1967), the advancing tsunami wave with its interpreted rumble falls far short of the 'strong roar' and 'people drowned 'characteristics associated with the K_0 V category of the modified Sieberg scale of seismic wave intensity. The category III+/−I (Light to Rather Strong) would appear to be more reasonable for this event on the limited information available, implying run-up values of 1–4 m. Conclusions gleaned from the available evidence would appear to be congruent with the interpretation of an intensity of a K_0 I–III tsunami.

The tsunami of 28 December 1908

The earthquake of 28 December 1908 was centred on the Straits of Messina, causing at least 120 000 fatalities (Borg *et al.* 2016) in and around the cities of Messina and Reggio di Calabria, the single greatest loss of life in recorded earthquake history at that time. Its effects were felt over a radius of 200 km from its epicentre. The associated tsunami made landfalls along 100 km of the Sicilian coast and 38 km of that of Calabria, with observed maximum run-ups of 11.7 and 9.7 m, respectively (Omori 1909; Platania 1909). Comerci *et al.* (2008) estimated that 2000 further deaths were attributable to the tsunami, which also destroyed many properties. A contemporary Maltese account of the Messina event is provided by Galea (1909).

In a thorough review, Pino *et al.* (2009) summarized studies of the 1908 event. The magnitude of the 1908 earthquake is estimated at $M_w = 7.1$. A majority opinion is that it was caused by strike-slip along a north-striking fault located within the Messina Strait (Amoruso *et al.* 2002), whilst Pino *et al.* (2009) affirmed that models to date do not wholly satisfy the observed data. Ridente *et al.* (2014), based on submarine geomorphology, concluded that a blind fault was most likely responsible and that submarine landslides may have locally distorted the run-up pattern, thus confounding attempts to reconstruct the event.

The tsunami impact on Malta, some 250 km from the epicentre, is widely recorded, including numerous contemporaneous newspaper reports (Fig. 3) (Borg *et al.* 2016). Official records of tide levels (Qarantena, Grand Harbour Valletta) and daily weather logs (Ġurdan Lighthouse, Gozo and British naval ships berthed in Grand and Marsamxett harbours, Valletta) are available for inspection at the National Archives of Malta, Gozo and the UK, respectively. Baratta (1910) provided witness observations from 11 localities throughout the Maltese Islands, summarized in Table 1. Not all characteristics were recorded for all stations but it is possible to identify some general patterns. The tsunami made landfall on 28 December, initially at Xlendi (Gozo) at 06.00 and down to Birżebbuġa at 06.50. Characteristic multiple tsunami waves were recorded at Mġarr and Msida. Disturbances of the sea apparently continued at Tigné until 16.00. There is considerable variation in maximum elevations (run-up) of the sea, according to local topographical circumstances, as is characteristic of tsunami waves. Maximum values of >1 m height were observed only at Msida and Birżebbuġa, and, within this limited dataset, the total range of sea-level variation exceeded 2 m only at Birżebbuġa. The single quantifiable record of inundation (onshore advance) relates to Marsaxlokk (Baratta 1910), suggesting that the tsunami inundated the land up to 300 m inland. Within the broader context of global tsunami events, this tsunami impact is of only modest proportion. Although reaching at least X on the Papadopoulos & Imamura (2001) 12-point tsunami intensity scale close to its epicentre at Messina, its intensity was substantially reduced to around point V by the time it arrived at Malta, consistent with a tsunami wave height of <1 m.

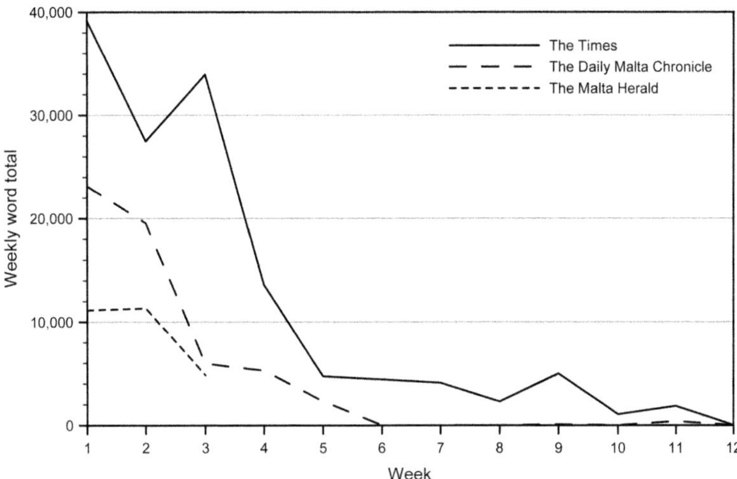

Fig. 3. Weekly newspaper wordage covering the Messina earthquake and tsunami disaster of 28 December 1908 in the immediate aftermath of the event.

It is noteworthy, and perhaps indicative of its limited impact, that no record of this tsunami event was recorded either by the weather observatory at Ġurdan lighthouse, NW Gozo, or in the logs of the many British naval ships at anchor in Grand and Marsamxett harbours, Valletta. The damage that the 1908 event induced in the Maltese Islands was apparently limited to the flooding of cellars in Sliema and Msida, and the stranding of boats at Qarantena and Marsaxlokk. Relatively minor disruption was caused by the fluctuations of sea level in the form of boats breaking from their moorings and ferry passengers having difficulty in accessing embarkation pontoons. No serious damage was reported and, perhaps, most significantly there is no record of human fatality or even injury.

The exposure of the Maltese populace to information regarding the 1908 Messina earthquake disaster, including the tsunami, can be assessed from contemporary press reports. These are represented in this study by the two leading Maltese English language newspapers, *The Daily Malta Chronicle* and *The Malta Herald*, both available in local archives. The volume of information was assessed in terms of weekly word totals for weeks following the earthquake and tsunami, estimated from the column length of daily articles and their average words

Table 1. *Summary of tsunami observations on Gozo and Malta from 1908 presented by Baratta (1910)*

Location	Time (h)	Rise (m)	Fall (m)	Range (m)	Waves (no.)	Run-up (m)	Inundation (m)
Xlendi	06.00	/	/				
Mġarr	06.10	+0.91	/		4		
Marsalforn		+0.72	−0.38	1.10			
Francesi	06.30	+0.32	−0.23	0.55			
Sliema	06.50	+0.92					
Msida (Valletta)		+1.40	/		5		
Qarantena (Valletta)	06.45	/	/				
Sliema Marina	07.45	0.92					
Tigné	<16.00	Vortices					
Birżebbuġa	06.30	+1.50	−1.20	2.70			
Marsaxlokk						+2.80*	300*

Values indicated by * were reconstructed by ground surveys in this study from descriptions by Baratta (1910). The locations are illustrated in Figure 2.

per line (Fig. 3). Newspapers at this time did not publish photographs but printed articles did, however, include descriptions of the event itself and its aftermath, survivors' tales, reports of ongoing rescue and relief work, charitable requests and donations, and obituary notices. It is evident from the material published that the Maltese were exposed to an abundant flow of detailed information following the event. The reportage of these two local papers can be evaluated by comparison with *The Times (of London)* for the same period (Fig. 3), in which individual news items are identified by word length in the online *The Times Digital Archive* (http://www.thetimes.co.uk/Digital_Archive).

Analysis shows that in the 3 weeks immediately following the disaster, approximately 96 000 words were published in the two Maltese papers, 60% of them in *The Daily Malta Chronicle*. The international scale of interest in the Messina disaster is shown by the fact that over a 12 week period, *The Daily Malta Chronicle* alone published a word volume amounting to some 56% of that of *The Times*. In terms of significance, if the total word content (all articles) within each paper at that time is taken into account the 'proportion' of news reporting on the 1908 event within *The Daily Malta Chronicle* is almost certainly higher than that of *The Times*.

The news coverage in the press would have been reinforced by visible activity of British naval movements in and out of Valletta in support of the relief effort, and of the arrival in Malta of Sicilian refugees. The Maltese population would have been in no doubt as to the magnitude of its impact on their geotectonically close neighbours along the Sicilian and Calabrian coasts, in sharp contrast with the minimal physical impact on their own island home, as would have also been the case in 1693 (Borg *et al.* 2016).

In summary, the two tsunami landfalls recorded within more than 300 years of recent Maltese history were consequent upon major seismic events in nearby Sicily, but, while each was responsible for an estimated >60 000 human fatalities in their source localities, their impact on Malta was minimal with apparently no human fatalities. Most notably, the legacy of this cumulative experience of local tsunami landfalls would not appear conducive to a belief in a significant tsunami hazard to the Maltese archipelago.

Geomorphological evidence of tsunami in Malta

The field evidence for historical extreme wave events in the Maltese Islands is concentrated along the east coast, clearly indicating wave attack from that direction (Mottershead *et al.* 2014, 2015). This is not to suggest that such events are exclusive to the east, but rather it offers topographies that act as traps for coastal sediment deposition. The cliffed nature of the upfaulted west coast, in contrast, offers few such opportunities and, to date, no clear evidence of historical extreme wave activity has been observed on that side of the archipelago.

A major difficulty in interpreting extreme wave deposits is to discriminate them from regular storm waves. Storm wave height is limited by the available fetch, in the case of Malta up to 2000 km, which, in turn, limits the maximum height of the storm waves themselves, modelled by Drago *et al.* (2013) as 5.5 m. The height of a tsunami, in contrast, is initially dependent on the magnitude of the event that creates it. At the point of landfall on a coastline, its height may be magnified up to 10-fold by the characteristics of local bathymetry and topography. A coastline will therefore consist of two zones: a lower zone, which receives run-up from both storm and weak tsunami events; and a higher zone beyond the reach of storm wave run-up, which is inundated only by strong tsunami waves. In the former case, both storm and tsunami sediments may accumulate together to form a composite deposit, whereas the latter zone captures and retains only tsunami sediments.

Extreme waves, including tsunami, may leave a wide range of both erosional and depositional signatures as evidence of their visitation to a coastline. Depositional signatures may take the form of fine-grained sediments deposited in the calm waters of coastal lagoons or estuaries to form sediment sequences susceptible to study by coring or trenching. On exposed terrestrial surfaces, coarser material of nearshore origin such as sand, gravel and boulders may be swept ashore to form spreads and, in some cases, landforms of measurable relief such as ridges and berms.

The range of extreme wave depositional landforms observed across sites widely distributed along the east coast of Malta is shown in Table 2. The majority of signatures are in the form of boulders, characteristic of the rocky shores and coastal platforms and ramps in the coralline limestones of Malta. They may occur as isolated individuals, clusters or scatters forming a line or berm marking a velocity threshold of a slowing wave riding up a coastal slope. Imbricate boulders, in a cluster or occasionally stacked in a line, are indicative of a linear flow of high velocity. Split boulders, which now consist of two or more closely spaced fragments, are indicative of deposition by a heavy force. Bio-encrusted boulders are indicative of a source within the intertidal or subtidal zone. Sediment sheets are characteristically formed by sand, whereas dump deposits tend to be diamictic. Distinct erosion features are formed by the detachment

Table 2. *Distribution of depositional extreme wave signatures at Maltese sites*

Location	Boulders					Sediment sheet	Dump deposit
	Isolated	Line/berm	Imbricate	Split	Bio-encrusted		
Aħrax	/	/	/			/	
Għemieri	/	/	/		/		/
N Comino coast	/		/		/		
Qorrot	/		/	/			/
Quarry	/	/	/				
Water Park	/		/		/		
White Tower	/		/		/	/	/
Xghajra	/		/	/	/		
Żonqor	/	/	/		/		

The locations are illustrated in Figure 2.

and removal of bedrock to create a range of features and forms also present along the east coast, as shown in Table 3. The majority of these forms have been described by Bryant & Young (1996) and further defined within the Maltese context by Mottershead *et al.* (2015). Broadly speaking, they are arranged in order of diminution of power required for their formation from sockets to swept terrain. Collapsed sea caves are interpreted on the basis that other signatures are indicative of very-high-energy events and so conceivably are linked to extreme cave-breaching events. The identification of a plunge pool is a unique case, at the base of a massive flow evidenced by other signatures (Għemieri, below).

It is doubtful that any single extreme wave signature is itself diagnostic of the activity of tsunami. It is the case, however, that the signatures occur in assemblages rather than individually, and in patterns relating both to each other and to topography and its interaction with an overwashing flow. Meaningful relationships can therefore be identified between different signatures, and between signatures and topography. This is particularly the case in areas of varied terrain such as the Maltese Islands. Examples of spatial relationships of this kind are shown in Figure 4 (Aħrax). The headland and the lower slopes facing the NE are the most exposed slopes of the peninsula and exhibit fields of sockets, from whence boulders have been extracted by the decompression effect of a large overwashing wave. These are interpreted as the sources of the boulders now lying deposited further to landwards. Two lines of large individual boulders form a funnel leading to a col at 7.3 m asl in the Aħrax peninsula in the northeasternmost corner of the island of Malta. Landwards of this line is a spread of other sediment, cobbles diminishing landwards to gravels and sand. The boulders possess a long-axis orientation indicating a movement from the coast (Fig. 4). This group

Table 3. *Distribution of erosional extreme wave signatures at Maltese sites (modified after Mottershead et al. 2015, with permission from John Wiley & Sons, Inc.)*

Location	Sockets	Eroded scarps	Clifftop detachment scar	Scoured terrain	Swept terrain	Collapsed sea cave	Plunge pool
Aħrax	/	/	/	/		/	
Għar Għana	/	/		/	/		
Għemieri	/	/	/	/		/	/
N Comino coast	/			/			
Qorrot		/			/		
Quarry					/		
St Paul's Island	/				/	/	
Water Park	/						
White Tower	/	/					
Xghajra		/			/		
Żonqor		/			/		

The locations are illustrated in Figure 2.

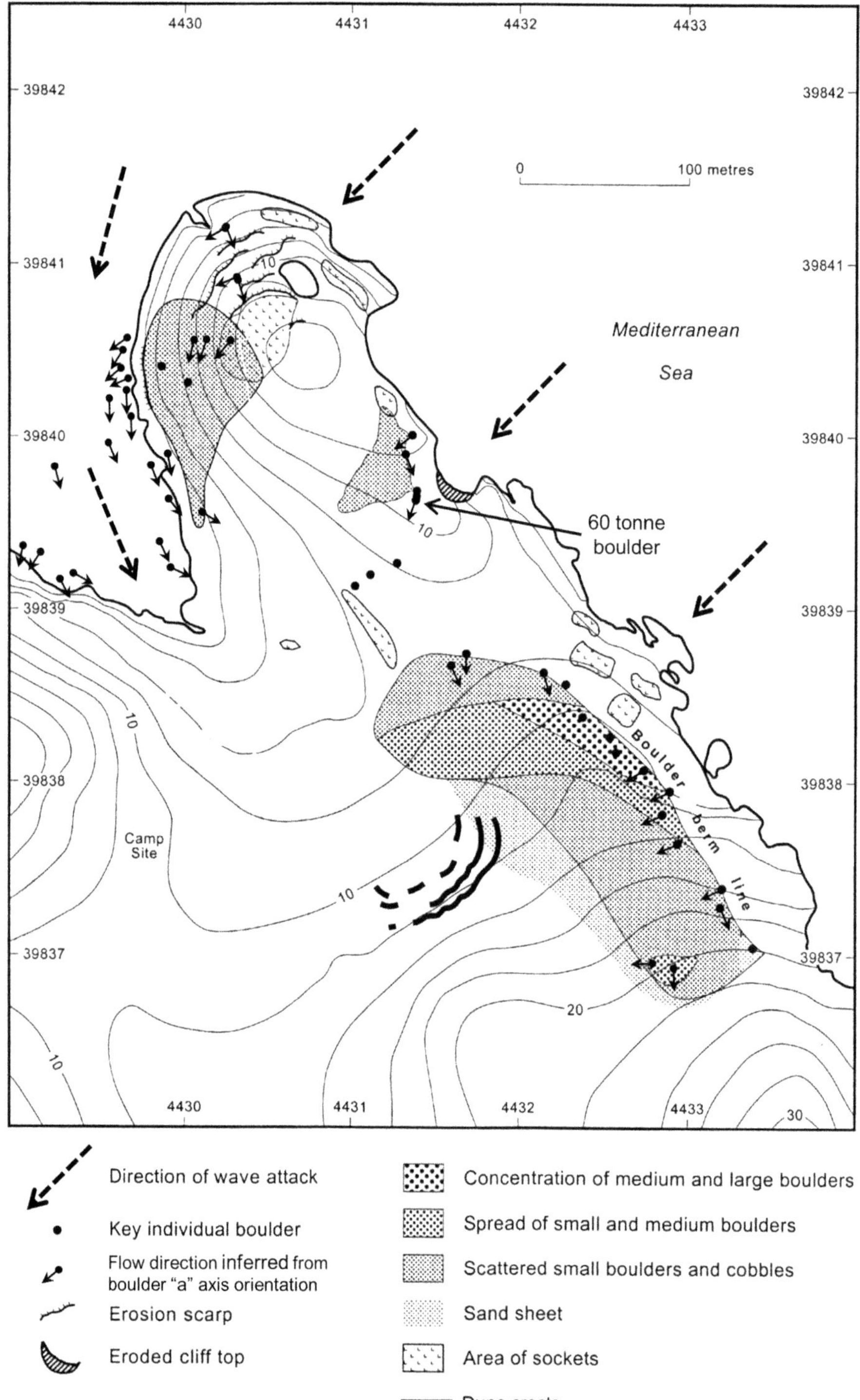

Fig. 4. Ahrax study site showing distributions of boulder deposits, erosional features and inferred directions of direct and refracted wave impacts. The location of this site is illustrated in Figure 2. Modified after Mottershead *et al.* (2014), with permission from Schweizerbart Science Publishers (http://www.schweizerbart.de).

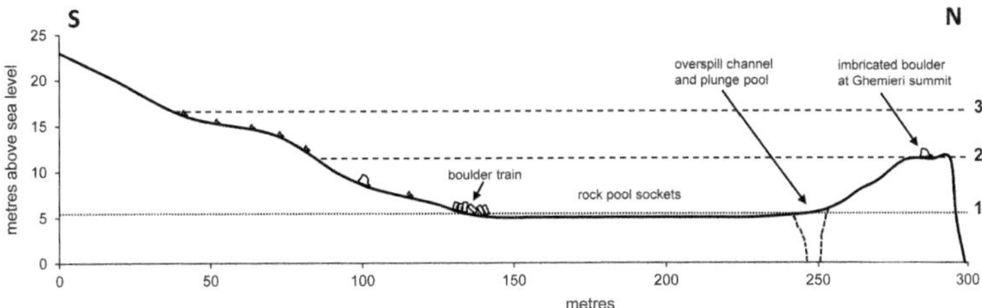

Fig. 5. Composite north–south section across Għemieri col, Comino, showing reconstructed levels of overwashing as follows: 1, maximum known storm wave of 5.5 m would marginally overtop the crest of the col; 2, a large imbricate boulder on Għemieri summit indicates a high-energy flow overwashing the peninsula to at least 11.5 m asl; 3, a scatter of small boulders on the southern slope suggests that the outlying overwashing flow attained a minimum level of 17 m asl. The location of this site is illustrated in Figure 2. Modified after Mottershead *et al.* (2015), with permission from John Wiley & Sons, Inc.

of signatures is interpreted as a landwards-fining sequence of deposition due to a decelerating sheet of water. The boulder line extends to an elevation of >20 m, indicating that the water which transported them attained at least that elevation. Boulders also occur submerged in the bay to the west of Aħrax headland, indicative of refraction of flow around the headland and into the bay. This observation can be linked to the occurrence inland of the bay of a sand sheet terminating in arcuate berms. This is interpreted as being deposited by the flow which entered the bay and ran up the valley to an elevation of circa 12 m inland of the bay.

In these ways, the patterns of erosional and depositional signatures reflect the interaction between the assailing wave and the terrain over which it flowed. Figure 5 shows a landscape cross-section across the col though the Għemieri peninsula at the NE point of the island of Comino. The evidence of erosional and depositional signatures shows that this also acted as a spillway for an overwashing flow. In this case, the flow overrode the 4 m-high sea cliff and passed over a col just behind the cliff itself from where it was able to flow without hindrance into the bay beyond. Despite the presence of this open routeway, the existence of depositional signatures up to 16.5 m asl on the summit to the north (marked N) and on the slope to the south (marked S) indicates a flow of water up to 10 m deep through this col. The evidence of these depositional signatures is reinforced by the presence of a plunge pool and other erosional signatures on the floor of the col itself, indicative of a local high-velocity thread of water concentrated by a canyon at the base of the overwashing flow.

A further perspective on the distribution of extreme wave patterns is shown in Figure 6, showing the altitudes attained by extreme wave deposits along the margins of the Comino Channel. At the eastern entrance to the Comino Channel, the upfaulted blocks of land forming the Aħrax peninsula, Comino Island and St Anthony's Head, Gozo, form an eastwards-facing bastion to extreme waves from that direction, rising to 35, 75 and 120 m, respectively. This would block the passage of an assailing wave, causing it to back up, thus increasing its hydraulic potential. The wave would then propagate though the limited gaps open to it, namely the North and South Comino Channels. Figure 6 demonstrates a steady decline in elevation over a distance of 4 km from east to west; from the highest signatures at 20 m asl, adjacent to the barriers represented by cliffs, down towards 2 m asl at more sheltered sites, declining westwards as the wave propagating along the channel diffracts and attenuates.

An alternative approach to the identification of historical events and the processes by which they shaped local landscapes lies in hydrodynamic considerations. Models have been developed both for tsunami run-up and for boulder detachment by wave action, whether storm or tsunami in origin. The run-up equation of Synolakis (1987) models the relationships between shoreline height of tsunami waves, onshore and offshore slope gradients, and run-up elevation. Observed values of extreme wave sedimentary signatures provide a minimum estimate of run-up height; this may then be used to retrodict tsunami wave height when bathymetric and topographical gradients are known. For the Maltese sites directly exposed to the NE, the Synolakis equation yields shoreline wave heights ranging from 1.37 to 3.83 m (Mottershead *et al.* 2014; note that this range now incorporates updating of data following more extensive searching at two sites).

Hydrodynamic considerations can also be applied both to processes of boulder detachment and transportation. These permit the identification of thresholds of force required for the removal of clasts in terms of the height of wave required to

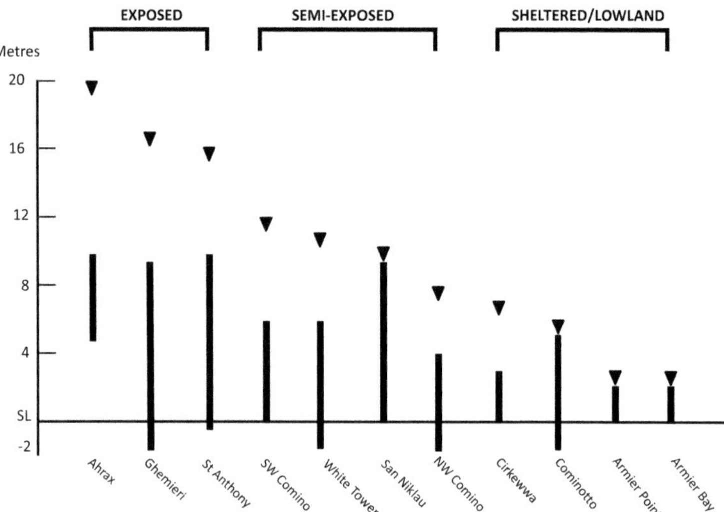

Fig. 6. Sites along the channels to the north and south of Comino plotted according to their relative exposure to the NE. (Triangles show highest depositional boulder signatures; bars show the height range of erosional signatures; SL, sea level.)

do so, whether storm or tsunami. In this way, discrimination is possible as to whether a storm wave of plausible height within the Mediterranean context would be capable of removing an observed boulder of specific size from its seating on or within a rock platform surface. If that boulder lies above the maximum storm wave threshold, then a more powerful wave must then be responsible, pointing to the conclusion that it was likely to be a tsunami wave. Applying the models of Nandasena *et al.* (2011) and Mottershead *et al.* (2014) showed that at two sites at the shoreline (Water Park and White Tower: Fig. 2) 29 and 16%, respectively, of the larger boulders exceed the capacity of the conservatively assumed maximum storm wave height of 9 m, and were thus interpreted as having been detached and transported to their present locations by tsunami waves. The largest boulders in this study implied tsunami wave heights at the shoreline of 4.3 and 3.6 m, respectively. Recent work by Biolchi *et al.* (2015) across a wider range of shoreline sites in the same area produced conclusions not inconsistent with this finding. They also noted that contemporary storms have been capable of moving some large boulders, whilst onshore boulders just above the shoreline show evidence of having been moved onshore from the intertidal zone at dates not inconsistent with known tsunami as far back as the past 1000 years. Although the elevations of the dated boulders are not specifically detailed, the dates appear to relate to shoreline boulders rather than those associated with the higher level (10–20 m asl) overwashing of the landscapes of Aħrax and Għemieri.

A special case is the model of Hansom *et al.* (2008), which describes clifftop boulder detachment. This was used by Mottershead *et al.* (2015) to model the wave required to detach an approximately 60 t boulder from a clifftop cornice at approximately 10 m asl (Fig. 4), and revealed the required storm wave to be between 9.1 and 28.4 m (depending on the original but unknown orientation of the boulder), which is clearly unfeasible at this elevation and with maximum sea waves of 5.5 m (Drago *et al.* 2013). However, the required power would be produced by a tsunami wave at 10 m asl of between 2.1 and 7.3 m height, which is a more feasible proposition, particularly for a site with extreme wave signatures present up to 22 m asl.

When evaluating the geomorphological evidence of historical tsunami run-up, it needs to be remembered that, for two very good reasons, the observed field evidence is likely to understate the magnitude of the actual event. First, at the point of maximum run-up, the tsunami wave itself has zero velocity and may only be present at that point for a matter of seconds. This means that by that stage it has minimum capacity for sediment transport and will have already deposited much of its sediment load, and during its brief presence at that point will offer minimal opportunity for fine sediment to fall out of suspension. Therefore, any deposition at this point is likely to be floating material such as driftwood and other biodegradable vegetation unlikely to leave any enduring signature. The sediments, especially large boulders, are likely to have been deposited at an early stage during run-up well short of the maximum limit, and will thus significantly underestimate the elevation of the latter. Second, agricultural, residential and industrial development has encroached on many coastal sites, obscuring and/or destroying relevant field evidence. Thus, the encroachment of terraced fields downslope at Żonqor has blurred the boundaries of coastal marine sedimentary deposits; the development of quarrying at Għar Dorf (NE Gozo) has destroyed any surfacial evidence that would corroborate similar deposits near St Anthony's Head; seafront development at Xghajra and Sliema has encroached on deposits evident on the shore platforms beneath; and the coastline flanking the urban concentrations of Sliema and Valletta has seriously obscured natural features of the coastline from Paceville to Xghajra over a distance of more than 7 km. It is, perhaps, ironic that development, the very phenomenon which obscures and thereby limits the evidence for tsunami attack, is the factor whose very presence actually increases the potential tsunami hazard.

Considering in tandem these two factors, the limited markers emplaced by tsunami at their run-up limit and the loss of depositional evidence caused by coastal development, point to the conclusion that the remaining surviving and observable evidence is likely to under-represent significantly the maximum run-up limit and may only serve as a minimum estimate of its true elevation.

The geomorphological evidence presented above contrasts markedly with the historical experiences and current perception of the tsunami history of Malta. The evidence of run-up to >10 m asl at most of the coastal sites directly exposed to wave approach from the NE greatly exceeds anything reported in relation to the 1693 and 1908 events. The notion of a substantial flow of water 2.1–7.3 m in depth flowing over Aħrax col and higher up the peninsula is not easy to comprehend in the field and stretches the imagination of the trained geomorphologist. It is more conceivable, however, that the evidence outlined above alludes to an event of substantially greater magnitude and threat than any in recorded history. It is entirely feasible that this great inundation occurred during an era when Malta was uninhabited or during the many centuries when the population lived well inland, with coastal lands far too dangerous to settle under the threat of seaborne marauders. Unfortunately, the authors have not yet been able to establish the dates of the higher-altitude deposits, which remains a significant research imperative in resolving the paradox between the historical record and the geomorphological evidence.

Maltese tsunami history in the central Mediterranean context

Recent research in the surrounding region offers important evidence regarding the tsunami history of the central Mediterranean zone in which Malta is located. This includes both historical tsunami landfalls on coasts opposed to Malta across the Mediterranean Sea (Mottershead *et al.* 2014) and marine sediments on the seabed.

The nearest coastline facing the Maltese archipelago is that of SE Sicily. Field evidence of tsunami landfall impacts on this coast has been summarized by de Martini *et al.* (2012), embracing the work of Scicchitano *et al.* (2007), Barbano *et al.* (2010), de Martini *et al.* (2010) and Gerardi *et al.* (2012), and spanning the entire east coast of Sicily from Gurna to Pantano Morghella (including a nearshore seabed study in Augusta Bay). A tsunami event assigned to AD 365 (see Shaw *et al.* 2008) occurred at four sites throughout this geographical range, although it is unclear to what extent the temporally clustered earthquakes of AD 361 in Sicily (M6.6) and 374 AD (M6.3) in Reggio, Calabria, may have confounded the interpretation of the sedimentary record. Of particular interest in this case is the study of Pantano Morghella in SE Sicily by Gerardi *et al.* (2012), for two geographical reasons. First, of the sites in the study, it is the nearest to Malta at only 90 km distant and may, therefore, serve as an analogue for Maltese events. Secondly, it is in the zone characterized by modelling efforts (Tinti *et al.* 2005; Lorito *et al.* 2008) as a focus of impact by high tsunami waves. Gerardi *et al.* (2012) identified two tsunami layers in a coastal wetland site, which they associate with the Hellenic far-field and Messina near-field sources. These are dated at AD 365 and AD 1908, respectively. The former reveals an inundation of up to 1200 m distance and the latter 380 m distance. The ingress of the 1908 event at this site was enhanced by an early twentieth century drainage canal, so a more realistic contrast to the AD 365 event is offered by nearby 1908 tsunami impacts of 1–3 m run-up and 10–15 m inundation. If these latter values are more representative of the 1908 event, then the AD 365 event assumes an even more outstanding relative magnitude.

Studies focused on the continuously accumulating marine sediments on the Mediterranean deep seabed have also revealed some significant insights to better understand the temporally and spatially fragmented terrestrial tsunami evidence. These have been studied both in the Ionian Sea immediately to the east of Malta (Köng *et al.* 2016; Polonia *et al.* 2016*a*, *b*; San Pedro *et al.* 2016) and in the broader east Mediterranean Basin beyond (Polonia *et al.* 2013, 2015, 2016*b*). Late Quaternary sedimentation in the central and east Mediterranean is dominated by turbidites, formed by seismically triggered mass flows of submarine sediments that are subsequently re-deposited. They occur both in perched basins, on the floors of submarine canyons, and in abyssal plains such as the 4000 m-deep Ionian abyssal plain where such beds may attain a thickness of 25 m. The trigger events for the flows may have been seismic shaking, volcanic collapse or the passage of a tsunami wave. First investigated by Kastens & Cita (1981), the turbidites are both consistent in nature and widespread across the Eastern Mediterranean. Recent studies by Polonia *et al.* (2013, 2016*b*) revealed that they extend from Crete to Sicily and from North Africa to Calabria, reaching as far north as 38° in the Ionian Sea, and extending over an area of $>150\ 000\ km^2$.

Polonia *et al.* (2013, 2016*b*) studied short (1–3 m) seabed cores that provide a record of late Holocene sedimentary events in the Ionian abyssal plain east of Malta. In particular, they reveal the presence, only 160 km from Malta, of a major megaturbidite unit, the Homogenite/Augias Turbidite (Polonia *et al.* 2013, hereafter HAT). This unit, on account of its uniformity and its breadth of occurrence, is interpreted to be a consequence of a single basin-wide event of magnitude unmatched elsewhere within the late Holocene. Both Polonia *et al.* (2016*b*) and San Pedro *et al.* (2016) presented models that demonstrated that each turbidite emplacement comprises up to four identifiable phases, consistent with the disturbance caused by the passage of a tsunami wave.

Radiometric dating evidence from a wide distribution of sites in differing topographical circumstances reveals a widespread synchronicity of deposition, confirming that emplacement was the consequence of a single widespread major event. Calibrated ^{14}C dating places the event in the interval AD 364–415. This associates it with the M_w 8.4 earthquake event of 21 July AD 365, the greatest seismic event known to have occurred in the Mediterranean domain during the historical period, and the associated tsunami. Polonia *et al.* (2013, 2015) identified a turbidite layer similar to the above in both its sedimentary character and wide distribution, which was interpreted in relation to a tsunami event of similar magnitude of impact to the HAT. This bed was dated as at least 15 ka BP. Köng *et al.* (2016) studied long (10–15 m) cores from the Ionian Basin, embracing a sedimentary record extending back 60 kyr. They identified a siliclastic megaturbidite, which they correlated with the HAT, and a further similar layer dated at 21 ka BP, which they interpreted as evidence of a tsunami. This latter deposit could relate to the post-15 ka BP basin-wide turbidite layer reported by Polonia *et al.* (2013) for which only a minimum date could be estimated.

In cores retrieved from canyons and basins in the western Ionian Sea below the Malta Escarpment, some 190 km east of Malta, Polonia *et al.* (2016*a*) were able to identify turbidites assigned to the 1693 and 1908 tsunami, confirming the capability of these events to displace submarine sediment some 200–300 km from the seismic event of their origins. In addition, Köng *et al.* (2016) have identified some 75 comparable turbidites within their cores on the Calabrian arc, dated within the past 60 kyr.

Nonetheless, these studies document regional (near-field) and not basin-wide events. In the case of the Messina 1908 event, it was estimated that seismic shaking caused several large submarine landslides that contributed significantly to the tsunami that was generated (Polonia *et al.* 2016*a*). Conceivably, if the 1908 tsunami were landslide-generated/augmented, then much of its energy would have comprised shorter period waves that would attenuate significantly with distance from their origin. This might provide one plausible explanation for the limited historical impacts of this tsunami on the Maltese Islands.

A further approach to interpreting AD 365 or equivalent palaeotsunami events, and their impacts, is via modelling efforts (Tinti *et al.* 2005; Lorito *et al.* 2008; Shaw *et al.* 2008; Pararas-Carayannis & Mader 2010; Pararas-Carayannis 2011). Such models uniformly show that the AD 365 tsunami could have reached as far west as the Sicily Channel, although without significantly penetrating the Western Mediterranean Basin. Modelled tsunami wave amplitude at the longitude of Malta ranges from 2 to 7 m. Lorito *et al.* (2008, p. 11), in commenting on local extreme values of absolute maximum water height (HMAX) attained during tsunami propagation, noted that shoaling effects (as they approach along their propagation path from the NE) can create 'local extreme HMAX values are observed at Malta and along the southeastern coasts of Sicily, Calabria and Apulia'. It is thus likely that tsunami waves approaching from the NE are transformed by shoaling as they pass over the major submarine obstacle represented by 3500 m-high Malta escarpment, as clearly shown by England *et al.* (2015).

Conclusion

A new perspective on the tsunami risk to Malta is now provided by recent studies of Holocene events in the Eastern Mediterranean Basin. It is evident that tsunami of near-field origin from the Calabrian Arc making landfall at Malta in historical times have caused little recorded damaging impact. There is certainly good public awareness of these events as a result of the 1908 earthquake and associated tsunami. Both devastated the cities of Messina and Reggio Calabria, yet the latter had only a modest impact on Malta. Earthquake events of similar magnitude along the Calabrian Arc are not uncommon, with an estimated recurrence interval of 100–700 years (Polonia *et al.* 2015) or 450–1000 years (Köng *et al.* 2016): yet, on the basis of recorded history, do not apparently constitute a significant threat to Malta.

Biolchi *et al.* (2015) showed that modern storms are capable of moving and depositing shoreline boulders on Malta and it is possible that some of these deposits may be associated with known historical tsunami. In contrast, however, we are concerned here with the patterns of signatures in Maltese coastal landscapes extending upwards to >20 m asl. Their recent recognition in Malta (Mottershead *et al.* 2014, 2015) represents a much more severe landfall impact and would appear to require a different explanation. The recent evidence of Polonia *et al.* (2013, 2016*a*, *b*) and Köng *et al.* (2016) implies stratigraphic and temporal support for the AD 365 tsunami and at least one previous basin-wide event originating in the Hellenic Arc, as potentially responsible agencies. As such, they represent low-frequency events of very high magnitude and a potentially highly damaging threat to the Maltese archipelago. However, since only two basin-wide events of this type separated by some 19 000 years have been identified from studies covering the past 60 kyr, it is very difficult to estimate a return period with any confidence.

While this paper has sought to reconcile the apparent paradox between the historical record and the geomorphological evidence, the dating of the higher-altitude landfall deposits is still a prerequisite to the verification of the AD 365 tsunami as the primary formative agent, and this remains an important near-term research objective. It is concluded, therefore, that the populated NE shores of the Maltese Islands are exposed to moderately frequent severe storms and infrequent near-field tsunami capable of major impacts along the shoreline. By contrast, evidence of high run-up and long inundation distances are tentatively linked to very infrequent, far-field basin-wide events with long, although as yet uncertain, return periods. If this type of event does represent the major tsunami hazard in the Maltese archipelago, then there may be some solace in the interpretation that it is likely to be of very low frequency of occurrence and, with its origin in the Western Hellenic Arc, would provide at least 60–70 min warning to Malta.

Diana Martin and Aaron Micallef are acknowledged for assistance with Italian and Maltese translations, respectively. Joanna Causon Deguara is thanked for field assistance. Derek Mottershead received a contribution to field expenses from the Department of Geography, University of Portsmouth. We are grateful to Jim Rose and an anonymous reviewer for their constructive comments for improving the manuscript.

Correction notice: The original version was incorrect. This was due to some errors in the references.

References

Ambraseys, N.N. 1962. Data for the investigation of the seismic sea-waves in the Eastern Mediterranean. *Bulletin of the Seismological Society of America*, **52**, 895–913.

Amoruso, A., Crescentini, L. & Scarpa, R. 2002. Source parameters of the 1908 Messina Straits, Italy, earthquake from geodetic and seismic data. *Journal of Geophysical Research*, **107**, 2080.

Baratta, M. 1910. *La Catastrofe Sismica Calabro-Messinese (28 Dicembre 1908)* [The Calabrio-Messinian seismic catastrophe (28 December 1908)]. Presso la Società Geografica Italiana, Rome.

Barbano, M.S., Pirrotta, C. & Gerardi, F. 2010. Large boulders along the south-eastern Ionian coast of Sicily: storm or tsunami deposits? *Marine Geology*, **275**, 140–154.

Biolchi, S., Furlani, S. et al. 2015. Boulder accumulations related to extreme wave events on the eastern coast of Malta. *Natural Hazards and Earth System Sciences Discussions*, **3**, 5977–6019.

Boccone, P. 1697. *Intorno il terremoto della Sicilia seguito l'anno 1693* [On the earthquake in Sicily in the year 1693]. Museo di Fisica, Venice.

Borg, R.P., Borg, R.C. & Borg, A.G. 2008. The seismic risk of buildings in Malta. *In*: Mazzolani, F., Mistakidis, E. et al. (eds) *Urban Habitat Constructions Under Catastrophic Events*. University of Malta, Msida, Malta, 430–441.

Borg, R.P., D'Amico, S. & Galea, P. 2016. Earthquakes and people: the Maltese experience of the 1908 Messina earthquake. *In*: D'Amico, S. (ed.) *Earthquakes and their Impact on Society*. Springer International, Cham, Switzerland, 533–562.

Boschi, E., Guidoboni, G., Ferrari, D., Mariotti, G., Valensise, A. & Gasperini, P. 2000. Catalogue of strong Italian earthquakes from 461 B.C. to 1997. *Annali di Geofisica*, **43**, 843–868. (CD-ROM.)

Bryant, E.A. & Young, R.W. 1996. Bedrock-sculpturing by tsunami, South Coast New South Wales, Australia. *Journal of Geology*, **104**, 565–582.

Camilleri, D.H. 2006. Tsunami construction risks in the Mediterranean – outlining Malta's scenario. *Disaster Prevention and Management*, **15**, 146–162.

Comerci, V., Blumetti, A.M. et al. 2008. One century after the 1908 Southern Calabria – Messina earthquake (southern Italy): a review of the geological effects. *Geophysical Research Abstracts*, **10**, EGU2008-A-09190.

De la Torre, F.R. 1995. Spanish sources concerning the 1693 earthquake in Sicily. *Annali di Geophysica*, **28**, 523–569.

De Martini, P.M., Barbano, M.S., Smedile, A., Gerardi, F., Pantosti, D., Del Carlo, P. &

PIRROTTA, C. 2010. A unique 4,000 year long geological record of multiple tsunami inundations in the Augusta Bay (eastern Sicily, Italy). *Marine Geology*, **276**, 42–57.

DE MARTINI, P.M., BARBANO, M.S., PANTOSTI, D., SMEDILE, A., PIRROTTA, C., DEL CARLO, P. & PINZI, S. 2012. Geological evidence for paleotsunamis along eastern Sicily (Italy): an overview. *Natural Hazards and Earth System Sciences*, **12**, 2569–2580.

DE SOLDANIS, G.P.F.A. 1746. *Il Gozo Antico-Moderno e Sacro-Profano (Gozo, Ancient & Modern, Religious & Profane)*. Rabat, Gozo. (In Italian) Unpublished. In National Library of Malta, Libr.145.

DE SOLDANIS, G.P.F.A. [1746] 1936. *Għawdex bil-ġrajja tiegħu (Il Gozo Antico-Moderno e Sacro-Profano)*. Government Press, Malta, 336pp. (Trans G. Farrugia from Italian to Maltese).

DE SOLDANIS, G.P.F.A. [1746] 1999. *Gozo, Ancient & Modern, Religious & Profane (Il Gozo Antico-Moderno e Sacro-Profano)*. Media Centre Publications, Malta, 226 pp. (Trans. A. Mercieca from Maltese to English).

DRAGO, A., AZZOPARDI, J., GAUCI, A., TARASOVA, R. & BRUSHY, A. 2013. Assessing the offshore wave energy potential for the Maltese Islands. *In*: *Sustainable Energy 2013: the ISE Annual Conference*, Thursday 21 March 2013, Qawra, Malta. Institute for Sustainable Energy, University of Malta, 16–27.

ENGLAND, P., HOWELL, A., JACKSON, J. & SYNOLAKIS, C. 2015. Palaeotsunamis and tsunami hazards in the Eastern Mediterranean. *Philosophical Transactions of the Royal Society A: Mathematical, Physical and Engineering Sciences*, **373**, https://doi.org/10.1098/rsta.2014.0374

GALEA, A.M. 1909. *It-Theżhiża ta'Messina; it-28 tax-Xahar tal Milied 1908*. [The earthquake at Messina on the 28th December 1908]. Il-Kotba tal-Moghdija taż-Żmien. [The passage of Time Books]. Printed by G. Muscat, Valletta.

GALEA, P. 2007. Seismic history of the Maltese islands and considerations on seismic risk. *Annals of Geophysics*, **50**, 725–740.

GERARDI, F., BARBANO, M.S., DE MARTINI, P.M. & PANTOSTI, D. 2008. Discrimination of tsunami sources (earthquake v. landslide) on the basis of historical data in eastern Sicily and southern Calabria. *Bulletin of the Seismological Society of America*, **98**, 2795–2805.

GERARDI, F., SMEDILE, A. *ET AL.* 2012. Geological record of tsunami inundations in Pantano Morghella (south-eastern Sicily) both from near and far-field sources. *Natural Hazards and Earth System Sciences*, **12**, 1185–1200.

GUIDOBONI, E., COMASTRI, A. & TRAINA, G. 1994. *Catalogue of Ancient Earthquakes in the Mediterranean Area up to the 10th Century*. ING-SGA, Rome.

GUTSCHER, M.-A., ROGER, J., BAPTISTA, M.-A., MIRANDA, J.M. & TINTI, S. 2006. Source of the 1693 earthquake and tsunami (Southern Italy): new evidence from tsunami modeling of a locked subduction fault plane. *Geophysical Research Letters*, **33**, L08309, https://doi.org/10.1029/2005gl025442

HANSOM, J.D., BARLTROP, N.D.P. & HALL, A.M. 2008. Modelling the processes of cliff-top erosion and deposition under extreme storm waves. *Marine Geology*, **253**, 36–50.

IIDA, K., COX, D. & PARARAS-CARAYANNIS, G. 1967. *Preliminary Catalog of Tsunamis Occurring in the Pacific Ocean*. Data Report No. 5. Hawaii Institute of Geophysics, University of Hawaii, Honolulu.

KASTENS, K.A. & CITA, M.B. 1981. Tsunami induced sediment transport in the abyssal Mediterranean Sea. *Geological Society of America Bulletin*, **89**, 591–604.

KÖNG, E., ZARAGOSI, S. *ET AL.* 2016. Untangling the complex origin of turbidite activity on the Calabrian Arc (Ionian Sea) over the last 60 ka. *Marine Geology*, **373**, 11–25.

LORITO, S., TIBERTI, M.M., BASILI, R., PIATANESI, A. & VALENSISE, G. 2008. Earthquake generated tsunamis in the Mediterranean Sea: scenarios of potential threats to Southern Italy. *Journal of Geophysical Research*, **113**, B01301, https://doi.org/10.1029/2007jb004943

MICALLEF, A., FOGLINI, F., LE BAS, T., ANGELETTI, L., MASELLI, V., PASUTO, A. & TAVIANI, M. 2013. The submerged paleolandscape of the Maltese Islands: morphology, evolution and relation to Quaternary environmental change. *Marine Geology*, **335**, 129–147.

MOTTERSHEAD, D., BRAY, M., SOAR, P. & FARRES, P.J. 2014. Extreme wave events in the central Mediterranean: geomorphic evidence of tsunami on the Maltese Islands. *Zeitschrift für Geomorphologie*, **58**, 385–411.

MOTTERSHEAD, D.N., BRAY, M.J., SOAR, P.J. & FARRES, P.J. 2015. Characterisation of erosional features associated with tsunami terrains on rocky coasts of the Maltese Islands. *Earth Surface Processes and Landforms*, **40**, 2093–2111.

NANDASENA, N.A.K., PARIS, R. & TANAKA, N. 2011. Reassessment of hydrodynamic equations: minimum flow velocity to initiate boulder transport by high energy events (storms, tsunamis). *Marine Geology*, **281**, 70–84.

OMORI, F. 1909. Preliminary report on the Messina–Reggio Earthquake of Dec. 28, 1908. *Bulletin of Imperial Earthquake Investigation Committee*, **3**, 37–46.

PAPADOPOULOS, G. 2009. Tsunamis. *In*: WOODWARD, J.C. (ed.) *The Physical Geography of the Mediterranean*. Oxford University Press, Oxford, 483–512.

PAPADOPOULOS, G.A. & IMAMURA, F. 2001. A proposal for a new Tsunami intensity scale. *In*: *Proceedings of the International Tsunami Symposium*, August 7–10, 2001, Seattle, Washington, USA, 569–576.

PAPADOPOULOS, G.A., GRÀCIA, E. *ET AL.* 2014. Historical and pre-historical tsunamis in the Mediterranean and its connected seas: geological signatures, generation mechanisms and coastal impacts. *Marine Geology*, **354**, 81–109.

PARARAS-CARAYANNIS, G. 2011. The earthquake and tsunami of July 21, 365 AD in the eastern Mediterranean Sea. Review of Impact on the Ancient World – Assessment of recurrence and future impact. *Science of Tsunami Hazards*, **30**, 253–292.

PARARAS-CARAYANNIS, G. & MADER, C.L. 2010. The earthquake and tsunami of 365 A.D. in the eastern Mediterranean Sea. *In*: *Proceedings of the 9th U.S. National and 10th Canadian Conference of Earthquake Engineering*, 25–29 July 2010, Toronto, Ontario, Canada, Volume **4**, 3284–3293, Paper No.

1846, Tsunami Society International, Honolulu, Hawaii, USA.

PIATANESE, A. & TINTI, S. 1998. A revision of the 1693 eastern Sicily earthquake and tsunami. *Journal of Geophysical Research*, **103**, 2719–2758.

PINO, N.A., PIATANESI, A., VALENSISE, G. & BOSCHI, E. 2009. The 28 December 1908 Messina Straits Earthquake (*Mw* 7.1): A Great Earthquake throughout a Century of Seismology. *Seismological Research Letters*, **80.2**, 243–259.

PLATANIA, G. 1909. I fenomini marittimi che accompagnarono il terremoto di Messina del 28 Dec 1908. [The maritime phenomena that accompanied the Messina earthquake of 28 Dec. 1908.] *Rivista Geografica Italiana*, **16**, 154–161.

POLONIA, A., BONATTI, E., CAMERLENGHI, A., LUCCHI, R.G., PANIERI, G. & GASPERINI, L. 2013. Mediterranean megaturbidite triggered by the AD 365 Crete earthquake and tsunami. *Scientific Reports*, **3**, 1285, https://doi.org/10.1038/srep01285

POLONIA, A., ROMANO, S. ET AL. 2015. Are repetitive slumpings during sapropel S1 related to paleo-earthquakes? *Marine Geology*, **361**, 41–52.

POLONIA, A., NELSON, C.H., ROMANOA, S., VAIANI, S.C., COLIZZA, E., GASPAROTTO, G. & GASPERINI, L. 2016*a*. A depositional model for seismo-turbidites in confined basins based on Ionian Sea deposits. *Marine Geology*, first published online 18th May 2016, https://doi.org/10.1016/j.margeo.2016.05.010

POLONIA, A., VAIANI, S.C. & DE LANGE, G.J. 2016*b*. Did the A.D. 365 Crete earthquake/tsunami trigger synchronous giant turbidity currents in the Mediterranean Sea? *Geology*, **44**, 19–22.

RIDENTE, D., MARTORELLI, E., BOSMAN, A. & CHIOCCI, F.L. 2014. High-resolution morpho-bathymetric imaging of the Messina Strait (Southern Italy). New insights on the 1908 earthquake and tsunami. *Geomorphology*, **208**, 149–159.

SAN PEDRO, L., BABONNEAU, N., GUTSCHER, M.-A. & CATTANEO, A. 2016. Origin and chronology of the Augias deposit in the Ionian sea (central Mediterranean sea), based on new regional sedimentological data. *Marine Geology*, https://doi.org/10.1016/j.margeo.2016.05.005

SCICCHITANO, G., MONACO, C. & TORTORICI, L. 2007. Large boulder deposits by tsunami waves along the Ionian coast of south-eastern Sicily (Italy). *Marine Geology*, **238**, 75–91.

SHAW, B., AMBRASEYS, N.N. ET AL. 2008. Eastern Mediterranean tectonics and tsunami hazard inferred from the AD 365 earthquake. *Nature Geoscience*, **1**, 268–276.

SHOWER, J. 1693. *Practical reflections on the late earthquakes in Jamaica, England, Sicily, Malta, & c. Anno 1692: With a particular, historical account of those, and divers other earthquakes.* Printed for John Salusbury at the Rising Sun in Cornhill and Abraham Chandler at the Chirurgion's Arms in Aldersgate Street, London.

SYNOLAKIS, C.E. 1987. The runup of solitary waves. *Journal of Fluid Mechanics*, **185**, 523–545.

TINTI, S., ARMIGLIATO, A., PAGNONI, G. & ZANIBONI, F. 2005. Scenarios of giant tsunamis of tectonic origin in the Mediterranean. *ISET Journal of Earthquake Technology*, **42**, 171–188, Paper No. 464.

VISINI, F., DE NARDIS, R., BARBANO, M.S. & LAVECCHIA, G. 2009. Testing the seismogenic sources of the January 11th 1693 Sicilian earthquake (Io X/XI): insights from macroseismic field simulations. *Bollettino della Società Geologica Italiana*, **128**, 147–156.

Cataloguing tsunami events in the UK

DAVE LONG

British Geological Survey, Edinburgh, EH14 4AP, UK

Present address: 22a Edgehead, Pathhead, Midlothian EH37 5RJ, UK, davelongmarinegeology@gmail.com

Abstract: Tsunami catalogues provide important datasets in assessing the risk from infrequent but potentially high-impact events. Although the UK is located away from subduction zones (the most common origin of tsunamis), tsunamis have struck its shores, most notably those triggered by the prehistoric Storegga submarine landslide and the 1755 Lisbon earthquake. Since the major events of 2004 (Indian Ocean) and 2011 (Japan) tsunamis are in the public psyche, even if the risks to UK coasts are not. Due to this heightened awareness, many reported events are claimed to be tsunamis and the potential for tsunamis is increasingly included in risk planning; understanding the true frequency of tsunamis is therefore important. Within the UK, the evidence for tsunamis includes tide gauge readings, reported visual observations and interpretation of sedimentological features. Catalogues need to consider whether the event is a true tsunami in order to avoid a plethora of claims that confound risk assessments; for example, recent well-documented events generated by weather systems (meteotsunamis) provide a possible explanation for some historical events. A detailed examination of the impact of tsunamis upon the UK coast is provided, including examples of events triggered by the three primary causes of tsunamis: seismicity, submarine landslides and coastal landslides.

One of the most dramatic and unpredictable marine threats to any coastal community is a tsunami. Tsunamis comprise translational waves that travel at great speed in deep water but, as they approach the coastline and enter shallow water, the wave slows down and increases dramatically in height. The waves can be very destructive as they strike the coast, both in their initial impact and also as they withdraw, dragging loose material out to sea. They are usually associated with large earthquakes, typical of active margins, but can be triggered by other sudden changes in submarine topography such as underwater landslides or volcanic eruptions. They can also be created by sudden impacts upon the water surface causing displacement, such as from a coastal landslide or a bolide impact.

Understanding the frequency of tsunami events is a key element in assessing the potential risk arising from their impacts on coastal communities and associated infrastructure. Establishing a database of events that have been attributed to tsunamis is therefore an important stage in assessing the tsunami hazard and risk to any country. In recent years, particularly following the Indian Ocean tsunami of 2004 and the Tohoku tsunami of 2011, the use of the word 'tsunami' and claims of tsunami events, past and present, has increased significantly. These claims need to be evaluated to improve risk assessment. Global databases such as the US National Geophysical Data Center NOAA/WDS (National Oceanic and Atmospheric Administration/World Data Service) Historical Tsunami database (https://www.ngdc.noaa.gov/hazard/tsu_ db.shtml) are an important source of information. However, when considering the tsunami risk to a country, reports found in a wider range of sources such as newspapers, personal documents and folklore need to be examined and all individual events recorded in this way evaluated.

In order to compile a relevant and reliable UK tsunami database, previous misinterpretations of events need to be removed. For example, events within the US National Geophysical Data Center NOAA/WDS Global Historical Tsunami Database that are known to be reports of seiches (a standing wave in an enclosed or semi-enclosed body of water due to resonance) have been excluded. Similarly, the Global Historical Tsunami Database lists an event for both Flanders and Norfolk on 31 January 1953 which was, in fact, a major storm surge with extensive flooding. This surge began when a northerly storm combined with spring tides to breach coastal defences, leading to 1836 people being killed in the Netherlands, 307 in England and 28 in Belgium (http://floodlist.com/europe/1953-north-sea-floods). This event has therefore been omitted from the UK tsunami database. Further, reports of tsunamis that have mispositioned the event have been ignored, for example the event of 31 March 1761 at Carrick in County Wexford, Ireland was misrecorded as occurring at Carrickfergus, UK.

From: Scourse, E. M., Chapman, N. A., Tappin, D. R. & Wallis, S. R. (eds) 2018. *Tsunamis: Geology, Hazards and Risks*. Geological Society, London, Special Publications, **456**, 143–165.
First published online June 29, 2017, https://doi.org/10.1144/SP456.10

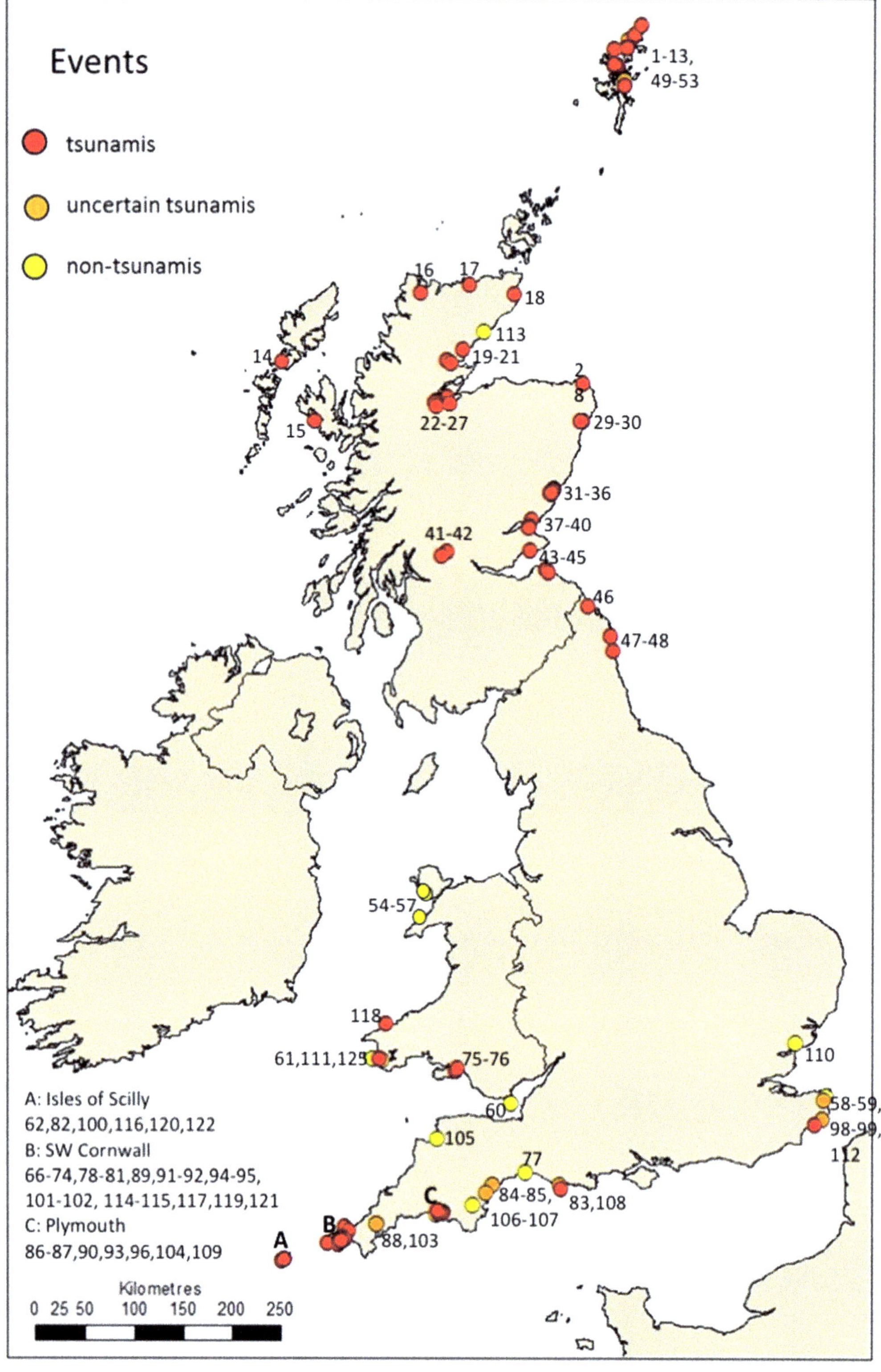
Events
tsunamis
uncertain tsunamis
non-tsunamis
1-13,
49-53
16
17
18
113
14
19-21
2
8
22-27
29-30
15
31-36
37-40
41-42
43-45
46
47-48
54-57
118
61,111,125
75-76
110
60
58-59,
98-99,
112
105
77
84-85,
106-107
83,108
C
B
88,103
A
A: Isles of Scilly
62,82,100,116,120,122
B: SW Cornwall
66-74,78-81,89,91-92,94-95,
101-102, 114-115,117,119,121
C: Plymouth
86-87,90,93,96,104,109
Kilometres
0 25 50 100 150 200 250

For the UK, it may be considered that the rarity of tsunamis means that they are not an important geohazard which requires to be considered in coastal risk planning. Notably, they are not included within the UK National Risk Register (2015). However, risk assessments for some infrastructure need to take account of events of very low frequency (perhaps once in 10 ka) because the impact could be very high upon critical infrastructure located on the coast such as industrial facilities like refineries, chemical plants and nuclear power stations, or on emergency services required to respond to any impact, particularly hospitals.

Creation of a UK catalogue

A catalogue of tsunami events for the UK was created for the EU TRANSFER project (Tsunami Risks and Strategies for the European Region) in the form of a database and report contributing to a Europe-wide review (Long & Wilson 2007). That study was founded on information gathered together by a previous EU study, GITEC (Genesis and Impact of Tsunamis on European Coasts), and used additional historical documents. The final product was incorporated into a GITEC-based Europe-wide compilation (Maramai *et al.* 2014). Copies of the UK database and report (Long & Wilson 2007) were widely distributed and made publicly available (e.g. NORA, http://nora.nerc.ac.uk/502969/). Several of the UK events reported were subsequently reviewed by Haslett & Bryant (2008). The catalogue was extended (Long 2015) as a component of the NERC Arctic Research Programme Landslide Tsunami project (http://arp.arctic.ac.uk/projects/landslide-tsunami; and a BGS internal report placed in open access (http://nora.nerc.ac.uk/513298/). These catalogues discuss in detail the individual events previously claimed to be tsunamis, evaluating the evidence for the claim of a tsunami triggered by geological processes. The database behind these catalogues has subsequently been extended by the inclusion of additional events and sites that have since been identified, and further information added to provide an up-to-date listing provided in Appendix 1.

The original database comprised three tables: events, locations and references. The locations table was the key dataset recording all sites where a tsunami-like event had been mentioned, and cross-correlated with tables of references and events. Plotting out these locations shows that the sites where a tsunami-like event has been suggested are primarily located in the SW of England and Wales or around the Scottish coast (Fig. 1). A few events have been claimed elsewhere but these have generally been shown to be non-tsunami events, with meteotsunamis being a frequent cause.

Meteotsunamis

Tsunami-like effects can be created by certain meteorological conditions such as cyclones, hurricanes and thunderstorms, or more distinct events such as atmospheric pressure jumps and atmospheric gravity waves. These processes can generate a wave with similar characteristics to tsunamis when the disturbance travels at the same speed as the surface wave that it has generated, thereby both sustaining and adding to the wave (Monserrat *et al.* 2006).

Several such events have been recognized from historical records around the UK (Table 1) and have been considered as an important potential coastal hazard having caused several fatalities (1929, 1932, 1964) (Haslett *et al.* 2009). Events in 2011 (Fig. 2) and 2015 have been examined in detail and correlated with particular meteorological conditions (Tappin *et al.* 2013; Sibley *et al.* 2016).

As meteotsunamis have been recognized over the last 100 years it must be assumed that similar events occurred previously, and may explain the origin of tsunamis reported around the British Isles for which no potential tsunami trigger has been identified. For example, an event in 1858 was examined in detail by Newig & Kelletat (2011) who suggested that a tsunami, sourced in the Atlantic, generated anomalous waves on both sides of the English Channel, extending into the North Sea as far east as Denmark. The observations of the waves were recorded at the same time that meteorological phenomena, including sudden severe thunderstorms, were noted; a meteotsunami therefore seems a more likely explanation for the dramatic waves reported.

These events are termed as meteotsunamis and should not be confused with tsunamis which have a geological (seismic or structural failure) trigger. However, in any assessment of risk to coastal communities and infrastructure, the threat from both geologically triggered tsunamis and meteotsunamis

Fig. 1. Map of locations around the UK with reported tsunami events. Modified from Long (2015). Locations colour coded according to event status with certain tsunamis overlying uncertain tsunamis overlying non-tsunamis where more than one event is reported at a single location. Locations are numbered to link with sites listed in Appendix 1. Note that, due to the number of events and locations in SW England, three areas (Isles of Scilly, SW Cornwall and Plymouth) are listed separately as A, B and C, respectively.

Table 1. *Probable meteotsunami events in the UK*

Year	Month	Day	Location	References
1761	July	28	Mount's Bay (Cornwall)	Borlase (1762)
1824	November	23	Chesil Beach (Dorset)	Haslett & Bryant (2009)
1847	May	23	Cornwall	Long (2015)
1848	July	7	Dorset	Roberts (1849)
1858	June	5	English Channel	Newig & Kelletat (2011)
1859	October	4	Cornwall and Devon	Dawson *et al.* (2000), Edmonds (1860, 1862)
1868	April	23	Dorset	Haslett & Bryant (2009)
1869	September	29	Cornwall and Devon	Dawson *et al.* (2000), Perrey (1872)
1883	October	17	Severn Estuary	Haslett & Bryant (2009)
1892	August	18	Devon and Cornwall	Haslett & Bryant (2009)
1910	December	16	Ilfracombe (Devon)	Haslett & Bryant (2009)
1929	July	20	Folkestone (Kent), Brighton (Sussex)	Haslett & Bryant (2009), Haslett *et al.* (2009)
1932	August	2	Aberavon (Glamorgan)	Haslett *et al.* (2009)
1938	August	5	Bridlington (Yorkshire)	Haslett *et al.* (2009)
1939	July	4	Milford Haven (Dyfed)	Haslett *et al.* (2009)
1939	July	5	Weymouth (Dorset)	Haslett *et al.* (2009)
1957	July	6	Bembridge (Isle of Wight)	Haslett *et al.* (2009)
1964	May	17	Arnside (Cumbria)	Haslett *et al.* (2009)
1966	July	31	Westward Ho! (Devon) and Pembrokeshire	Haslett & Bryant (2009), Haslett *et al.* (2009)
1979	February	13	Bristol Channel and western part of English Channel	Haslett & Bryant (2009)
2008	May	28	Grampian	Sibley *et al.* (2016)
2011	June	27	English Channel	Tappin *et al.* (2013)
2015	July	1	Grampian	Sibley *et al.* (2016)

Fig. 2. Screen shot from a video of 27 July 2011 meteotsunami in the Yealm estuary, Cornwall (by permission of Simon Fitch) (Tappin *et al.* 2013).

should be considered. The fact that the frequency of the latter in the UK appears to be greater than the frequency of tsunamis *sensu stricto*, and that they have been known to cause fatalities, means that meteotsunamis are important phenomena which should be included in any risk assessment for the UK coastline. A table of these events was included in Long (2015), but further examination of nineteenth-century newspapers suggests additional events (Table 1). A greater occurrence during summer months is apparent, when thunderstorms are more common. The frequency of meteotsunamis may increase in the future as rising temperatures due to climate change are likely to be associated with increased storminess, including thunderstorms (Diffenbaugh *et al.* 2013).

Examination of the evidence

Modern tsunami events are observed and often recorded on eye witnesses' videos, and transmitted globally almost instantly. These observations and recordings can provide a wealth of information to be used to reliably describe and quantify the event in terms of location, time, extent of run-up, number of waves, etc. However, there is less information available to review historic events. Within the database (Appendix 1), three categories have been used to classify the types of evidence for describing an event as a tsunami, namely 'tide gauges', 'observations' or 'deposits'.

Tide gauges

Tide gauges provide a record of the changing height of the sea at a fixed location, overcoming the limitations of one-off measurements. Self-recording devices to measure the height of the sea were developed in the 1830s. They are often located on piers, providing a record of changes in sea level around the coast. Some sites around the UK have records extending back continuously for more than 100 years, with the records preserved by the British Oceanography Data Centre (BODC; https://www.bodc.ac.uk/data). The stations were established to record low-frequency processes such as tides and storm surges. However, tide gauges can also record tsunami events that are smaller in amplitude than those likely to be noted by human observations; they also provide a useful record of events occurring at times when there are few or no observers, for example during the hours of darkness, providing that high-frequency recording is available.

Originally, tide gauges recorded the height of the sea on a roll of paper, often on a daily, weekly or monthly basis. Since the 1970s, charts have been replaced by electronic sampling at most UK sites at intervals averaging 15 minutes, although there are plans to reduce the period to 5 minutes (P.L. Woodworth, pers. comm., 2008). There are 44 tide gauges around the coast of the UK, run by the Tide Gauge Inspectorate, which can provide near-real-time data distributed by BODC. Data from another 12 tide gauges are also available in near-real time from the Channel Coastal Observatory (http://www.channelcoast.org). However, these real-time readings may not be of high enough frequency to extract an optimum dataset to describe tsunami waves. There are only a few high-frequency (0.1 Hz) tide gauges in the UK; most only record every 15 minutes (*c.* 1.1 mHz) which is too infrequent for analysing individual waves.

Information on past events can be examined in the form of the analogue records preserved. For example, several small tsunamis initiated by earthquakes offshore Iberia have been detected on UK tide gauges (Dawson *et al.* 2000) and can be shown to extend over several hours (Fig. 3). Even the 2004 Indian Ocean tsunami was detected at several UK gauges (Woodworth *et al.* 2005), although the 15 minute sampling underestimated the amplitude of the 2004 tsunami impact at Newlyn by about one-third compared with digitization of paper records at 5 minute intervals. These recorded events are listed in Appendix 1.

Observations

Although the scientific term 'tsunami' has now entered common parlance, prior to the 2004 Indian Ocean event a range of other terms was often used. Within the UK, these included 'tidal wave' and 'seismic sea wave' and, even further back in time, the term 'sea-bore' was used to describe such events in historical documents. Sometimes there was simply a description of unusual movements of the sea, often under the term 'phenomenon', that may subsequently be interpreted as a tsunami-like event.

Few documents or reports were published prior to the eighteenth century, and the recognition of potential tsunami events was rarely clear, with possible conflation with other events such as storms. However, by the nineteenth century, newspaper reporting of events was more reliable and letters/reports from local agents were distributed to many newspapers with some reports reprinted in ten or more publications, usually *verbatim*. Newspaper reports of events in 1802, 1821, 1843, 1847, 1858 and 1869 are listed in Appendix 1.

For example, the 1858 event at Folkestone was reported in the Kentish Gazette on 8 June 1858, page 6, as follows:

> Violent Thunder Storm. – On Saturday morning about seven o'clock, a heavy tempest broke over our town, which was one of the most terrific we have witnessed for some years past, a very extraordinary phenomenon occurred in the harbour. About eight o'clock the storm

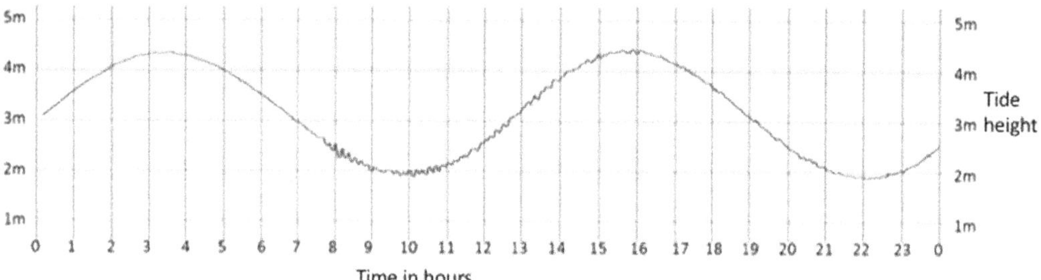

Fig. 3. Tide gauge for St Mary's, Isles of Scilly on 28 Feb 1969 showing a small tsunami (maximum amplitude trough to crest 0.25 m) arriving at about 0735, generated by 7.3 M_s earthquake west of Portugal (site 116 in Appendix 1). Drawn from scan of original record provided by BODC.

> had reached its height; the thunder keeping up a continuous roll, interrupted only by the heavy peals which followed the most vivid lightning; the rain falling in torrents for half an hour, accompanied by hail. The wind at this time being E.N.E. suddenly veered round to W.N.W., and the tide which had ebbed sufficiently for the steamers to ground suddenly returned, and rushed into the harbour at the rate of ten knots an hour, and with such force as instantly to fleet the steamers, rising at least three feet, snapping their hawsers like pack-thread, and causing such commotion among the vessels and boats as to create great alarm. The Princess Mary carried away her bowsprit and figurehead in fouling the Princess Helena, which was thrown almost on her beam ends; a number of large fishing boats were cast adrift, one being capsized, and drawn out to sea, where it sunk by the reflux, which was as sudden and almost as strong as the influx, and as dangerous to the shipping, great fears being entertained of the drift vessels being carried out to sea. The phenomenon was several times repeated, but each time with less force; the most singular feature was, that outside the harbour was a dead calm. An occurrence of this nature is beyond the memory of the oldest seaman.

This event (site 97 in Appendix 1) has been claimed to be a tsunami with an unknown source in the Atlantic (Newig & Kelletat 2011). However, numerous newspaper reports of dramatic thunderstorms, both here and at sites across England and France, suggest that a meteotsunami is a more likely explanation for the damaging waves.

For larger-scale events, most notably the 1755 Lisbon earthquake, scientific institutions gathered reports of observations (Borlase 1755; Huxham 1755); these provide a valuable resource when examining the effects of tsunamis and their impact on wider stretches of coastline.

Deposits

The recognition of sediments within coastal deposits as evidence for tsunami events allows databases to be extended backwards in time prior to modern tidal gauges and historical records; events can therefore be identified where no records of observations exist, either due to the lack of any preserved written record or the absence of any observer or tide gauge. Interpreting sediments as tsunamigenic requires an understanding of the position of the coastline when a possible event struck; this means that only sediments of Holocene age can be interpreted as there is too much uncertainty regarding the coastal geometry prior to around 10 ka ago. These deposits usually comprise a layer of sediment showing evidence of a marine origin within a sequence of terrestrial deposits. They may include a chaotic assemblage of diverse material or can be well-sorted, depending on the source material of the deposit. It is important to be able to distinguish these deposits from the fluctuations in sea level reflecting isostatic/eustatic changes that have affected the UK coastline over the last 10 ka. It is also important to distinguish tsunami deposits from storm deposits. The latter are usually quite variable in thickness and distribution, whereas the former may contain a wide variety of material from mud clasts to boulders and usually extend much further inland (Tuttle *et al.* 2004; Morton *et al.* 2007).Within the UK, tsunami deposits are most easily demonstrated by products of the Storegga tsunami of *c.* 8150 BP (Smith *et al.* 2004; Fig. 4), but deposits have also been located associated with the 1755 Lisbon tsunami (Foster *et al.* 1993). Within the database (Appendix 1) sedimentary deposits are further classified as to whether they have been recognized in boreholes or in exposed sections. Where they are exposed, they provide an opportunity to examine spatial relationships more easily than that possible in boreholes. However, because of their coastal setting, exposed sections are vulnerable to post-event erosion.

Other than the evidence provided by layers of fine-grained marine deposits, Haslett & Bryant (2007) suggested that boulder trains in Anglesey and the nearby Lleyn Peninsula are another type of sediment deposit formed by a tsunami and may even have been triggered by a bolide impact with a

Fig. 4. Sand layer within coastal peat deposits at Whalefirth, Yell, Shetland, formed by the Storegga tsunami *c.* 8150 years BP (site 5 in Appendix 1) (from Long 2015; adapted from Tappin *et al.* 2015).

date of AD 1014. They pointed to the presence of imbricated boulder trains comprising very large boulders (the largest *c.* 193 tonnes), suggesting that they could not have been moved by storms but required the energy of a tsunami to move them to their present elevated position. No earthquake or earthquake-like event is included for 1014 in Musson's (2008) review of historic earthquakes for the British Isles. As these boulders appear comparable to megaclasts within the glacial deposits that form the cliff sequences in this part of NW Wales, it is possible that they have simply been eroded out of the local glacial sequence with their positions reflecting glacial processes, particularly as the orientation of the boulder trains mimic the direction of ice movement across Anglesey (Phillips *et al.* 2010). These boulder trains are therefore not considered to be evidence of a tsunami event. Elsewhere, comparable-sized boulder trains have been ascribed to storm deposits and shown not to be formed by a tsunami (Cox *et al.* 2012).

Event classification

The catalogue produced in 2007 for the EU-funded project TRANSFER included three levels of event classification:

- tsunami event;
- uncertain tsunami event; or
- non-tsunami event.

These classes were applied to each event as listed in the NOAA and GTDB databases, and a comment was provided in the catalogue giving a reason for the classification (Long & Wilson 2007). These classifications were also applied to the additional events in the subsequent catalogue (Long 2015), and are summarized in Appendix 1.

Although many events are claimed as tsunamis, it is often difficult to support such claims without evidence of a triggering mechanism such as an earthquake or a submarine landslide. Alternative explanations therefore have to be considered, most notably meteotsunamis as discussed above. If an alternative explanation is the most plausible, then the event should be classified as a non-tsunami event.

In some cases, anomalous waves have been noted in semi-enclosed water bodies, including harbours, and may be seiches triggered by earthquakes. These are not strictly tsunamis but are hazards that need to be considered. Checking details on the timing of events is important in evaluating such reports. For example, there are many reports of water movement associated with the 1755 Lisbon earthquake; however, those noted in the morning of 1 November are clearly associated with the earthquake and are seiches created by the passage of the seismic waves, whereas those noted from mid-afternoon onwards are related to the subsequent tsunami.

Extensive coastal flooding reported in the Severn Estuary in 1607 has been considered as evidence for a major tsunami (Bryant & Haslett 2002), although it is difficult to identify a potential cause. However, examination of oceanographic conditions and meteorological reports over a wider area suggests that it was a combination of exceptionally high tides and strong winds, producing a storm surge that caused the flooding (Horsburgh & Horritt 2006). This event is therefore considered to be a non-tsunami.

Key tsunami events in the UK

Three tsunami events and their impact upon the UK coast are described in detail as examples of events triggered by different primary sources of tsunami, namely seismicity, submarine landslides and coastal landslides.

Storegga slide

Numerous sites have been identified with a thin continuous layer of marine sediments, usually sands, around Shetland and along the eastern and northern coasts of mainland Scotland (Fig. 4; Smith *et al.* 2004), with a further three sites located in NE England (Fig. 5). A similar horizon has been detected at sites along much of the western coast of Norway (Bondevik *et al.* 1997) and at sites in Denmark, the Faroes, Iceland and Greenland. This event is attributed to the failure of 3500 km^3 of sediments on the mid-Norwegian margin known as the Holocene Storegga Slide (Dawson *et al.* 1988; Long *et al.* 1989).

The Holocene Storegga Slide (Bugge *et al.* 1987; Jansen *et al.* 1987) is dated offshore to 7250 $\pm$ 250 ^{14}C years BP (Haflidason *et al.* 2005) and its associated tsunami deposit is dated onshore to 7300 $\pm$ 20 ^{14}C years BP (Bondevik *et al.* 2012). This is approximately 8150 calendar years ago, during the 8.2 ka Holocene cold event. The geological model developed for the Ormen Lange project indicated that the Holocene Storegga Slide was just the most recent of a series of megaslides (>2000 km^2) that occurred offshore mid-Norway from the end of the Pliocene, with a frequency of roughly once every 100 ka over the last 0.5 Ma (Solheim *et al.* 2005). Current studies are suggesting that their frequency may be double that previously estimated, as a new Late Pleistocene megaslide has been identified from the Storegga area (Watts *et al.* 2016). The geological model for these events suggests that, following a major glacial episode and the deposition on the continental slope of a thick pile of unconsolidated sediment upon lithologically contrasting interglacial sediments, enhanced seismic activity associated with the unloading and stress release caused by the melting of ice sheets triggered slope failure (Solheim *et al.* 2005).

The tsunami deposits generally consist of fine- to medium-grained sand, often showing a fining-upwards sequence that can be repeated multiple times, interpreted as sedimentation by different flooding episodes within the complete tsunami event (Shi 1995; Smith *et al.* 2007). The tsunami deposit layer is generally less than 10 cm in thickness but may be more than 1 m thick (sites 21, 23 and 43 in Appendix 1). The sands include microfossils that are indicative of shallow marine conditions. These fossils are commonly broken, suggesting transport in turbulent conditions. Where the layer has been deposited within coastal peats, occasional ripped clasts of peat are incorporated within the tsunami deposit (e.g. at Maryton near Montrose, site 36 in Appendix 1). At the Maggie Kettle's Loch section (site 11 in Appendix 1), it can be shown clearly that the peat clasts are associated with the second wave. This suggests that the coastal peats were eroded by the first wave, causing blocks of peat to be floating when they became incorporated within the deposits of the second wave (Bondevik *et al.* 2003).

There is a general decrease in run-up heights from north to south. The highest sediment run-ups occur in inlets (*c.* 20 m) at sites around Sullom Voe (sites 7–11 in Appendix 1), a large north-facing inlet (Bondevik *et al.* 2003), reducing to a few metres run-up in the vicinity of the Firth of Forth and along the Northumberland coast (Fig. 6). South of Northumberland the former shoreline is now offshore, and any tsunami deposits from this event would have been vulnerable to erosion and reworking during the subsequent marine transgression. However, as several of the sites in Shetland show, the layer can be preserved beneath later sediments below the present-day sea level, even with subsequent marine transgression. The extent of inundation, both horizontally and vertically, can be difficult to estimate but, by examining how the deposit occurs within lake basins, lake thresholds provide altitude control. In coastal sequences where the deposit transgresses from within intertidal muds into coastal peats, such as at Fullerton or Creich (sites 35 and 20, respectively, in Appendix 1), the transgression provides a position for the high-water mark on the day the wave struck (Long *et al.* 1989; Smith *et al.* 2004). The actual run-up of the wave would have exceeded the extent of preserved sediments, so these provide only a minimum inundation.

Examination of clasts within the deposit can also indicate the season of the event as well as providing suitable material for dating. The stage in the development of buds (Bondevik *et al.* 1997), mosses (Rydgren & Bondevik 2015) and fruit (Dawson & Smith 2000), and also the size of fish bones (Bondevik *et al.* 1997) entrapped within the tsunami deposits at sites in Norway and Scotland, suggest that the tsunami struck in late autumn.

Two further sand layers have been found within coastal peats on the east coast of Shetland (sites 49–53 in Appendix 1) and have been attributed to tsunami events suggested at 5500 and 1500 years BP (Bondevik *et al.* 2005; Dawson *et al.* 2006). It was considered that a late stage landslide in the Storegga area may have caused the 5500 BP event (Bondevik *et al.* 2005); however, no likely source for this tsunami event has been identified to date. Dawson *et al.* (2006) suggested that the younger event may have been triggered by a local submarine landslide off the eastern coast of Shetland. However, existing regional morphological seafloor information is not sufficiently detailed to test this hypothesis, but detailed multibeam submarine surveys between Yell and Fetlar, near the entrance of Basta Voe, do not show any evidence of a submarine landslide (Dick 2015). The areas where this feature is noted in Basta Voe include areas of reported peat sliding

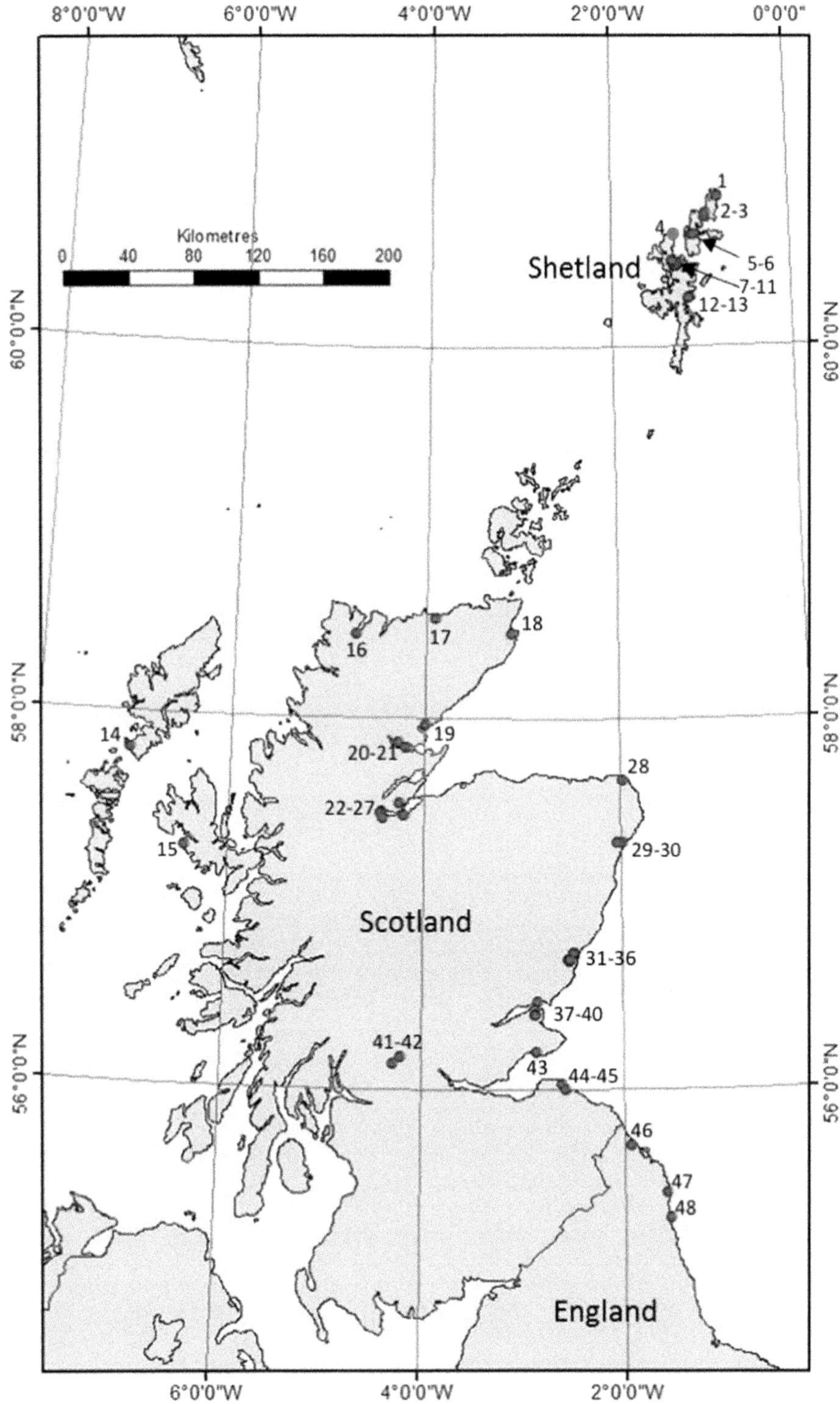

Fig. 5. Map of locations where sedimentary features associated with the Storegga tsunami have been reported. The numbers correlate with site numbers given in Appendix 1. Modified from Long (2015).

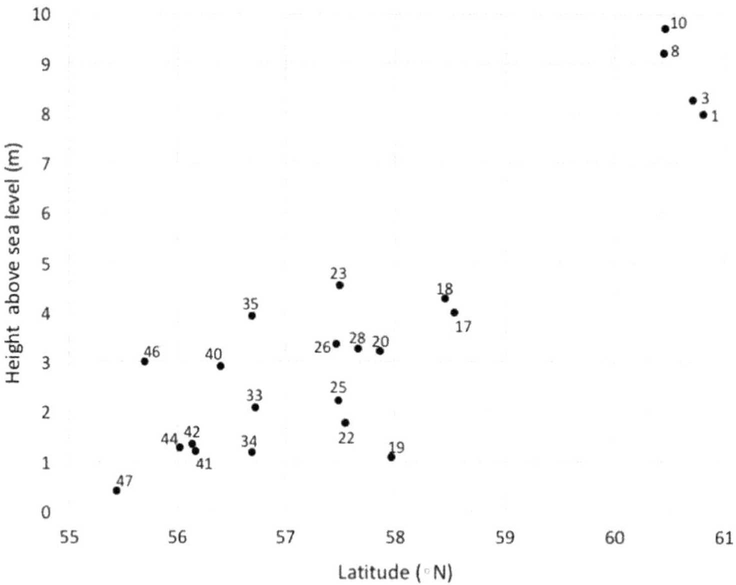

Fig. 6. Plot showing the variation in altitude (metres) of tsunami deposit against latitude at transects from *c.* 55.5°N (NE England) to *c.* 60.5°N (Shetland), using the lowest and highest recorded tsunami sediment basal altitudes in the transect. Derived primarily from data in Smith *et al.* (2004). The numbers correlate with site numbers given in Appendix 1.

(Flinn 1994) that may have incorporated and transported basal sands, as noted elsewhere on Shetland (Dykes & Warburton 2008).

Lisbon Earthquake

The largest seismic event to have struck Europe in the last few hundred years occurred on 1 November 1755 and was noted from Scotland to Austria and also in north Africa, either felt as ground motion or seen as seiches in baths, lakes and enclosed harbours. However, the greatest damage occurred in western Iberia and Morocco. The earthquake is estimated to have had a magnitude M_s of 8.5 and was probably centred on the Azores–Gibraltar boundary between the African and Eurasian plates (Baptista *et al.* 1998). It struck at 09:50 (local time), destroying many buildings in Lisbon, Cadiz and Morocco, and was followed by a tsunami 20 minutes later at Lisbon, compounding the destruction inflicted upon that city. The event is often referred to as the Lisbon Earthquake. The tsunami was observed along the western Iberian coast and Morocco, and in the UK and Ireland, and even across the Atlantic in the Caribbean and South America. The earthquake and tsunami killed between 60 000 and 100 000 people. Reports of the event in Britain were gathered together by the Royal Society in London and provide a very useful record. They are predominantly of the seiche noted about 11:00 in the morning in various harbours around the UK, as well as in many lakes and ponds; however, there were also several reports from the SW coast of England and Wales (Fig. 7) and southern Ireland in the afternoon and evening, when a series of waves was noted.

The observations described dramatic movements of the water. Borlase (1755) described the arrival of the waves in Mount's Bay, Cornwall:

> A little after 2 o'clock in the afternoon, about half an hour after ebb, the sea was observed at the Mounts-bay pier to advance suddenly from the eastward. It continued to swell and rise for the space of 10 minutes; it then began to retire, running to the west and southwest, with a rapidity equal to that of a millstream descending to an undershot-wheel; it ran so for about 10 minutes, till the water was 6 feet lower than when it began to retire. The sea then began to return, and in 10 minutes it was at the before-mentioned extraordinary height; in 10 minutes more it was sunk as before; and so it continued alternately to rise and fall between 5 and 6 feet, in the same space of time. The 1st and 2nd fluxes and refluxes were not so violent at the Mount pier as the 3rd and 4th, when the sea was rapid beyond expression, and the alterations continued in their full fury for 2 hours; they then grew fainter gradually, and the whole commotion ceased about low water, $5\frac{1}{2}$ hours after it began.

Further along the coast Huxham (1755) described events at Plymouth:

> . . .the tide had made a very extraordinary out (or recess) almost immediately after high water (about 4pm) left both the passage-boats, with some horses, and several persons, at once quite dry in the mud, though the minute

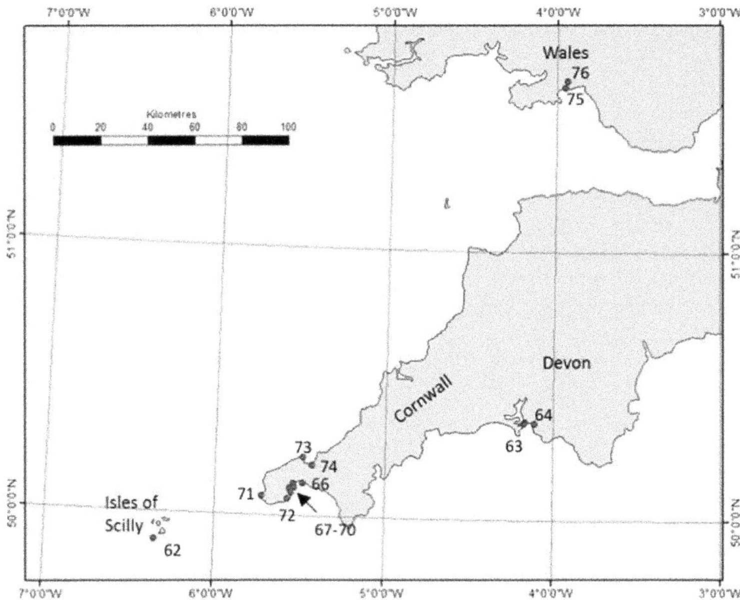

Fig. 7. Sites in SW England and Wales where the 1755 Lisbon earthquake tsunami was observed. The numbers correlate with site numbers given in Appendix 1. Modified from Long (2015).

> or two before, in four or five feet water: in less than eight minutes the tide returned with the utmost rapidity, and floated both the boats again, so that they had near five feet water. The sea sunk and swelled, though in a much less degree, for near half an hour longer.

Sedimentation attributed to this tsunami event has been reported (Foster *et al.* 1991, 1993) from a series of cores taken at Big Pool on St Agnes in the Isles of Scilly (site 62 in Appendix 1). The particle size analysis of the main sand unit, which exceeds 1 m in thickness in places, suggests a high-magnitude, low-frequency depositional sequence, with a number of large waves spread over a period of several hours, in order to produce the fining-upward sequences observed. Radiocarbon dating suggests a mid-eighteenth century age (Foster *et al.* 1993) with a similar age estimate from optically stimulated luminescence dating for the deposits (Banerjee *et al.* 2001). Based on the morphology of the deposit, it is attributed to the 1755 tsunami rather than a storm event. Edmonds (1843) reported that blocks of granite 3 m above the high water mark were moved by the waves.

The event of 1755 is the largest from the Azores–Gibraltar area during historic times, with smaller earthquakes triggering tsunamis detected on UK tide gauges in 1941, 1969 and 1975 (sites 114, 116–120 in Appendix 1). It is possible that tsunamis comparable to that of 1755 have struck the UK previously, as studies in SW Iberia indicate that similar large-scale tsunami events occurred with a frequency of 1000–2000 years (Luque *et al.* 2001). There is therefore a possibility that a prehistoric tsunami could have struck the southwestern part of the UK and the southern part of Ireland. As in the case of the 1755 Lisbon Earthquake, the Azores–Gibraltar area appears to be the most likely source area for future tsunami waves that might strike the British Isles.

The 1911 cliff fall at Folkestone

A tsunami of a different origin occurred on 31 December 1911 at Folkestone. The chalk cliffs of southern England are subject to erosion and frequently collapse. In 1911 a large landslide at Abbot's Cliff, east of Folkestone, caused a wave in Folkestone Harbour that was most notable in the shallow areas. Ropes and stanchions were damaged as hawsers snapped. The event was described in the *Folkestone, Hythe, Sandgate and Cheriton Herald* of Saturday 6 January 1912 by a seaman (Charles Harrison) who was standing at the harbour as follows:

> The great rush of water swept up towards the harbour from the east without the slightest warning. I have never witnessed anything like it. The sea was choppy, but not rough, previously. All at once there seemed to be a long heaving, lifting motion, as if the Channel were being raised in the air. The vessels in sight 'danced' violently, the colliers were tossed pell mell, and the crews on various steamers lying around shouted at the tremendous shock. It was an exciting moment – for it passed in a moment – and left us all wondering.

The newspaper article notes that vessels were lifted two to three feet (0.6–0.9 m) by the wave. The effects in the outer (deeper) harbour were described as minimal.

As this event occurred in the evening, while dark, it was some time before the waves were correlated with a large cliff fall located approximately 5 km NE of Folkestone Harbour that had occurred at Abbot's Cliff (Fig. 8). This comprised a large volume of chalk; figures of 800 000 tons were quoted. The cliff fall was reported in the *Whitstable Times and Herne Bay Herald* on Saturday 6 January 1912 as follows:

> From the nearest coastguard station the noise made, as the avalanche of chalk swept down from the face of the cliff into the sea, is described as sounding like heavy guns booming. An examination on Monday showed that the chalk extended like a causeway some 400 yards to sea. It is about 200 yards wide, and at some places 30ft deep. The displacement of water caused by this immense mass entering the sea set up conditions similar to a tidal wave at Folkestone. The water rose several feet, and several colliers broke their cables and ran adrift, causing great alarm in the outer harbour. Fishing smacks danced like corks. The noise of the approaching wave was heard a mile distant, and not within living memory has such a one been seen.

The cliff fall consisted of the middle and upper parts of the Chalk sequence. It was speculated that heavy rains during the preceding months had weakened the cliff, causing the rockfall and depositing it at the foot of the cliff. Abbot's Cliff was previously 348 feet (106 m) high at the failure site (McDakin 1912). Hutchinson (2002) gave a volume of 500 000 m^3 for this fall.

Coastal landslips in this part of the English Channel are a common occurrence (Hutchinson 2002), with impact on local infrastructure (Birch & Warren 2007). Some are slow moving such as the 1915 slide at nearby Folkestone Warren; cliff falls of the Chalk can be sudden however and, if they include a sufficient volume of material entering the sea, they can trigger a tsunami, thereby extending the area at risk from that immediately adjacent to the cliff fall. Modelling of potential cliff falls to explain the 1580 possible tsunami event at Dover (Neilson *et al.* 1984) suggested that volumes of 10^6 m^3 could generate a tsunami (Roger *et al.* 2012). Either a single very large or multiple smaller coastal landslides offers a possible explanation of the reported observations, albeit that Musson (2008) rebutted both of the reported instances.

Discussion

In any assessment of the risks from a tsunami striking a coast, the return period of the triggering mechanism needs to be evaluated. The three tsunami events described above exhibit differing return periods.

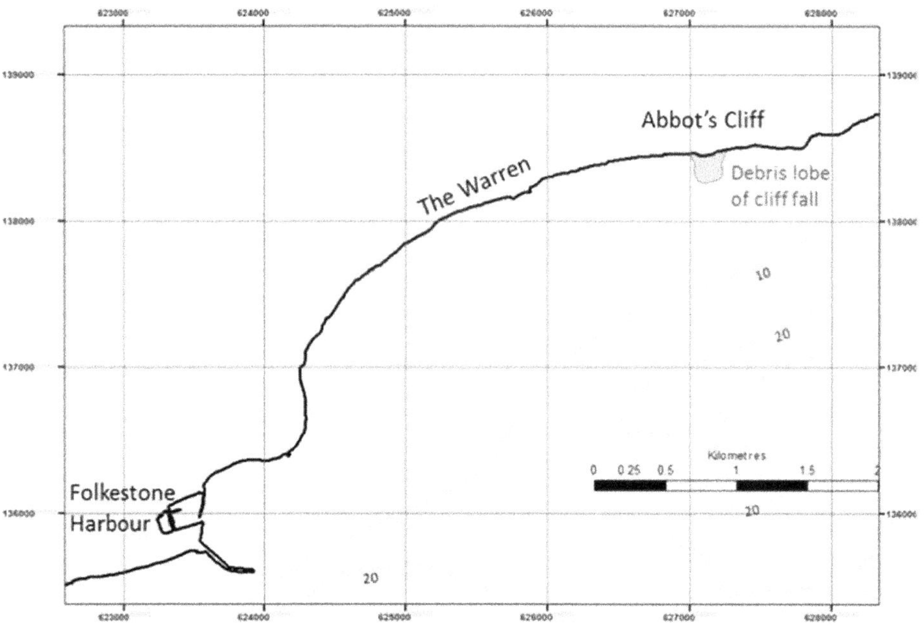

Fig. 8. Map showing location of coastal cliff fall at Abbot's Cliff and Folkestone Harbour where the consequent tsunami was observed on 31 December 1911 (site 112 in Appendix 1). Modified from Long (2015).

The Storegga Slide is the most recent megaslide to have affected the continental margin offshore mid-Norway. Studies have shown that major slides have occurred on this section of the Norwegian margin approximately every 100 ka (Solheim *et al.* 2005), although they may be more frequent (Watts *et al.* 2016). These occur when major glaciations rapidly deposit large quantities of sediments that become unstable when subjected to the neotectonic stresses associated with changes in ice loading that happen during the transition from glacial to interglacial conditions, or when temperature rises change stresses within sediments containing gas hydrates. Other megaslides have been noted in the Arctic with sufficient volume to create a tsunami large enough to reach the UK, depending on the style of failure. The exact trigger mechanisms and frequency of large megaslides in this area are still to be determined. It is also possible that future changes in climate may influence the frequency of megaslides in other parts of the North Atlantic and Arctic Oceans, most notably the Greenland margin, if significant changes in ice-loading occur resulting from global temperature rise.

As noted previously the UK is located in an area of low seismicity compared with other parts of the world due to its position away from plate boundaries, the focus of larger seismic events. The nearest plate boundary is located at the Mid-Atlantic Ridge. Spreading centres tend to have an upper limit of magnitude 7 for seismic events. Higher magnitudes tend to occur at destructive margins. The nearest location is along the Azores–Gibraltar alignment, where the African and European plates meet, and is regularly the site of earthquakes in excess of magnitude 7 that have initiated tsunamis. Studies of Holocene coastal sequences in Portugal and Spain have shown evidence of several destructive events comparable with that associated with the 1755 Lisbon tsunami (Luque *et al.* 2001). This suggests that a significant tsunami striking the UK coast from offshore Portugal can be expected every few thousand years or so.

Determining the frequency of locally triggered tsunamis from sources such as cliff falls requires an evaluation of the varying instability of the wide range of rock types around the UK. As this instability often depends on changes in moisture levels, incorporating climatic factors is important when assessing the frequency of coastal landslide-induced tsunami events.

The occurrence of meteotsunamis around the coast of the UK is another hazard that should be included in risks to the UK, particularly as they have caused fatalities in past. They may explain many of the reports of an 'anomalous wave' or 'phenomenon' published in nineteenth-century newspapers. It is considered that these events are under-reported in the UK, but recent analyses of events in 2011 and 2015 (Sibley *et al.* 2016) show that meteorological monitoring may assist in risk assessment.

Conclusions

The UK has been struck by tsunamis in prehistoric and historic times. They are a hazard that should be considered in strategic planning and risk assessment for key infrastructure in a coastal setting. However, it must be recognized that tsunamis are infrequent events and it is possible that planning for other hazards that occur more frequently, such as storm surges, may provide some of the mitigation that is appropriate in planning for tsunamis. Planning for infrastructure with extremely long lifetimes, such as nuclear waste repositories, has to take account of a range of geological processes that are not normally considered and should include the possibility of tsunami events.

The tsunami events that threaten the UK exhibit a wide range of triggering mechanisms. These encompass earthquakes, underwater landslides and cliff falls into the sea. Evaluation of these potential threats needs to cover an extensive geographical area, well beyond the UK. The example of the 1755 Lisbon tsunami highlights the fact that earthquakes thousands of kilometres from the UK can pose a threat. Similarly, the failure of very large sections of the continental margin outwith the UK can have a severe impact upon British coastlines, as demonstrated by the Storegga Slide.

Meteotsunamis are a further hazard and, as past events have incurred fatalities, these also need to be considered in coastal planning. The fact that the frequency of the latter in the UK appears to be greater than the frequency of tsunamis *sensu stricto*, and that they have been known to cause fatalities, means that meteotsunamis are important phenomena which should be included in any risk assessment for the UK coastline.

Studies on establishing a catalogue of tsunamis in the UK began under the EU project TRANSFER and were completed as part of the NERC project Arctic Research Programme Landslide Tsunami (http://arp.arctic.ac.uk/projects/landslide-tsunami). Thanks are due to BODC for providing scans of original tide gauge records. The initial catalogue was published with permission of the Executive Director of BGS. Thanks are given to Professor Dave Tappin for comments on an earlier draft of this paper. Two anonymous referees are thanked for their diligent comments that have greatly improved the clarity of this paper.

Appendix 1

Table of UK tsunami events, based upon the tables presented in Long & Wilson (2007) and Long (2015).

Appendix 1. *Locations of tsunami evidence in the UK*

Site	Event	Confidence	Location	Longitude	Latitude	Evidence	Date	Time	Wave height	Comment	References
1	1	T	Norwick	−0.8035	60.8051	B	~8150 yr BP			Grey sand with clasts and organic fragments	Smith 1993; Smith *et al.* 2004
2	1	T	Snarravoe,	−0.9579	60.6923	B	~8150 yr BP			Lacustrine cores show sand layer with gravel including silt clasts	Bondevik *et al.* 2005
3	1	T	Burragarth	−0.9499	60.7138	B	~8150 yr BP			Grey sand layer up to 0.75 m thick thinning landward with twig fragments	Smith 1993; Smith *et al.* 2004
4	1	T	Loch of Flugarth	−1.3399	60.5972	B	~8150 yr BP			Dated sand layer with marine diatoms within gyttja sequence	Bondevik *et al.* 2002
5	1	T	Whalefirth	−1.1193	60.6048	S	~8150 yr BP			Extensive layer in coastal peats at head of loch, easy access from road	Cascalho *et al.* 2016
6	1	T	Mid Yell	−1.0849	60.6056	S	~8150 yr BP			Sand layer at base of peats	Costa *et al.* 2015
7	1	T	Scatsta Voe	−1.2805	60.4367	S	~8150 yr BP			Run up probably >20 m above palaeo sealevel	Birnie 1981; Smith 1993; Smith *et al.* 2004
8	1	T	Garth's Voe	−1.2565	60.4492	S	~8150 yr BP			Sand layer with fragments of vegetation. Run up probably >20 m above palaeo sealevel	Birnie 1981; Smith 1993; Smith *et al.* 2004
9	1	T	Otter Loch	−1.3168	60.4369	B	~8150 yr BP			Run up probably >20 m above palaeo sealevel	Bondevik *et al.* 2002
10	1	T	The Houb, Sullom Voe	−1.3364	60.4550	S	~8150 yr BP			150 m outcrop showing sand layer with gravel. Run up probably >20 m above palaeo sealevel	Bondevik *et al.* 2002, 2003, 2005
11	1	T	Maggie Kettle's Loch	−1.3326	60.4631	S	~8150 yr BP			Sand layer with clasts of peat, reaches 9.2 msl probably >20 m above palaeo sea level	Bondevik *et al.* 2002, 2003
12	1	T	Garth Loch	−1.1504	60.2663	B	~8150 yr BP			15 mm thick layer of very coarse sand and fine gravel and rip-up clasts	Bondevik *et al.* 2005
13	1	T	Loch of Benston	−1.1595	60.2645	B	~8150 yr BP			Coarse sand with twig fragments	Bondevik *et al.* 2005
14	1	T	Northton	−7.0596	57.7975	B	~8150 yr BP			Sudden short incursion of marine sediments within freshwater sequence	Jordan *et al.* 2010
15	1	T	Talisker Bay	−6.4574	57.2850	B	~8150 yr BP			Inundation of brackish waters into freshwater site, possible erosion of beach barrier	Selby & Smith 2015
16	1	T	Loch Eriboll	−4.7433	58.4485	S	~8150 yr BP			Dating suggests an erosion surface with sandy sediment cover	Long *et al.* 2016

17	1	T	Strath Halladale	−3.9069	58.5377	B	~8150 yr BP		Intraclast suggests an autumn event	Dawson & Smith 1997; Dawson & Smith 2000
18	1	T	Wick River	−3.1276	58.4535	B	~8150 yr BP		Fine sand layer thinning inland	Dawson & Smith 1997
19	1	T	Smithy House	−4.0100	57.9656	B	~8150 yr BP		Fine sand layer	Smith *et al.* 1992
20	1	T	Creich	−4.2777	57.8685	B	~8150 yr BP		Sand layer transgressing from within the carse clay to within peat	Smith *et al.* 1992
21	1	T	Dounie	−4.1971	57.8456	B	~8150 yr BP		Sand layer up to a 1 m thick tapering inland	Smith *et al.* 1992
22	1	T	Munlochy Bay	−4.2609	57.5462	B	~8150 yr BP		Thin fine to medium sand layer	Firth 1984
23	1	T	Bellevue	−4.4417	57.4997	B	~8150 yr BP		Sand layer locally 1.1 m thick	Firth 1984
24	1	T	Tomich	−4.4611	57.4912	B	~8150 yr BP		Sand layer	Firth 1984
25	1	T	Barnyards	−4.4610	57.4894	B	~8150 yr BP		Silty fine sand layer, minimum 2 m run-up above palaeo high water mark	Haggart 1982
26	1	T	Moniack	−4.4308	57.4621	B	~8150 yr BP		Fining upward grey sand layer. Minimum 4 m run-up above palaeo high water mark	Haggart 1982; Smith *et al.* 1985
27	1	T	Castle St., Inverness	−4.2217	57.4786	B	~8150 yr BP		Archaeological site 'white sand' layer	Wordsworth 1985; Dawson *et al.* 1990
28	1	T	Water of Philorth	−1.9765	57.6670	B	~8150 yr BP		Fine to medium sand layer thinning inland	Smith *et al.* 1982, 2004
29	1	T	Waterside	−1.9884	57.3310	B	~8150 yr BP		Grey, micaceous fine sand layer	Smith *et al.* 1983
30	1	T	Tarty Burn	−2.0299	57.3346	B	~8150 yr BP		Silty sand layer tapering inland	Smith *et al.* 1999
31	1	T	Dryleas	−2.4806	56.7363	B	~8150 yr BP		Fine sand layer	Smith & Cullingford 1985
32	1	T	Dubton	−2.4904	56.7326	B	~8150 yr BP		Fine sand layer	Smith & Cullingford 1985
33	1	T	Puggieston	−2.4937	56.7335	B	~8150 yr BP		Grey, micaceous, silty fine sand	Smith & Cullingford 1985
34	1	T	Old Montrose	−2.5438	56.6992	B	~8150 yr BP		Fine sand layer	Smith & Cullingford 1985
35	1	T	Fullerton	−2.5307	56.6947	B	~8150 yr BP		Fine sand layer	Smith *et al.* 1980
36	1	T	Maryton	−2.5177	56.7002	S	~8150 yr BP		Sand layer with clasts of peat in cliff section	Smith *et al.* 1980
37	1	T	Broughty Ferry	−2.8539	56.4710	S	~8150 yr BP		Mesolithic archaeological site with sand layer	Hutcheson 1886; Lacaille 1954; Smith *et al.* 2004
38	1	T	Craigie	−2.8833	56.4070	B	~8150 yr BP		Sand layer	Haggart 1978
39	1	T	St Michael's Wood	−2.8864	56.4043	B	~8150 yr BP		Sand layer	Haggart 1978
40	1	T	Silver Moss	−2.8847	56.4007	B	~8150 yr BP		Sand layer thinning inland	Chisholm 1971
41	1	T	Goodie Water	−4.2168	56.1764	B	~8150 yr BP		Silty fine sand layer	Holloway 2002
42	1	T	Over Easter Offerance	−4.2902	56.1373	B	~8150 yr BP		Sand layer	Sissons & Smith 1965
43	1	T	Cocklemill Burn	−2.8672	56.1977	S	~8150 yr BP		1.77 m thick layer of coarse sand with some gravel, sand clasts and organic material	Tooley & Smith 2005
44	1	T	Lochhouses	−2.6162	56.0303	B	~8150 yr BP		Sand layer with marine diatoms within peat sequence fining and thinning landward	Newey 1965; Smith *et al.* 1991
45	1	T	Hedderwick	−2.5773	55.9999	S	~8150 yr BP		Fine sand and shell hash layer	D.E. Smith pers comm

(Continued)

Appendix 1. (*Continued*)

Site	Event	Confidence	Location	Longitude	Latitude	Evidence	Date	Time	Wave height	Comment	References
46	1	T	Broomhouse Farm	−1.9411	55.7003	B	~8150 yr BP			Sand layer at 2.5–2.7 mOD	Horton *et al.* 1999; Shennan *et al.* 2000
47	1	T	Howick	−1.5917	55.4406	B	~8150 yr BP			Typical beach pebbles and cobbles in a coarse sand-silt matrix at −3.1 mOD	Boomer *et al.* 2007
48	1	T	Low Hauxley	−1.5542	55.3095	B	~8150 yr BP			Archaeological site	Waddington 2014
49	2	uT	Loch of Benston	−1.1595	60.2645	B	~5500 yr BP			Sand layer with gyttja clasts and twigs within Holocene gyttja sequence	Bondevik *et al.* 2005
50	2	uT	Mid Yell	−1.0849	60.6056	S	~5500 yr BP			Sand layer within peats above prominent birch horizon	Costa *et al.* 2015
51	2	uT	Garth Loch	−1.1504	60.2663	B	~5500 yr BP			Fine to medium sand with clasts of gyttja	Bondevik *et al.* 2005
52	3	uT	Basta Voe	−1.0648	60.6698	S	~1500 yr BP			Thin sand layer within peat seen along 200 m cliff section thinning inland to a line of sand grains 2 km away	Bondevik *et al.* 2005; Dawson *et al.* 2006; Toothill 1994
53	3	uT	Dury Voe	−1.1653	60.3229	S	~1500 yr BP			Sand layer thinning inland to 5.5 m above high tide	Bondevik *et al.* 2002, 2005
54	4	nT	Porth Cwyfan	−4.4917	53.1848	S	~1014 AD			Boulder train	Haslett & Bryant 2007
55	4	nT	Porth Terfyn	−4.4706	53.1816	S	~1014 AD			Boulder train	Haslett & Bryant 2007
56	4	nT	Llanddwyn Island	−4.4156	53.1360	S	~1014 AD			Boulder train	Haslett & Bryant 2007
57	4	nT	Porth Dinllaen	−4.5673	52.9458	S	~1014 AD			Boulder train	Haslett & Bryant 2007
58	5	uT	Sandwich	1.3501	51.2817	O	06/04/1580			Most likely a harbour seiche	Neilson *et al.* 1984
59	5	uT	Dover	1.3111	51.1181	O	06/04/1580			Most likely a harbour seiche but may be due to associated cliff falls	Neilson *et al.* 1984
60	6	nT	Bristol Channel	−3.1493	51.3335	O	30/01/1607	900		Extensive flooding	Bryant & Haslett 2002; Haslett & Bryant 2004; Horsburgh & Horritt 2006
61	7	nT	Dale, Miford Haven	−5.1683	51.7075	O	02/07/1749	1100		Sea runs up and down seven times in 45 minutes	Anon 1749
62	8	T	Isles of Scilly, Big Pool	−6.3480	49.8954	B	01/11/1755			Layer of marine sand between peat up to 1 m thick	Banerjee *et al.* 2001; Dawson *et al.* 1991, 2000; Foster *et al.* 1993
63	8	T	Stonehouse Creek, Plymouth	−4.1642	50.3651	O	01/11/1755	1600		Tearing up of mud and sand banks in a very alarming manner	Huxham 1755; Dawson *et al.* 2000
64	8	T	Creston, Plymouth	−4.1091	50.3608	O	01/11/1755	1600		Sea withdraw 4–5 ft, sea returns in 8 mins.	Huxham 1755

65	8	T	Crunhill, Plymouth	−4.1738	50.3596	O	01/11/1755	1600		Sea withdraws and returns, breaks cable	Huxham 1755
66	8	T	St Mount's Bay	−5.4738	50.1174	O	01/11/1755	after 1400		Sudden advance and retreat of the sea, took 5.5 hrs to settle	Borlase 1755
67	8	T	Penzance	−5.5284	50.1176	O	01/11/1755	1445	2.4	Sea rose 8 ft	Borlase 1755
68	8	T	Newlyn	−5.5509	50.0998	O	01/11/1755		3.1	Sea rose 10 ft	Borlase 1755
69	8	T	Mousehole	−5.5413	50.0839	O	01/11/1755			Similar to Newlyn	Borlase 1755
70	8	T	Gwavas Lake	−5.5245	50.1024	O	01/11/1755			Boat estimated sea velocity at 7 mph	Borlase 1755
71	8	T	Lands End, Cornwall	−5.7177	50.0685	O	01/11/1755			Agitation perceived	Borlase 1755
72	8	T	Larmorna Cove, Cornwall	−5.5647	50.0607	O	01/11/1755			Large blocks of granite deposited ten feet above high water	Edmonds 1843, 1845
73	8	T	St Ives	−5.4767	50.2154	O	01/11/1755		2.7	On north side sea rose 8–9 ft	Borlase 1755
74	8	T	Hayle	−5.4214	50.1874	O	01/11/1755	after 1500	2.1	Surge 7 ft high	Borlase 1755
75	8	T	Swansea	−3.9485	51.6129	O	01/11/1755	1845		Agitation	Borlase 1755
76	8	T	Whiterock, Swansea	−3.9333	51.6352	O	01/11/1755	1700		Floating of beached vessels, vessels turned onto river bank, lasted 2 hrs	Borlase 1755
77	9	nT	Lyme Regis	−2.9351	50.7233	O	31/05/1759			Sea flowed in and out three times during an hour	Perrey 1849; Roberts 1849
78	10	T	Penzance	−5.5284	50.1176	O	31/03/1761	1700	1.8	Sea rose 6 feet	Borlase 1762
79	10	T	Mousehole	−5.5413	50.0839	O	31/03/1761			Great agitation	Borlase 1762
80	10	T	Newlyn	−5.5509	50.0998	O	31/03/1761			Sea rose almost as much as at Penzance	Borlase 1762
81	10	T	St Michael's Mount	−5.4738	50.1174	O	31/03/1761	1700	1.2	Tide rose and fell 4 ft at pier	Borlase 1762
82	10	T	Isles of Scilly	−6.3165	49.9172	O	31/03/1761		1.2	Sea rose 4 feet and agitation lasted 2 hours	Borlase 1762
83	11	uT	Weymouth	−2.4523	50.6093	O	09/08/1802	0600	0.35	Tide rose instantaneously, ebbed again, and repeated doing so three times within an hour	Anon 1802*a*; Dawson *et al.* 2000
84	11	uT	Exmouth	−3.4191	50.6156	O	10/08/1802		0.9	The tide suddenly rose three feet in height, remained a short period, suddenly retreated	Anon 1802*a*
85	11	uT	Teignmouth	−3.4961	50.5421	O	10/08/1802	0800	0.6	Sea instantaneously rose and fell nearly two feet several times in the space of ten minutes	Anon 1802*b*; Dawson *et al.* 2000
86	12	nT	Plymouth	−4.1094	50.3661	O	31/05/1811		2.4	Coincides with widespread gales	Dawson *et al.* 2000
87	13	uT	Sutton Harbour, Plymouth	−4.1330	50.3682	O	13/09/1821	1400	1.2	'Sea-boar' observed, several vessels adrift. Event also noted in Cherbourg	Anon 1821
88	13	uT	Truro	−5.0446	50.2554	O	13/09/1821	1400		Similar occurrence to that seen at Plymouth	Anon 1821
89	14	nT	Penzance	−5.5284	50.1176	O	05/07/1843	1130		Strong currents, waves returning at 20 m intervals	Anon 1843*a*; Edmonds 1843, 1845
90	14	nT	Plymouth	−4.1094	50.3661	O	05/07/1843	1100		Noted in association with sudden storms	Edmonds 1845

(*Continued*)

Appendix 1. (*Continued*)

Site	Event	Confi-dence	Location	Longitude	Latitude	Evidence	Date	Time	Wave height	Comment	References
91	14	nT	Newlyn	−5.5509	50.0998	O	05/07/1843	1200	1	Sudden withdrawal, four waves at 10–15 minute intervals	Anon 1843*a*, *b*; Edmonds 1843
92	15	nT	St Mount's Bay	−5.4738	50.1174	O	23/05/1847	0500	1.5	Waves 0.9–1.5 m noted all day following slight tremor felt night before. Max waves 5 pm	Edmonds 1869
93	15	nT	Plymouth	−4.1094	50.3661	O	23/05/1847			Waves similar to St Mount's Bay, max waves noted 8–9 pm	Edmonds 1869
94	15	nT	Newlyn	−5.5509	50.0998	O	23/05/1847	1730	0.6	Sudden rush of the sea	Blight 1847
95	15	nT	Mousehole	−5.5413	50.0839	O	23/05/1847		1.2	Waves continued until 9 pm max amplitude 8 ft	Blight 1847
96	15	nT	Anderton	−4.2080	50.3487	O	23/05/1847		1.2	Several boats damaged	Anon 1847
97	16	nT	Folkestone	1.1862	51.0788	O	05/06/1858	0800	0.9	Boats cast adrift and one capsized and sank	Newig & Kelletat 2011; Anon 1858*a*
98	16	nT	Dover	1.3111	51.1181	O	05/06/1858			Considered same as Boulogne with 2.4 m wave	Newig & Kelletat 2011
99	16	nT	Pegwell Bay	1.3854	51.3171	O	05/06/1858	0915		Tide receded about 200 yards and returned to its former position after about 20 minutes	Newig & Kelletat 2011; Anon 1858*b*
100	17	nT	Isles of Scilly	−6.3165	49.9172	O	29/09/1869			Water disturbed similar to seen at Penzance	Anon 1869*a*; Perrey 1872
101	17	nT	Newlyn	−5.5509	50.0998	O	29/09/1869			Water disturbed similar to seen at Penzance	Anon 1869*a*; Perrey 1872
102	17	nT	Penzance	−5.5284	50.1176	O	29/09/1869	0600	0.75	Waves at 20 min intervals for up to 4 hrs, succession of bores affecting vessels	Anon 1869*a*, *b*; Perrey 1872
103	17	nT	Truro	−5.0446	50.2554	O	29/09/1869	0600	0.6	Several 'tidal bores' observed over 5 hrs	Anon 1869*a*; Perrey 1872
104	17	nT	Plymouth	−4.1094	50.3661	O	29/09/1869		0.75	Water swept in suddenly then withdrew	Anon 1869*c*
105	17	nT	Bideford	−4.2042	51.0189	O	29/09/1869		0.9	Tide receded and reappeared 3 or 4 times, vessels refloated	Anon 1869*d*
106	17	nT	Totnes	−3.6825	50.4304	O	29/09/1869			Sudden flow on the river	Anon 1869*e*
107	17	nT	Exmouth	−3.4191	50.6156	O	29/09/1869	1000		Wave suddenly rose and rushed 20 ft up beach	Anon 1869*f*
108	18	T	Portland	−2.4392	50.5831	T	28/08/1883	1015	0.025	May be seiching from air-waves due to eruption of Krakatau	Berninghausen 1968
109	18	T	Devonport	−4.1872	50.3833	T	28/08/1883	1045	0.1	May be seiching from air-waves due to eruption of Krakatau	Berninghausen 1968
110	19	nT	East Mersea Island	0.9784	51.7914	O	22/04/1884			Newspaper report after Colchester earthquake	Haslett & Bryant 2008

111	20	nT	Milford Haven	−5.0370	51.6936	O	18/08/1892			Waves noted coincident with Pembrokeshire earthquake	Davison 1897
112	21	T	Folkestone	1.1862	51.0788	O	31/12/1911	1830	0.9	Ropes and stanchions damaged by wave following large cliff fall	Anon 1912*a*, 1912*b*
113	22	nT	Helmsdale	−3.6483	58.1154	O	24/01/1927			Large rollers noted following an earthquake	Tyrrell 1932
114	23	T	Newlyn	−5.5413	50.1028	T	25/11/1941	2215	0.2	7 tsunami waves over ~4 hrs, period 15 min	Dawson *et al.* 2000
115	24	T	Newlyn	−5.5413	50.1028	T	23/05/1960		0.025	Not clear	Van Dorn 1987
116	25	T	St Mary's, Isle of Scilly	−6.3165	49.9172	T	28/02/1969	0735	0.12	Oscillations (peak to trough max 0.25 m) seen for at least ten hours	BODC
117	25	T	Newlyn	−5.5413	50.1028	T	28/02/1969			Gauge record shows heavy seiching on a day with calm sea conditions	Dawson *et al.* 2000
118	25	T	Fishguard	−4.9837	52.0132	T	28/02/1969	1200	0.03	Oscillations only weakly evident on tide gauge	BODC
119	26	T	Newlyn	−5.5413	50.1028	T	26/05/1975	1525	0.06	Eight tsunami waves over nearly 4 h with an average period of circa 12 min	Dawson *et al.* 2000
120	26	T	St Mary's, Isle of Scilly	−6.3165	49.9172	T	26/05/1975	1420	0.05	Oscillations extend over seven hours	BODC
121	27	T	Newlyn	−5.5413	50.1028	T	27/12/2004	1400	0.15	Variability	Woodworth *et al.* 2005
122	27	T	St Mary's, Isle of Scilly	−6.3165	49.9172	T	27/12/2004			Less certain than at Newlyn	Woodworth *et al.* 2005
123	27	T	Milford Haven	−5.0613	51.7034	T	27/12/2004	0938	0.16	May be confused by a storm surge	Woodworth *et al.* 2005

Site: Sequential number for evidence considered to be a tsunami.
Event: Sequential number for events that have been considered to be a tsunami.
Confidence: Opinion of event to be a geologically triggered tsunami event (derived from Long, 2015). T = tsunami, nT = non-tsunami, uT = uncertain tsunami.
Evidence: B = Sedimentary deposit reported in a borehole; O = Visual observation; S = Sedimentary deposit reported in an exposed section; T = tide gauge.

References

Anon, 1749. *Scots Magazine*, Friday 9th July 1749, page 348.

Anon, 1802*a*. *The Gloucester Journal*, Monday 16th August 1802.

Anon, 1802*b*. *The Salisbury and Winchester Journal*, Monday 16th August 1802.

Anon, 1821. *The Windsor and Eton Express*, Sunday 16th September 1821, page 3.

Anon, 1843*a*. *The Cornwall Royal Gazette*, Friday 21st July 1843, page 2.

Anon, 1843*b*. *The Exeter and Plymouth Gazette*, Saturday 22nd July 1843, page 3.

Anon, 1847. *The Cornwall Royal Gazette*, Friday 28th May 1847, page 3.

Anon, 1858*a*. *The Kentish Gazette*, Tuesday 8th June 1858, page 6.

Anon, 1858*b*. *The Isle of Wight Observer*, Saturday 12th June 1858, page 4.

Anon, 1869*a*. *The Cornubian and Redruth Times*, Friday 1st October 1869, page 4.

Anon, 1869*b*. *The Exeter and Plymouth Gazette*, Friday 1st October 1869, page 5.

Anon, 1869*c*. *The London Daily News*, Saturday 2nd October 1869, page 5.

Anon, 1869*d*. *The North Devon Journal*, Thursday 7th October 1869, page 8.

Anon, 1869*e*. *The Totnes Times and Dartmouth Gazette*, Saturday 2nd October 1869, page 4.

Anon, 1869*f*. *The Western Times*, Tuesday 5th October 1869, pages 3 and 8.

Anon, 1912*a*. *The London Daily News*, Tuesday 2nd January 1912.

Anon, 1912*b*. *The Folkestone, Hythe, Sandgate and Cheriton Herald*, Saturday 6th January 1912.

Banerjee, D., Murray, A.S. & Foster, I.D.L. 2001. Scilly Islands, UK: optical dating of a possible tsunami deposit from the 1755 Lisbon earthquake. *Quaternary Science Reviews*, **20**, 715–718.

Baptista, M.A., Heitor, S., Miranda, J.M., Miranda, P. & Mendes Victor, L. 1998. The 1755 Lisbon Tsunami; evaluation of the tsunami parameters. *Journal of Geodynamics*, **25**, 143–157.

Berninghausen, W.H. 1968. *Tsunamis and Seismic Seiches reported from western North and South Atlantic and the coastal waters of northwestern Europe*. Naval Oceanographic Office, Informal Rep. No. 68–85, Washington, D.C. 20396, p 41.

Birch, G. & Warren, C.D. 2007. Landslide and chalk fall management on the Folkestone to Dover railway line, Kent. *In*: McInnes, R., Jakeways, J., Fairbank, H. & Mathie, E. (eds) *Landslides and Climate Change*. Taylor & Francis Group, London.

Birnie, J.F. 1981. *Environmental Changes in Shetland since the end of the Last Glaciation*. Unpublished Ph.D. Thesis, University of Aberdeen.

Blight, R. 1847. *The Cornwall Royal Gazette*, Friday 28th May 1847, page 2.

Bondevik, S., Dawson, A.G., Dawson, S., Drange, I., Lohne, O., Mangerud, J. & Svendsen, J.-I. 2002. Records of palaeotsunamis around the Norwegian Seas. Norsk Hydro Contract: NHT-B44-07822-00, 124pp.

Bondevik, S., Mangerud, J., Dawson, S., Dawson, A. & Lohne, Ø. 2003. Record-breaking height for 8000-year-old tsunami in the North Atlantic. *Eos*, **84**, 289–293.

Bondevik, S., Mangerud, J., Dawson, S., Dawson, A. & Lohne, Ø. 2005. Evidence for three North Sea tsunamis at the Shetland Islands between 8000 and 1500 years ago. *Quaternary Science Reviews*, **24**, 1757–1775.

Bondevik, S., Stormo, S.K. & Skjerdal, G. 2012. Green mosses date the Storegga tsunami to the chilliest decades of the 8.2 ka cold event. *Quaternary Science Reviews*, **45**, 1–6.

Bondevik, S., Svendsen, J.I., Johnsen, G., Mangerud, J. & Kaland, P.E. 1997. The Storegga tsunami along the Norwegian coast, its age and run-up. *Boreas*, **26**, 29–53.

Boomer, I., Waddington, C., Stevenson, T. & Hamilton, D. 2007. Holocene coastal change and geoarchaeology at Howick, Northumberland, UK. *The Holocene*, **17**, 89–104.

Borlase, W. 1755. On the coast of Cornwall. *Philosophical Transactions of the Royal Society of London*, **49**, 373–379.

Borlase, W. 1762. Some Account of the extraordinary Agitation of the Waters in Mount's-bay, and other Places, on the 31st of March 1761: In a Letter for the Reverend Dr. C Lyttelton. *Philosophical Transactions of the Royal Society*, **52**, 418–431.

Bryant, E.A. & Haslett, S.K. 2002. Was the AD 1607 coastal flooding event in the Severn Estuary and Bristol Channel (UK) due to a tsunami? *Archaeology in the Severn Estuary*, **13**, 163–167.

Bugge, T., Befring, S. *et al.* 1987. A giant three-stage submarine slide off Norway. *Geo-Marine Letters*, **7**, 191–198.

Cascalho, J., Costa, P., Dawson, S., Milne, F. & Rocha, A. 2016. Heavy mineral assemblages of the Storegga tsunami deposit. *Sedimentary Geology*, **334**, 21–33.

Chisholm, J.I. 1971. The stratigraphy of the post-glacial marine transgression of NE Fife. *Bulletin of the Geological Survey of Great Britain*, **37**, 91–107.

Costa, P.J., Andrade, C., Cascalho, J., Dawson, A.G., Freitas, M.C., Paris, R. & Dawson, S. 2015. Onshore tsunami sediment transport mechanisms inferred from heavy mineral assemblages. *The Holocene*, **25**, 795–809.

Cox, R., Zentner, D.B., Kirchner, B.J. & Cook, M.S. 2012. Boulder ridges on the Aran Islands (Ireland): Recent movements caused by storm waves, not tsunamis. *Journal of Geology*, **120**, 249–272.

Davison, C. 1897. On the Pembroke Earthquakes of August, 1892, and November, 1893. *Quarterly Journal of the Geological Society*, **53**, 157–176.

Dawson, S. & Smith, D.E. 1997. Holocene relative sea-level changes on the margin of a glacio-isostatically uplifted area: an example from northern Caithness, Scotland. *The Holocene*, **7**, 59–77.

Dawson, S. & Smith, D.E. 2000. The sedimentology of Middle Holocene tsunami facies in northern Sutherland, Scotland, UK. *Marine Geology*, **170**, 169–179.

Dawson, A.G., Dawson, S. & Bondevik, S. 2006. A late Holocene tsunami at Basta Voe, Yell, Shetland Isles. *Scottish Geographical Journal*, **122**, 100–108.

DAWSON, A.G., FOSTER, I.D.L., SHI, S., SMITH, D.E. & LONG, D. 1991. The identification of tsunami deposits in coastal sediment sequences. *Science of Tsunami Hazards*, **9**, 73–82.

DAWSON, A.G., LONG, D. & SMITH, D.E. 1988. The Storegga Slides: evidence from eastern Scotland for a possible tsunami. *Marine Geology*, **82**, 271–276.

DAWSON, A.G., MUSSON, R.M.W., FOSTER, I.D.L. & BRUNSDEN, D. 2000. Abnormal historic sea-surface fluctuations, SW England. *Marine Geology*, **170**, 59–68.

DAWSON, A.G., SMITH, D.E. & LONG, D. 1990. Evidence for a tsunami from a mesolithic site in Inverness. *Journal of Archaeological Science*, **17**, 509–512.

DICK, R. 2015. *Geomorphological mapping of the seabed in North-East Shetland through interpretation of multibeam and seismic profile data.* British Geological Survey Internal Report, IR/15/012.

DIFFENBAUGH, N.S., SCHERER, M. & TRAPP, R.J. 2013. Robust increases in severe thunderstorm environments in response to greenhouse forcing. *PNAS*, **110**, 16361–16366.

DYKES, A.P. & WARBURTON, J. 2008. Characteristics of the Shetland Islands (UK) peat slides of 19 September 2003. *Landslides*, **5**, 213–226.

EDMONDS, R. 1843. An account of an extraordinary movement of the sea in Cornwall, in July, 1843, with notices of similar movements in previous years, and also of earthquakes which have occurred in Cornwall. *Transactions of the Royal Geological Society of Cornwall*, **13**, 111–121.

EDMONDS, R. 1845. On earthquakes and extraordinary movements of the sea; and on remarkable lunar periodicities in earthquakes, oscillations of the sea, and great atmospherical changes. *Edinburgh New Philosophical Journal*, **38**, 271–279.

EDMONDS, R. 1860. An account of the extraordinary agitations of the sea in the west of England, on the 25 and 26 June, and the 4 October; with notices of the earthquake shocks in Cornwall on the 11th of November, and the 21st October 1859. *Edinburgh New Philosophical Journal*, **12**, 1–15.

EDMONDS, R. 1869. On extraordinary agitations of the sea not produced by winds or tides. *Report and Transactions of the Devonshire Association*, **3**, 144–152.

FIRTH, C.R. 1984. *Raised shorelines and ice limits in the inner Moray Firth and Loch Ness areas, Scotland.* Unpublished Ph.D. Thesis, Council for National Academic Awards, Coventry (Lanchester) Polytechnic.

FLINN, D. 1994. *Geology of Yell and Some Neighbouring Islands in Shetland.* British Geological Survey (Scotland), HMSO for the British Geological Survey, Memoir **130**, 119.

FOSTER, I.D.L., ALBION, A.J. *ET AL.* 1991. High energy coastal sedimentary deposits: an evaluation of depositional processes in southwest England. *Earth Surface Processes and Landforms*, **16**, 341–356.

FOSTER, I.D.L., DAWSON, A.G., DAWSON, S., LEES, J. & MANSFIELD, L. 1993. Tsunami sedimentation sequences in the Scilly Isles, southwest England. *Science of Tsunami Hazards*, **11**, 35–46.

HAFLIDASON, H., LIEN, R., SEJRUP, H.P., FORSBERG, C.F. & BRYN, P. 2005. The dating and morphometry of the Storegga Slide. *Marine and Petroleum Geology*, **22**, 123–136.

HAGGART, B.A. 1978. *A Pollen and Stratigraphical Investigation into a Peat Deposit in St. Michael's Wood, near Leuchars, Fife.* Unpublished M.A. Thesis, University of St. Andrews.

HAGGART, B.A. 1982. *Flandrian sea-level changes in the Moray Firth area.* Unpublished Ph.D. Thesis, University of Durham.

HASLETT, S.K. & BRYANT, E.A. 2007. Evidence for historic coastal high-energy wave impact (tsunami?) in North Wales, United Kingdom. *Atlantic Geology*, **43**, 137–147.

HASLETT, S.K. & BRYANT, E.A. 2008. Historic tsunami in Britain since AD 1000: a review. *Natural Hazards and Earth System Sciences*, **8**, 587–601.

HASLETT, S.K. & BRYANT, E.A. 2009. Meteorological tsunamis in southern Britain: an historical review. *The Geographical Review*, **99**, 146–163.

HASLETT, S.K., MELLOR, H.E. & BRYANT, E.A. 2009. Meteo-hazard associated with summer thunderstorms in the United Kingdom. *Physics and Chemistry of the Earth*, **34**, 1016–1022.

HOLLOWAY, L.K. 2002. *Sedimentary environments in the western Forth valley during the Late Devensian and early Holocene.* Unpublished Ph.D. Thesis, Coventry University.

HORSBURGH, K. & HORRITT, M. 2006. The Bristol Channel floods of 1607 – reconstruction and analysis. *Weather*, **61**, 272–277.

HORTON, B.P., INNES, J.B., SHENNAN, I., GEHRELS, W.R., LLOYD, J.M., MCARTHUR, J.J. & RUTHERFORD, M.M. 1999. The Northumberland Coast. *In*: BRIDGLAND, D.R., HORTON, B.P. & INNES, J.P. (eds) *The Quaternary of North-East England. Field Guide*, Quaternary Research Association, 146–165.

HUTCHESON, A. 1886. Notice of the discovery of a stratum containing worked flints at Broughty-Ferry. *Proceedings of the Society of Antiquaries of Scotland*, **20**, 166–169.

HUTCHINSON, J.N. 2002. Chalk flows from coastal cliffs of northwest Europe. *In*: EVANS, S.G. & DEGRAFF, J.V. (eds) *Catastrophic Landslides: Effects, Occurrence and Mechanisms.* Geological Society of America, Boulder, Colorado, Reviews in Engineering Geology, **XV**, 257–302.

HUXHAM, J. 1755. Letter XIV, In Devonshire and Cornwall, at Plymouth, Mounts Bay, Penzance etc. *Philosophical Transactions of the Royal Society*, **49**, 371–373.

JANSEN, E., BEFRING, S., BUGGE, T., EIDVIN, T., HOLTEDAHL, H. & SEJRUP, H.P. 1987. Large submarine slides on the Norwegian Continental Margin: Sediments, transport and timing. *Marine Geology*, **78**, 77–107.

JORDAN, J.T., SMITH, D.E., DAWSON, S. & DAWSON, A.G. 2010. Holocene relative sea-level changes in Harris, Outer Hebrides, Scotland, UK. *Journal of Quaternary Science*, **25**, 115–134.

LACAILLE, A.D. 1954. *The Stone Age in Scotland.* Oxford University Press, Oxford.

LONG, D. 2015. *A catalogue of tsunamis reported in the UK.* British Geological Survey Internal Report, IR/15/043.

LONG, D. & WILSON, C.K. 2007. *A catalogue of tsunamis in the UK*. British Geological Survey Commissioned Report, CR/07/077.

LONG, D., SMITH, D.E. & DAWSON, A.G. 1989. A Holocene tsunami deposit in eastern Scotland. *Journal of Quaternary Science*, **4**, 61–66.

LONG, A.J., BARLOW, N.L.M., DAWSON, S., HILL, J., INNES, J.B., KELHAM, C., MILNE, F.D. & DAWSON, A. 2016. Lateglacial and Holocene relative sea-level changes and first evidence for the Storegga tsunami in Sutherland, Scotland. *Journal of Quaternary Science*, **31**, 239–255.

LUQUE, L., LARIO, J., ZAZO, C., GOY, J.L., DABRIO, C.J. & SILVA, P.J. 2001. Tsunami deposits as paleoseismic indicators: examples from the Spanish coast. *Acta Geologica Hispanica*, **36**, 197–211.

MARAMAI, A., BRIZUELA, B. & GRAZIANI, L. 2014. The Euro-Mediterranean Tsunami Catalogue. *Annals of Geophysics*, **57**, https://doi.org/10.4401/ag-6437

MCDAKIN, J.G. 1912. Geological notes. *Reports and Transactions of the East Kent Scientific and Natural History Society, Series II*, **11**, 15.

MONSERRAT, S., VILIBIC, I. & RABINOVICH, A.B. 2006. Meteotsunamis: atmospherically induced destructive ocean waves in the tsunami frequency band. *Natural Hazards Earth System Science*, **6**, 1035–1051.

MORTON, R.A., GELFENBAUM, G. & JAFFE, B.E. 2007. Physical criteria for distinguishing sandy tsunami and storm deposits using modern examples. *Sedimentary Geology*, **200**, 184–207.

MUSSON, R.M.W. 2008. *The seismicity of the British Isles to 1600*. British Geological Survey Open Report, OR/08/049.

NATIONAL RISK REGISTER FOR CIVIL EMERGENCIES 2015. National Risk Register for Civil Emergencies – 2015 edition, available from https://www.gov.uk/government/publications/national-risk-register-for-civil-emergencies-2015-edition

NEILSON, G., MUSSON, R.M.W. & BURTON, P.W. 1984. The 'London' earthquake of 1580, April 6, *Engineering Geology*, **20**, 113–141.

NEWEY, W.W. 1965. *Post-glacial vegetational and climatic changes in part of south-east Scotland as indicated by the pollen-analysis and stratigraphy of some of its peat and lacustrine deposits*. Unpublished Ph.D. Thesis, University of Edinburgh.

NEWIG, J. & KELLETAT, D. 2011. The North Sea Tsunami of June 5, 1858. *Journal of Coastal Research*, **27**, 931–941.

PERREY, A. 1849. Note sur les tremblements de terre ressentis en 1848, *Memoirs Academie, Dijon* (part 2, pp. 1–40).

PERREY, A. 1872. Note sur les tremblements de terre en 1869, avec supplements pour les annees anterieures de 1843 a 1868, *Memoires cour. Bruxelless*, 22, no. 4, 116pp.

PHILLIPS, E.R., EVEREST, J. & DIAZ-DOCE, D. 2010. Bedrock controls on subglacial landform distribution and geomorphological processes: Evidence from the Late Devensian Irish Sea Ice Stream. *Sedimentary Geology*, **232**, 98–118.

ROBERTS, G. 1849. *Remarkable tide in the British Channel, Friday, July 7, 1848, as it appeared at Lyme Regis, Dorset.*, Annual report of the British Association for the Advancement of Science for 1848, transactions of the sections 37.

ROGER, J., GUNNELL, Y., KRIEN, Y. & RAY, P. 2012. Potential for landslide-related tsunami in the Dover Strait area (English Channel) based on numerical modeling. *Proceedings of the Fifteenth World Conference on Earthquake Engineering*, Lisbon, Portugal, 2012.

RYDGREN, K. & BONDEVIK, S. 2015. Moss growth patterns and timing of human exposure to a Mesolithic tsunami in the North Atlantic. *Geology*, **43**, 111–114.

SELBY, K.A. & SMITH, D.E. 2015. Holocene relative sea-level change on the Isle of Skye, inner hebrides, Scotland. *Scottish Geographical Journal*, **132**, 42–65.

SHENNAN, I., HORTON, B., INNES, J., GEHRELS, R., LLOYD, J., MCARTHUR, J. & RUTHERFORD, M. 2000. Late Quaternary sea-level changes, crustal movements and coastal evolution in Northumberland, UK. *Journal of Quaternary Science*, **15**, 215–237.

SHI, S. 1995. *Observational and Theoretical Aspects of Tsunami Sedimentation*. Unpublished Ph.D. thesis, Coventry University.

SIBLEY, A., COX, D., LONG, D., TAPPIN, D. & HORSEBURGH, K. 2016. Meteorologically generated tsunami-like waves in the North Sea on 1/2 July 2015 and 28 May 2008. *Weather*, **71**, 68–74.

SISSONS, J.B. & SMITH, D.E. 1965. Peat bogs in a post-glacial sea and a buried raised beach in the western part of the Carse of Stirling. *Scottish Journal of Geology*, **1**, 247–255.

SMITH, D.E. 1993. Norwick, Unst; Burragarth, Unst; Sullom Voe, Mainland. *In*: BIRNIE, J., GORDON, J., BENNETT, K. & HALL, A. (eds) *The Quaternary of Shetland*. Quaternary Research Association Field Guide, 52–56.

SMITH, D.E. & CULLINGFORD, R.A. 1985. Flandrian relative sea-level changes in the Montrose Basin Area. *Scottish Geographical Magazine*, **101**, 91–104.

SMITH, D.E., MORRISON, J., JONES, R.L. & CULLINGFORD, R.A. 1980. Dating the main postglacial shoreline in the Montrose area, Scotland. *In*: CULLINGFORD, R.A., DAVIDSON, D.A. & LEWIN, J. (eds) *Timescales in Geomorphology*. Wiley, Chichester, 225–245.

SMITH, D.E., CULLINGFORD, R.A. & SEYMOUR, W.P. 1982. Flandrian relative sea-level changes in the Philorth valley, North-East Scotland. *Transaction of the Institute of British Geographers (New Series)*, **7**, 321–336.

SMITH, D.E., CULLINGFORD, R.A. & BROOKS, C.L. 1983. Flandrian relative sea level changes in the Ythan valley, northeast Scotland. *Earth Surface Processes and Landforms*, **8**, 423–438.

SMITH, D.E., CULLINGFORD, R.A. & HAGGART, B.A. 1985. A major coastal flood during the Holocene in eastern Scotland. *Eiszeitalter und Gegenwart*, **35**, 109–118.

SMITH, D.E., TURBAYNE, S.C., DAWSON, A.G. & HICKEY, K.R. 1991. The temporal and spatial variability of major floods around European coasts. Final report of Coventry Polytechnic. Contract EV4C 0047 UK (H). European Commission, Brussels.

SMITH, D.E., FIRTH, C.R., TURBAYNE, S.C. & BROOKS, C.L. 1992. Holocene relative sea-level changes and shoreline displacement in the Dornoch Firth area, Scotland. *Proceedings of the Geologists' Association*, **103**, 237–257.

SMITH, D.E., FIRTH, C.R., BROOKS, C.L., ROBINSON, M. & COLLINS, P.E.F. 1999. Relative sea-level rise during the Main Postglacial Transgression in NE Scotland, UK. *Transactions of the Royal Society of Edinburgh: Earth Science*, **90**, 1–27.

Smith, D.E., Shi, S. *et al.* 2004. The Holocene Storegga Slide tsunami in the United Kingdom. *Quaternary Science Review*, **23**, 2291–2321.

Smith, D.E., Foster, I.D.L., Long, D. & Shi, S. 2007. Reconstructing the pattern and depth of flow onshore in a palaeotsunami from associated deposits. *Sedimentary Geology*, **200**, 362–371.

Solheim, A., Berg, K., Forsberg, C.F. & Bryn, P. 2005. The Storegga Slide complex: repetitive large scale sliding with similar cause and development. *Marine and Petroleum Geology*, **22**, 97–107.

Tappin, D., Sibley, A., Horseburgh, K., Daubord, C., Cox, D. & Long, D. 2013. The English Channel 'tsunami' of 27 June 2011 – a probable meteorological source. *Weather*, **68**, 144–152.

Tappin, D.R., Long, D. & Carter, G.D.O. 2015. *Shetland Islands Field Trip May 2014 – Summary of Results*. British Geological Survey Open Report, OR/15/017.

Tooley, M.J. & Smith, D.E. 2005. Relative sea-level change and evidence for the Holocene Storegga Slide tsunami from a high-energy coastal environment: Cocklemill Burn, Fife, Scotland, UK. *Quaternary International*, **133–134**, 107–119.

Toothill, N.P. 1994. *Holocene storm surge or tsunami deposit at Basta Voe, Shetland Isles*, unpublished BSc thesis, Dept of Geography, Coventry University 98pp.

Tuttle, M.P., Ruffman, A., Anderson, T. & Jeter, H. 2004. Distinguishing tsunami from storm deposits in eastern North America: the 1929 Grand Banks Tsunami v. the 1991 Halloween Storm. *Seismological Research Letters*, **75**, 117–131.

Tyrrell, G. 1932. Recent Scottish earthquakes. *Transactions of the Geological Society of Glasgow*, **19**, 1–41.

Van Dorn, W.G. 1987. Tide gage response to tsunamis. Part II: Other oceans and smaller seas. *Journal of Physical Oceanography*, **17**, 1507–1516.

Waddington, C. 2014. *Rescued from the Sea: an archaeologist's tale*. Northumberland Wildlife Trust.

Watts, M., Talling, P., Hunt, J., Xuan, C. & van Peer, T. 2016. A new date for a large pre-Holocene Storegga Slide. *Geophysical Research Abstracts*, **18**, 17055.

Woodworth, P.L., Blackman, D.L. *et al.* 2005. Evidence for the Indonesian tsunami in British tidal records. *Weather*, **60**, 263–267.

Wordsworth, J. 1985. The excavation of a Mesolithic horizon at 13–24 Castle Street, Inverness. *Proceedings of the Society of Antiquaries of Scotland*, **115**, 89–103.

The application of microtextural and heavy mineral analysis to discriminate between storm and tsunami deposits

PEDRO J. M. COSTA[1]*, G. GELFENBAUM[2], S. DAWSON[3], S. LA SELLE[2], F. MILNE[3], J. CASCALHO[1,4], C. PONTE LIRA[1], C. ANDRADE[1], M. C. FREITAS[1] & B. JAFFE[2]

[1]*Instituto Dom Luiz and Departamento de Geologia, Faculdade de Ciências, Universidade de Lisboa, 1749-016 Lisboa, Portugal*

[2]*United States Geological Survey, Pacific Coastal and Marine Science Center, 400 Natural Bridges Drive, Santa Cruz, CA 95060, USA*

[3]*Department of Geography, University of Dundee, Tower Building, Nethergate, Dundee DD1 4HN, UK*

[4]*Museu Nacional de História Natural e da Ciência (MUHNAC-UL), Rua da Escola Politécnica 56/58, 1250-102, Lisboa, Portugal*

**Correspondence: ppcosta@fc.ul.pt*

Abstract: Recent work has applied microtextural and heavy mineral analyses to sandy storm and tsunami deposits from Portugal, Scotland, Indonesia and the USA. We looked at the interpretation of microtextural imagery (scanning electron microscopy) of quartz grains and heavy mineral compositions. We consider inundation events of different chronologies and sources (the AD 1755 Lisbon and 2004 Indian Ocean tsunamis, the Great Storm of 11 January 2005 in Scotland, and Hurricane Sandy in 2012) that affected contrasting coastal and hinterland settings with different regional oceanographic conditions. Storm and tsunami deposits were examined along with potential source sediments (alluvial, beach, dune and nearshore sediments) to determine provenance.

Results suggest that tsunami deposits typically exhibit a significant spatial variation in grain sizes, microtextures and heavy minerals. Storm deposits show less variability, especially in vertical profiles. Tsunami and storm quartz grains had more percussion marks and fresh surfaces compared to potential source material. Moreover, in the studied cases, tsunami samples had fewer fresh surfaces than storm deposits.

Heavy mineral assemblages are typically site-specific. The concentration of heavy minerals decreases upwards in tsunamigenic units, whereas storm sediments show cyclic concentrations of heavy minerals, reflected in the laminations observed macroscopically in the deposits.

The sedimentology of storm and tsunami deposits: present knowledge

A key problem in reconstructing chronologies of coastal hazards is distinguishing between storm and tsunami deposits. The sedimentological differentiation between these deposits has been debated over the years and, so far, no unequivocal criteria have been established despite different generation mechanisms and wave characteristics. Storms are produced by meteorological disturbances that generate both high-energy waves and elevated sea levels, while tsunamis are typically generated from earthquakes that rupture the seafloor and cause significant vertical displacements that result in waves with very long periods. Submarine landslides, meteorite impacts, and volcanic edifice failures and other large-scale volcanic processes may also generate tsunamis.

When sandy deposits from storms and tsunamis are present in the same location, the thickness and number of layers within the deposit can be used to differentiate between a tsunami and a storm genesis (e.g. Williams & Hutchinson 2000). Sandy tsunami deposits tend to have few, normally graded beds, suggesting deposition from suspension by successive tsunami waves or successive depositional pulses (Paris *et al.* 2009; Jaffe *et al.* 2012; Shanmugam 2012; Szczuciński *et al.* 2012). Storm deposits tend to have thinner and more numerous laminations, with sediments transported and deposited by higher-frequency but lower-energy storm waves. In areas with a higher frequency of tsunami events, the greatest inland extent of tsunami deposits, when

From: Scourse, E. M., Chapman, N. A., Tappin, D. R. & Wallis, S. R. (eds) 2018. *Tsunamis: Geology, Hazards and Risks*. Geological Society, London, Special Publications, **456**, 167–190.
First published online February 23, 2017, https://doi.org/10.1144/SP456.7

compared with storm deposits, may be used to identify a tsunami source.

Over the last decade, a number of papers revisited the issue of distinguishing tsunami and storm deposits: for example, Goff *et al.* (2004), Tuttle *et al.* (2004), Kortekaas & Dawson (2007), Morton *et al.* (2007), Switzer & Jones (2008) and Costa *et al.* (2012*a*). Typically, tsunami deposits are present in a stratigraphic sequence of a coastal depositional environment as conspicuous sediment units bearing compositional or textural signatures in contrast to the under- and overlying materials accumulated under the normal sedimentation regime. It is therefore crucial to fully understand local, seasonal and mesoscale background sedimentation processes prior to attaching a tsunamigenic or storm origin to any sedimentary unit or feature within the geological record.

Typical storm-related washovers consist of fan- or sheet-like subaerial to subaqueous accumulations of sand supplied by overwash from the nearshore areas and supratidal sediment resident within mainland surfaces (i.e. beach and coastal dune) (Costa 2012). Factors that influence the dimensions and planform of fans and related overwash bedforms include wind-induced surge elevation, astronomical tide, wave set-up and run-up, the intensity and duration of channelled flow, the presence of topographical obstructions, and the volumes of sand in the shoreface, beach and dunes that are available for erosion and subsequent deposition by overwash (Donnelly *et al.* 2001; Morton *et al.* 2007). Storm deposits can also form graded or homogeneous deposits of sand, shell, gravel and mud within the prominently clastic sediments in coastal lagoons – their identification also relies on the contrast with low-energy background sedimentation (Otvos & Carter 2008).

Several studies stress the similarities between tsunami and storm deposits (Nanayama *et al.* 2000; Goff *et al.* 2004; Tuttle *et al.* 2004; Morton *et al.* 2007; Switzer & Jones 2008). Overall, all of these studies suggest that while no single criteria may exist to definitively discriminate between tsunami and storm origin, the combined study of sediment composition, texture, grading, stratification, thickness, geometry and landscape conformity may be used to determine the generation mechanism of a specific unit.

In conclusion, storm deposits share many textural, palaeontological, geochemical and geomorphological characteristics with tsunami deposits, making their differentiation a complex task. This is also further complicated by studies addressing storm activity but not clearly separating storm surge from storm wave impacts and sedimentation. The diversity of topographical, hydrodynamic and sedimentological settings, as well as post-depositional aspects, may create major difficulties in correlating individual sand layers with specific events.

In this work, results from a variety of locations and chronologies affected by either storm or tsunami events are presented. We try to improve the understanding of storm and tsunami sedimentological signatures using results from microtextural and heavy mineral analysis, and discuss the contribution of these sedimentological techniques to the differentiation of storm and tsunami deposits. The differentiation of tsunami and storm deposits is a critical determinant in the hazard; therefore, the development and improvement of sedimentary criteria is a scientific requisite to accurate tsunami and storm hazard assessment.

The application of microtextural and heavy mineral studies to the discrimination of storm and tsunami deposits

In recent years, microtextural and heavy mineral analysis have demonstrated their potential in the study of the depositional imprints of tsunami and, to a lesser extent, storm layers. As a summary, and based in the literature described above, it can be said that microtextural and heavy mineral analysis have proven to be a reliable provenance tool, even if its capability to unequivocally differentiate tsunami and storm deposits is still a matter of debate.

The routine use of scanning electron microscopy (SEM) in sedimentary studies was in large part developed by Krinsley & Donahue (1968) and Mahaney (2002). The use of SEM photographs has been applied mostly to quartz grains, but has also been successfully used for other minerals (Mahaney 2002; Dahanayake & Kulasena 2008; Lakshmi *et al.* 2010). A recent summary of microtextural features of quartz grains can be found in Costa (2012).

Scanning electron microscopy of quartz grains is used to establish provenance relationships between source material and tsunami deposits (e.g. Bruzzi & Prone 2000; Dahanayake & Kulasena 2008; Costa *et al.* 2009; Mahaney & Dohm 2011). Bruzzi & Prone (2000) compared microtextural signatures of quartz grains deposited by the AD 1755 tsunami in Boca do Rio (Portugal) and quartz grains deposited by a storm in the Rhone delta in November 1997. They found several microtextural features were associated with tsunamis, including a greater number of upturned plates, and fractures and other mechanical marks.

Comparing tsunami and storm deposits, Costa *et al.* (2012*a*) were not able to assign a specific microtextural signature to high-energy inundations, because generation of microtextures is dependent on specific local conditions, such as available sediment source, flow conditions (velocity and depth) and sediment concentration during transport. Furthermore, the characteristics of each study site (i.e. topography, coastal configuration and sediment sources) are determining factors of the type of microtextures observed in tsunamigenic grains (Mahaney *et al.* 2010; Mahaney & Dohm 2011). For example, higher sediment concentrations flows have less space between grains and this leads to a greater number of grain impacts but with less kinetic energy, thus producing more percussion marks (Costa *et al.* 2012*a*). The authors concluded that, in general, tsunami grains presented the greatest microtextural variance relative to potential source material, as well as the greatest number of fresh surfaces and percussion marks, including near total resurfacing in some grains.

More recently, Bellanova *et al.* (2015) were unable to identify a distinctive microtextural feature in the 2010 tsunami deposits from a bank profile of the Tirúa river in Chile. Grains from the tsunami deposits showed many fresh surfaces and percussion marks, but these features were also present in the non-tsunamigenic deposits and three source samples (beach, dune and river). Therefore, the tsunamigenic deposits did not exhibit statistically significant differences in characteristics and abundances in all microtextural families. This could be attributed to one or a combination of the following reasons: (a) the small dataset analysed (seven samples); (b) the fact that dissolution might have masked other microtextural imprints; (c) fluvial grains are typically rich in percussion marks (Mahaney 2002); (d) the tsunami event was not able to significantly imprint grains from this specific site; and (e) local hydrodynamic and climatic conditions that might not favour the preservation of mechanical microtextural features. To address the issues raised here, only an extensive and detailed analysis of several tsunami deposits will provide a robust assessment on the worldwide applicability of microtextural analysis to tsunami and storm studies.

Another sedimentological technique used to identify tsunami deposits is the study of heavy mineral assemblages (minerals denser than 2.89 g cm^{-3}). This technique has been successfully used to establish provenance of tsunamigenic sediment or to infer phases of tsunami inundation (Switzer *et al.* 2005; Babu *et al.* 2007; Bahlburg & Weiss 2007; Morton *et al.* 2007; Narayana *et al.* 2007; Higman & Bourgeois 2008; Jagodziński *et al.* 2009, 2012; Nakamura *et al.* 2012; Cascalho *et al.* 2016).

Jagodziński *et al.* (2009) used heavy mineral assemblages to establish the sediment source of a deposit in Thailand from the 2004 Indian Ocean tsunami. The authors mainly described tsunami samples that were richer in mica – especially in the upper part of the tsunamigenic unit – and less tourmaline than beach and pre-tsunami soils. The abundance of mica in the tsunamigenic deposit was attributed to its lower density, shape and transport mode. Nakamura *et al.* (2012) detected a high percentage of heavy minerals (up to 89%) in the tsunami deposits from the 2011 Tohoku-oki tsunami on the Misawa coast (Japan). The authors detected an overall decrease in heavy mineral content, due to waning flow energy as the wave moved inland.

Costa *et al.* (2015*a*) studied heavy mineral assemblages of tsunami deposits from a wide range of locations and ages (the 1.5 ka BP Basta Voe tsunami in Scotland, the AD 1755 tsunami in Portugal and the 2004 Indian Ocean tsunami in Indonesia). In the Portuguese sites, palaeotsunami samples shared fewer similarities with nearshore materials than they did with local dune and beach sediments, suggesting that the beach and dunes were the likely source areas. In the Indonesian sites, the authors identified distinct depositional episodes (inundation and backwash) based on the presence of rounded or euhedral zircon. The euhedral zircon was associated with an inland source and therefore had to have been transported seawards during the backwash phase.

In storm deposits, it is common to observe several individual laminations or subunits segregated into multiple discrete lamina sets that commonly exhibit either normal or inverse grading (Schwartz 1975; Leatherman & Williams 1983; Morton *et al.* 2007). However, detailed heavy mineral studies in storm deposits are scarce and mainly consist of macroscopic analysis. In fact, they do not focus on the analysis of heavy mineral assemblages or establishing provenance relationships. It has also been observed that the concentration of heavy minerals decreases upwards in tsunamigenic units, whereas storm sediments tends to display cyclic concentrations of heavy minerals, reflected in the laminations observed in the deposits (Costa *et al.* 2015*a*).

Event and studied areas

In this work, sandy deposits from two different storm and two different tsunami events are studied and compared. The events studied are from diverse locations, inundation events of different chronologies and source mechanisms (the Great Storm of 11 January 2005 and Hurricane Sandy 2012 in Scotland and the USA, respectively; the AD 1755 and 26 December 2004 tsunamis in Portugal and Indonesia, respectively) that affected distinct coastal settings with different regional oceanographic conditions. Storm and tsunami deposits were compared with present-day analogues (alluvial, beach, dune and nearshore sediments) in an attempt to determine their provenance. A full list of samples from all locations and results are presented in Tables 1–3.

Table 1. *Heavy mineral (HM) data from the four studied deposits*

Sample Name	Country	Sed Env	Comment	Number of grains counted	HM grains counted, except Op + Alt + NI	% of HMs in the total sediment fraction	% of transparent HMs	Zircon	Tourmaline	Rutile	Staurolite
14	Portugal	Tsunami	Tsunami profile inland	308	115	0.393	62.66	0.00	49.57	0.00	19.13
LV10a	Portugal	Tsunami	Tsunami vertical variation	309	134	0.511	56.63	0.00	58.96	0.00	17.16
LV10b	Portugal	Tsunami	Tsunami vertical variation	351	152	0.403	56.70	0.00	48.03	3.29	15.13
LV10c	Portugal	Tsunami	Tsunami vertical variation	326	152	0.335	53.37	0.00	57.89	1.32	15.13
LV10d	Portugal	Tsunami	Tsunami vertical variation	236	128	0.291	45.76	0.00	49.22	0.78	13.28
LV10e	Portugal	Tsunami	Tsunami vertical variation	318	135	0.319	57.55	0.00	49.63	0.74	16.30
LV7a	Portugal	Tsunami	Tsunami vertical variation	337	179	0.357	46.88	0.00	54.19	0.00	15.64
LV7b	Portugal	Tsunami	Tsunami vertical variation	329	201	0.043	38.91	0.00	47.76	0.00	16.92
20	Portugal	Tsunami	Tsunami profile inland	125	87	0.251	30.40	0.00	35.63	0.00	16.09
26	Portugal	Tsunami	Tsunami profile inland	312	169	0.015	45.83	0.00	47.93	1.78	18.34
N9	Portugal	Nearshore	Source sed	330	137	0.092	58.48	0.00	56.93	0.73	1.46
N13_5	Portugal	Nearshore	Source sed	316	39	0.091	66.09	0.00	58.97	0.00	7.69
N6	Portugal	Nearshore	Source sed	337	149	0.075	52.85	0.00	58.39	2.01	5.37
N18	Portugal	Nearshore	Source sed	227	87	0.047	61.67	0.00	50.00	4.65	5.81
N18_5	Portugal	Nearshore	Source sed	311	141	0.029	54.66	0.71	51.77	0.71	6.38
N_15	Portugal	Nearshore	Source sed	115	161	0.019	52.23	0.00	52.17	2.48	6.83
D4	Portugal	Dune	Source sed	342	94	0.426	72.51	0.00	50.00	0.00	13.83
D1	Portugal	Dune	Source sed	353	163	0.279	53.82	0.00	57.06	1.23	7.98
D6	Portugal	Dune	Source sed	321	138	0.276	57.01	0.00	63.04	0.00	8.70
DC1	Portugal	Dune	Source sed	261	90	0.245	65.52	0.00	51.11	0.00	12.22
D2	Portugal	Dune	Source sed	325	122	0.229	62.46	0.00	46.72	1.64	7.38
D3	Portugal	Dune	Source sed	323	162	0.106	49.85	0.00	59.88	4.32	6.79
D5	Portugal	Dune	Source sed	242	105	0.101	56.61	0.00	60.00	0.95	9.52
DC2	Portugal	Dune	Source sed	219	88	0.075	59.82	0.00	52.27	0.00	13.64
B4	Portugal	Beach	Source sed	348	186	0.192	46.55	0.00	62.90	4.84	3.23
B1	Portugal	Beach	Source sed	324	164	0.185	49.38	0.00	50.61	6.71	2.44
B3	Portugal	Beach	Source sed	342	98	0.182	71.35	0.00	50.00	0.00	9.18
B2	Portugal	Beach	Source sed	122	58	0.077	52.46	0.00	67.24	1.72	3.45
2A	Indonesia	Tsunami	Tsunami profile inland	228	102	0.187	55.26	2.94	0.00	0.00	0.00
2G	Indonesia	Tsunami	Backwash	201	80	0.062	60.20	7.50	0.00	0.00	1.25
7B	Indonesia	Tsunami	Tsunami profile inland	241	59	0.244	70.95	0.00	0.58	0.00	0.58
7C	Indonesia	Tsunami	Tsunami profile inland	233	104	0.271	55.36	1.92	0.96	0.00	0.96
7D	Indonesia	Tsunami	Tsunami profile inland	282	72	0.003	42.20	0.00	0.84	0.00	3.36
7E	Indonesia	Tsunami	Tsunami profile inland	95	27	0.047	35.79	2.94	8.82	0.00	2.94
9A	Indonesia	Tsunami	Tsunami profile inland	302	123	0.352	59.27	4.07	0.81	0.00	0.00
LAG6	Indonesia	Tsunami	Lagoon (backwash)	289	85	0.149	73.36	0.94	0.47	0.00	0.47
11	Indonesia	Tsunami	Backwash	310	24	0.149	57.74	4.47	2.79	0.56	1.12
15A	Indonesia	Tsunami	Tsunami profile inland	117	71	0.028	29.91	11.43	0.00	0.00	0.00
15B	Indonesia	Tsunami	Tsunami profile inland	269	164	0.245	40.15	4.63	0.93	0.00	2.78
18	Indonesia	Lagoon	Tsunami profile inland	318	88	0.193	48.43	1.83	0.00	0.00	1.83
N21	Indonesia	Nearshore	Nearshore (backwash)	316	189	0.195	84.49	0.37	0.75	0.00	1.50
HB1_(1–2)	Scotland	Storm deposit	Vertical and horizontal profile	345	320	0.087	92.75	0.58	0.29	0.00	0.00
HB1_(5_6)	Scotland	Storm deposit	Vertical and horizontal profile	328	295	0.088	89.94	0.00	0.30	0.00	0.00
HB1_(10–11)	Scotland	Storm deposit	Vertical and horizontal profile	347	322	0.132	92.80	0.00	0.00	0.00	0.00

Andalusite	Silimanite	Kyanite	Garnet	Monazite	Amphiboles	Pyroxene	Epidote	Sphene	Mica	Titanite	Olivine	Brookite	Anatase	Alanite	Olivine
31.30	0.00	0.00	0.00	0.00	0.00	0.00	0.00	0.00	0.00	0.00	0.00	0.00	0.00	0.00	0.00
23.13	0.00	0.00	0.00	0.00	0.75	0.00	0.00	0.00	0.00	0.00	0.00	0.00	0.00	0.00	0.00
30.92	0.00	0.00	0.00	0.00	1.32	0.66	0.66	0.00	0.00	0.00	0.00	0.00	0.00	0.00	0.00
25.66	0.00	0.00	0.00	0.00	0.00	0.00	0.00	0.00	0.00	0.00	0.00	0.00	0.00	0.00	0.00
35.16	0.00	0.00	0.00	0.00	0.78	0.78	0.00	0.00	0.00	0.00	0.00	0.00	0.00	0.00	0.00
31.11	0.00	0.00	0.00	0.00	0.74	0.74	0.74	0.00	0.00	0.00	0.00	0.00	0.00	0.00	0.00
27.93	0.00	0.00	0.00	0.00	1.12	1.12	0.00	0.00	0.00	0.00	0.00	0.00	0.00	0.00	0.00
34.83	0.00	0.00	0.00	0.00	0.50	0.00	0.00	0.00	0.00	0.00	0.00	0.00	0.00	0.00	0.00
47.13	0.00	0.00	0.00	0.00	0.00	0.00	0.00	0.00	0.00	0.00	0.00	1.15	0.00	0.00	0.00
30.77	0.00	0.00	0.00	0.00	0.00	1.18	0.00	0.00	0.00	0.00	0.00	0.00	0.00	0.00	0.00
38.69	0.00	0.00	0.73	0.00	0.73	0.00	0.73	0.00	0.00	0.00	0.00	0.00	0.00	0.00	0.00
33.33	0.00	0.00	0.00	0.00	0.00	0.00	0.00	0.00	0.00	0.00	0.00	0.00	0.00	0.00	0.00
29.53	0.67	0.00	0.00	0.00	1.34	1.34	1.34	0.00	0.00	0.00	0.00	0.00	0.00	0.00	0.00
27.91	1.16	0.00	1.16	0.00	5.81	0.00	3.49	0.00	0.00	0.00	0.00	0.00	0.00	0.00	0.00
36.17	1.42	0.00	0.00	0.00	1.42	0.00	1.42	0.00	0.00	0.00	0.00	0.00	0.00	0.00	0.00
29.19	1.24	0.00	2.48	0.00	2.48	1.86	0.00	0.62	0.00	0.00	0.00	0.00	0.62	0.00	0.00
31.91	0.00	0.00	1.06	0.00	2.13	1.06	0.00	0.00	0.00	0.00	0.00	0.00	0.00	0.00	0.00
31.29	0.00	0.00	0.00	0.00	1.23	0.61	0.61	0.00	0.00	0.00	0.00	0.00	0.00	0.00	0.00
27.54	0.00	0.00	0.00	0.00	0.00	0.00	0.72	0.00	0.00	0.00	0.00	0.00	0.00	0.00	0.00
36.67	0.00	0.00	0.00	0.00	0.00	0.00	0.00	0.00	0.00	0.00	0.00	0.00	0.00	0.00	0.00
41.80	0.00	0.00	0.00	0.00	1.64	0.00	0.82	0.00	0.00	0.00	0.00	0.00	0.00	0.00	0.00
26.54	0.00	0.00	0.62	0.00	0.62	1.23	0.00	0.00	0.00	0.00	0.00	0.00	0.00	0.00	0.00
25.71	0.95	0.00	0.00	0.00	2.86	0.00	0.00	0.00	0.00	0.00	0.00	0.00	0.00	0.00	0.00
31.82	0.00	0.00	0.00	0.00	2.27	0.00	0.00	0.00	0.00	0.00	0.00	0.00	0.00	0.00	0.00
27.42	0.00	0.00	0.00	0.00	0.00	0.54	0.54	0.54	0.00	0.00	0.00	0.00	0.00	0.00	0.00
38.41	0.00	0.00	0.00	0.00	0.00	0.61	1.22	0.00	0.00	0.00	0.00	0.00	0.00	0.00	0.00
36.73	0.00	0.00	0.00	0.00	2.04	1.02	1.02	0.00	0.00	0.00	0.00	0.00	0.00	0.00	0.00
27.59	0.00	0.00	0.00	0.00	0.00	0.00	0.00	0.00	0.00	0.00	0.00	0.00	0.00	0.00	0.00
35.29	0.00	0.00	0.00	0.00	51.96	7.84	1.96	0.00	0.00	0.00	0.00	0.00	0.00	0.00	0.00
38.75	0.00	0.00	2.50	0.00	47.50	2.50	0.00	0.00	0.00	0.00	0.00	0.00	0.00	0.00	0.00
1.17	0.00	0.00	0.58	0.00	23.39	4.09	2.34	0.00	0.58	0.58	0.58	0.00	0.00	0.00	0.00
12.50	0.00	0.00	1.92	0.00	74.04	0.96	3.85	0.00	0.00	0.00	2.88	0.00	0.00	0.00	0.00
4.20	0.00	0.00	0.00	0.00	40.34	5.04	5.04	0.00	0.84	0.84	0.00	0.00	0.00	0.00	0.00
0.00	0.00	0.00	0.00	0.00	47.06	2.94	14.71	0.00	0.00	0.00	0.00	0.00	0.00	0.00	0.00
13.82	0.00	0.00	2.44	0.00	76.42	1.63	0.81	0.00	0.00	0.00	0.00	0.00	0.00	0.00	0.00
0.00	0.47	0.00	0.94	0.00	18.87	16.51	1.89	0.00	0.47	0.47	0.00	0.00	0.00	0.00	0.00
1.12	0.00	0.00	0.56	0.00	29.05	3.35	3.91	0.00	0.56	0.00	0.00	0.00	0.00	0.00	0.00
20.00	0.00	0.00	0.00	0.00	34.29	2.86	0.00	0.00	0.00	0.00	0.00	0.00	0.00	0.00	0.00
6.48	0.00	0.00	0.00	0.00	39.81	2.78	1.85	0.00	1.85	3.70	0.93	0.00	0.00	0.00	0.00
7.93	0.00	0.00	0.00	0.00	85.98	0.61	1.83	0.00	0.00	0.00	0.00	0.00	0.00	0.00	0.00
1.87	0.00	0.00	1.12	0.00	38.58	25.09	0.75	0.00	0.75	0.00	0.00	0.00	0.00	0.00	0.00
1.16	0.00	0.00	2.61	0.87	79.71	7.54	0.29	0.00	0.00	0.00	0.00	0.00	0.00	0.29	0.00
17.07	0.30	0.00	5.18	0.00	61.89	3.96	0.00	0.00	0.00	0.30	0.00	0.00	0.00	0.91	0.00
2.31	0.00	0.00	3.46	0.58	76.95	8.36	0.29	0.00	0.00	0.00	0.00	0.00	0.00	0.00	0.86

(Continued)

Table 1. *Heavy mineral (HM) data from the four studied deposits* (*Continued*)

Sample Name	Country	Sed Env	Comment	Number of grains counted	HM grains counted, except Op + Alt + NI	% of HMs in the total sediment fraction	% of transparent HMs	Zircon	Tourmaline	Rutile	Staurolite
HB1_(15–16)	Scotland	Storm deposit	Vertical and horizontal profile	257	247	0.1277	96.11	0.00	0.00	0.00	0.00
HB2_(11–12)	Scotland	Storm deposit	Vertical and horizontal profile	174	158	0.1193	90.80	0.00	0.00	0.00	0.00
HB2_(18–19)	Scotland	Storm deposit	Vertical and horizontal profile	249	227	0.1625	91.16	0.00	0.00	0.00	0.00
HB2_(27–28)	Scotland	Storm deposit	Vertical and horizontal profile	369	359	0.1123	97.29	0.00	0.27	0.00	0.00
HB3_(9–10)	Scotland	Storm deposit	Vertical and horizontal profile	176	165	0.1134	93.75	0.00	0.00	0.00	0.00
HB3_(14–15)	Scotland	Storm deposit	Vertical and horizontal profile	142	132	0.1595	92.96	0.00	0.00	0.00	0.00
HB3_(21–22)	Scotland	Storm deposit	Vertical and horizontal profile	168	162	0.1021	96.43	0.00	0.00	0.00	0.00
HB4_(4–5)	Scotland	Storm deposit	Vertical and horizontal profile	128	123	0.1622	96.09	0.00	0.00	0.00	0.00
HB4_(9–10)	Scotland	Storm deposit	Vertical and horizontal profile	344	320	0.1144	93.02	0.00	0.00	0.00	0.00
HB5_(4–5)	Scotland	Storm deposit	Vertical and horizontal profile	303	275	0.0263	90.76	0.00	0.00	0.00	0.00
HB5_(8–9)	Scotland	Storm deposit	Vertical and horizontal profile	328	295	0.1052	89.94	0.00	0.30	0.00	0.00
HB5_(11–12)	Scotland	Storm deposit	Vertical and horizontal profile	386	356	0.0139	92.23	0.52	0.00	0.00	0.00

OP, opaque grains; NI, non-identified grains; Alt, altered minerals.

Tsunami

Salgados (AD 1755 tsunami). The AD 1755 tsunami inundation and associated deposits have been described in the Salgados lagoonal basin (Costa *et al.* 2012*b*) (Fig. 1). The AD 1755 tsunami sandy layer is in strong contrast with the under- and overlying muddy sediments representing the background low-energy sedimentary environment. The deposit overlays an unconformity surface interpreted as an erosive contact (Costa *et al.* 2012*b*). This deposit was assigned to the AD 1755 tsunami event based in its spatial distribution, as well as on results from textural, micropalaeontological, microtextural and age-constraining analysis. The tsunamigenic unit sediment source was identified as the dune and beach (Costa *et al.* 2012*b*).

Lhok Nga (2004 Indian Ocean tsunami). Lhok Nga Bay is a flat 5 km large coastal embayment (25 km^2) opened to the west and delimited by calcareous steep slopes (Fig. 1). The coast prior to the 26 December 2004 Indian Ocean tsunami consisted of a continuous beach, breached by small estuaries during the wet season (Paris *et al.* 2009). The 2004 tsunami deposits in this area consist of medium–coarse, greyish to yellowish sands, which display lateral variations in thickness and grain size, and vertical variations in texture. Several subunits were identified in the tsunami deposit (Paris *et al.* 2007), corresponding to inundation and backwash phases. The lower contact of the tsunami unit with the underlying sand is abrupt or erosional.

Tsunami samples used in this study were collected a few weeks after the tsunami event (Paris *et al.* 2007). Tsunami samples consist of vertical and horizontal profiles up to 3 km inland. The vertical analysis included samples representing inundation and backwash (Paris *et al.* 2007; Costa *et al.* 2012*a*).

Storm

Stoneybridge (Great Storm of January 2005). The 11 January 2005 storm strongly affected the Benbecula and South Uist coastlines of Scotland (Fig. 1), producing changes in the coastal landscape along the entire western seaboard and, in particular, on headlands (Angus & Rennie 2007). According to

Andalusite	Silimanite	Kyanite	Garnet	Monazite	Amphiboles	Pyroxene	Epidote	Sphene	Mica	Titanite	Olivine	Brookite	Anatase	Alanite	Olivine
1.56	0.00	0.00	2.33	0.78	87.55	3.50	0.00	0.00	0.00	0.00	0.00	0.00	0.00	0.00	0.39
4.60	0.57	0.00	4.02	0.00	74.71	6.90	0.00	0.00	0.00	0.00	0.00	0.00	0.00	0.00	0.00
11.65	0.00	0.00	1.20	0.40	67.07	10.44	0.40	0.00	0.00	0.00	0.00	0.00	0.00	0.00	0.00
4.07	0.00	0.00	1.90	0.00	81.03	9.21	0.27	0.00	0.00	0.00	0.00	0.00	0.00	0.54	0.00
11.36	0.00	0.00	1.70	0.57	69.32	10.23	0.00	0.00	0.00	0.57	0.00	0.00	0.00	0.00	0.00
3.52	0.00	0.00	2.82	0.00	80.99	5.63	0.00	0.00	0.00	0.00	0.00	0.00	0.00	0.00	0.00
13.10	0.00	0.00	2.98	0.00	76.79	2.98	0.60	0.00	0.00	0.00	0.00	0.00	0.00	0.00	0.00
36.72	0.00	0.00	1.56	1.56	53.13	2.34	0.00	0.00	0.00	0.78	0.00	0.00	0.00	0.00	0.00
31.10	0.00	0.00	2.91	0.58	53.20	4.94	0.29	0.00	0.00	0.00	0.00	0.00	0.00	0.00	0.00
17.16	0.00	0.00	1.98	0.00	68.65	2.64	0.00	0.00	0.00	0.33	0.00	0.00	0.00	0.00	0.00
17.07	0.30	0.00	5.18	0.00	61.89	3.96	0.00	0.00	0.00	0.30	0.00	0.00	0.00	0.91	0.00
15.03	0.00	0.00	1.55	0.26	71.24	3.37	0.26	0.00	0.00	0.26	0.00	0.00	0.00	0.00	0.26

Dawson *et al.* (2007), the coastal response to the storm varied depending on sedimentary environment. In some areas, coastal dunes retreated. In other areas, shingle/gravel and even boulders were deposited.

A series of trenches and cores were investigated at Stoneybridge (Fig. 1) along a horizontal transect (on a flat surface) perpendicular to the shoreline at approximately 6 m ordnance datum (OD) (Fig. 2) that extended approximately 300 m inland. Lithostratigraphic logs were compiled and stratigraphic correlations between sample sites were made. Nineteen samples were prepared to observe heavy minerals under the microscope.

Sand was deposited as a sheet, with distinct layers of coarser and lighter sand in sharp contact with layers of finer and darker sand units (Fig. 2). Overall, a dark-grey fine basal sand (subunit 3) is overlaid by intercalations of light-grey medium sand and light-grey coarse sand (subunit 2) throughout the middle section, while the top layers are darker fine sands (subunit 1). The thickness of the central coarser section decreases in a landwards direction, forming a landwards-tapering wedge of coarser material within the tapering storm deposit itself.

Fire Island (2012 Hurricane Sandy). Hurricane Sandy impacted the NE coast of the USA on 29 October 2012 and is the largest historical storm on record. Barrier islands in the region were strongly affected by the hurricane, which caused severe erosion of dunes and beaches. On Fire Island, a barrier island along the southern coast of Long Island, New York, sandy lobes were deposited by the storm waves and changed the geomorphology of the area (Hapke *et al.* 2013; La Selle *et al.* 2015) (Figs 1 & 3). A field investigation 1 year after the storm measured profiles perpendicular to the coast, documented characteristics of the overwash deposit and collected samples for laboratory analyses (La Selle *et al.* 2016).

Methodology

The laboratory protocols applied in the microtextural and heavy mineral analysis used in this work are detailed below.

Table 2. *Digital image analysis results for grain-shape characteristics from samples from Fire Island*

Location	Sample	Sediment environment	Minor axis (mm)	Major axis (mm)	Diameter (mm)	Circularity	Shape factor	Ratio (minor/major)	Mean (micra)	Sorting (phi units)	Skewness	Kurtosis
Fire Island	FI13-E4-S1-Beach	Beach	0.430	0.533	0.488	0.931	0.826	0.00064	776.1	2.504	0.437	1.893
Fire Island	FI13-E4-S3-Dune	Dune	0.324	0.438	0.376	0.919	0.801	0.00045	444.1	1.407	−0.008	0.980
Fire Island	FI13-E4-T1-2-5	Grey dirty sand with roots	0.296	0.421	0.346	0.889	0.771	0.00041	250.1	1.362	−0.013	1.067
Fire Island	FI13-E4-T1-11-13	Medium–coarse sand	0.346	0.495	0.408	0.923	0.775	0.00049	514.6	1.776	0.305	0.952
Fire Island	FI13-E4-T2-5-8	Grey sand	0.388	0.489	0.437	0.917	0.796	0.00060	411.7	1.379	−0.002	1.031
Fire Island	FI13-E4-T3-hm	Heavy mineral lamina	0.441	0.576	0.492	0.940	0.824	0.00062	271.6	1.454	0.073	0.953
Fire Island	FI13-E4-T3-5-10	Rooted faintly laminated	0.244	0.353	0.284	0.914	0.815	0.00041	335.3	1.405	0.026	1.017
Fire Island	FI13-E4-T3-40-45	Top of laminated layer	0.436	0.625	0.530	0.915	0.799	0.00058	416.9	1.497	−0.033	1.058
Fire Island	FI13-E4-T3-50-53	Aeolian	0.362	0.529	0.426	0.892	0.779	0.00049	395.8	1.542	−0.019	1.160
Fire Island	FI13-E4-T4-45-50	Top laminated sand	0.307	0.446	0.377	0.914	0.778	0.00043	359.5	1.381	−0.059	1.043
Fire Island	FI13-E4-T4-63-65	Loose aeolian	0.394	0.583	0.459	0.918	0.832	0.00052	458.6	1.714	0.104	1.034
Fire Island	FI13-E4-T7-22-26	Massive fine–medium sand	0.387	0.485	0.431	0.934	0.799	0.00052	374.0	1.420	−0.102	1.093
Fire Island	FI13-E4-T8-2-5	Very fine sand slightly grey	0.276	0.387	0.320	0.916	0.794	0.00041	253.6	1.392	−0.004	1.089
Fire Island	FI13-E4-T9-1-5	Very fine sand (rooty)	0.273	0.407	0.331	0.904	0.759	0.00039	264.0	1.407	−0.041	1.081

Table 3. *Microtextural data from all studied sites from 0 (absent) to 5 (covering > 75% of the grain's surface)*

Location	Sample	Comment	Dissolution	Adhering particles	Fresh surfaces	Percussion marks	Roundness
Fire Island	FI13-E4-S1-Beach	Beach	2	1	3	2	4.000
Fire Island	FI13-E4-S3-Dune	Dune	2	2	1	1	4.000
Fire Island	FI13-E4-T1-2-5	Grey dirty sand with roots	1.5	1	3	2	3.000
Fire Island	FI13-E4-T1-11-13	Medium–coarse sand	1	1	3	2	3.500
Fire Island	FI13-E4-T2-5-8	Grey sand	2.5	1	1	1	4.000
Fire Island	FI13-E4-T3-hm	Heavy mineral lamina	1	0	2	1	4.000
Fire Island	FI13-E4-T3-5-10	Rooted faintly laminated	1	1.5	3	2	4.000
Fire Island	FI13-E4-T3-40-45	Top of laminated layer	1	1	2	2	4.000
Fire Island	FI13-E4-T3-50-53	Aeolian	3.5	2	1	1	4.000
Fire Island	FI13-E4-T4-45-50	Top laminated sand	0	0	3	2	4.000
Fire Island	FI13-E4-T4-63-65	Loose aeolian	1	1	3	2	4.000
Fire Island	FI13-E4-T7-22-26	Massive fine–medium sand	1.5	0	2.5	2	4.000
Fire Island	FI13-E4-T8-2-5	Very fine sand slightly grey	2	1	1.5	1.5	4.000
Fire Island	FI13-E4-T9-1-5	Very fine sand (rooty)	1	2	3	2	3.000
Stoneybridge	HB_2_3	Storm	4.643	4.500	1.643	0.286	3.143
Stoneybridge	HB_5_6	Storm	3.071	3.429	2.500	0.643	3.643
Stoneybridge	HB_7_8	Storm	3.000	3.500	2.000	0.083	3.250
Stoneybridge	HB_9_10	Storm	3.857	4.429	0.929	0.357	2.214
Salgados	ArmP_SG_17m	Nearshore	3.000	3.357	0.214	1.071	1.214
Salgados	SG_16_5m	Nearshore	3.692	4.000	0.308	0.010	1.692
Salgados	SG_16m	Nearshore	3.933	3.800	0.667	2.000	1.000
Salgados	ArmP_SG_16m	Nearshore	2.273	3.273	1.364	0.818	1.818
Salgados	ArmP_SG_15_1m	Nearshore	3.083	3.667	0.417	1.083	2.250
Salgados	ArmP_SG_15m	Nearshore	2.143	3.571	0.857	1.429	2.714
Salgados	ArmP_SG_14m	Nearshore	3.000	3.231	1.231	0.923	2.462
Salgados	SG_13m	Nearshore	3.167	3.056	0.778	1.500	2.000
Salgados	SG_11_5m	Nearshore	3.750	2.700	0.400	1.150	1.900
Salgados	ArmP_SG_11m	Nearshore	2.615	3.231	0.923	1.385	2.385
Salgados	SG_10m	Nearshore	2.769	3.769	1.462	0.538	2.154
Salgados	ArmP_SG_9m	Nearshore	2.636	2.455	1.818	1.000	3.000
Salgados	SG_7m	Nearshore	2.500	2.438	2.000	2.000	2.125
Salgados	SG_4m	Nearshore	1.700	2.450	2.650	1.600	2.550
Salgados	SG_Praia	Beach	0.333	0.250	3.167	2.750	3.250
Salgados	SG_Praia_TdV	Beach	1.533	0.667	2.600	1.800	1.800
Salgados	SG_Berma	Beach	1.583	2.417	2.000	1.333	2.750
Salgados	SG_Esc_Bar_Base	Beach	1.333	2.000	2.583	0.917	3.083

(*Continued*)

Table 3. *Microtextural data from all studied sites from 0 (absent) to 5 (covering > 75% of the grain's surface)* (*Continued*)

Location	Sample	Comment	Dissolution	Adhering particles	Fresh surfaces	Percussion marks	Roundness
Salgados	SG_Barra	Beach	1.000	0.429	3.143	1.071	2.929
Salgados	SG_Esc_Bar_Top	Beach	1.364	1.545	2.909	1.818	2.636
Salgados	SG_Zona_Bar	Beach	0.583	1.167	3.583	1.083	3.000
Salgados	SG_Duna	Dune	3.333	3.417	1.250	2.333	2.250
Salgados	SG_A_Duna	Dune	2.333	3.400	1.800	1.800	2.000
Salgados	SG_D5	Dune	2.538	2.308	2.231	2.231	2.000
Salgados	SG_Duna_F	Dune	2.364	2.636	1.727	2.091	3.091
Salgados	SG_Crista_Duna	Dune	2.214	2.929	2.071	1.643	2.500
Salgados	SG_Duna_R	Dune	1.750	2.667	1.917	2.083	2.083
Salgados	SG_Duna_2F	Dune	2.545	3.091	1.364	2.364	2.545
Salgados	SG_Duna_2Crista	Dune	1.933	3.133	2.467	1.000	3.000
Salgados	SG_Duna_2R	Dune	2.308	2.615	1.231	1.538	2.385
Salgados	SG_Duna_2RR	Dune	2.583	3.000	0.750	1.500	2.167
Salgados	SG_14_(0.20–0.28)	Storm	0.083	1.000	2.667	3.167	2.750
Salgados	SG_14_(0.40–0.82)	Tsunami	0.636	0.545	2.091	3.727	1.909
Salgados	SG_20_menor	Tsunami	1.444	2.667	3.778	0.963	3.481
Salgados	SG_20	Tsunami	1.286	1.500	1.143	1.321	2.000
Salgados	SG_20_Maior	Tsunami	1.800	1.300	1.600	2.400	2.100
Salgados	SG_22	Tsunami	1.320	1.240	2.000	1.840	2.480
Salgados	SG_26	Tsunami	0.706	0.941	2.882	0.882	2.471
Salgados	SG_LV7_(0.40–0.41)	Tsunami	2.048	0.286	2.048	1.524	1.476
Lhok Nga	IND_Praia	Beach	1.000	4.667	2.000	0.833	3.500
Lhok Nga	IND_Nearshore	Nearshore	3.833	4.333	0.778	0.778	2.500
Lhok Nga	NGA_2A	Tsunami	0.875	1.938	1.813	0.250	2.750
Lhok Nga	NGA_2E	Tsunami	0.692	2.692	3.000	0.615	3.692
Lhok Nga	NGA_2G	Tsunami backwash	0.857	2.286	3.571	0.786	3.714
Lhok Nga	NGA_7C	Tsunami	1.069	2.966	3.276	0.759	3.552
Lhok Nga	NGA_7F	Tsunami backwash	0.857	1.357	3.429	1.643	3.143
Lhok Nga	NGA_9A	Tsunami	1.621	2.828	2.931	0.621	4.000
Lhok Nga	NGA_18	Tsunami	2.833	2.583	0.917	0.010	2.500

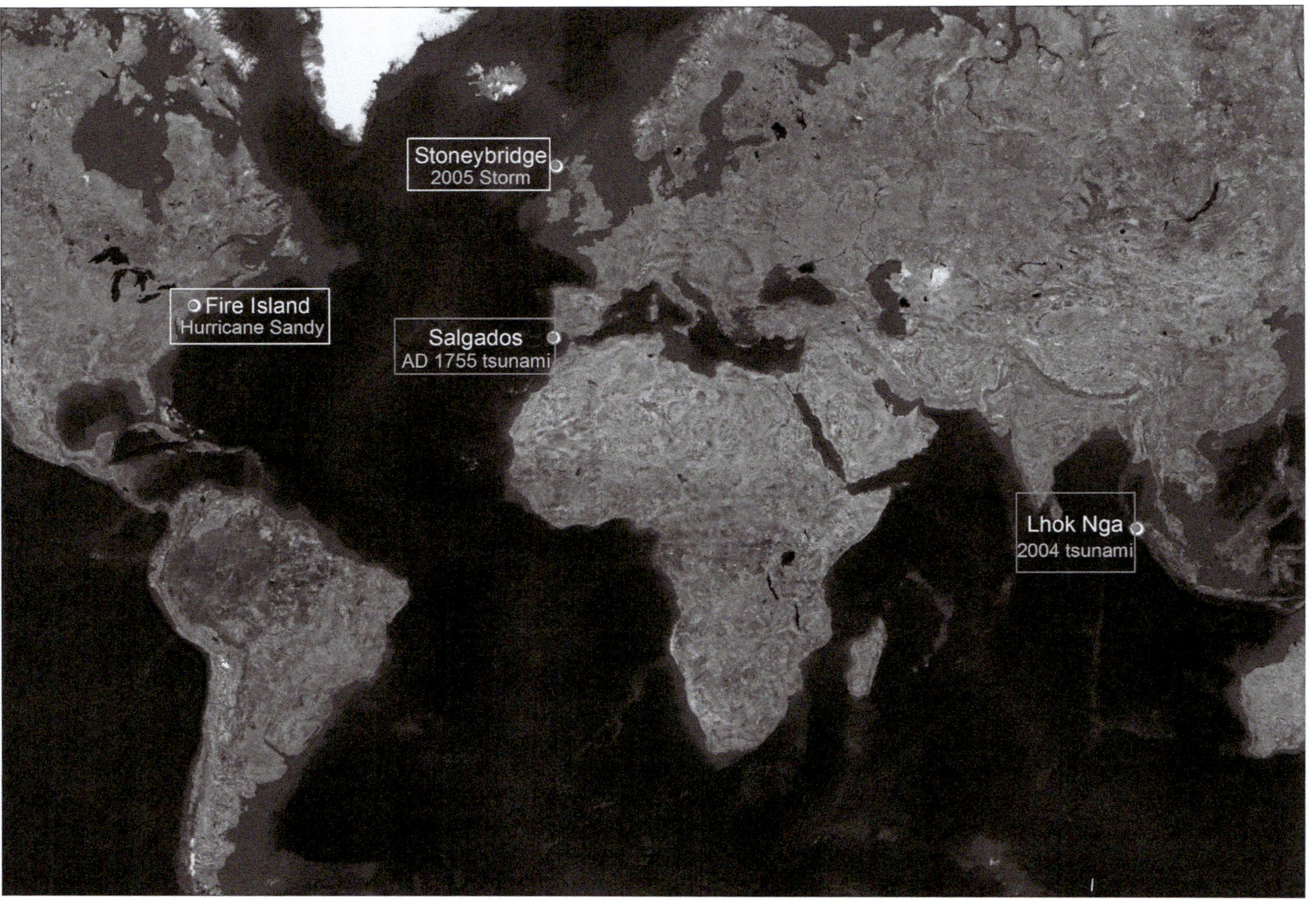

Fig. 1. Overview and geographical location map of the studied events and field sites.

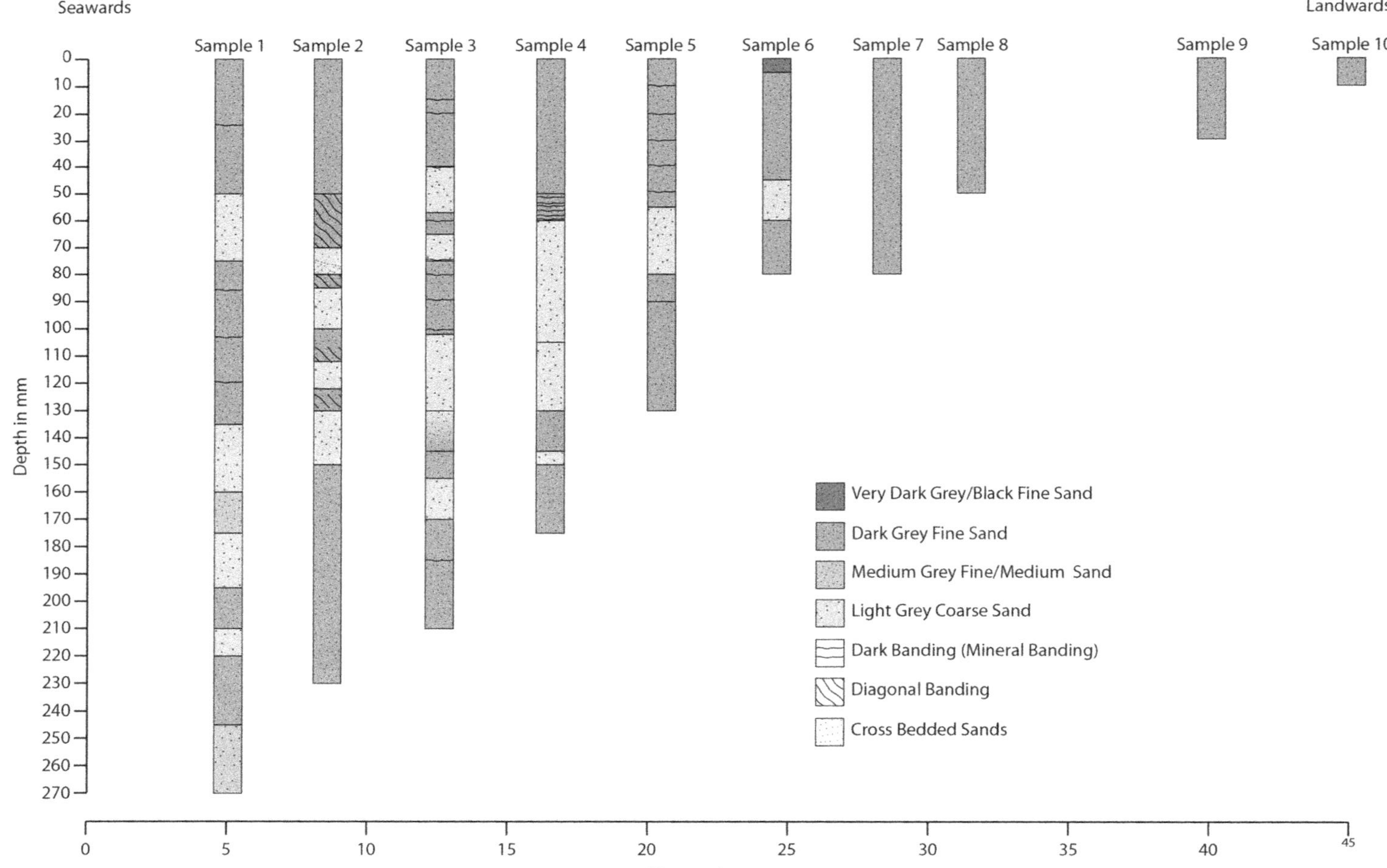

Fig. 2. Lithostratigraphic profile at Stoneybridge. Core 1 is to seawards and 10 to landwards of the present coastline. Note the coarser sand wedge within the overall storm deposit in cores 1–6.

Fig. 3. Push core (left) and U-channel (right) sampling Hurricane Sandy deposit in the trench in Fire Island.

Grain-size analysis

Grain-size analysis is conducted using a set of sieves at 0.5ϕ interval. Initially, the bulk samples are washed with tapwater and two fractions separated (>63 and <63 µm) using a 4ϕ mesh. The finer fraction (<63 µm) is analysed using the laser granulometer Malvern Mastersizer Hydro 2000 MU after deflocculating with 30% sodium hexametaphosphate. The coarser fraction (>63 µm) is analysed exclusively through sieving, with the exception of samples that have a coarser fraction weighing less than 10 g. Statistical parameters of the grain-size distribution (e.g. mean grain size, ϕ10, ϕ90, median, mode, standard deviation, kurtosis and skewness) of the samples are calculated using the graphic method (Folk & Ward 1957) in the GradiStat software (Blott & Pye 2001).

Microtextural analysis

Grain-surface microtexture (also named exoscopy: i.e. the micromorphological analysis of the surface of grains) is an analysis usually reserved for sand-sized quartz particles. Many quartz sand grains develop microtextural features that can often be referred to specific transport processes or specific sedimentary environments. In order to recognize surface microtextures, a scanning electron microscope was used to obtain high-resolution images (JEOL JSM 5200 LV (University of Lisbon)).

Laboratory procedure. The number of grains necessary for sedimentological interpretation varies within a wide range (cf. Mahaney 2002; see Costa *et al.* 2012*a* for a summary and examples). A minimum of 25 quartz grains per sample from the 1ϕ to 3ϕ (125–500 µm) size fraction were randomly selected under the binocular microscope and prepared for SEM analysis. This involves the mineralization of the particles with gold to avoid electrical inferences when obtaining the images. The grains were then taken to the SEM laboratory and photographed. Visual analyses of each enlarged grain allowed classification based on a group of microtextural characteristics (see below).

Microtextural classification. The microtextural classification was established based on selected references (e.g. Mahaney 2002; Costa *et al.* 2012*a*). Each grain is characterized using a number of descriptive microtextural features revealed in SEM

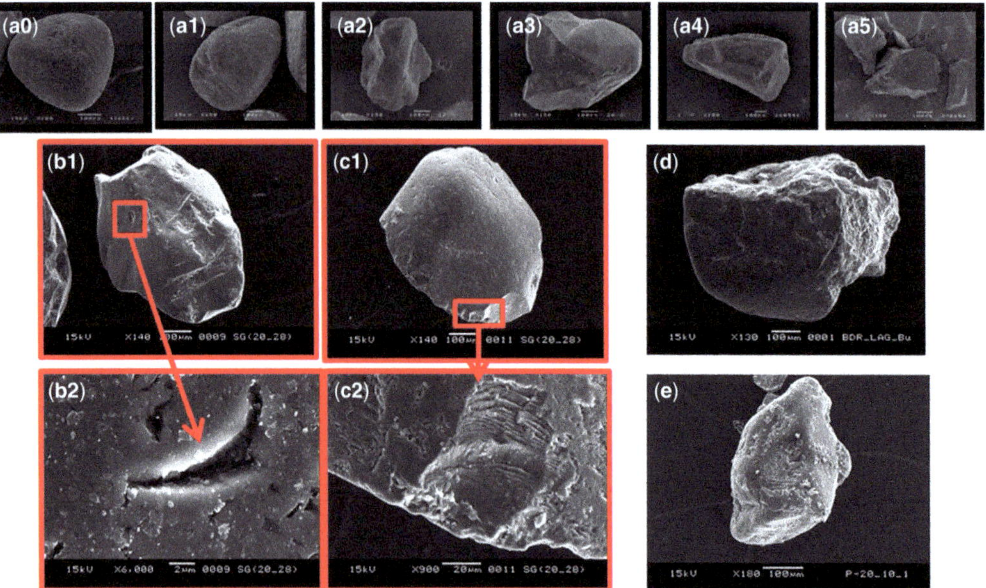

Fig. 4. Five main microtextural features. (**a0**)–(**a5**) angularity (very rounded to very angular); (**b1**) percussion marks; (**b2**) detailed view of a percussion mark; (**c1**) fresh surface (in this case, sharp edge, fractures and steps); (**c2**) detailed view of a fresh surface; (**d**) dissolution (especially visible on the right face of the grain); and (**e**) adhering particles (more visible in the centre of the grain) (Costa *et al.* 2012*a*).

microphotographs (e.g. fresh surfaces, percussion marks, dissolution features, angularity and adhering particles). The microtextures (Fig. 4) can be characterized as follows:

- Angularity (Fig. 4a0–a5) – classified using Powers scale (Powers 1953).
- Fresh surfaces (Fig. 4b1 & b2) – characterized by the presence of mechanical marks (i.e. fractures, abrasion marks, sharp edges) that are responsible for the recent exposure of part, or the totality, of the grain surface. These are also habitually characterized by the absence of dissolution or precipitation in those new surfaces.
- Percussion marks (Fig. 4c1 & c2) – characterized by the presence of, usually, V-shaped depressions, typically the result of grain collision.
- Dissolution (Fig. 4d) – microtexture of chemical nature indicating the degree of dissolution on the surface of individual grains. The effects of dissolution are noticed by the destruction of fresh surfaces and sharp edges on the surface of the grains, and by the formation of grooves.
- Adhering particles (Fig. 4e) – characterized by the presence of microparticles on the surface of quartz grains. Usually, these microparticles are within grooves and many are attached to the surfaces of individual grains.

A semi-quantitative approach to the microtextural classification of each grain is applied, based upon the proportion of grain surface occupied by each feature: 0 (absent) to 5 (>75% of the grain surface). The angularity of the grains was classified from 0 to 5 (very rounded to very angular). Median values for each variable are calculated for each sample and used in the subsequent statistical analysis.

Shape analysis. To decrease the human operator subjectivity, grain-shape analysis was conducted, in samples from Fire Island, using a fully automated routine based on image analysis procedures and mathematical morphology techniques. This procedure quantifies five different shape parameters: (1) minor and major axis; (2) equivalent diameter; (3) circularity; (4) form factor; and (5) rolling index. The technique calculates each parameter by analysing the 2D projected shape of each individual particle (Fig. 5). In the specific case of the rolling index, the parameter is evaluated by an automated procedure that relates the visual roundness chart developed by Powers (1953) and Shepard (1973) with the shape of the analysed particle, thus assigning the matching rolling index. The methodology is completely described in Lira (2011), and follows the works of Wadell (1932, 1934), Powers (1953), Sneed & Folk (1958), Zunic & Hirota (2008) and Lira & Pina (2007, 2009).

Furthermore, we emphasize that we only applied this new digital image analysis approach to samples from Fire Island, New York, USA to provide more

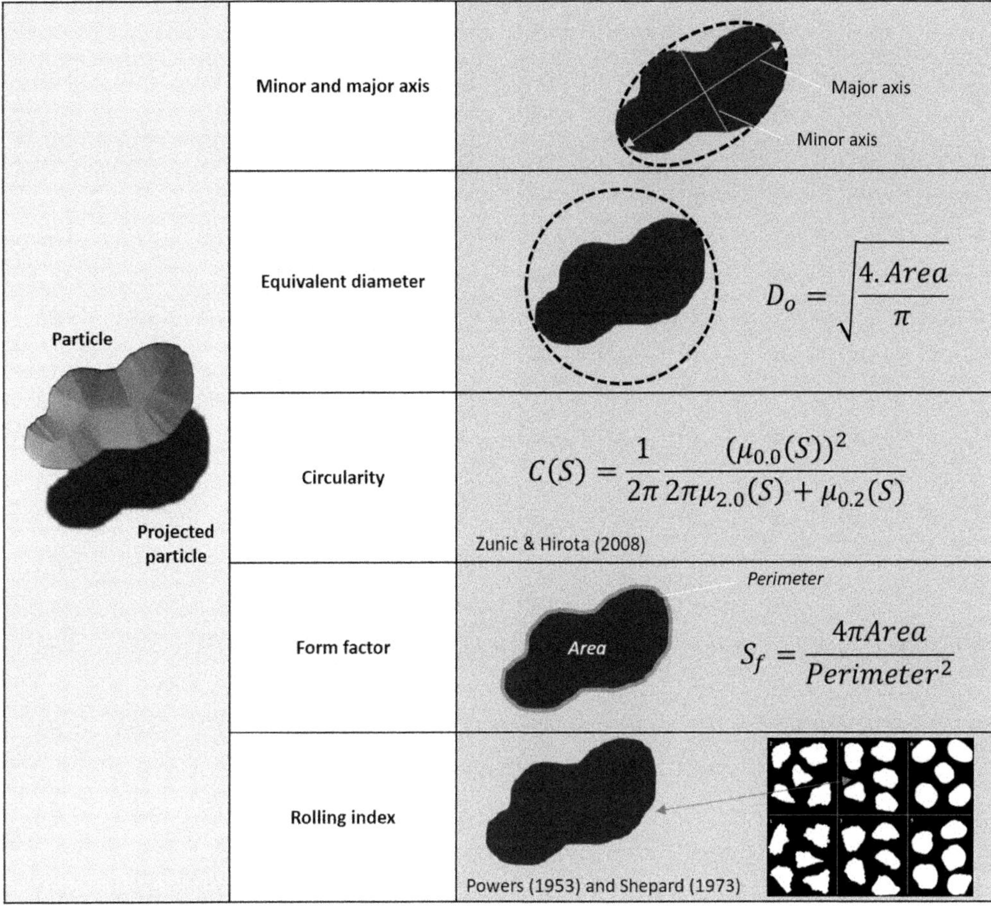

Fig. 5. Schematic representation of five different parameters measured using an automated image analysis of the grain shape.

quantitative (grain-shape microtextural) data from a contemporaneous storm (hurricane) event. We have done this to explore what we consider to be a very promising tool to develop microtextural analysis.

Heavy mineral analysis

For heavy mineral analysis, the approach used was similar to the one described in Costa *et al.* (2015*a*). Sediment samples were grouped in the 1ϕ–3ϕ fraction and prepared for observation on slides under the petrographical microscope. Heavy minerals were separated using bromoform and then mounted using Entelan resin on glass slides. For each sample, about 300 heavy minerals per slide were identified and counted under an Olympus BX40 petrographical microscope. The percentage of heavy minerals was recalculated to only consider transparent grains.

In the interpretation of the results, we used mean values for the density for each mineral species. These criteria were applied based on the knowledge of the more common types of minerals for each heavy mineral species/group and in the usual approach used on heavy mineral studies (Mange & Maurer 1992).

Heavy mineral analysis was not conducted on samples from Fire Island.

Results

Tsunami

Salgados (Portugal). The microtextural analysis conducted in Salgados (Costa *et al.* 2012*a*) is summarized in Table 3, which shows that tsunamigenic samples have high values of fresh surfaces, the highest values of percussion marks (thus suggesting energetic hydrodynamic conditions before deposition) and the widest range of values observed in all the microtextural features. On the other hand, beach grains presented the highest values for fresh surfaces and a strong presence of percussion marks, while dissolution and adhering particles are almost absent. Dune grains revealed a well-balanced representation of all the microtextural features, whereas nearshore samples presented high values of dissolution and adhering particles, and a moderate presence of percussion marks and fresh surfaces.

The study and comparison of the heavy mineral content from different sedimentary environments and the tsunamigenic unit (Table 1) revealed that the assemblage is dominated by andalusite, tourmaline and staurolite (combined, they represent up to 90% of the heavy mineral assemblage) (Costa *et al.* 2015*a*).

Results suggest that the tsunamigenic samples present very similar percentage values of heavy mineral as dune samples (i.e. andalusite, tourmaline and staurolite), the only exception being staurolite which presents higher values (between 13.2 and 19.1% in tsunamigenic samples). Staurolite in dune samples varies between 7 and 14%. Beach samples were distinguishable by the higher percentage of other minerals (e.g. amphiboles, zircon, apatite) and by the lowest percentage of staurolite. On the other hand, nearshore samples were differentiated by the higher content in amphiboles and other minerals.

Lhok Nga (Indonesia). In Lhok Nga (Indonesia), according to Costa *et al.* (2012*a*), microtextural results solely revealed vertical variations within the tsunamigenic unit in fresh surfaces and percussion marks (increasing to the top of the tsunamigenic unit), and that the highest values for these microtextures were identified in the sediment deposited by backwash (Table 3).

The study of the heavy mineral assemblage of the Lhok Nga tsunamigenic samples revealed the dominance of amphiboles (with values up to 85%) (Costa *et al.* 2015*a*) (Table 1). Other relevant (>5% of the assemblage) heavy minerals observed were andalusite and pyroxene. Samples collected in subunits associated with the backwash phase revealed more zircon and garnets. Costa *et al.* (2015*a*) also detected a higher number of euhedral zircon grains in the backwash samples, which was in contrast with inundation samples that presented more rounded zircon grains (implying reworking in the coastal area) (Fig. 6).

Storm

Stoneybridge (Scotland). Microtextural classification was conducted on samples from the storm deposit at Stoneybridge (Table 3). Angularity exhibited high values (*c.* 3.1–3.6), except for the base of the unit (with 2.2). The highest values of fresh surfaces were observed in samples HB_(0.05–0.06)

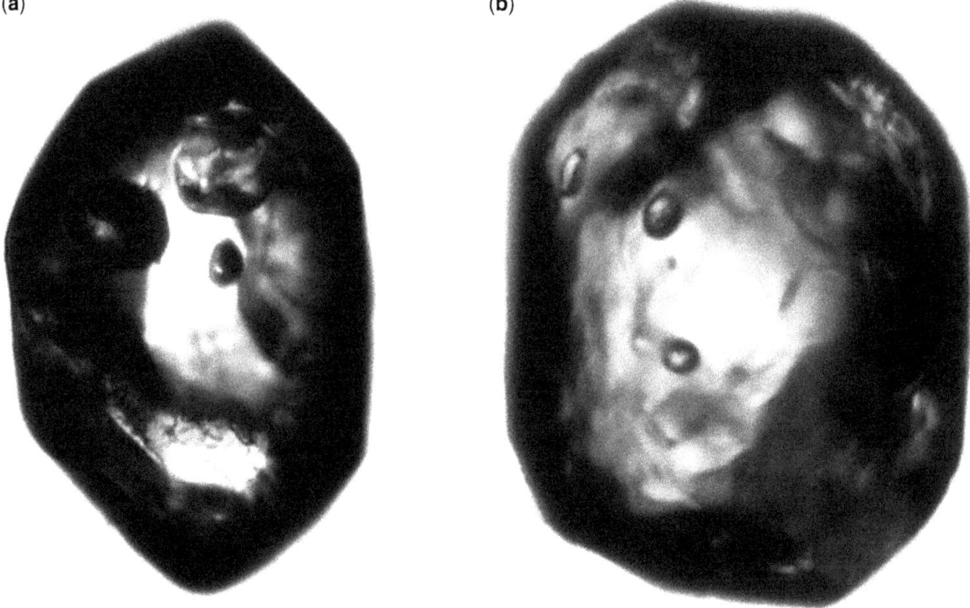

Fig. 6. Images of euhedral (**a**) and rounded (**b**) zircon obtained in run-in and backwash samples, respectively, collected in Lhok Nga (Costa *et al.* 2015*a*).

and HB_(0.07–0.08) with results of 2.5 and 2.0. Fewer fresh surfaces were measured in the basal sample HB_(0.09–0.10). On the other hand, dissolution presented very high values (all above 3.0, and the highest at the top and base of the storm deposit). Regarding adhering particles, they present values of approximately 3.5 in the middle samples, while the top and base of the deposit revealed values of around 4.5. Percussion marks present its highest value (0.643) in sample HB_(0.05–0.06), while the lowest result of 0.083 was observed in sample HB_(0.07–0.08).

Microtextural data were not useful in establishing a vertical trend for the storm deposit, although it was clear that the base and top of the deposit exhibited contrasting results with the other storm deposit samples (except for percussion marks) probably related with incorporation of *in situ* finer material.

In terms of the percentage of heavy minerals in total sediment fraction, no vertical trend was observed (Table 1). The presence of opaque grains within the studied thin sections is exceptionally low (<1.5%) in the assemblage (Table 1).

The storm deposit is mainly composed of amphiboles, pyroxenes, andalusite and garnet (forming >90% of the heavy mineral assemblage). Other minerals like monazite, epidote, allanite, tourmaline and zircon were also detected but in very small percentages (Table 1). The sum of the percentage of the four main heavy mineral species tends to slightly decrease towards the top of each core. Amphiboles are the dominant mineral species with a presence, on average, of 73.5%: with a maximum value of 87.55% and a minimum of about 53%. In compositional terms, no vertical trend was clearly established based on percentage variation of the main mineralogical components of these samples: however, horizontal variations were detected and are discussed below (e.g. andalusite and garnet).

Fire Island (USA). Here, we also present new microtextural, shape and grain-size analyses performed from beaches, dunes and washover deposits on Fire Island (New York, USA) after the 2012 Hurricane Sandy. Generally, the microtextural results obtained revealed a mostly homogenous assemblage with minor differences – mainly related with sediment source (aeolian v. beach) or compositional (heavy mineral layers) or textural (fine vs coarser samples) – between samples (see the summary in Tables 2 & 3). However, some discrepancies provided interesting insights into the micromorphological and compositional sedimentological impacts of the Hurricane Sandy on Fire Island.

In this assemblage, samples with higher dissolution values were mainly from an aeolian source (Table 3). A similar pattern was also observed in terms of adhering particles with higher values in aeolian samples in contrast with lower values for the heavy mineral laminae (Table 1). In what concerns fresh surfaces, lower values (i.e. 1) were observed in samples with an aeolian origin, while most of the remaining samples presented values around of 2.5–3 (Table 3).

Samples analysed typically present a value of 2 for percussion marks. However, lower values were detected in four samples: FI13-E4-T3-hm (heavy mineral lamina), FI13-E4-T2-5-8 (grey sand), FI13-E4-S3-Dune and FI13-E4-T3-50-53 (aeolian).

In the roundness classification, all samples present a median value of 4 (sub-angular), except for three more angular samples: FI13-E4-T9-1-5 (very fine sand (rooty)), FI13-E4-T1-2-5, (grey dirty sand with roots) and FI13-E4-T1-11-13 (medium–coarse sand). Confirming the pattern exhibited in the roundness, circularity results show small discrepancies between the different samples, with values varying solely from 0.892 to 0.940. In what concerns the shape ratio, all values are within a narrow interval of 0.759 and 0.32. The ratio between minor and major axis exhibits, interestingly, the highest values in samples FI13-E4-T3-hm (heavy mineral lamina) and FI13-E4-S1-Beach, and the lowest value in sample FI13-E4-T9-1-5 (very fine sand (rooty)).

Discussion

Microtextural signatures in storm and tsunami deposits

Several attempts (cf. Costa *et al.* 2012*a* for a summary) have been made to characterize surface microtextural signatures in quartz grains from distinct sedimentary environments. The results proposed have been controversial, and far from unanimous in the identification of microtextures, interpretation of imprinting mechanisms or processes, and association of specific microtextural signatures with specific sedimentary environments. These difficulties are amplified in the case of storm-driven deposits because of the short transport (both distance and time) and also because the sediment source is typically less diverse than other extreme marine inundations (e.g. tsunamis). Difficulties in the microtextural analysis of tsunami deposits were observed by Bellanova *et al.* (2015), who were unable to detect a clear pattern or microtextural signature in tsunamigenic samples from Chile. This result stressed the importance of background studies and the relevance of considering local geomorphological and sedimentological factors before conducting microtextural analysis. We might consider that a diverse sediment source is a prerequisite for the microtextural technique to be successfully applied. We can speculate that due to the physical differences (namely wavelength) between storm and tsunami waves, and if the source is diverse enough, we expect to see a greater diversity of microtextural features in tsunami deposits compared to storms.

The fact that five different microtextural features (roundness or angularity, percussion, fresh surfaces, dissolution and adhering particles) were sufficient to discriminate sedimentary environments and establish the tsunamigenic sedimentary source in Salgados (Fig. 7) and Lhok Nga (Costa *et al.* 2012*a*) does not necessary mean that these groups can be applied worldwide. In fact, in the Chilean samples, a strong presence of dissolution features probably masked mechanical imprints: thus, it might be advantageous to study a larger number of microtextural features to differentiate source and tsunami samples. The microtextural features have been summarized by Mahaney (2002) and Costa (2012), and it is advised to test a combination of groups of microtextures and select the ones that reveal a stronger discriminating factor for each study area. In theory, as suggested by Costa *et al.* (2012*a*), these associations of microtextures should involve the contrast between mechanical (percussion marks and fresh surfaces) and chemical features (dissolution and adhearing particles), as well as more permanent/long-term characteristics (such as roundness/angularity).

Furthermore, the development and application of digital image analysis to the study of quartz grains microtextures (as demonstrated here in the case of Fire Island) might provide more objective parameters and decrease the observer influence in this analysis. Although we have not yet applied this approach to other studied sites, we think this methodology is certainly a way to advance this technique that can also be quantified and used to discriminate grains from different sedimentary environments.

Moreover, previous studies (e.g. Costa 2012) have suggested that storm deposits share more similarities with beach samples. The results obtained for both Stoneybridge and Fire Island seem to confirm this pattern. For example, if aeolian, dune and beach samples are considered as the potential source material of the Hurricane Sandy deposit, one can observe that samples deposited by this event present microtextural features compatible with the beach samples and discrepancies with the aeolian (reworked overwash) and dune samples – the latter two presenting higher values for dissolution and adhering particles, and lower values for percussion marks and fresh surfaces – suggesting that either the sand in the overwash fans were essentially derived from the beach sediment or that dune material was reworked and strongly reprinted (Fig. 8; Tables 2 & 3). The simpler process is the former, which makes it the more likely to occur in nature. Furthermore, roundness, circularity, shape factor and ratio of minor/major axis did not present major differences between samples (Table 2), which once more suggest these characteristics as long-term processes that required more time to produce noticeable changes in the grains. Rounding could have occurred because of differences in sources or by rapid smoothening or removal of edges and corners induced by effective grain–grain collisions during transport. The latter hypothesis raises questions in terms of the transport duration necessary to transform sand-sized angular quartz grains on more rounded ones. The issue has been debated for larger particles, and there is a wide consensus that high-energy and turbulent transport is able to produce rapid changes of surface features in quartz particles

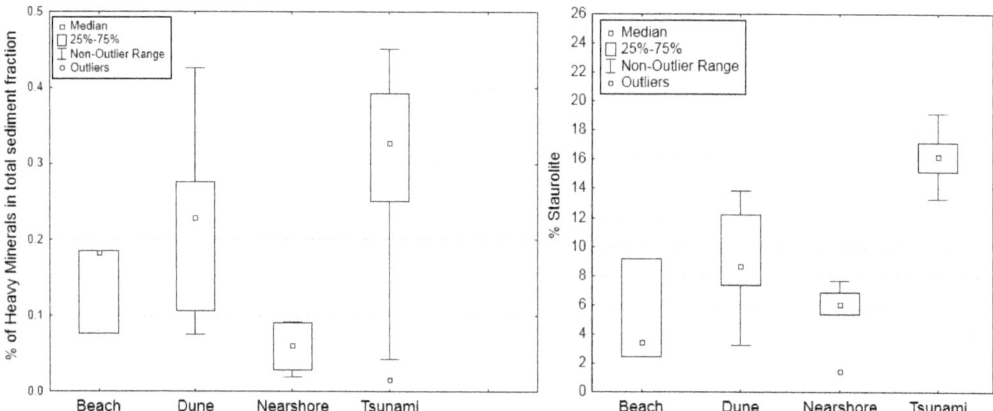

Fig. 7. Percentage of heavy minerals in the total sediment fraction (left image); and the percentage of staurolite (right image) observed in beach, dune, nearshore, tsunami and storm samples retrieved from Salgados.

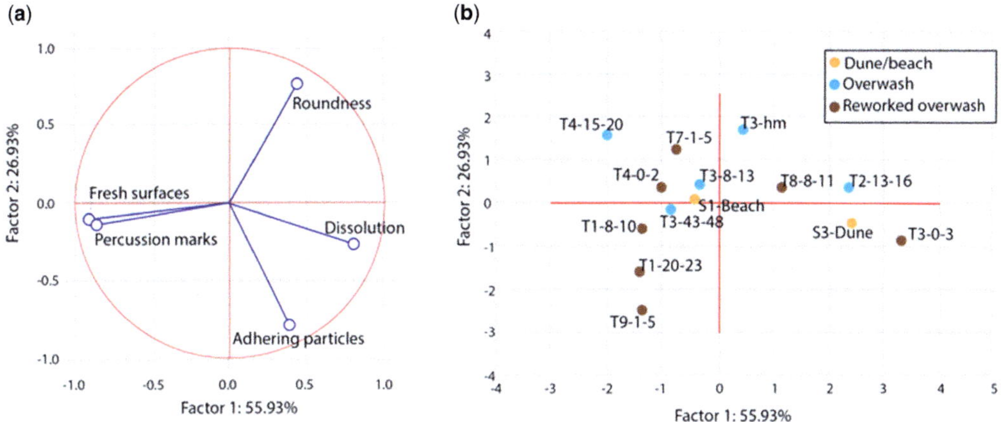

Fig. 8. Principal component analysis of microtextural features of Fire Island. (**a**) Projection of the variables (microtextural features) on the principal components. (**b**) Projection of the samples on the principal components. Blue circles, sediments that were most likely to have been primarily transported during the hurricane (by overwash); brown circles, sediments on the overwash fan that appear to have been reworked by post-Sandy aeolian processes; orange circles, surface samples from the beach and dune (La Selle *et al.* 2016).

(i.e. short-distance and evolution of roundness); however, in shallow-marine environments, the roundness characterizing quartz sand may essentially be acquired by pressure-induced near-bottom vibration induced by travelling waves, rather than impact between particles moving alongshore or cross-shore and entrained, re-suspended and deposited by coastal currents.

In the case of the Fire Island samples, roundness (and related microtextural shape features) seems to be unvarying between different samples (i.e. aeolian, dune, beach and 'storm' deposits), which suggest that the brief duration of the event was not capable of causing significant changes in the shape of the grains.

Microtextural results are further supported by mean grain size, also suggesting that beach sediments are the main source. The beach sample is the coarsest sample and exhibits the highest sorting, while dune and aeolian samples are (very) well sorted and present much finer grain-size curves. In fact, dune and aeolian samples present finer grain-size distribution than several of the hurricane samples, supporting the reasoning that the main source was the beach sediment because it is highly unlikely that finer source material can produce coarser depositional samples, especially if we assume that samples were collected in a region where transport by density dominated (this can also be confirmed by the grain-size horizontal variation of the hurricane sampling profile). Moreover, our reasoning is in agreement with McLaren & Bowles (1985) which concluded that sedimentary deposits may become either: (a) coarser, better sorted and more positively skewed (in high-energy transport) or finer, better sorted (the case presented here if we consider beach as the main sediment source); or (b) more negatively skewed (in a low-energy regime).

This study on sediment samples laid down during a hurricane marine invasion suggests that these samples were very likely to have been sourced in the beach sediment available in the area at the time of the event. Even if geomorphological evidence suggests some dune material was transported inland, the core of the hurricane-driven deposit was very probably mainly, but not exclusively, composed of beach sand.

A contrasting pattern was established for the tsunamigenic samples. For instance, in Salgados, it was clear that the tsunamigenic sediments share fewer similarities with both alluvial and nearshore materials, and more similarities with dune and, to a lesser degree, beach deposits: thus suggesting the latter as the probable sedimentary sources of the tsunamigenic units.

Based in two examples (Fire Island and Salgados), somewhat preliminary, microtextural results, one might speculate about the dominance of different sedimentary source environments for storm and tsunami deposits. However, both the cases of Stoneybridge and Lhok Nga did not reveal conclusive microtextural results. Only further studies will clarify if this discrepancy was caused by the lower number of samples (when compared with Salgados or Fire Island) or if, due to sedimentological constraints, the five microtextural features are insufficient to discriminate the source in Stoneybridge and Lhok Nga, and other microtextural imprints should be sought (similar to what was suggested for the Chilean samples, see Bellanova *et al.* 2015).

The results might imply that storms and tsunami may have different main sediment sources. A possible explanation lies on the different duration of the events and on its erosional effects on dune fields. During storm events, the water rises over hours and the dunes are essentially eroded from the seawards side until they finally collapse, so large portions of dune sediment were washed offshore when overwash and inland deposition finally occurs. By contrast, during a tsunami, the large waves could overtop the dune, eroding the dune from the top, carrying dune sand landwards – recent modelling exercises provide ground to support this reasoning (Costa *et al.* 2015*b*).

Heavy mineral analysis in storm and tsunami deposits

Although sedimentological similarities between the different studied deposits were salient, some differences were noted – all deposits (except Salgados) presented stratification/laminations distinctly separating depositional phases. The tsunamigenic deposit in Salgados was macroscopically massive, which is probably associated with either the depositional specificity associated with turbulent flow or with the alteration of the sedimentary structures. Related with this is the presence of mud-drapes separating different subunits in Lhok Nga and their absence in the deposits of Salgados. The deposition and preservation of mud-drapes on sand is favoured by a large sand-wave asymmetry, a high bottom concentration of suspended mud and large time-velocity asymmetry (Allen 1982) conditions that were probably registered in Lhok Nga but apparently did not prevail in Salgados.

The Indonesian samples presented lower values of heavy mineral content in the total sediment fraction and no clear vertical trend was detectable. This was reflected in the heavy mineral results (Table 1) by the maximum horizontal and vertical variation in the proportion characterizing each heavy mineral species (although without alterations in the mineral spectrum) having been found in Indonesia and attributed to preservation of distinct depositional episodes including run-up and backwash, and selective density-controlled transport. Similar behaviour was observed in the storm samples analysed here (Table 1), thus likely to be associated with the cyclic presence of heavy minerals laminae under- and overlaid by quartz and bioclast-dominated layers.

By contrast, the Portuguese tsunami sediments are more homogenous and, in some cores, display a vertical trend (higher concentration of heavy minerals at the base, see Table 1). These results are in accordance with Bahlburg & Weiss (2007), who detected higher heavy mineral concentrations at the base of individual sand layers inferred to have been laid down by different waves from the same event, and, broadly speaking, also in line with Morton *et al.* (2007), who concluded that storm and tsunami deposits can contain heavy mineral laminae at the base and within the deposit because the heavy mineral assemblage is source dependent.

The formation of these laminae was described by Komar & Wang (1984) on the Oregon coast. The authors describe the development of small black-sand deposits during single storm events and suggest that the concentrating processes takes place within the zone of wave swash on the beach face. Swash processes are ideal for the development of heavy mineral concentrates by the selective winnowing of low-density quartz and feldspar, leaving behind a lag deposit of heavy minerals. The waves and currents in the outer portions of beaches are not as conducive to mineral sorting as in the swash zone. The breaker zone is characterized by high levels of turbulence, which carry sand in suspension, thus having the effect of mixing of the sediments rather than sorting them according to their densities. Particles with low settling rates will tend to drift offshore from the breaker zone, and this may result in some degree of concentration of heavy minerals on the beach as a whole.

In July 2003, at Faro beach (Portugal), Cascalho & Taborda (2006) monitored a beach face depositional imprint (over a 9 h period) and described how the process responsible for the heavy mineral concentration may work according to the following description: first, the erosion phase takes place promoted by the wave swash. Sand grains are winnowed and carried back down the beach face. This phase can be characterized by intense erosion, high offshore transport rate and no substantial grain sorting. Second, after the maximum erosion phase, the profile begins to grow and the swash hydrodynamic conditions promote the generation of the upper beach face deposits enriched in heavy minerals. The grain sorting is intense and very efficient. After this phase, the profile develops into a high-speed growth, and the conditions of sorting between the light and heavy minerals are again weaker. The Stoneybridge storm deposit may have been formed by the processes described in this conceptual model. Indeed, the present-day mechanisms of heavy mineral deposition were observed by the authors in the beach under normal wave and weather conditions, and it was possible to detect that the heavy minerals tend to concentrate at the limit of the swash zone. If we transpose this mechanism to the storm of 2005, the higher heavy mineral sediment concentration inland suggests an analogous process. Another peculiar aspect is the greater presence of garnet (densest of the most common heavy minerals of the assemblage) in Sample 7 (Fig. 2), again this probably mirrors the present-day swash transport processes. Moreover, this might be enhanced by a selective transport of the less dense minerals during the wave retreat favouring the concentration of the denser mineral grains (e.g. garnets).

In Stoneybridge, heavy mineral assemblages did not create clear vertical patterns within the whole deposit. However, in terms of their horizontal distribution, they contributed to the discussion of transport mechanisms responsible, for instance, for the higher concentration of garnets in the most inland sample.

One pattern common to all the studied deposits is that they show contrasts in sediment texture, vertical size grading, thickness of the deposit and sharpness of boundaries, with over- and underlying units grading into overlying sediments. Other common aspects briefly characterized above include an erosional/sharp basal contact, thinning inland, fining upwards and the presence of rip-up clasts. More important, it is clear that both storm and tsunami deposits also present a higher concentration of heavy minerals when compared with the source material. These characteristics reflect the suddenness of a unique high-energy marine inundation responsible for the deposition of the studied units. The tsunami deposits studied revealed overall, an upwards decrease in the percentage of heavy minerals in the total sediment fraction and in the concentration of the densest of the denser minerals (e.g. staurolite for Salgados, and zircon and garnets for Lhok Nga).

Heavy mineral data were useful, indicating likely sediment sources for the tsunami deposits (e.g. dune for Salgados: Fig. 9) or different inundation phases (e.g. shape of zircon in Lhok Nga: Fig. 6). Despite the higher heavy mineral concentrations in storm deposits, the usefulness of heavy mineral assemblages as a discriminating or provenance tool was somewhat reduced.

Final remarks

In this work, we were able to detect that tsunami deposits tend to display a relevant and more extensive spatial variation, reflected by a wider spectrum of grain sizes, microtextures and heavy minerals. Storm deposits show less variability, especially in the vertical profiles, and, as field evidence has shown, their inland extent is less than the tsunamis studied.

Based on microtextural results, it was observed that tsunami and storm grains generally had more percussion marks and fresh surfaces than their potential source material. Based in the limited number of cases presented in this work, the discrimination between the two types of deposits mainly relies on the different primary sources established (dune for tsunami samples and beach for storm samples). This aspect must be considered in future studies in regions where dunes are absent and microtextural differences in the deposits might not be so evident.

Although heavy mineral assemblages are typically site-specific, a concentration of heavy minerals was observed in both types of event deposits. In the tsunami deposit of Salgados, the concentration of heavy minerals decreases upwards, whereas both storm deposits and Lhok Nga show cyclic concentrations of heavy minerals (i.e. laminations associated with different inundation phases and backwash subunit: Costa *et al.* 2015*a*). Overall,

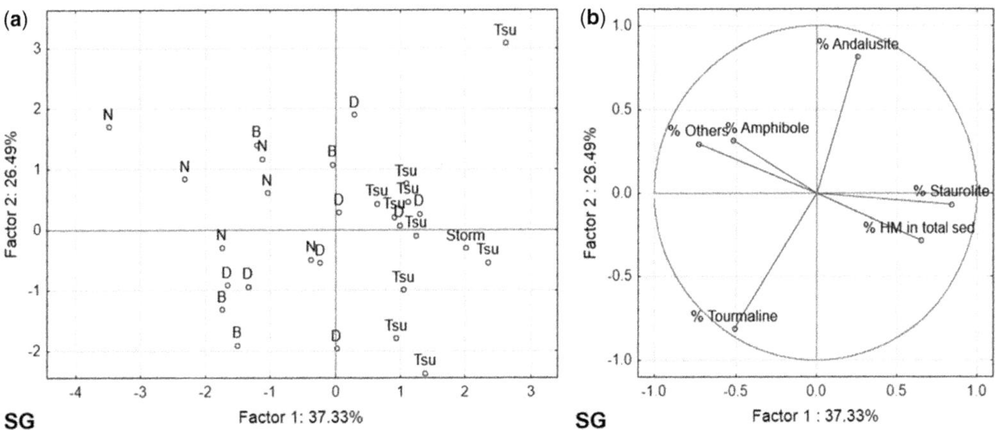

Fig. 9. Principal component analysis on heavy minerals analysed in the Salgados samples. Note in the left-hand image: A, alluvial sample; B, beach sample; D, dune samples; N, nearshore samples; Tsu, tsunami samples.

tsunami deposits presented a more diverse range of minerals and exhibited a higher concentration of the densest minerals of the studied heavy mineral assemblages. Heavy mineral analysis was also a useful provenance tool, especially for tsunami deposits.

Although results discussed here are insufficient for definitive conclusions, it is clear that there is a potential to use microtextural and heavy mineral analysis as useful sedimentological tools (and also complementing existing tools, such as grain size or compositional/morphoscopic analysis) in the establishment of source material, and as complementary tools in the discrimination of storm and tsunami deposits. As a conclusion, we can stress that discrimination at individual sites is possible, but cannot be extended to form the basis for identifying absolute criteria.

The authors wish to thank: Arthur Holmes Travel Grant 2015; NERC project 'Will climate change in the Arctic increase the landslide-tsunami risk to the UK?'; FCT project GETS PTDC/CTEGEX/65948/2006; FCT Post-Doctoral Fellowship SFRH/BPD/84165/2012; and all those who helped during field, laboratory or office work. Funding for GG, SLS and BJ came from the United States Geological Survey, Coastal and Marine Geology Program. The authors wish to express their deep gratitude to Editor David Tappin for his enormous input to improve the initial version of this manuscript.

References

ALLEN, J.R.L. 1982. *Sedimentary Structures: Their Character and Physical Basis*. Elsevier, Oxford.

ANGUS, S. & RENNIE, A.F. 2007. *An Ataireachd Ard: The Great Sea Surge. The Natural Heritage Impact of the Storm of 11 January 2005 in the Uists and Barra, Outer Hebrides*. Scottish Natural Heritage commissioned Report Scottish Natural Heritage, Perth, UK.

BABU, N., SURESH BABU, D.S. & MOHAN DAS, P.N. 2007. Impact of tsunami on texture and mineralogy of a major placer deposit in southwest coast of India. *Environmental Geology*, **52**, 71.

BAHLBURG, H. & WEISS, R. 2007. Sedimentology of the December 26, 2004, Sumatra Tsunami deposits in eastern India (Tamil Nadu) and Kenya. *International Journal of Earth Sciences*, **96**, 1195–1209.

BELLANOVA, P., BAHLBURG, H. & NENTWIG, V. 2015. Test of the microtextural analysis of quartz grains of tsunami and non-tsunami deposits in Tirúa (Chile) – an unsuitable method for a valid tsunami identification. Abstract presented at the *AGU Fall Meeting*, 14–18 December 2015, San Francisco, CA, USA.

BLOTT, S.J. & PYE, K. 2001. GRADISTAT: a grain size distribution and statistics package for the analysis of unconsolidated sediments. *Earth Surfaces Processes and Landforms*, **26**, 1237–1248.

BRUZZI, C. & PRONE, A. 2000. A method of sedimentological identification of storm and tsunami deposits: exoscopic analysis, preliminary results. *Quaternaire*, **11**, 167–177.

CASCALHO, J. & TABORDA, R. 2006. Heavy mineral placer formation – An example from Algarve, Portugal. *Journal of Coastal Research*, **SI39**, 246–249.

CASCALHO, J., COSTA, P.J.M., DAWSON, S., MILNE, F. & ROCHA, A. 2016. Heavy mineral assemblages of the Storegga tsunami deposit. *Sedimentary Geology*, **334**, 21–33.

COSTA, P.J.M. 2012. *Sedimentological signatures of extreme marine inundations*. PhD thesis, University of Lisbon, Lisbon, Portugal.

COSTA, P.J.M., ANDRADE, C., FREITAS, M.C., OLIVEIRA, M.A. & JOUANNEAU, J.M. 2009. Preliminary results of exoscopic analysis of quartz grains deposited by a Palaeotsunami in Salgados Lowland (Algarve, Portugal). *Journal of Coastal Research*, **56**, 39–43.

COSTA, P.J.M., ANDRADE, C., DAWSON, A.G., MAHANEY, W.C., PARIS, R., FREITAS, M.C. & TABORDA, R. 2012*a*. Microtextural characteristics of quartz grains transported and deposited by tsunamis and storms. *Sedimentary Geology*, **275–276**, 55–69.

COSTA, P.J.M., ANDRADE, C. ET AL. 2012*b*. A tsunami record in the sedimentary archive of the central Algarve coast, Portugal: characterizing sediment, reconstructing sources and inundation paths. *The Holocene*, **22**, 899–914.

COSTA, P.J.M., ANDRADE, C., CASCALHO, J., DAWSON, A.G., FREITAS, M.C., PARIS, R. & DAWSON, S. 2015*a*. Onshore tsunami sediment transport mechanisms inferred from heavy mineral assemblages. *The Holocene*, **25**, 795–809, https://doi.org/10.1177/0959683615569322

COSTA, P.J.M., GELFENBAUM, G., LA SELLE, S., COSTAS, S., ANDRADE, C., CASCALHO, J. & FREITAS, M.C. 2015*b*. Modelling tsunami sedimentation associated with the AD 1755 event in Algarve (Portugal). Abstract presented at the *AGU Fall Meeting*, 14–18 December 2015, San Francisco, CA, USA.

DAHANAYAKE, K. & KULASENA, N. 2008. Recognition of diagnostic criteria for recent- and paleo-tsunami sediments from Sri Lanka. *Marine Geology*, **254**, 180–186.

DAWSON, A.G., DAWSON, S. & RITCHIE, W. 2007. Historical climatology, coastal change associated with the 'Great Storm' tof January 2005, South Uist, Benbecula, Scottish Outer Hebrides. *Scottish Geographical Journal*, **123**, 135–149.

DONNELLY, J.P., ROLL, S., WENGREN, M., BUTLER, J., LEDERER, R. & WEBB, T., III. 2001. Sedimentary evidence of intense hurricane strikes from New Jersey. *Geology*, **29**, 615–618.

FOLK, R.L. & WARD, W.C. 1957. Brazos river bar: a study of significant of grain size parameters. *Journal of Sedimentary Petrology*, **27**, 3–26.

GOFF, J., MCFADGEN, B.G. & CHAGUÉ-GOFF, C. 2004. Sedimentary differences between the 2002 Easter storm and the 15th-century Okoropunga tsunami, southeastern North Island, New Zealand. *Marine Geology*, **204**, 235–250.

HAPKE, C.J., BRENNER, O., HEHRE, R. & REYNOLDS, B.J. 2013. *Coastal Change from Hurricane Sandy and the 2012–13 Winter Storm Season – Fire Island,*

New York. United States Geological Survey, Open-File Report, **2013-1231**.

Higman, B. & Bourgeois, J. 2008. Deposits of the 1992 Nicaragua tsunami. *In*: Shiki, T., Tsuji, Y., Yamazaki, T. & Minoura, K. (eds) *Tsunamiites – Features and Implications*. Elsevier, Amsterdam, 81–103.

Jaffe, B.E., Goto, K., Sugawara, D., Richmond, B.M., Fujino, S. & Nishimura, Y. 2012. Flow speed estimated by inverse modeling of sandy tsunami deposits: results from the 11 March 2011 tsunami on the coastal plain near the Sendai Airport, Honshu, Japan. *Sedimentary Geology*, **282**, 90–109, https://doi.org/10.1016/j.sedgeo.2012.09.002

Jagodziński, R., Sternal, B., Szczuciński, W. & Lorenc, S. 2009. Heavy minerals in 2004 tsunami deposits on Kho Khao Island, Thailand. *Polish Journal of Environmental Studies*, **18**, 103–110.

Jagodziński, R., Sternal, B., Szczuciński, W., Chagué-Goff, C. & Sugawara, D. 2012. Heavy minerals in the 2011 Tohoku-oki tsunami deposits – insights into sediment sources and hydrodynamics. *Sedimentary Geology*, **282**, 57–64, https://doi.org/10.1016/j.sedgeo.2012.07.015

Komar, P.D. & Wang, C. 1984. Processes of selective grain transport and the formation of placers on beaches. *Journal of Geology*, **92**, 637–655.

Kortekaas, S. & Dawson, A.G. 2007. Distinguishing tsunami and storm deposits: an example from Martinhal, SW Portugal. *Sedimentary Geology*, **200**, 208–221.

Krinsley, D.H. & Donahue, J. 1968. Environmental interpretation of sand grain surface textures by electron microscopy. *Geological Society of America Bulletin*, **79**, 743–748, https://doi.org/10.1130/0016-7606(1968)79[743:eiosgs]2.0.co;2

La Selle, S., Gelfenbaum, G., Costa, P., Jaffe, B. & Lunghino, B. 2015. Hurricane Sandy deposits on Fire Island, NY: Using washover deposit stratigraphy to understand sediment transport during large storms. Abstract presented at the *AGU Fall Meeting*, 14–18 December 2015, San Francisco, CA, USA.

La Selle, S.M., Lunghino, B.D., Jaffe, B.E., Gelfenbaum, G. & Costa, P.J.M. 2016. *Hurricane Sandy Washover Deposits on Fire Island, NY*. United States Geological Survey, Open-File Report, **2016-XXXX**, In print.

Lakshmi, C.S.V., Srinivasan, P., Murthy, S.G.N., Trivedi, D. & Nair, R.R. 2010. Granularity and textural analysis as a proxy for extreme wave events in southeast coast of India. *Journal of Earth System Science*, **119**, 297–305.

Leatherman, S.P. & Williams, A.T. 1983. Vertical sedimentation units in a barrier island washover fan. *Earth Surface Processes and Landforms*, **8**, 141–150.

Lira, C. 2011. *Development of new image analysis methodologies for morphometric characterization of sediments*. PhD thesis, IST, Universidade Técnica de Lisboa, https://doi.org/10.13140/RG.2.1.4685.9681 [in Portuguese].

Lira, C. & Pina, P. 2007. Sedimentological analysis of sands. *In*: Martí, J., Benedí, J.M., Mendonça, A.M. & Serrat, J. (eds) *Pattern Recognition and Image Analysis. Third Iberian Conference, IbPRIA 2007, Girona, Spain, June 6–8, 2007, Proceedings, Part II*. Lecture Notes in Computer Science, **4478**. Springer, Berlin, 388–396.

Lira, C. & Pina, P. 2009. Automated grain shape measurements applied to beach sands. *Journal of Coastal Research*, **SI 56**, 1527–1531.

Mahaney, W.C. 2002. *Atlas of Sand Grain Surface Textures and Applications*. Oxford University Press, Oxford.

Mahaney, W.C. & Dohm, J.M. 2011. The 2011 Japanese 9.0 magnitude earthquake: test of a kinetic energy wave model using coastal configuration and offshore gradient of Earth and beyond. *Sedimentary Geology*, **239**, 80–86.

Mahaney, W.C., Dohm, J.M., Costa, P. & Krinsley, D.H. 2010. Tsunamis on Mars: earth analogues of projected Martian sediment. *Planetary and Space Science*, **58**, 1823–1831.

Mange, M.A. & Maurer, H.F.W. 1992. *Heavy Minerals in Colour*, 1st edn. Chapman & Hall, London.

McLaren, P. & Bowles, D. 1985. The effects of sediment transport on grain-size distribution. *Journal Sedimentary Petrology*, **55**, 457–470.

Morton, R.A., Gelfenbaum, G. & Jaffe, B.E. 2007. Physical criteria for distinguishing sandy tsunami and storm deposits using modern examples. *Sedimentary Geology*, **200**, 184–207.

Nakamura, Y., Nishimura, Y. & Putra, P.S. 2012. Local variation of inundation, sedimentary characteristics, and mineral assemblages of the 2011 Tohoku-oki tsunami on the Misawa coast, Aomori, Japan. *Sedimentary Geology*, **282**, 216–227.

Nanayama, F., Shigeno, K., Satake, K., Shimokawa, K., Koitabashi, S., Miyasaka, S. & Ishii, M. 2000. Sedimentary differences between the 1993 Hokkaido-nansei-oki tsunami and the 1959 Miyakojima typhoon at Taisei, southwestern Hokkaido, northern Japan. *Sedimentary Geology*, **135**, 255–264.

Narayana, A.C., Tatavarti, R., Shinu, N. & Subeer, A. 2007. Tsunami of December 26, 2004 on the southwest coast of India: post-tsunami geomorphic and sediment characteristics. *Marine Geology*, **242**, 155–168.

Otvos, E.G. & Carter, G.A. 2008. Hurricane degradation–barrier development cycles, northeastern Gulf of Mexico. Landform evolution and island chain history. *Journal of Coastal Research*, **24**, 463–478.

Paris, R., Lavigne, F., Wassmer, P. & Sartohadi, J. 2007. Coastal sedimentation associated with the December 26, 2004 tsunami in Lhok Nga, west Banda Aceh (Sumatra, Indonesia). *Marine Geology*, **238**, 93–106.

Paris, R., Wassmer, P. *et al.* 2009. Tsunamis as geomorphic crisis: lessons from the December 26, 2004 tsunami in Lhok Nga, west Banda Aceh (Sumatra, Indonesia). *Geomorphology*, **104**, 59–72.

Powers, M.C. 1953. A new roundness scale for sedimentary particles. *Journal of Sedimentary Research*, **23**, 117–119.

Schwartz, R.K. 1975. *Nature and Genesis of Some Storm Washover Deposits*. United States Army Corps of Engineers, Coastal Engineering Research Center, Technical Memorandum, 61.

Shanmugam, G. 2012. Process-sedimentological challenges in distinguishing paleo-tsunami deposits. *Natural Hazards*, **63**, 5–30.

Shepard, F.P. 1973. *Submarine Geology*. Harper and Row, New York.

Sneed, E.D. & Folk, R.L. 1958. Pebbles in the lower Colorado River, Texas, a study in particle morphogenesis. *Journal of Geology*, **66**, 114–150.

Switzer, A.D. & Jones, B.G. 2008. Large-scale washover sedimentation in a freshwater lagoon from the southeast Australian coast: tsunami or exceptionally large storm? *The Holocene*, **18**, 787–803.

Switzer, A.D., Pucillo, K., Haredy, R.A., Jones, B.G. & Bryant, E.A. 2005. Sea-level, storms or tsunami; enigmatic sand sheet deposits in sheltered coastal embayment from southeastern New South Wales Australia. *Journal of Coastal Research*, **21**, 655–663.

Szczuciński, W., Kokociński, M., Rzeszewski, M., Chagué-Goff, C., Cachão, M., Goto, K. & Sugawara, D. 2012. Sediment sources and sedimentation processes of 2011 Tohoku-oki tsunami deposits on the Sendai Plain, Japan – Insights from diatoms, nannoliths and grain size distribution. *Sedimentary Geology*, **282**, 40–56.

Tuttle, M.P., Ruffman, A., Anderson, T. & Hewitt, J. 2004. Distinguishing tsunami from storm deposits in eastern North America. The 1929 Grand Banks tsunami v. the 1991 Halloween storm. *Seismological Research Letters*, **75**, 117–131.

Wadell, H.A. 1932. Volume, shape and roundness of rock particles. *Journal of Geology*, **40**, 443–451.

Wadell, H.A. 1934. Some new sedimentation formulas. *Physics*, **5**, 281–291.

Williams, H.F.L. & Hutchinson, I. 2000. Stratigraphic and microfossil evidence for late Holocene tsunamis at Swantown Marsh, Whidbey Island, Washington. *Quaternary Research*, **54**, 218–227.

Zunic, J. & Hirota, K. 2008. Measuring shape circularity. *In*: Ruiz-Shulcloper, J. & Kroopstasch, W. (eds) *Progress in Pattern Recognition, Image Analysis and Applications*. Springer, Berlin, 94–101.

Risk-informed tsunami warnings

GORDON WOO

RMS, 30 Monument Street, London EC3R 8NB, UK
Gordon.Woo@rms.com

Abstract: Tsunami warning traditionally has been perceived within the scientific community as an essentially deterministic process. This may be satisfactory in regions where tsunami monitoring is very extensive and the uncertainty in tsunami forecasting is low. However, where there is significant uncertainty, the worthy aspiration to achieve a zero-tolerance policy on tsunami fatalities is set against the reduction in public alert compliance associated with previous false alerts. A risk-informed approach is presented, which outlines a decision process that allows for a decline in alert compliance due to false alerts. This risk-informed approach assesses probabilistically the tsunamigenicity of fault rupture, using data not just from historical catalogues but also the geological record. Prior probabilities could be automatically updated in real time by a Bayesian Belief Network from observations obtained in real time: earthquake depth information; pressure sensor recordings, tide gauge measurements; or visual reports of coastal inundation. A decision to issue a general evacuation warning in a specific region might be made provided the updated posterior probability reached a critical threshold.

The challenge of false alerts

Traditional tsunami warning for coastal populations has relied upon the perception of earthquake ground shaking itself. Over centuries of experience, native populations exposed to tsunami hazard have learned to move quickly to higher ground on sensing ground motion. This type of hazard warning is optimal in respect of psychological response: human beings respond better when they can sense the danger for themselves. However, as a warning system, the perception of seismic ground motion is of limited effectiveness because the epicentre may either be too close to take evasive action or too distant for the earthquake to be much felt.

Primitive though this warning system is, false alerts are comparatively rare because felt earthquakes are uncommon, and the disruptive cost of moving to higher ground is typically quite modest for rural populations. For any modern scientific warning system, the process of warning (e.g. a siren) is remote from the hazard itself, and is prone to technical malfunction and forecasting error, which can induce false alerts due to the overestimation of the tsunami threat. The lack of individual sensory perception of the threat, such as precursory ground shaking, can lead to scepticism about the importance of a specific alert, and the value of tsunami alerts in general. This may erode public trust and diminish alert compliance.

One of the common shortcuts in human response to a piece of information is to adopt the principle of social proof: decide to do what other people like us are doing (Cialdini 2007). This often makes good sense, since an action that is popular in a particular situation may well also be functional and appropriate. People view a behaviour as more correct in a given situation to the extent that we see others performing it. In November 2006, tsunami warnings were shown repeatedly on television after an M8.1 earthquake occurred in the Russian-held islands north of Hokkaido. The tsunamis ended up being only about 40 or 50 cm high at their highest and few people, even those living in the area deemed most dangerous, heeded the warnings and headed for higher ground or shelters.

In January 2007, an M8.2 earthquake occurred off the coast of Hokkaido. Tens of thousands of residents were ordered to seek higher ground in the event of a tsunami. Tsunami warnings were issued on television and radio but were widely ignored. The tsunami that hit Hokkaido was only around 20 cm at its highest; it had been predicted to be 1 m high.

Following a tsunami alert in Whakatane, Bay of Plenty, New Zealand, in December 2013, only a few families evacuated up the hill. Social proof induces a highly non-linear response outcome: the fewer who take notice of an alert, the less likely others will follow. The concern of civic officials in Whakatane is that a genuine alert will not be taken seriously in future, the sound of a tsunami siren being linked with another computer failure. The problem of computer false alerts is only exacerbated by increasing the number of sirens along the eastern Bay of Plenty coastline to increase the audibility range of a tsunami alert.

In his book on the 'cry wolf' syndrome, Breznitz (1984) has urged that warning systems must protect their decision-makers from being personally accountable for false alerts. The pressure on the shoulders of individual decision-makers is illustrated by

From: Scourse, E. M., Chapman, N. A., Tappin, D. R. & Wallis, S. R. (eds) 2018. *Tsunamis: Geology, Hazards and Risks*. Geological Society, London, Special Publications, **456**, 191–197.
First published online January 23, 2017, https://doi.org/10.1144/SP456.3

repeated false tsunami alert evacuations by one Pacific shore community; it was made clear to the community leader that he had better take care not to waste their time and resources again. To cope better with the 'cry wolf' syndrome, decision-makers need the assistance of decision support systems, such as the risk-informed approach presented here.

In Japan, JMA (Japan Meteorological Agency) seismologists felt obliged to apologize to 1.5 million coastal residents for overestimating tsunami heights associated with the M8.8 Maule, Chile, earthquake of 27 February 2010. Experts defended the agency's decision to warn that waves of 1 m or more might strike Japan's Pacific coast after the Maule earthquake. They acknowledged the risk of making residents blasé about the danger next time. The key point made at the Pacific Tsunami Warning Center (PTWC) in Hawaii was that failure to warn of a devastating tsunami was not an option. Recognizing they overstated the threat from the tsunami triggered by the Maule earthquake, the PTWC defended their actions, saying they took the proper steps and learned the lessons of the 2004 Indian Ocean tsunami.

Behavioural response to hazard warnings

The issuance of warnings for tsunamis and other natural hazards is an interdisciplinary task involving not just seismologists, geologists and other Earth scientists, but also social scientists. Accordingly, Earth scientists should have some broad familiarity with studies of the behavioural response of populations to hazard warnings.

The wide-ranging studies of human judgements undertaken by the sociologist Paul Slovic (2010) have contributed much to elucidate the attitudes of ordinary individuals to risk. Public attitudes towards untimely death are quite finely structured; death is not just a simple statistic. Deaths are differentiated according to whether they are voluntary (as in a skiing or skydiving accident), caused by an industrial accident, by an act of terrorism or the consequence of a natural disaster. Mindful of the intricacies and anomalies of the public perception of major risks, Slovic has long been an ardent critic of policy which neglects human nature, the choices people make and their societal consequences.

The human brain has two different processing systems: the experiential processing system, which controls survival behaviour and is the source of emotions and instincts (e.g. feeding, fighting, fleeing); and the analytical processing system, which controls analysis of scientific information. To the extent that emotion is a key driver of attitudes towards risk, studies of emotion are crucial for understanding human behaviour to hazard warnings. Emotion is a universal human behavioural trait that is exercised in the decision-making process. In psychology, 'affect' is defined as feelings or emotions experienced about a stimulus, and is a regular aspect of brain functioning. Slovic *et al.* (2002) developed the affect heuristic concept. People consult their emotions in making decisions; rather than think about a question, they check their feelings about it. The affect heuristic is a form of mental substitution: how one feels about something substitutes for thinking about it.

In emergency crisis situations, such as may arise with the sudden onset of natural hazard events like a tsunami, reflex reaction may have to replace thinking. An affect system for reacting to stimuli is of obvious evolutionary significance. The affect system simplifies things by directing attention to one stimulus at a time. This is especially important if that thing, be it a wild animal or a force of nature, poses an immediate danger to life and limb.

The neuroscientist Antonio Damasio (1994) proposed that the emotional evaluation of outcomes, and the bodily states and the approach and avoidance tendencies associated with them, all play a central role in guiding decision-making. Damasio and co-workers observed that if people fail to display the appropriate emotions before making a decision, the ability to make good decisions may be impaired. An inability to experience a sensible degree of fear is a notable emotional drawback. In the context of hazard warnings, incapability of developing a healthy fear of physical harm may delude people into ignoring them. Public education has to form part of any tsunami alert system.

In reacting to a warning, an individual employs some form of mental model to perform his or her own cost–benefit analysis (Green 2009). The three main components are: the perceived cost of compliance; perception of the danger level; and perceived personal, social and cultural factors. Since all of these are personal and subjective, general sociological studies are required to understand public attitudes to warnings.

Dennis Mileti, the foremost pioneering social scientist of natural disasters, has pointed out that public reaction to warnings of impending disaster is not well characterized by a simple stimulus–response model. If this were the case, a warning would need only to be heard to be followed by protective action. Instead, people who receive warnings first typically go through a social psychological process of evaluating the risk, and then formulate ideas about what to do, before they take a protective action. Mileti (1995) has summarized the five principal factors underlying behavioural response. For each of these five factors, negative aggravating experience of past false tsunami alerts could de-sensitize and mentally block an individual from making the required response to a genuine alert.

Hearing a warning

The first stage in the social psychological process of public response to hazard warnings is hearing that there is an emergency, typically through a public warning like a siren or a worded warning message (Mileti & Sorensen 1987). Even when it is physically possible for people to hear a warning, various factors may inhibit a message from actually being heard. People may fail to listen because of habituation: for example, they never really listen attentively to radio or television, or rarely check their phone, e-mail or social media messages.

Understanding what was really meant by the warning

Once it is heard, the information in a warning must be understood. This means more than just understanding what is heard, personal meaning must then be attached to the message. Risk communication is a two-way interactive process. Meaning or understanding varies from person to person, and these varied understandings may or may not conform to the meaning intended by those who issued the warning. Just as messages may become strangely garbled, so also can understanding.

Developing belief in the warning

Protective public action is also encouraged and improved if people develop a belief that the warning is real, and that the contents of the message are accurate. The empirical literature indicates that the greater the perceived threat, the greater the probability of evacuation (Perry 1982). Threat perception is especially dependent on these three factors: warning content; prior education or training; and the warning source. A warning that contains detailed information about the hazard and its consequences is more likely to engender the sense of fear that enhances belief that the warning should be taken seriously.

Personalizing the risk

People also consider the implications of warnings for themselves and their groups (e.g. their family). The literature indicates a direct positive relationship between an individual's perceived level of personal risk and the probability of evacuation. If people mistakenly do not imagine that emergency information was actually intended for them, but rather meant for others, they are likely to ignore it. This might be a misunderstanding for inland residents, who might imagine they are not exposed to tsunami hazard. However, if people think they are, indeed, the intended targets of emergency information, they may act accordingly and not ignore the information. Personalization can lead to both under-response and over-response in emergencies.

Deciding what, if anything, to do

Finally, when a person has heard the emergency information; then formed an understanding of what is being said; then defined a level of belief in what is being said; and determined a level of risk personalization, then his or her behaviour would follow based on the personal perceptions formed (Quarantelli 1980). A person typically goes through these distinct phases each time that new warning information is received. Public warning systems that recognize these phases can be very effective in helping those at risk find safety before disasters strike. By contrast, warning systems that are not designed to take the social psychology of public warning response into account are much less likely to foster pre-disaster public protective actions.

Decision-making under uncertainty

Given the pressure on tsunami decision-makers, the methodology for raising alerts needs to have these three attributes: it should be rational, equitable and defensible (Woo 2016). The methodology has to be rational in the sense that there is a reasoned and considered balance between the economic costs of a false alert and the safety benefits of a timely alert and evacuation. It has to be equitable to all stakeholders, without discrimination against any group at risk. Furthermore, it has to be defensible both in a court of law and in the court of public opinion. This requirement has become obligatory since the L'Aquila earthquake trial in Italy, where leading Italian scientists were indicted for manslaughter, over a matter of public risk communication.

The task of estimating run-up heights along a coast is fraught with a diverse range of uncertainties: for example, earthquake source parameters, Earth model and bathymetry. There are various alternative ways by which decision-makers can deal with such uncertainties, which may be considerable. The simplest approach is to make a conservative estimate of the tsunami run-up heights. Where this exceeds a minimum threshold, an alert would be issued as a practical transparent expression of a strict policy of zero-tolerance of casualties. Such a policy would be intrinsically more tenable for tsunamis, which have an inherently short hazard duration, than for other geological hazards such as earthquakes or volcanic eruptions, for which the period of heightened threat might last for weeks or even months.

According to the European Commission, the precautionary principle may be invoked when a phenomenon, product or process may have a dangerous effect, identified by a scientific and objective evaluation, if this evaluation does not allow the risk to be determined with sufficient certainty. Tsunami warning systems are not widely deployed in Europe. Even where tsunami detection systems are deployed, the uncertainty in risk estimation may still be substantial. Dangerous tsunamis are rare in Europe, although they are liable to affect southern Europe in particular. Given the rarity of potential alerts, a precautionary approach to tsunami warning might be appropriate and justified in some areas of southern Europe.

Risk-informed warnings

A more sophisticated approach would be to estimate the likelihood of dangerous run-up heights, and base a decision on alert issuance on the balance between the expected reduction in casualties and the expected alert disruption costs. This approach is elaborated here. The cost of an alert involves three main factors: (a) expenses and hardship associated with evacuation; (b) disruption of business and societal dysfunction; and (c) loss of public trust in the warning system, leading to future warnings being ignored by some proportion of the population at risk.

Suppose that an alert might save lives at risk along a threatened stretch of coastline, if a dangerous tsunami occurs with run-up height above some given threshold. Denote by P, the probability that the tsunami run-up height along this stretch of coastline exceeds this danger threshold. Then an optimal cost–benefit decision is to sound an alert if this probability exceeds the ratio of the expected alert cost to the loss if no alert is issued. The loss, L, here is measured simply in terms of the number of fatalities resulting from failure to evacuate.

The cost of an unnecessary evacuation arises predominantly from (c). If there is no significant tsunami, the evacuation time and the period of business disruption is measured in minutes or hours, but is a tiny fraction of human lifespan. There are some professionals, such as surgeons, for whom a tsunami alert might be very inconvenient and especially disruptive. However, hospitals should have in place contingency plans, so that the cost of disruption would still be far less than that of the loss of life. As with fire and earthquake drills, there may be some educational and training advantage in having a brief tsunami alert, even if no tsunami materializes.

Suppose that a proportion, λ, of the population at risk are liable to ignore any future real alerts if they themselves are aggrieved by being subject to a false alert. This proportion needs to be assessed by sociologists from past false alert experience combined with attitude surveys of the population at tsunami risk. This proportion would be expected to vary according to whether the tsunami threat is near-field or far-field. Even though there may be hours of warning, far-field tsunami warnings are still prone to substantial uncertainty. Hawaii coastal residents have been resentful at false alerts in the past. A risk-informed approach to tsunami warnings provides the scientific rationale for estimating the proportion λ, which otherwise would be an exercise in social studies without clear practical purpose.

Assume pessimistically that those made averse to responding to future alerts may eventually succumb to tsunami hazard, and that economic discounting of such future loss can be ignored. Then a conservative upper estimate of the expected cost of a false alert may be expressed succinctly as: $C = \lambda L(1 - P)$. The conservative nature of this estimate implicitly absorbs the other evacuation economic costs (a) and (b).

The expected loss PL from not sounding an alert exceeds the false alert cost C if $PL > \lambda L(1 - P)$ or $P > \lambda/(1 + \lambda)$. The threshold values of P are tabulated in Table 1 for different values of proportion λ of the population at risk, who are liable not to comply with any future real alerts if they suffer a false alert.

If false alerts had very little impact on future compliance with tsunami alerts (i.e. if λ is very small), then the threshold probability for issuing an alert might well be taken to be very low. This corresponds to the situation where there is little cost associated with a false alert. A precautionary approach might then be adopted, whereby whenever there is some perceived prospect of a dangerous tsunami, an alert is issued. This is similar to the common situation with office fire alarms, where there is quite a high ratio of false alarms to actual fires.

Table 1. *Dependence of threshold alert probabilities* (P) *on the population proportion* (λ) *who would be inclined not to comply with future tsunami warnings*

λ	0.1	0.2	0.3	0.4	0.5	0.6	0.7	0.8	0.9	1.0
P	0.09	0.17	0.23	0.29	0.33	0.38	0.41	0.44	0.47	0.50

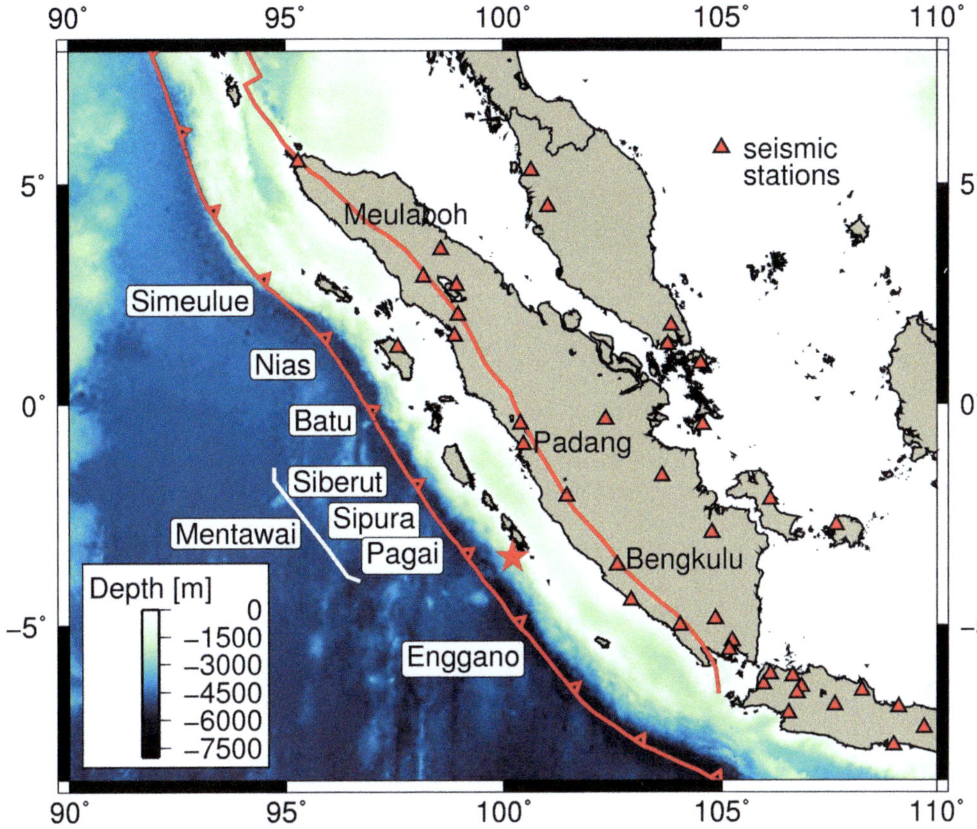

Fig. 1. Map of Sumatra showing the Sunda subduction trench. Seismic stations are indicated by red triangles. (From Blaser *et al.* 2011.)

However, if false alerts are known to reduce future alert compliance, or if strong public objections have been raised over past experience of false alerts, a more circumspect considered approach to alert issuance is required: lives are effectively put at risk by false alerts. For example, suppose there were a 20% reduction in compliance, then the probability threshold for alert issuance would need to be as high as 0.17. If there were 30% reduction in compliance, the probability threshold for

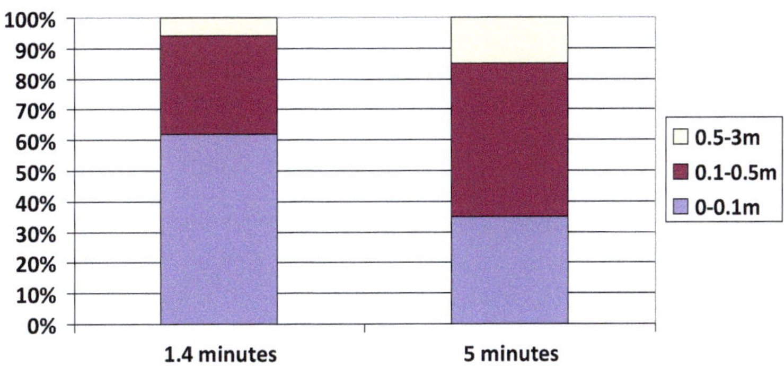

Fig. 2. Cumulative probability distribution for tsunami heights in metres at Meulaboh. Two elapse time periods are indicated: 1.4 and 5 min. (Data from Blaser *et al.* 2011.)

alert issuance would need to be increased to 0.23. If there were 40% reduction in compliance, the probability threshold for alert issuance would need then to be 0.29.

If as much as half of the population at risk might not respond subsequently to a tsunami alert, then the threshold probability for issuing an alert would be raised to one-third. In the limit that the entire population becomes sceptical of all future tsunami alerts, then the threshold probability had better be evens or more. If this probability were less than evens, the subsequent casualties from future tsunamis would be too great a loss potential for it to be worthwhile to issue a tsunami alert in the current circumstances. In principle, the degree of alert non-compliance might be elicited through a public opinion survey. However, it is well known to social scientists that what people say they would do under hypothetical hazard circumstances may be quite different from what they actually do. Bearing in mind the principle of social proof (i.e. that alert non-compliance might be contagious), it would be prudent to assume that λ might be at least 0.1. In this case, careful thought should be given to issuing an alert if the likelihood of a tsunami was below 0.09. If a tsunami seems very unlikely, it would be better not to 'cry wolf'.

Case study: Sumatra earthquake of 6 April 2010

Because the predominant approach to tsunami forecasting remains deterministic, there have been few attempts made to quantify the many sources of uncertainty by calculating the probability distribution for tsunami wave heights at coastal locations. A comprehensive probabilistic approach for aggregating the various sources of uncertainty involves the construction of Bayesian Belief Networks (BBNs) (Woo 2011). A BBN is a graphical representation of causal relationships in a domain, and provides a very efficient language for building models of domains with inherent uncertainty (Jensen 1996). Other less formal methods for estimating the probability distribution for tsunami heights might also be used.

BBN applications to earthquakes and volcanic eruptions have been developed, and Blaser *et al.* (2011) have expanded the domain of natural hazard application to tsunami forecasting. For a case study, they have analysed the M7.8 Sumatra earthquake of 6 April 2010. The regional tectonic setting is shown in Figure 1. In the north of Sumatra is the city of Meulaboh, in Aceh province, which was among the hardest hit areas by the 2004 Indian Ocean earthquake and tsunami, being the closest town, just 150 km from the earthquake epicentre. About one-third of the population of 120 000 perished.

The first real-time parameter estimates for hypocentre location and magnitude arrived 1.4 min after the earthquake of 6 April 2010 occurred. Owing to the complex bathymetry, a site-specific tsunami threat analysis is needed. At Meulaboh, the BBN analysis yields a 62% chance of a negligible tsunami height of less than 0.1 m, and a 32% chance of a minor tsunami height of between 0.1 and 0.5 m. More significantly, the BBN analysis yields a 6% chance of a tsunami of between 0.5 and 3 m.

After 5 minutes, more seismological information was available to refine the estimate of hypocentre and magnitude, and update the BBN. At Meulaboh, there was then a 35% chance of a negligible tsunami height of less than 0.1 m, and a 50% chance of a minor tsunami height of between 0.1 and 0.5 m. More significantly, the BBN analysis yields a 15% chance of a tsunami height of between 0.5 and 3 m. The cumulative probability distributions at Meulaboh after 1.4 and 5 min are charted in Figure 2.

At Meulaboh, the population at risk liable to ignore real alerts, λ, was likely to be low in 2010, given the catastrophic tragic experience of December 2004. Taking 9% as the probability threshold for sounding a tsunami alert of a minor tsunami of height above 0.1 m, then an alert would have been sounded early after 1.4 min. If the threshold for a major tsunami of 0.5 m or more was considered more relevant, the BBN results would suggest that the alert be delayed beyond 1.4 min, at which time the probability of attaining this threshold was just 6%. Being about 230 km towards the north of the epicentre, timely alerts for Meulaboh could have been issued using a risk-informed approach.

As it turned out, the local authorities released a tsunami warning that was withdrawn 1 h later. The actual wave height recorded at the Meulaboh tide gauge was 0.44 m, which was within the tsunami height range of 0.1–0.5 m. As always with any hazard decision support tool, the output from the tsunami BBN does not itself automatically trigger an alert. Rather, it may be used to help the decision-maker arrive at a good decision on issuing an alert that is rational, equitable and defensible.

This work was partly supported by the REAKT project of the EC's Seventh Framework Programme (FP/2007–13) under grant agreement No. 282862.

References

BLASER, L., OHRNBERGER, M., RIGGELSEN, C., BABEYKO, A. & SCHERBAUM, F. 2011. Bayesian networks for tsunami early warning. *Geophysical Journal International*, **185**, 1431–1443.

BREZNITZ, S. 1984. *Cry Wolf: The Psychology of False Alarms*. Lawrence Erlbaum, Hillsdale, NJ.

CIALDINI, R.B. 2007. *The Psychology of Persuasion*. Collins Business, New York.

DAMASIO, A.R. 1994. *Descartes' Error: Emotion, Reason and the Human Brain*. Putnam, New York.

GREEN, M. 2009. *The Psychology of Warnings*, http://www.visualexpert.com/Resources/psychwarnings.html

JENSEN, F. 1996. *An Introduction to Bayesian Networks*. Taylor & Francis, London.

MILETI, D.S. 1995. Factors related to flood warning response. Paper presented at the *US–Italy Workshop on the Hydrometeorology Impacts and Management of Extreme Floods*, 13–17 November 1995, Perugia, Italy.

MILETI, D.S. & SORENSEN, J.H. 1987. Why people take precautions against natural disasters. *In*: WEINSTEIN, N. (ed.) *Taking Care: Why People Take Precautions*. Cambridge University Press, Cambridge, 296–320.

PERRY, R.W. 1982. *The Social Psychology of Civil Defense*. Lexington Books, Lexington, MA.

QUARANTELLI, E.L. 1980. Some research emphases for studies on mass communications systems and disasters. *In*: *Disasters and Mass Media*. National Academy of Sciences, Washington, DC, 293–299.

SLOVIC, P. 2010. *The Feeling of Risk: New Perspectives on Risk Perception*. Earthscan, London.

SLOVIC, P., FINUCANE, M., PETERS, E. & MACGREGOR, D.G. 2002. The affect heuristic. *In*: GILOVICH, T. & GRIFFIN, D. & KAHNEMAN, D. (eds) *Heuristics and Biases: The Psychology of Intuitive Judgment*. Cambridge University Press, Cambridge, 397–420.

WOO, G. 2011. *Calculating Catastrophe*. Imperial College Press, London.

WOO, G. 2016. Participatory decision-making on hazard warnings. *In*: GARDONI, P., MURPHY, C. & ROWELL, A. (eds) *Risk Analysis of Natural Hazards*. Risk, Governance and Society, **19**. Springer, New York, 221–240.

The New Zealand Probabilistic Tsunami Hazard Model: development and implementation of a methodology for estimating tsunami hazard nationwide

WILLIAM POWER[1]*, XIAOMING WANG[1], LAURA WALLACE[1,2], KATE CLARK[1] & CHRISTOF MUELLER[1]

[1]*GNS Science, 1 Fairway Drive, Avalon, Lower Hutt, New Zealand*

[2]*Institute for Geophysics, University of Texas at Austin, Austin, TX 78713, USA*

**Correspondence: w.power@gns.cri.nz*

Abstract: For New Zealand, a country straddling the Pacific 'Ring of Fire', effective mitigation of the risks posed by tsunamis is an urgent priority. Mitigation measures include evacuation mapping, land-use planning and engineering of tsunami resilient buildings and infrastructure; but for these to be effective, a quantitative estimate of the tsunami hazard is needed. For this purpose we present the New Zealand Probabilistic Tsunami Hazard Model (NZPTHM). The model uses a Monte Carlo method for sampling from the geophysical parameters that constrain the magnitude–frequency distributions of the earthquake sources that can cause tsunamis affecting New Zealand. The sampled parameters are used to construct synthetic catalogues of the source events and the subsequent tsunami heights. Processing of these synthetic catalogues produces hazard curves, describing maximum tsunami height as a function of return period, which include 'error bars' (confidence intervals) as determined by the Monte Carlo model. Most practical mitigation measures require inundation modelling, and for this purpose we propose using de-aggregation, a process by which a small set of scenarios can be extracted from the NZPTHM for the purpose of detailed inundation modelling.

Motivation

This paper describes the first implementation of a probabilistic tsunami hazard model for all New Zealand coasts. The model considers the hazard posed by distant, regional and local tsunamis caused by earthquakes.

There are many ways in which the risks caused by natural hazards can be mitigated; in the case of tsunamis, these include early warning systems, evacuation mapping, public education in self-evacuation, land-use zoning and engineered sea defences (Eisner 2005). However, these techniques must be used appropriately to ensure that mitigation measures are effective in their operation and are suitably prioritized relative to mitigation of other natural and man-made hazards.

A probabilistic assessment of risk, defined as an estimate of the probable economic losses or human casualties in a period of time, is generally considered the best way to make comparisons across multiple hazards. The relationship between risk, hazard, exposure and vulnerability is, in general terms, defined as (Kron 2002):

$$\text{Risk} = \text{Hazard} \times \text{Exposure} \times \text{Vulnerability}.$$

Mitigation measures reduce the exposure or the vulnerability to the hazard. The reduction in risk is then a measure of the effectiveness of mitigation (e.g. see Bründl *et al.* 2009).

The purpose of the model described here is to quantitatively estimate the tsunami hazard around the New Zealand coast, so the results may be applied to the estimation of risk and to the development of appropriate mitigation measures. While the focus of the analysis presented here is tsunami hazard in New Zealand, the methods presented are general in nature and potentially applicable to other areas of the world exposed to tsunami hazard.

Setting

New Zealand occupies a small portion of the mostly submerged continent of Zealandia (Fig. 1). Presently, Zealandia is bisected by the plate boundary between the Pacific and Australian tectonic plates. The Hikurangi subduction zone lies offshore the North Island's east coast, and accommodates westwards subduction of the Pacific Plate beneath the eastern North Island along the Hikurangi Trough (Wallace *et al.* 2009). The Hikurangi subduction

From: Scourse, E. M., Chapman, N. A., Tappin, D. R. & Wallis, S. R. (eds) 2018. *Tsunamis: Geology, Hazards and Risks*. Geological Society, London, Special Publications, **456**, 199–217.
First published online March 3, 2017, https://doi.org/10.1144/SP456.6

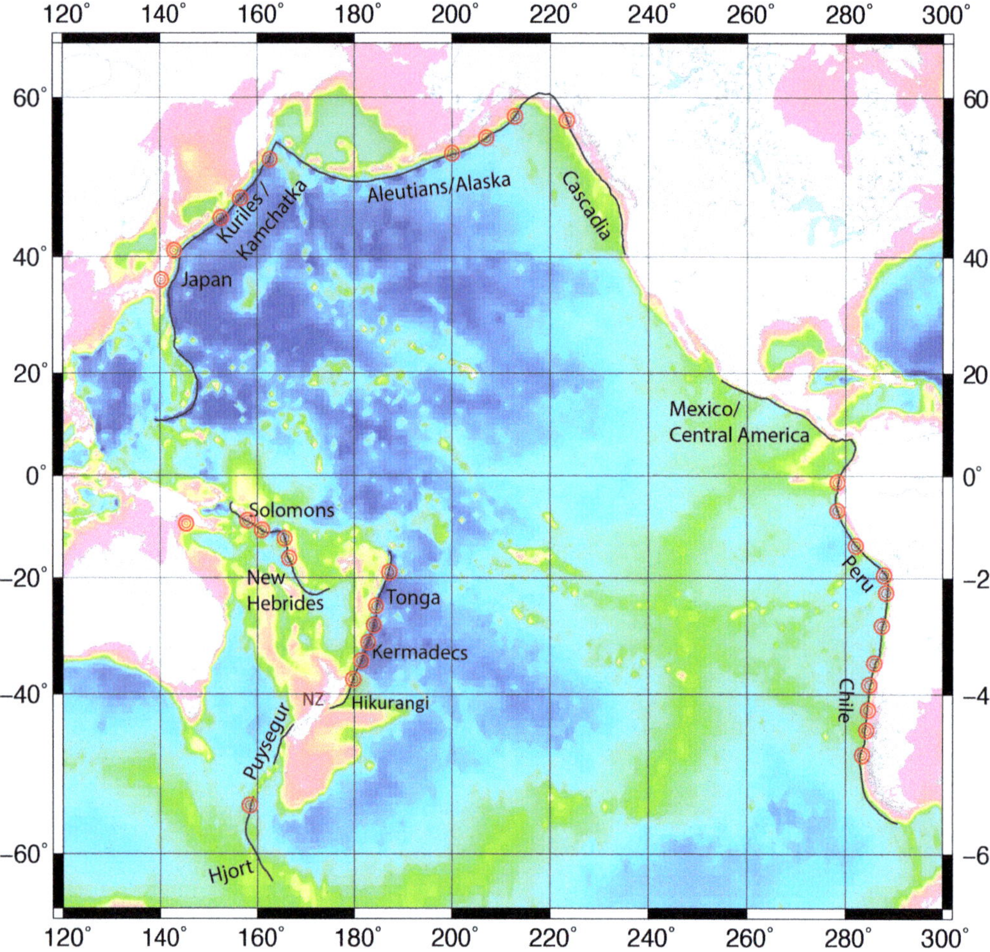

Fig. 1. Subduction margins in the circum-Pacific region, location of New Zealand (labelled with NZ) within the Pacific Basin, and locations of distant or regional earthquakes that have caused tsunami affecting New Zealand (1835–2011) indicated by red rings.

zone connects further north to the Kermadec–Tonga Trench (Fig. 1) (Power *et al.* 2012). In the northern South Island, the plate boundary becomes dominated by strike-slip in the Marlborough Fault System, which transitions southwards into a transpressive boundary along the Alpine Fault (Wallace *et al.* 2012). Southwest of the South Island, eastwards subduction of the Australian Plate beneath the Fiordland region occurs along the Puysegur Trench, which continues offshore for about 500 km (Hayes & Furlong 2010). The Hikurangi and Puysegur subduction interfaces and their associated upper plate and outer rise faults are potential sources of future tsunamis, and in some cases are known to be sources of historical or prehistorical tsunamis. There are also smaller offshore fault systems in the plate boundary zone that are not directly associated with subduction, such as the Kapiti–Manawatu Fault System, the fault systems in Pegasus Bay and the Bay of Plenty, and faults along the west coast of the South Island.

Regionally, in the SW Pacific, there are two major subduction zones north of New Zealand: the Tonga–Kermadec Trench and the Solomon Islands–New Hebrides Trench (Fig. 1). To the south, there is incipient subduction along the Hjort Trench in addition to the already mentioned Puysegur Trench. Historically, the Tonga–Kermadec and Solomon Islands–New Hebrides Trench are known to have experienced earthquakes that have caused tsunamis (McAdoo *et al.* 2008; Beavan *et al.* 2010; Lay *et al.* 2010), while the potential for large subduction earthquakes on the Hjort Trench is generally regarded as low (Meckel *et al.* 2003; Berryman

2005); although, admittedly, very little is known about the potential for earthquakes on this feature.

As tsunamis can travel long distances with little attenuation of energy, New Zealand is also susceptible to tsunamis generated around much of the Pacific 'Rim of Fire', the ring of plate boundaries – mostly subduction zones – around the edge of the Pacific tectonic plate, which, for the most part, mark the edge of the Pacific Basin that covers about one-third of the Earth's surface area. The Peru–Chile Trench, where the Nazca Plate subducts beneath South America, is a well-known source of tsunamis for New Zealand. However, for the other subduction zones in the north and eastern Pacific (e.g. Middle America Trench, Cascadia, Aleutians/Alaska), their potential of generating tsunamis impacting New Zealand is not well known.

Historical events

New Zealand has been affected by at least 80 tsunamis during the historical period (since *c.* AD 1835). The New Zealand Tsunami Database contains a record of tsunami observations in New Zealand from 1835 to the present day (NZTD 2015), the database uses many data sources: early observations of tsunamis were typically recorded in newspapers, journals and letters; while instrumental recordings of tsunamis on tide gauges at major ports are generally available from about 1960 onwards. The rates of recorded tsunamis are relatively steady at approximately four or five per decade up until about 2000. From 2000 onwards, the frequency of tsunamis increases, but this is largely attributed to an increase in data collection from tide gauges which record fluctuations caused by smaller tsunamis that would not have been noticed by human observation alone.

Of the 80 tsunamis to have affected New Zealand in historical times (post-1835):

- 27 were from distant sources (>3 h tsunami travel time);
- 12 were from regional sources (1–3 h tsunami travel time);
- 28 were from local sources (<1 h tsunami travel time);
- 13 were from unknown sources.

The ring of subduction zones around the Pacific Ocean is responsible for most of the distant-source tsunamis to affect New Zealand (Fig. 1). Tsunamis from the South American margin along Peru and Chile are most frequent, but New Zealand has also been affected by tsunamis from the Alaska–Aleutian margin, the Kamchatka–Kuril–Japan margin, and the south Pacific subduction zones of the Solomon Islands and the Tonga–Kermadec Trench (Fig. 1). Tsunamis generated at the Sumatra subduction zone (the M_w 9.3 Indian Ocean tsunami, 2004) and by the Krakatau volcanic eruption (1883) were recorded in New Zealand but did not cause any significant damage.

Regional-source tsunamis are typically from the Puysegur Trench, SW of New Zealand, and the Tonga–Kermadec Trench, NE of New Zealand (depending on the distance from New Zealand, tsunamis generated on the Tonga–Kermadec Trench can be classified as distant or regional source). Local-source tsunamis have been predominantly associated with upper plate faults or the plate interface along the Hikurangi subduction zone or the Fiordland–Puysegur subduction zone.

According to the New Zealand Tsunami Database (NZTD 2015), the five largest historical tsunamis in New Zealand were generated by: the M_W 8.2 Wairarapa earthquake in 1855 (Grapes & Downes 1997), a M_W 7.1 earthquake 50 km offshore of Gisborne in March 1947 (Bell *et al.* 2014), and distant earthquakes in South America in 1868, 1877 and 1960 (Power *et al.* 2007). The distribution of run-up heights for these events is shown in Figure 2. The local-source tsunami have generated damaging tsunami with run-up heights up to 10 m in localized areas (within *c.* 100 km of the earthquake epicentre). Distant-source tsunami have generated run-up heights of up to 4 m (but generally <1 m) over much of the eastern coastline of mainland New Zealand: however, the 1868 South America tsunami generated run-up of 7.6 m at one location on Banks Peninsula and run-up of 6 m (and possibly up to 10 m) on the Chatham Islands (Fig. 2).

Palaeotsunami

Palaeotsunami records can extend the tsunami record much further back in time than the historical and instrumental record, thereby improving our knowledge of tsunami hazard. Given New Zealand's relatively short historical period (*c.* 180 years), the extension of tsunami records back in time is particularly important as the full spectrum of tsunami behaviour will not have been captured during the historical period. Evidence of palaeotsunami is typically obtained from geological records of tsunami sediment and debris deposit in the coastal zone, and also from tsunami erosional landforms, archaeological sites and oral records. In New Zealand, palaeotsunami have been identified at many places around the coastline. Most palaeotsunami records have been compiled in the New Zealand Palaeotsunami Database (NZPD) (Goff 2008; Goff *et al.* 2010*c*). This database describes 293 observations around the New Zealand coastline of likely to possible palaeotsunami (Fig. 3), which are related to between 35 and 40 palaeotsunami (i.e. there are multiple observations that are attributed to the same event). The NZPD utilizes many data sources

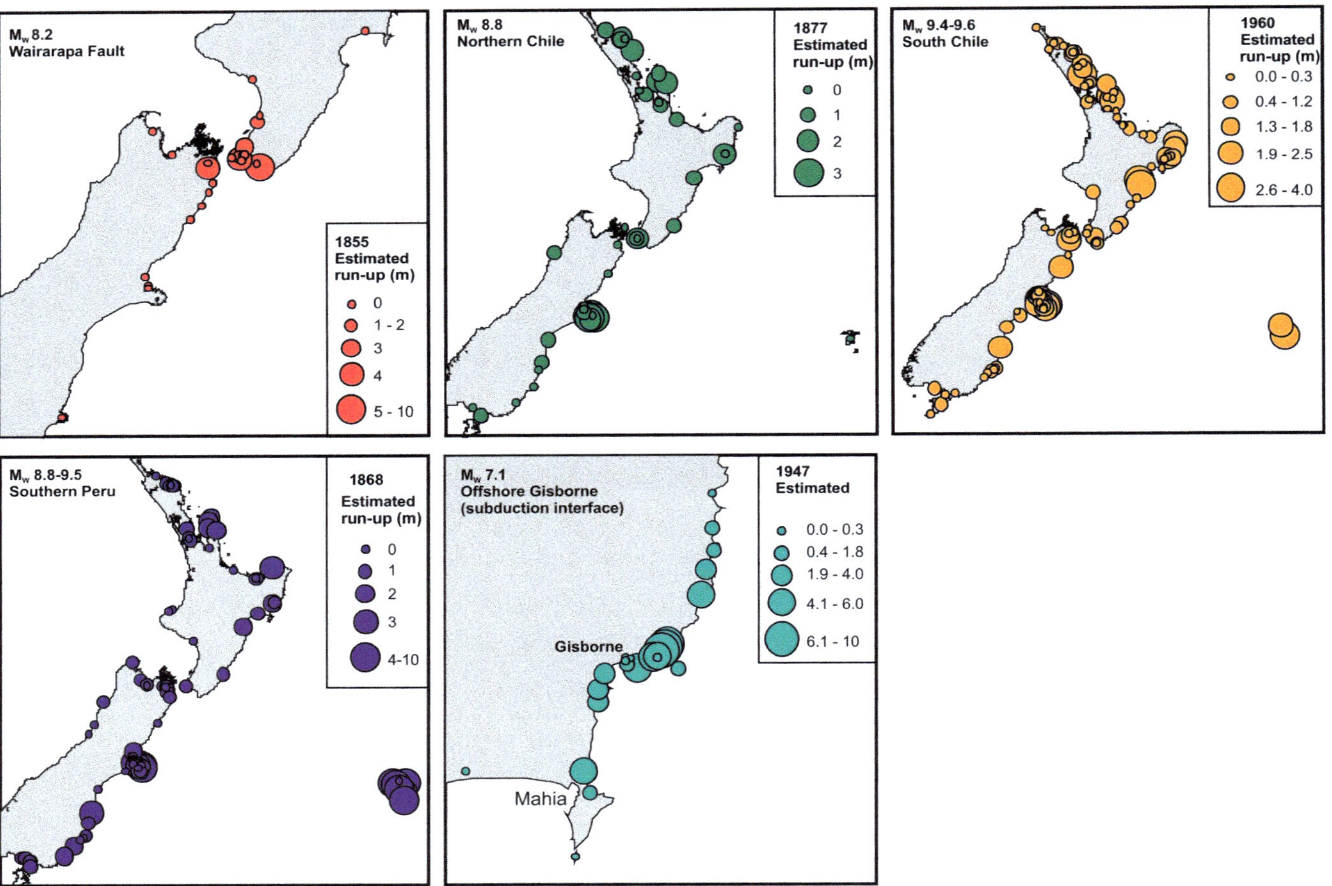

Fig. 2. Estimated tsunami run-up values for the five largest tsunami in New Zealand between 1835 and 2015. Note the scale varies between boxes. Local-source tsunami have high run-ups, but are typically smaller in spatial extent; distant-source tsunami have widespread effects but lower run-ups. The run-up values are from the New Zealand Tsunami Database (NZTD 2015), and include a combination of instrumentally recorded wave heights and estimates based on descriptions given in newspaper (and other) reports.

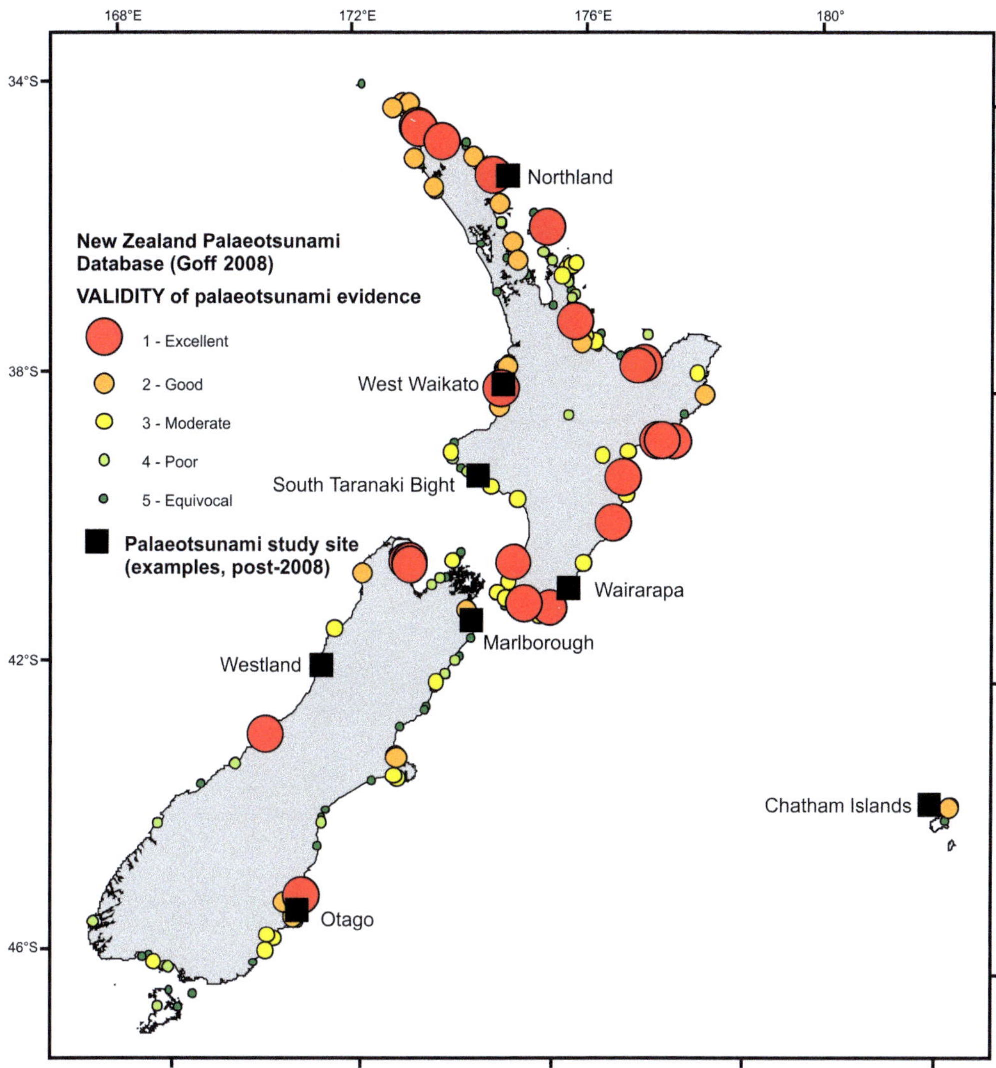

Fig. 3. Distribution of palaeotsunami deposits in the New Zealand Palaeotsunami Database (Goff 2008), labelled according to assessed validity, along with the locations of selected recent (post-2008) palaeotsunami study sites mentioned in the text (black boxes).

from multi-proxy, peer-reviewed studies to poorly verified unpublished observations. Most well-verified entries in the NZPD are located along the east coast of the North Island, in the Wellington region, along parts of the east coast of the South Island, and on the Chatham Islands. Palaeotsunami evidence also exists along the west coast of New Zealand but is less abundant than east coast records. The spatial distribution of palaeotsunami evidence is approximately consistent with the distribution of historical tsunami observations and proximal to New Zealand's most active offshore faults (e.g. the Hikurangi margin). Much evidence of palaeotsunami (*c.* 160 entries in the database) is estimated to be related to tsunamis that occurred between AD 1300 and AD 1600, which indicates either a particularly intense period of tsunami activity with many several tsunamis from multiple sources or, perhaps, one or two very large tsunami with the spread in age reflecting poor age control rather than many discrete tsunamis.

Since publication of the NZPD in 2008, many new studies have been made of palaeotsunami at various locations around New Zealand (e.g. Otago coast: Goff *et al.* 2009; Chatham Islands: Goff *et al.* 2010*a*; Nichol *et al.* 2010; Northland: Goff *et al.* 2010*b*; Wairarapa coast: Berryman *et al.* 2011; Marlborough: Clark *et al.* 2015; and South Taranaki Bight and Westland: Goff & Chagué-Goff 2015). Studies of particular note include the correlation of a tsunami deposit on the Chatham Islands to an AD 1604 earthquake in Peru (Goff *et al.* 2010*a*), palaeotsunami deposits in Northland dated at approximately 6.5 and 2.8 ka BP that were inferred to be from a Tonga–Kermadec Trench earthquake (Goff *et al.* 2010*b*), and a palaeotsunami deposit on the Marlborough coastline that is coincident with sudden subsidence of the coastline and inferred to be from a subduction earthquake on the southern Hikurangi margin at approximately 850 years BP (Clark *et al.* 2015). This sampling of recent palaeostunami studies demonstrates the variability in age, location and inferred source of New Zealand palaeotsunami.

As palaeotsunami research in New Zealand continues to progress beyond 'reconnaissance-level' studies and into detailed multi-proxy, multi-site investigations, the reliability and quality of the palaeotsunami record will improve substantially. Increasingly, sufficient data are being gathered so that the tsunami source can be identified. A future challenge will be to bring our knowledge of palaeotsunami up to a standard where source models can be reliably calibrated using the inland extent and elevation of palaeotsunami deposits, thus ensuring tsunami source and inundation models, and the tsunami hazard and risk assessments based on them, are dependable.

Probabilistic tsunami hazard analysis

A probabilistic approach to evaluating tsunami hazard was first developed by Rikitake & Aida (1988), and the last decade has seen many studies and advances in methodology (e.g. Geist & Parsons 2006; Burbidge *et al.* 2008; González *et al.* 2009; Sørensen *et al.* 2012). The essential approach behind Probabilistic Tsunami Hazard Analysis (PTHA) is closely related to that of Probabilistic Seismic Hazard Analysis (PSHA: e.g. Cornell 1968; McGuire 2004). The same four key steps are involved (Downes & Stirling 2001): identification of the relevant sources; quantification of the magnitude–frequency of those sources; estimation of the relationship between source properties (typically earthquake magnitude and location) and consequences at the site of interest (e.g. peak ground acceleration for earthquakes or peak tsunami amplitude for tsunamis); and calculation of the hazard using these pieces of information.

PTHA studies in New Zealand include: a risk study for New Zealand cities compiled by Berryman (2005) which is a direct precursor of the study presented here; an analysis of tsunami hazard at New Zealand coasts from South American tsunamis (Power *et al.* 2007); and analysis of tsunami hazard in the Auckland region caused by earthquakes on the Kermadec–Tonga Trench (Lane *et al.* 2013; Power *et al.* 2013). The national PTHA model presented here was originally developed as part of a review of New Zealand's tsunami hazard compiled by Power (2013*a*), and was the first PTHA model for New Zealand to cover all earthquake sources and all coasts. In this paper we make reference to the technical report by Power (2013*a*) for some specific details, while providing additional context, explanation and discussion of the hazard model.

Methods

Outline

The approach used for estimating tsunami hazard in this report is based on a Monte Carlo modelling process (Musson 1999). The method aims to estimate the maximum tsunami height that can be expected over a specified interval of time within sections of the New Zealand coast that are approximately 20 km long. As is the case in most areas of science, an estimate of tsunami hazard is of little value without an assessment of the associated uncertainty and, consequently, the estimation of uncertainties is an important component of this study.

The methodology of this study makes a clear distinction between (aleatory) variability and (epistemic) uncertainty (Abrahamson & Bommer 2005). Variability refers to the natural variations that occur between different events. For instance, the magnitude of earthquakes on a fault naturally varies from one earthquake to the next. Uncertainty, on the other hand, is a measure of our lack of knowledge about things which are constant in time. For example, while the shape of a fault is fixed (at least within the time frames we are interested in), its shape is not known exactly, and the uncertainty is a measure of how well it is known.

Our Monte Carlo analysis operates on two levels (Fig. 4). On the inner level, we assume that we have perfect knowledge of the uncertain parameters that do not vary over time, and carry out a hazard assessment using Monte Carlo sampling of those properties that naturally vary between events. On the outer level, we perform Monte Carlo sampling of the uncertain parameters, and use this to build up a set of hazard estimates that differ from those calculated for the inner level. The spread of these estimates represents the uncertainty in the hazard.

Typically, in PSHA and PTHA, uncertainties are described using logic trees (e.g. Annaka *et al.* 2007), the approach here is essentially equivalent except that the uncertain variables are sampled

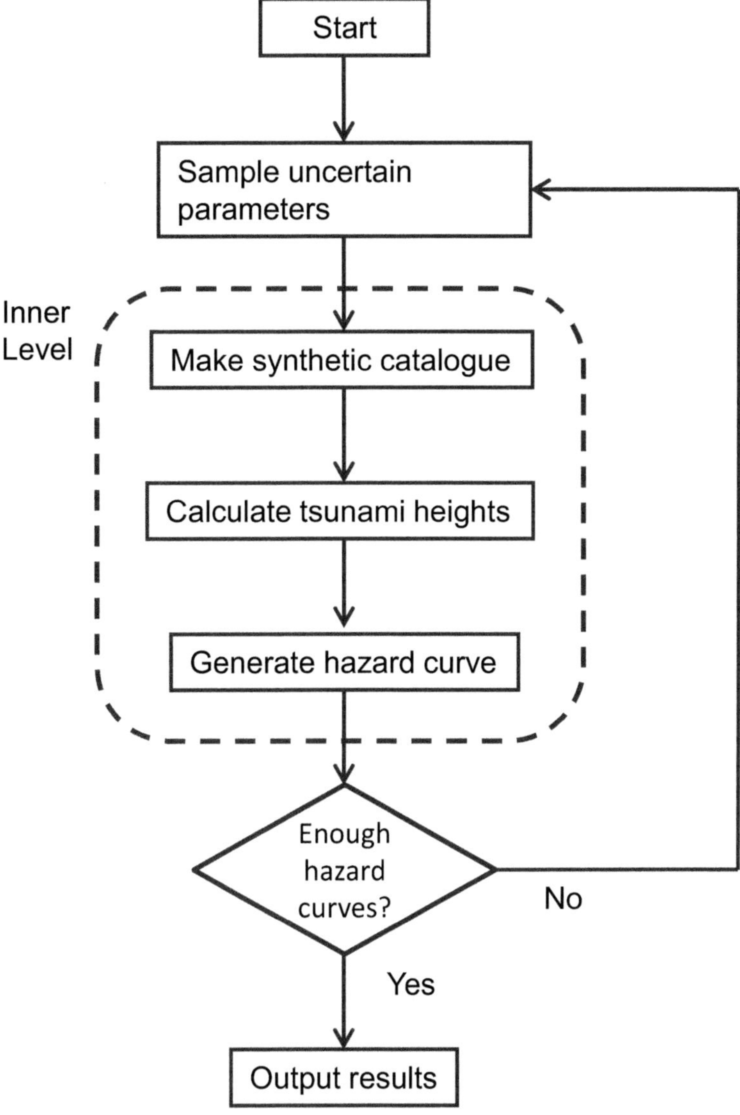

Fig. 4. Simplified flowchart representation of the Monte Carlo modelling scheme.

from continuous distributions rather than as discrete 'branches' of a logic tree.

A more detailed representation of the method is shown in Figure 5; in this figure each row going across the chart describes the steps used to construct one tsunami hazard curve. These steps are repeated many times using different samples of the uncertain parameters, and from these it is possible to assign 'error bars' (i.e. confidence intervals) to the tsunami hazard curves.

Each hazard curve describes the maximum tsunami height reached within a coastal section, as a function of return period. By sampling from the uncertain parameters, and creating multiple hazard curves, it is possible to estimate the uncertainty in the tsunami hazard (Fig. 6).

The uncertainties and variabilities fall into two broad categories: those associated with the source earthquake and those associated with the modelling process. For earthquakes, the primary uncertainty is the true form of the magnitude–frequency distribution of the faults (i.e. knowing how often earthquakes of varying magnitude occur along a fault), although it also encompasses such things as uncertainty in the geometry of the faults. The earthquake variabilities represent the variation in magnitude

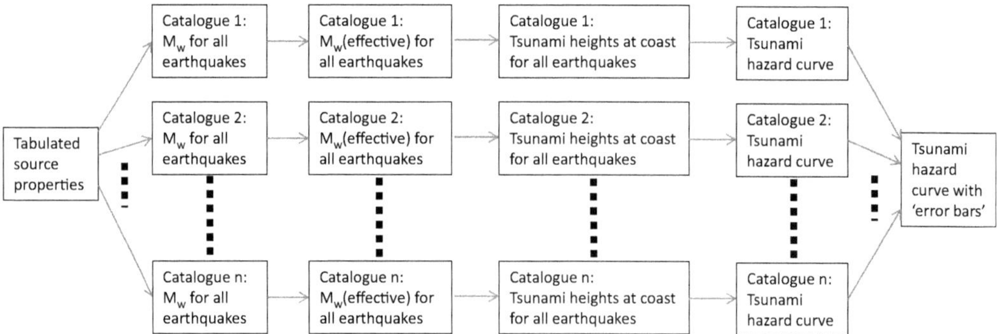

Fig. 5. Representation of the Monte Carlo modelling scheme.

from event to event on a particular fault, and also the variation in the distribution of slip (even among earthquakes of the same magnitude). Modelling uncertainty, on the other hand, reflects the inability of the model to fully capture the physics of tsunami generation and propagation, and uncertainties in bathymetric data. A further breakdown of different types of uncertainty and variability can be found in appendix 7.1 of Power (2013*a*).

Fault magnitude frequency

An essential input into the probabilistic hazard model is a definition of the physical and statistical properties of the various tsunami sources. The model described here is intended to estimate tsunami hazard within time frames of up to 2500 years. On these timescales, the major contribution to tsunami hazard is assessed to arise from both distant and local earthquakes. For some regions of the country, submarine landslides (e.g. see Tappin *et al.* 1999; Ward 2001; Harbitz *et al.* 2006) may contribute to the tsunami hazard in these time frames as well, and initial steps towards estimating potential landslide contributions are the focus of current research (Mueller *et al.* 2016; Wang *et al.* 2016).

The definition of tsunami sources from subduction zone earthquakes, which constitute all distant earthquake sources and the most important local ones, drew heavily on work that has been done for the Global Earthquake Model (GEM: Crowley *et al.* 2013). The assumed parameters of subduction zone earthquakes used for this report are summarized in Table 1, full details are in appendix 3 of Power (2013*a*).

The starting point for defining tsunami sources for local non-subduction zone earthquakes was the New Zealand Seismic Hazard Model (NZSHM: Stirling *et al.* 2012). The faults in the NZSHM were filtered to exclude those with characteristic magnitudes below 6.5 (which are too small to generate enough displacement to cause a tsunami), those with strike-slip mechanisms and those that are entirely onshore. Additional fault sources were added in the Hikurangi Outer Rise, the Taranaki Basin and along the west coast of the South Island (see appendix 5 of Power 2013*a*).

The creation of synthetic earthquake catalogues from the tabulated fault and subduction zone properties is illustrated in Figure 7. Subduction earthquake events in the synthetic catalogue were generated using a truncated Guttenberg–Richter (G–R) distribution (a sharp truncation was applied in the incremental G–R distibution, which results is a smooth tapering of the cumulative distribution: see Geist & Parsons 2014 for an analysis of issues associated with sampling from power-law distributions), magnitudes sampled from a characteristic distribution were used for intra-plate faults.

Magnitude alone is not enough to determine the size of tsunami that will be produced by an earthquake. It has been shown that the distribution of slip on a fault also plays an important role. Geist (2002) found that the peak amplitude of nearshore tsunamis varied by over a factor of 3 depending on the slip distribution, and Mueller *et al.* (2012, 2015) demonstrated great variation in the extent of inundation as a result of variable slip.

Mueller *et al.* (2015) introduced the concept used here of 'effective magnitude' to approximate the effects of non-uniform slip. In this approximation, the effect of non-uniform slip is treated as if it has the effect of altering the effective magnitude of the earthquake. By adding a normally distributed variation to the magnitudes in the synthetic earthquake catalogue, we create a new catalogue of 'effective magnitudes' that represent the consequences of the variable slip. It may be argued that this is not a true representation of the effects of variable slip, since variable slip may enhance the tsunami at some locations while reducing it at others,

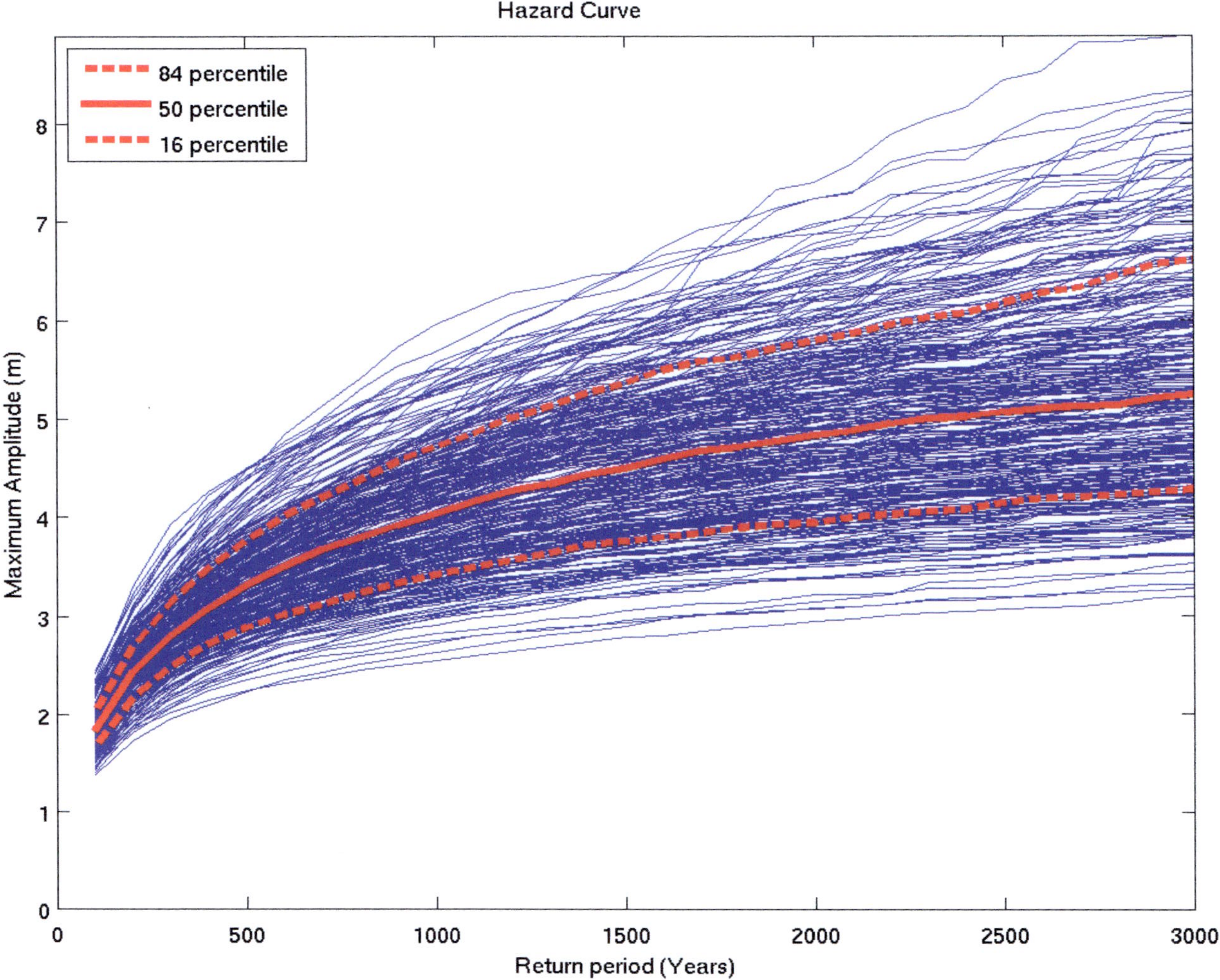

Fig. 6. Hazard curves for 300 samples of the uncertain parameters, illustrating how the 16th, 50th and 84th percentiles of uncertainty are calculated for one coastal section.

Table 1. *Assumed properties of subduction zone sources*

Subduction zone	M_{max} – pref	M_{max} – min	M_{max} – max	*C* – pref	*C* – min	*C* – max	*B*-value – pref	*B*-value – min	*B*-value – max
Alaska	9.60	9.50	9.70	0.50	0.30	0.70	0.75	0.50	1.00
Cascadia	9.00	8.80	9.20	0.80	0.70	0.90	0.75	0.50	1.00
Japan	9.07	9.00	9.14	0.70	0.60	0.90	0.75	0.50	1.00
Kanto	8.22	8.00	8.43	0.90	0.80	1.00	0.75	0.50	1.00
Nankai	8.73	8.50	8.95	0.90	0.80	1.00	0.75	0.50	1.00
Kurile–Kamchatka	9.35	9.00	9.70	0.80	0.70	0.90	0.75	0.50	1.00
Ryukyu	8.54	8.00	9.09	0.20	0.10	0.70	0.75	0.50	1.00
Izu–Bonin	8.20	7.20	9.21	0.20	0.10	0.70	0.75	0.50	1.00
Marianas	8.34	7.20	9.48	0.20	0.10	0.70	0.75	0.50	1.00
North Yap	8.11	7.20	9.01	0.20	0.10	0.70	0.75	0.50	1.00
Palau-South Yap	8.04	7.20	8.88	0.20	0.10	0.70	0.75	0.50	1.00
Hikurangi	8.50	8.00	9.00	0.54	0.40	0.70	0.75	0.50	1.00
Kermadec	8.74	8.10	9.39	0.30	0.20	0.75	0.75	0.50	1.00
Tonga	8.57	8.00	9.14	0.20	0.10	0.70	0.75	0.50	1.00
Puysegur	8.43	7.80	9.07	0.70	0.50	0.80	0.75	0.50	1.00
Hjort	7.78	7.20	8.36	0.50	0.30	0.70	0.75	0.50	1.00
Solomon NW	8.36	8.10	8.62	0.70	0.60	0.80	0.75	0.50	1.00
Solomon SE	8.58	8.10	9.06	0.70	0.60	0.80	0.75	0.50	1.00
New Hebrides North	8.01	7.60	8.43	0.25	0.15	0.70	0.75	0.50	1.00
New Hebrides Central	8.49	8.30	8.69	0.70	0.60	0.80	0.75	0.50	1.00
New Hebrides South	8.11	7.60	8.62	0.25	0.15	0.70	0.75	0.50	1.00
New Hebrides Mat. Hunt.	8.40	8.00	8.49	0.25	0.15	0.70	0.75	0.50	1.00
New Britain	8.41	8.00	8.82	0.70	0.60	0.80	0.75	0.50	1.00
New Guinea Trench East	8.27	7.60	8.93	0.70	0.60	0.80	0.75	0.50	1.00
New Guinea Trench West	8.64	8.20	9.07	0.70	0.60	0.80	0.75	0.50	1.00
Manus East	8.30	7.50	9.10	0.50	0.30	0.70	0.75	0.50	1.00
Manus West	8.33	7.50	9.17	0.50	0.30	0.70	0.75	0.50	1.00
Equador–Colombia	9.15	8.80	9.51	0.80	0.70	0.90	0.75	0.50	1.00
Peru	9.35	9.00	9.70	0.80	0.70	0.90	0.75	0.50	1.00
Northern Chile	9.04	8.60	9.48	0.80	0.70	0.90	0.75	0.50	1.00
Central Chile	9.51	9.50	9.51	0.80	0.70	0.90	0.75	0.50	1.00
Patagonia North	8.52	8.00	9.04	0.50	0.30	0.70	0.75	0.50	1.00
Patagonia South	8.74	8.00	9.49	0.50	0.30	0.70	0.75	0.50	1.00
Mexico Jalisco	8.33	8.20	8.45	0.50	0.30	0.70	0.75	0.50	1.00
Mexico Michoa	8.58	8.00	9.17	0.70	0.50	0.90	0.75	0.50	1.00
Central America ElSalv	8.29	8.00	8.58	0.30	0.10	0.70	0.75	0.50	1.00
Central America Costa Rica	8.22	7.70	8.74	0.50	0.30	0.70	0.75	0.50	1.00
Philippine	8.43	7.60	9.25	0.25	0.10	0.75	0.75	0.50	1.00
East Luzon Trough	7.86	7.30	8.43	0.50	0.30	0.70	0.75	0.50	1.00
Cotabato Trench	8.21	8.00	8.42	0.50	0.30	0.70	0.75	0.50	1.00

M_{max} is the maximum value of M_w, *C* is the coupling coefficient and *B*-value is the Gutenberg–Richter *B*-value. The *B*-value range was set to include the possibility that subduction plate interfaces may have low *b*-values relative to the global *B*-value of 1 (Bayrak *et al.* 2002). M_{max} – max is based on the assumption that the only ultimate constraint on the maximum magnitude is the length of the subduction zone (McCaffrey 2007), up to a global limit of M_w 9.7. M_{max} – min is based on the magnitude of the largest known historical earthquake or palaeotsunami. Pref indicates the preferred value of a parameter, and – min and – max indicate the assessed minimum and maximum of the plausible range of values of a parameter.

whereas our approach sees the effective magnitude of the earthquake increase (or decrease) in the same way at all locations. This would be a problem if we were to look at correlated hazard across multiple locations: however, as long as we view the hazard on a 'one site at a time' basis, this approximation appears valid.

This approach, of creating a catalogue of 'effective magnitudes', also provides a convenient way to incorporate the effects of modelling uncertainties. We regard the effects of modelling approximations, and of limited data on source geometry and ocean bathymetry, as having a similar effect to (usually small) increases or decreases in the magnitude of the source earthquake. Table 2 summarizes the parameters used for this purpose.

An 'effective magnitude' is calculated by applying the parameters that describe the uncertainties

For each subduction zone:

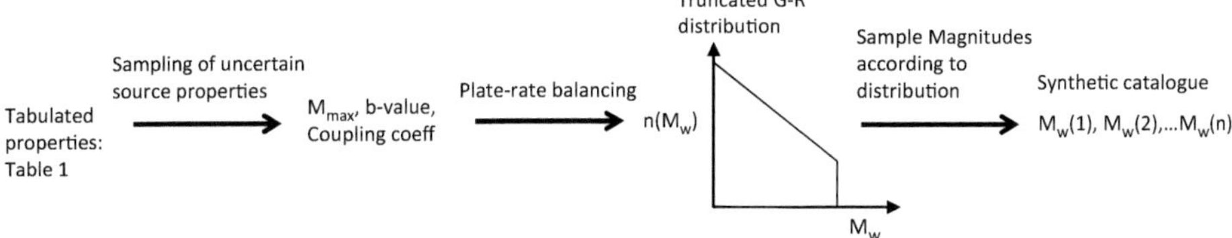

For each crustal fault:

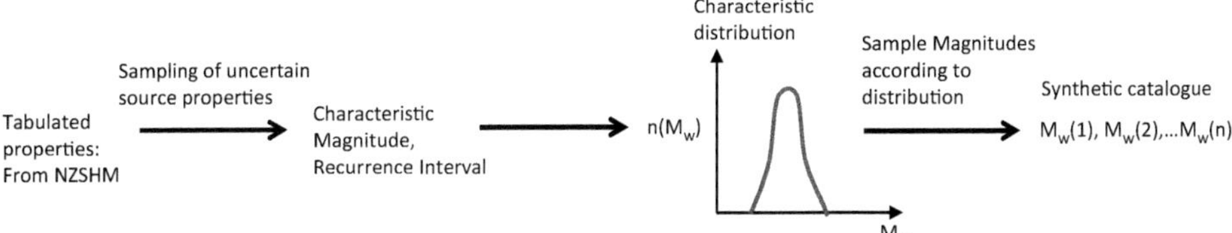

Fig. 7. Illustration of the steps by which the tabulated fault properties are used to create synthetic earthquake catalogues. This process corresponds to the leftmost set of arrows in Figure 5.

and variabilities that affect tsunami heights, using the following equation:

$$Mw_{ijk}(\text{effective}) = Mw_{ijk} + \sigma_v N(0,1)_{ijk} + \sigma_u N(0,1)_{jk} + \sigma_b N(0,1)_k \quad (1)$$

where i represents individual earthquakes on fault j, described in synthetic catalogue k. $N(0,1)$ represents a number sampled from the normal distribution with a mean of zero and a standard deviation of 1. The subscript to $N(0,1)$ describes the set over which individual samples are made: for example, $N(0,1)_{jk}$ is sampled for each fault in each catalogue, but has the same value for all earthquakes on a particular fault in a particular catalogue. This calculation of an 'effective magnitude' corresponds to the second step (going left to right) in Figure 8.

The choice of parameters in Table 2 was based on interpretation of available studies and expert judgement. The reasoning behind the values used is summarized in appendix 7.3 of Power (2013*a*). We regard these values as assumptions of the PTHA model, as further work is needed to better scientifically constrain their values.

Tsunami height estimation

The Monte Carlo method requires us to estimate the maximum tsunami height for each section of the coast following every event in a synthetic catalogue of earthquakes. A variety of techniques can be used to estimate tsunami heights, but it is important that the calculation can be performed quickly, since it

Table 2. *Standard deviations associated with random adjustments to the synthetic catalogue to create a catalogue of 'effective magnitudes'*

	Local crust fault (empirical model)	Local subduction zone (numerical model)	Distant subduction zone (numerical model)
Variability (e.g. non-uniform slip): σ_v	0.25	0.25	0.1
Modelling uncertainty (fault specific): σ_u	0.2	0.1	0.1
Modelling uncertainty (method bias): σ_b	0.14	0.05	0.05

The fault-specific uncertainty covers uncertainties that are specific to the modelling of each fault, while the method bias covers uncertainties that cause a systematic bias across all faults. Units are in the M_w scale.

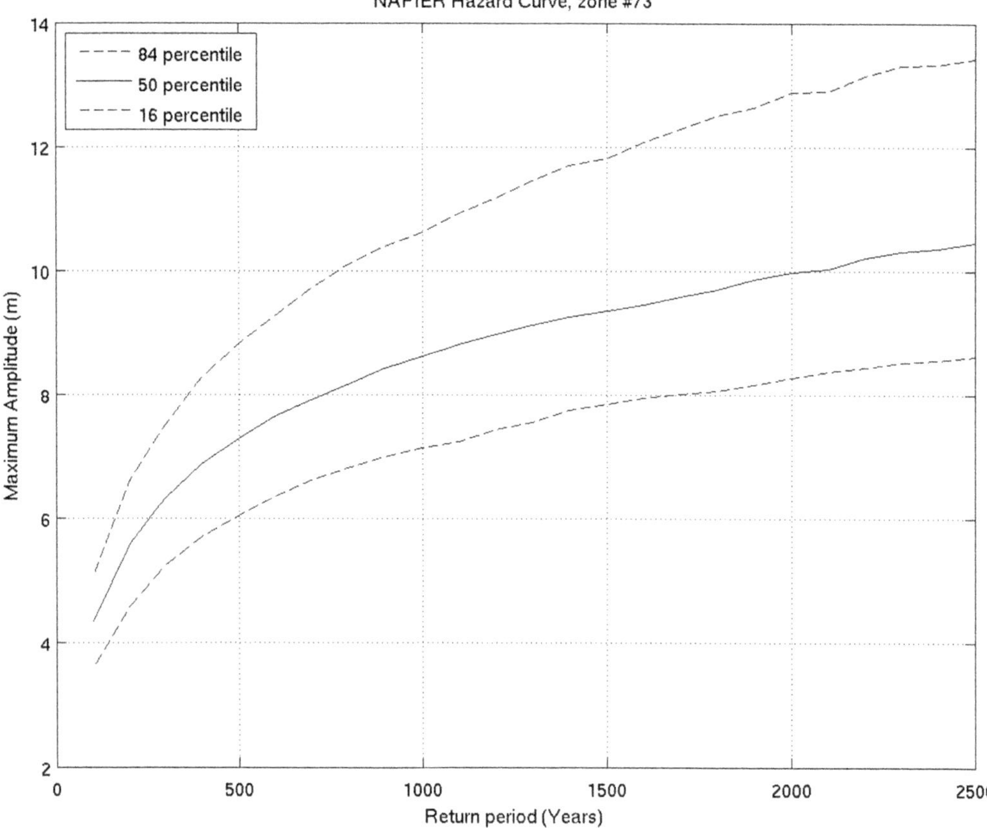

Fig. 8. Tsunami hazard curve for the city of Napier. The solid line shows the 50th percentile of epistemic uncertainty, and the dashed lines show the 16th and 84th percentiles.

is necessary to model many events to produce robust statistics. Three different methods were used in this study:

- Finding the closest available model (in terms of location and magnitude) in a pre-computed catalogue, and then applying scaling to the model results to match the synthetic catalogue magnitude.
- Using a collection of pre-calculated models of tsunamis from a particular source region to estimate coefficients in a semi-empirical scaling relationship.
- Using an empirically determined scaling relationship based only on the magnitude and distance of historical earthquakes that have caused tsunami.

In general, the accuracy of results diminishes down this list of methods, as does the work required to implement them for any particular source. The first method has been applied to subduction zone sources close to New Zealand, specifically the Hikurangi, Kermadec and Puysegur trenches, where the location of the earthquake within the source region plays a very major role in determining the consequences for particular sections of the New Zealand coast. The second method has been applied to all other Pacific subduction zones (i.e. those at regional or distant locations from New Zealand); the tsunami consequences of earthquakes at these distances are less sensitive to the precise location of the source. This method uses the empirical approach of Abe (1979), except that numerical results from the New Zealand forecast database (Power & Gale 2011: database contents updated in 2013) were used instead of historical catalogue data. The third method applies the empirical modelling approach developed by Abe (1995), and is applied to estimating the tsunamis caused by local faults other than the subduction zones. Details of the application of these methods are described in appendix 7.4 of Power (2013*a*).

Tsunami height is defined here as the maximum height that the tsunami would reach against an imaginary vertical wall at the coast, relative to the background sea level at the time of the tsunami. This choice of criteria permits us to reuse the modelling used for the New Zealand forecast database, where reflective boundary conditions were applied at the coast. In many situations where the tsunami does not penetrate far inland (i.e. less than several kilometres), this represents a reasonable approximation to the expected run-up height; although in a small number of locations where a tsunami is focused by small-scale topographical features, the run-up may locally reach about twice this height. For mitigation measures requiring estimates of inundation extents and depths, the recommended approach is to use deaggregation (see the 'Results' and 'Discussion and conclusions' sections later in this paper) to determine the degree to which different sources contribute to the hazard, and use this to select specific scenarios for detailed inundation modelling.

In the case of the empirical equation used for local upper plate and outer rise faults, Abe (1995) related the predicted tsunami height to the average run-up height measurement, rather than to the maximum height against an imaginary vertical wall at the shore. Here, these two quantities are treated as equivalent, but there is considerable uncertainty about this relationship; this uncertainty contributes to the corresponding bias parameter in Table 2.

The numerical models used for this study were developed using the COMCOT code (Wang & Liu 2006, 2007). A series of nested bathymetric grids were developed, ranging in size from the entire Pacific Ocean to small regions of New Zealand (the purpose of the nesting scheme is to maintain sufficient grid cells relative to the wavelength of the modelled tsunamis to allow an accurate representation of the wave form, and thereby avoid artificial attenuation of the modelled wave that can result from using a grid that is too coarse; as the tsunami wavelength reduces in shallow water, this means that finer grids are required for accurate modelling in these areas). Models from the New Zealand tsunami forecast database were used for the distant subduction zone sources. The local subduction zones were modelled using the same nested grid configuration in order to maintain consistency. In this grid set-up, the non-linear shallow-water wave equations were used to model the grids closest to New Zealand, where the water depths are such that the non-linear effects may be significant, and the linear shallow-water equations were used for all of the outer grids. The modelling included a wave-breaking scheme (Kennedy *et al.* 2000; Lynett 2002; Wang & Power 2011) and a Coriolis force term. Non-dispersive shallow-water equations were used but with a numerical dispersion scheme applied (Wang & Liu 2011), and no bottom friction was assumed. Simulation time steps were determined using the Courant–Friedrichs–Lewy (CFL) stability condition taking into account the nested grid set-up (Wang & Power 2011).

Processing

The Monte Carlo analysis of epistemic uncertainty was made using 300 samples of the uncertain parameters. For each of these 300 samples, a 100 000 year synthetic catalogue of earthquakes was constructed. Re-running the analysis using these same parameters and a different set of random numbers demonstrated good repeatability of the results, with variations in the hazard curves that were small compared to the cumulative effect of other sources of uncertainty. The probabilistic tsunami hazard model in this report does not include modelling of tides.

The process described in the preceding subsections allows the construction of tsunami hazard curves for individual sections of coast. These curves, which will be described in detail in the 'Results' section, indicate the height of tsunami that may be expected in a given time frame. On their own, these curves do not provide a measure of the extent of inundation, only the maximum height at the coast.

Deaggregation is a process for providing specific examples of events that correspond to a particular hazard level, and has been widely used in PSHA studies (McGuire 1995). Here, we use deaggregation to establish the extent to which different tsunami sources contribute to the probabilistic tsunami hazard. The intended purpose of the deaggregation methodology presented here is to establish a particular set of scenarios whose inundation can be modelled to give an approximation of the onshore tsunami hazard at a particular level of probability (i.e. return period) and confidence.

The probabilistic hazard analysis described here involves the generation of a large number ($N = 300$) of synthetic catalogues of effective earthquake magnitude. Each catalogue represents a sequence of earthquakes generated assuming a particular sampling of the uncertain parameters. For a selected return period R (500 and 2500 years have been used), the median tsunami height $H(R)$ was found from the corresponding hazard curve for the site of interest. Each synthetic catalogue was searched to find the three earthquakes that produced tsunamis at the site which were closest in height to $H(R)$. The proportion of these $3 \times N$ events coming from particular faults was calculated, and this was used to generate the pie charts described in the 'Results' section. In addition, an estimate was made of the median effective magnitude of the selected earthquakes from each of the identified faults.

Results

The coast of New Zealand was divided into 268 sections, each approximately 20 km long as measured along the open coast. Within each section, the model produces a hazard curve that illustrates the expected maximum tsunami height as a function of return period. A series of hazard curves were generated for each section: an example is shown in Figure 8 for the city of Napier on the east coast of the North Island, here the solid line indicates the best-estimate hazard curve and the dashed lines are 'error bars' indicating the 16th and 84th percentile of uncertainty (the full set of these curves is available online, see Power 2013*b*). During a tsunami, the peak water levels will vary considerably, even across a 20 km section of coast: in the curves shown here, the 'maximum amplitude' should be interpreted as the tsunami height measured at the location within the section where it is highest; the median tsunami height within the section may be significantly lower (e.g. see Power *et al.* 2010).

Deaggregation of the tsunami hazard for Napier at the 500 year return period, and 50th percentile of epistemic uncertainty is shown in Figure 9 (similar pie charts for other sites and return periods are available from Power 2013*a*, *b*). The pie chart in Figure 9 shows the breakdown of the relative contribution of different fault sources to the hazard: the area of each slice of the pie indicates the proportion of the hazard for which a particular fault is responsible – the larger the area, the more frequently that source is expected to produce tsunamis of the size corresponding to the return period.

The Figure 9 pie chart indicates the six tsunami sources that most frequently generate tsunamis at the median height for the 500 year return period. It

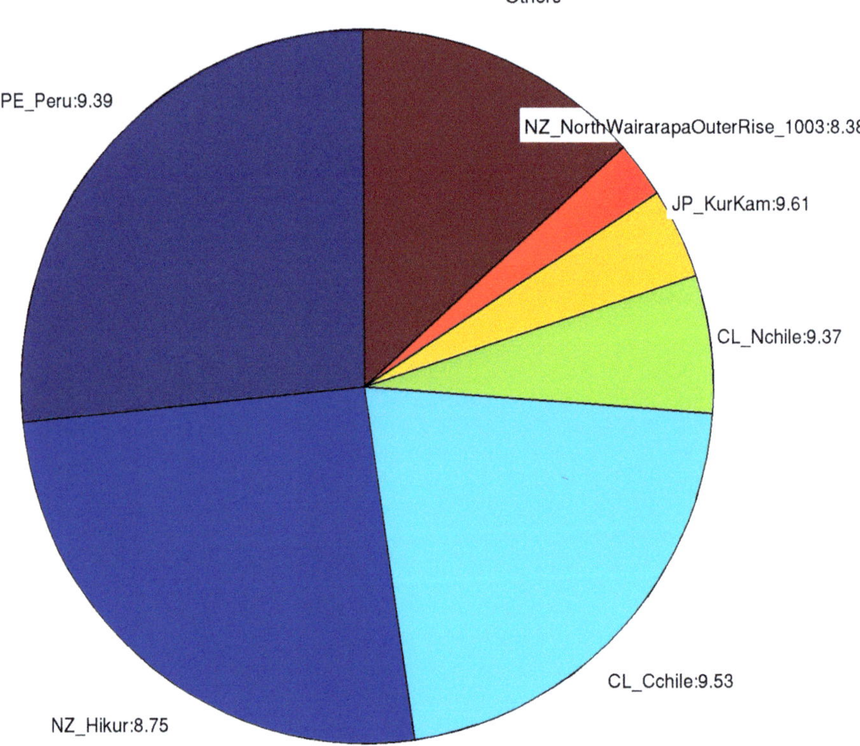

Fig. 9. Deaggregation of tsunami sources for the city of Napier. PE_Peru, Peru subduction zone; NZ_Hikur, Hikurangi subduction zone; CL_CChile, central Chile subduction zone; CL_NChile, northern Chile subduction zone; JP_KurKam, Kurile–Kamchatka subduction zone; NZ_NorthWairarapaOuterRise_1003, the North Wairarapa Outer Rise Fault (see Power 2013*a*). The numbers following the colon refer to the 'effective magnitude' (see the 'Methods' section of this paper) of these sources.

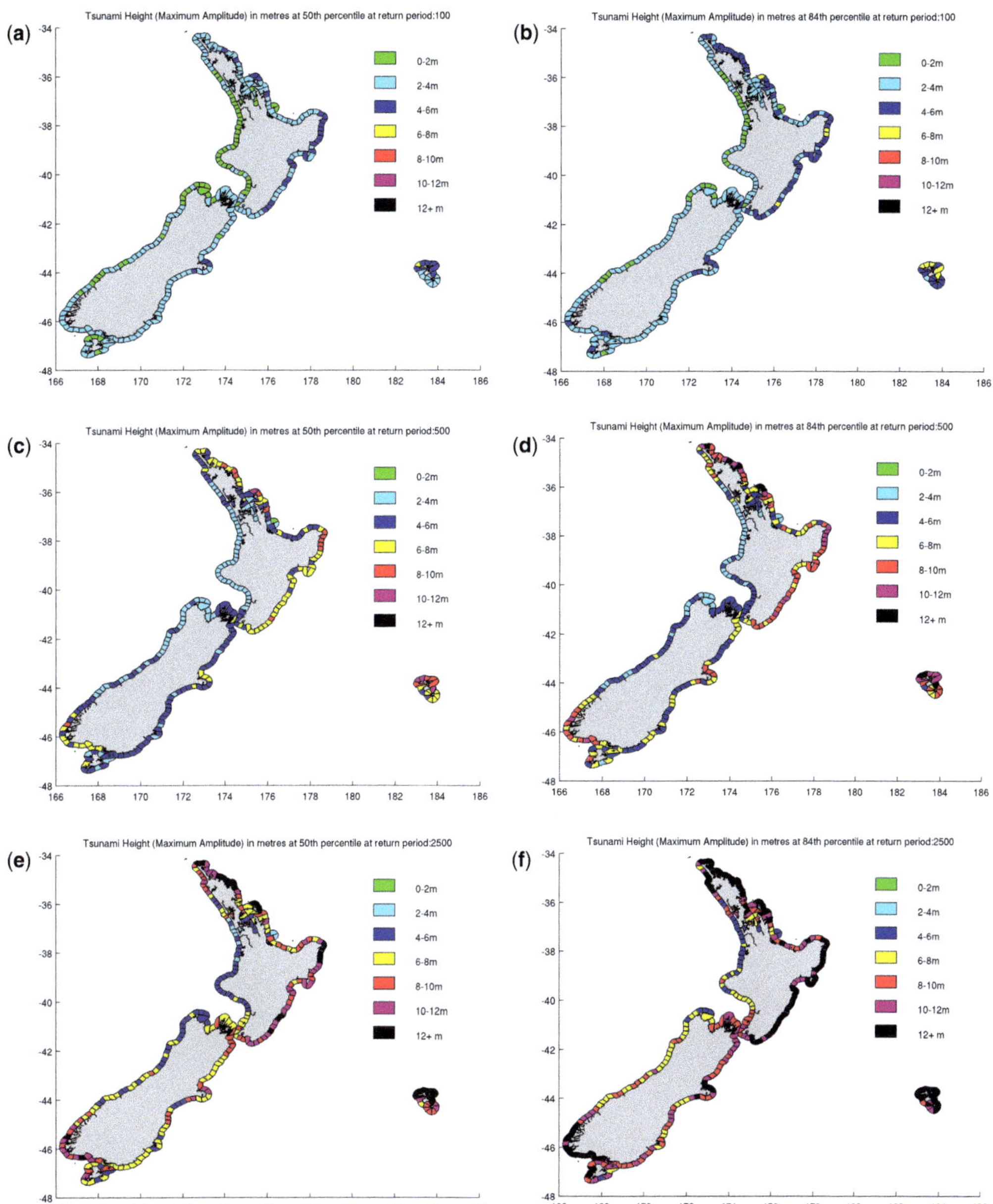

Fig. 10. Expected maximum tsunami height in metres at: (**a**) 100 year return period 50th percentile; (**b**) 100 year return period 84th percentile; (**c**) 500 year return period 50th percentile; (**d**) 500 year return period 84th percentile; (**e**) 2500 year return period 50th percentile; and (**f**) 2500 year return period 84th percentile of epistemic uncertainty.

also shows the effective magnitude of earthquakes on these faults that are necessary to generate a tsunami of the specified height. While these events are estimated to produce tsunamis of the same height at the coast, the extent of inundation is expected to vary between sources according to factors such as the number and period of waves (this is in addition to the variation in extent that naturally arises between areas of different physiography). For probabilistic inundation modelling we anticipate using a weighted average of the set of deaggregated scenarios.

In order to compare the hazard at different sites, the hazard at various return periods can be illustrated in a map view. Examples of these maps for return periods of 100, 500 and 2500 years, and at the 50th and 84th percentile of epistemic uncertainty, are shown in Figure 10.

Discussion and conclusions

The nationwide distribution of tsunami hazard in Figure 10 is generally consistent with expectations, showing a higher hazard in those areas of the coast directly exposed to local subduction zones and an overall trend for the east coast to be exposed to a higher tsunami hazard than the west coast.

While the results presented here represent a first iteration of an all-coast tsunami hazard model for New Zealand, there are many limitations and areas for further improvement which we summarize below.

The probabilistic model does not currently take into account variations in geophysical properties within subduction zones. This is an important issue for the Hikurangi Trench, where the northern portions experience weaker coupling and faster convergence than the southern portions (Wallace *et al.* 2009). Also, the probabilistic model currently does not treat 'tsunami earthquakes' (Kanamori 1972; for a New Zealand example, see Bell *et al.* 2014) on the shallowest parts of subduction interfaces as being distinct from other subduction interface earthquakes.

The method used to model tsunami caused by upper plate and outer rise faults is a simple attenuation rule that introduces uncertainties into the model, and biases in some situations such as by not incorporating the effects of constrictions (such as the Cook Strait that lies between the North and South Islands) on tsunamis that pass through. Improving the modelling of these sources is a key target of future research.

The division of the Pacific Rim into distinct subduction zones (Table 1) is in some cases based on well-defined geophysical changes, but in some locations the boundaries between subduction zones are more artificial. In some regions, subduction earthquakes may have ruptures that span more than one zone, a situation not represented in the current model.

The addition of landslide sources into the probabilistic model is an area requiring further research. Very large submarine landslides have been found around the New Zealand coast (e.g. Collot *et al.* 2001), but these have ages of multiple hundreds of thousands of years. Magnitude–frequency statistics are required for smaller landslides in order to include their contribution on the return times considered here. Work towards this goal is currently being carried out in the Cook Strait region (Lane *et al.* 2016; Mueller *et al.* 2016; Wang *et al.* 2016).

Another area of potential improvement is by reducing the length of the coastal sections used, from 20 km currently, as tsunami impacts may vary considerably even on this scale. It would be particularly helpful to be able to scale the hazard analysis to define separate coastal sections for the interior of the Waitemata (Auckland) and Port Nicholson (Wellington) harbours in which there are large populations and important infrastructure. This would require refining the associated numerical modelling grids in order to more accurately represent the harbour entrances.

Next steps for future work will include the use of the deaggregation charts to provide source scenarios for inundation modelling. Using a weighted average of the inundation properties (such as flow depths), with weights according to the size of contribution to the deaggregation, we intend to estimate the inundations corresponding to particular return periods. With this information, it should be possible to produce estimates of losses by combining estimates of flow depth with fragility curves, and estimates of exposed populations and buildings. This can then, in turn, be used to evaluate the reduction in risk that comes from mitigation options such as land-use planning and the engineering of tsunami-resistant structures.

The authors gratefully acknowledge the contributions of Kelvin Berryman, Philip Barnes, Nicola Litchfield, Andy Nicol, Martin Reyners, Aggeliki Barberopoulou, Stuart Fraser and Joshu Mountjoy to the selection and definition of tsunami sources. They thank Ursula Cochran for discussion of historical events and palaeotsunami, Biljana Lukovic for assistance with figure and data preparation, and Stefan Reese for discussion of tsunami risks. This work was supported by the New Zealand Ministry of Civil Defence and Emergency Management, and the New Zealand Natural Hazards Research Platform.

References

ABE, K. 1979. Size of great earthquakes of 1837–1974 inferred from tsunami data. *Journal of Geophysical Research*, **84**, 1561–1568.

ABE, K. 1995. Estimate of tsunami run-up heights from earthquake magnitudes. *In*: TSUCHIYA, Y. & SHUTO, N. (eds) *Tsunami: Progress in Prediction, Disaster Prevention and Warning*. Kluwer Academic, Dordrecht, The Netherlands, 21–35.

ABRAHAMSON, N.A. & BOMMER, J.J. 2005. Probability and uncertainty in seismic hazard analysis. *Earthquake Spectra*, **21**, 603–607.

ANNAKA, T., SATAKE, K., SAKAKIYAMA, T., YANAGISAWA, K. & SHUTO, N. 2007. Logic-tree approach for probabilistic tsunami hazard analysis and it applications to the Japanese coasts. *Pure and Applied Geophysics*, **164**, 577–592.

Bayrak, Y., Yilmaztürk, A. & Öztürk, S. 2002. Lateral variations of the modal (a/b) values for the different regions of the world. *Journal of Geodynamics*, **34**, 653–666.

Beavan, J., Wang, X. *et al.* 2010. Near-simultaneous great earthquakes at Tongan megathrust and outer rise in September 2009. *Nature*, **466**, 959–963.

Bell, R., Holden, C., Power, W., Wang, X. & Downes, G. 2014. Hikurangi margin tsunami earthquake generated by slow seismic rupture over a subducted seamount. *Earth and Planetary Science Letters*, **397**, 1–9.

Berryman, K. (compiler). 2005. *Review of Tsunami Hazard and Risk in New Zealand*. Client Report **2005/104**. Institute of Geological & Nuclear Sciences, Wellington, New Zealand.

Berryman, K., Ota, Y. *et al.* 2011. Holocene paleoseismic history of upper-plate faults in the southern Hikurangi subduction margin, New Zealand, deduced from marine terrace records. *Bulletin of the Seismological Society of America*, **101**, 2064–2087.

Bründl, M., Romang, H.E., Bischof, N. & Rheinberger, C.M. 2009. The risk concept and its application in natural hazard risk management in Switzerland. *Natural Hazards and Earth System Science*, **9**, 801–813.

Burbidge, D., Cummins, P.R., Mleczko, R. & Thio, H.K. 2008. A probabilistic tsunami hazard assessment for Western Australia. *Pure and Applied Geophysics*, **165**, 2059–2088.

Clark, K.J., Hayward, B.W., Cochran, U.A., Wallace, L.M., Power, W.L. & Sabaa, A.T. 2015. Evidence for past subduction earthquakes at a plate boundary with widespread upper plate faulting: southern Hikurangi margin, New Zealand. *Bulletin of the Seismological Society of America*, **105**, https://doi.org/10.1785/0120140291

Collot, J.Y., Lewis, K., Lamarche, G. & Lallemand, S. 2001. The giant Ruatoria debris avalanche on the northern Hikurangi margin, New Zealand: Result of oblique seamount subduction. *Journal of Geophysical Research: Solid Earth*, **106**, 19271–19297.

Cornell, C.A. 1968. Engineering seismic risk analysis. *Bulletin of the Seismological Society of America*, **58**, 1583–1606.

Crowley, H., Pinho, R., Pagani, M. & Keller, N. 2013. Assessing global earthquake risks: the Global Earthquake Model (GEM) initiative. *In*: Tesfamariam, S. & Goda, K. (eds) *Handbook of Seismic Risk Analysis and Management of Civil Infrastructure Systems*. Woodhead Publishing, Cambridge, 815–838.

Downes, G.L. & Stirling, M.W. 2001. Groundwork for development of a probabilistic tsunami hazard model for New Zealand. *In*: *Proceedings of International Tsunami Symposium, Session 1*, National Oceanic and Atmospheric Administration, Seattle, USA, 293–301.

Eisner, R.K. 2005. Planning for tsunami: reducing future losses through mitigation. *In*: Bernard, E.N. (ed.) *Developing Tsunami-Resilient Communities*. Springer, Dordrecht, The Netherlands, 155–162.

Geist, E.L. 2002. Complex earthquake rupture and local tsunamis. *Journal of Geophysical Research*, **107**, 2086, https://doi.org/10.1029/2000JB000139

Geist, E.L. & Parsons, T. 2006. Probabilistic analysis of tsunami hazards. *Natural Hazards*, **37**, 277–314.

Geist, E.L. & Parsons, T. 2014. Undersampling power-law size distributions: effect on the assessment of extreme natural hazards. *Natural hazards*, **72**, 565–595.

Goff, J.R. 2008. *The New Zealand Palaeotsunami Database*. NIWA Technical Report No. 131. National Institute of Water & Atmospheric Research, Auckland, New Zealand.

Goff, J.R. & Chagué-Goff, C. 2015. Three large tsunamis on the non-subduction, western side of New Zealand over the past 700 years. *Marine Geology*, **363**, 243–260, https://doi.org/10.1016/j.margeo.2015.03.002

Goff, J.R., Lane, E. & Arnold, J. 2009. The tsunami geomorphology of coastal dunes. *Natural Hazards and Earth System Sciences*, **9**, 847–854.

Goff, J., Nichol, S., Chagué-Goff, C., Horrocks, M., McFadgen, B. & Cisternas, M. 2010*a*. Predecessor to New Zealand's largest historic trans-South Pacific tsunami of 1868 AD. *Marine Geology*, **275**, 155–165.

Goff, J., Pearce, S., Nichol, S.L., Chagué-Goff, C., Horrocks, M. & Strotz, L. 2010*b*. Multi-proxy records of regionally-sourced tsunamis, New Zealand. *Geomorphology*, **118**, 369–382.

Goff, J.R., Nichol, S.L. & Kennedy, D. 2010*c*. Development of a palaeotsunami database for New Zealand. *Natural Hazards*, **54**, 193–208, https://doi.org/10.1007/s11069-009-9461-5

González, F.I., Geist, E.L. *et al.* 2009. Probabilistic tsunami hazard assessment at Seaside, Oregon, for near-and far-field seismic sources. *Journal of Geophysical Research: Oceans*, **114**, C11023.

Grapes, R. & Downes, G. 1997. The 1855 Wairarapa, New Zealand, earthquake: analysis of historical data. *Bulletin of the New Zealand National Society for Earthquake Engineering*, **30**, 271–368.

Harbitz, C.B., Lovholt, F., Pedersen, G. & Masson, D.G. 2006. Mechanisms of tsunami generation by submarine landslides: a short review. *Norsk Geologisk Tidsskrift*, **86**, 255.

Hayes, G.P. & Furlong, K.P. 2010. Quantifying potential tsunami hazard in the Puysegur subduction zone, south of New Zealand. *Geophysical Journal International*, **183**, 1512–1524.

Kanamori, H. 1972. Mechanism of tsunami earthquakes. *Physics of the Earth and Planetary Interiors*, **6**, 346–359.

Kennedy, A.B., Chen, Q., Kirby, J.T. & Dalrymple, R.A. 2000. Boussinesq modeling of wave transformation, breaking, and runup. part i: 1d. *Journal of Waterway, Port, Coastal and Ocean Engineering*, **126**, 39–47.

Kron, W. 2002. Keynote lecture: Flood risk = hazard × exposure × vulnerability. *In*: Wu, B., Wang, Z.-Y., Wang, G., Huang, G.G.H., Fang, H. & Huang, J. (eds) *Flood Defence 2002*. Science Press, New York, 82–97.

Lane, E.M., Gillibrand, P.A., Wang, X. & Power, W. 2013. A probabilistic tsunami hazard study of the Auckland region, Part II: inundation modelling and hazard assessment. *Pure and Applied Geophysics*, **170**, 1635–1646.

Lane, E.M., Mountjoy, J.J., Power, W.L. & Popinet, S. 2016. Initialising landslide-generated tsunamis for probabilistic tsunami hazard assessment in Cook Strait. *The International Journal of Ocean and Climate Systems*, **7**, 4–13.

Lay, T., Ammon, C.J., Kanamori, H., Rivera, L., Koper, K.D. & Hutko, A.R. 2010. The 2009 Samoa-Tonga great earthquake triggered doublet. *Nature*, **466**, 964–968.

Lynett, P. 2002. *A multi-layer approach to modelling generation, propagation, and interaction of water waves*. PhD thesis, Cornell University, Ithaca, NY.

McAdoo, B.G., Fritz, H. et al. 2008. Solomon Islands tsunami, one year later. *Eos, Transactions of the American Geophysical Union*, **89**, 169–170.

McCaffrey, R. 2007. The next great earthquake. *Science*, **315**, 1675.

McGuire, R.K. 1995. Probabilistic seismic hazard analysis and design earthquakes: closing the loop. *Bulletin of the Seismological Society of America*, **85**, 1275–1284.

McGuire, R.K. 2004. *Seismic Hazard and Risk Analysis*. Earthquake Engineering Research Institute, Oakland, CA.

Meckel, T.A., Coffin, M.F., Mosher, S., Symonds, P., Bernardel, G. & Mann, P. 2003. Underthrusting at the Hjort Trench, Australian–Pacific plate boundary: incipient subduction? *Geochemistry, Geophysics, Geosystems*, **4**, 1099.

Mueller, C., Power, W., Fraser, S. & Wang, X. 2012. Towards a robust framework for Probabilistic Tsunami Hazard Assessment (PTHA) for local and regional tsunami in New Zealand. *In*: Pittari, A. & Hansen, R.J. (eds) *Abstracts, Geosciences 2012 Conference, Hamilton, New Zealand*. Geoscience Society of New Zealand, Miscellaneous Publication 134A, 64.

Mueller, C., Power, W.L., Fraser, S.A. & Wang, X. 2015. Effects of rupture complexity on local tsunami inundation: implications for probabilistic tsunami hazard assessment by example. *Journal of Geophysical Research. Solid Earth*, **120**, 488–502, https://doi.org/10.1002/2014JB011301

Mueller, C., Mountjoy, J., Power, W., Lane, E. & Wang, X. 2016. Towards a spatial probabilistic submarine landslide hazard model for submarine canyons. *In*: Lamarche, G., Mountjoy, J. et al. (eds) *Submarine Mass Movements and Their Consequences*. Advances in Natural and Technological Hazards Research, **41**. Springer International Publishing, Cham, Switzerland, 589–597.

Musson, R.M.W. 1999. Determination of design earthquakes in seismic hazard analysis through Monte Carlo simulation. *Journal of Earthquake Engineering*, **3**, 463–474.

Nichol, S.L., Chagué-Goff, C., Goff, J.R., Horrocks, M., McFadgen, B.G. & Strotz, L.C. 2010. Geomorphology and accommodation space as limiting factors on tsunami deposition: Chatham Island, southwest Pacific Ocean. *Sedimentary Geology*, **229**, 41–52.

NZTD 2015. *New Zealand Tsunami Database: Historical and Modern Records*. GNS Science Database. GNS Science, Lower Hutt, New Zealand, http://data.gns.cri.nz/tsunami/index.html [last accessed December 2015].

Power, W. (compiler). 2013*a*. *Review of Tsunami Hazard in New Zealand (2013 Update)*. GNS Science Consultancy Report **2013**. GNS Science, Lower Hutt, New Zealand, http://www.civildefence.govt.nz/cdem-sector/ cdem-research-/mcdem-research-projects-and-resources/review-of-tsunami-hazard-in-new-zealand/

Power, W.L. 2013*b*. *Tsunami hazard curves and deaggregation plots for 20 km coastal sections, derived from the 2013 National Tsunami Hazard Model*. GNS Science Consultancy Report **2013/59**. GNS Science, Lower Hutt, New Zealand, http://www.gns.cri.nz/static/WillPower/SR_2013-059.pdf

Power, W. & Gale, N. 2011. Tsunami forecasting and monitoring in New Zealand. *Pure and Applied Geophysics*, **168**, 1125–1136.

Power, W., Downes, G. & Stirling, M. 2007. Estimation of tsunami hazard in New Zealand due to South American earthquakes. *In*: Satake, K., Okal, E.A. & Borrero, E.A. (eds) *Tsunami and Its Hazards in the Indian and Pacific Oceans*. Birkhäuser, Basel, Switzerland, 547–564.

Power, W., Wallace, L., Wang, X. & Reyners, M. 2012. Tsunami hazard posed to New Zealand by the Kermadec and southern New Hebrides subduction margins: an assessment based on plate boundary kinematics, interseismic coupling, and historical seismicity. *Pure and Applied Geophysics*, **169**, 1–36.

Power, W., Wang, X., Lane, E. & Gillibrand, P. 2013. A probabilistic tsunami hazard study of the Auckland region, part I: propagation modelling and tsunami hazard assessment at the shoreline. *Pure and Applied Geophysics*, **170**, 1621–1634.

Power, W.L., Gale, N.H., Lukovic, B., Gledhill, K.R., Clitheroe, G., Berryman, K.R. & Prasetya, G. 2010. *Use of Numerical Models to Inform Distant-Source Tsunami Warnings*. GNS Science Report 2010/11. GNS Science, Lower Hutt, New Zealand.

Rikitake, T. & Aida, I. 1988. Tsunami hazard probability in Japan. *Bulletin of the Seismological Society of America*, **78**, 1268–1278.

Sørensen, M.B., Spada, M., Babeyko, A., Wiemer, S. & Grünthal, G. 2012. Probabilistic tsunami hazard in the Mediterranean Sea. *Journal of Geophysical Research: Solid Earth*, **117**, B01305.

Stirling, M., McVerry, G. et al. 2012. National seismic hazard model for New Zealand: 2010 update. *Bulletin of the Seismological Society of America*, **102**, 1514–1542.

Tappin, D.R., Matsumoto, T. et al. 1999. Sediment slump likely caused 1998 Papua New Guinea tsunami. *Eos, Transactions of the American Geophysical Union*, **80**, 329–340.

Wallace, L.M., Reyners, M. et al. 2009. Characterizing the seismogenic zone of a major plate boundary subduction thrust: Hikurangi Margin, New Zealand. *Geochemistry, Geophysics, Geosystems*, **10**, Q10006.

Wallace, L.M., Barnes, P. et al. 2012. The kinematics of a transition from subduction to strike-slip: an example from the central New Zealand plate boundary. *Journal of Geophysical Research: Solid Earth*, **117**, B02405.

WANG, X. & LIU, P.L.-F. 2006. An analysis of 2004 Sumatra earthquake fault plane mechanisms and Indian Ocean tsunami. *Journal of Hydraulic Research*, **44**, 147–154.

WANG, X. & LIU, P.L.-F. 2007. Numerical simulations of the 2004 Indian Ocean tsunamis – coastal effects. *Journal of Earthquake and Tsunami*, **1**, 273–297.

WANG, X. & LIU, P.L.-F. 2011. An explicit finite difference model for simulating weakly nonlinear and weakly dispersive waves over slowly varying water depth. *Coastal Engineering*, **58**, 173–183.

WANG, X. & POWER, W.L. 2011. *COMCOT: A Tsunami Generation Propagation and Run-up Model.* GNS Science Report **2011/43**. GNS Science, Lower Hutt, New Zealand.

WANG, X., MOUNTJOY, J., POWER, W.L., LANE, E.M. & MUELLER, C. 2016. Coupled modelling of the failure and tsunami of a submarine debris avalanche offshore central New Zealand. *In*: LAMARCHE, G., MOUNTJOY, J. ET AL. (eds) *Submarine Mass Movements and Their Consequences*. Advances in Natural and Technological Hazards Research, **41**. Springer International Publishing, Cham, Switzerland, 599–606.

WARD, S.N. 2001. Landslide tsunami. *Journal of Geophysical Research: Solid Earth*, **106**, 11 201–11 215.

A global probabilistic tsunami hazard assessment from earthquake sources

GARETH DAVIES*[1], JONATHAN GRIFFIN[1], FINN LØVHOLT[2], SYLFEST GLIMSDAL[2], CARL HARBITZ[2], HONG KIE THIO[3], STEFANO LORITO[4], ROBERTO BASILI[4], JACOPO SELVA[4], ERIC GEIST[5] & MARIA ANA BAPTISTA[6,7]

[1]*Community Safety Branch, Geoscience Australia*

[2]*Norwegian Geotechnical Institute*

[3]*AECOM*

[4]*Istituto Nazionale di Geofisica e Vulcanologia*

[5]*Pacific Coast and Marine Science Centre, United States Geological Survey*

[6]*Instituto Superior de Engenharia de Lisboa, Instituto Politécnico de Lisboa*

[7]*Instituto Português do Mar e da Atmosfera*

**Correspondence: gareth.davies@ga.gov.au*

Abstract: Large tsunamis occur infrequently but have the capacity to cause enormous numbers of casualties, damage to the built environment and critical infrastructure, and economic losses. A sound understanding of tsunami hazard is required to underpin management of these risks, and while tsunami hazard assessments are typically conducted at regional or local scales, globally consistent assessments are required to support international disaster risk reduction efforts, and can serve as a reference for local and regional studies. This study presents a global-scale probabilistic tsunami hazard assessment (PTHA), extending previous global-scale assessments based largely on scenario analysis. Only earthquake sources are considered, as they represent about 80% of the recorded damaging tsunami events. Globally extensive estimates of tsunami run-up height are derived at various exceedance rates, and the associated uncertainties are quantified. Epistemic uncertainties in the exceedance rates of large earthquakes often lead to large uncertainties in tsunami run-up. Deviations between modelled tsunami run-up and event observations are quantified, and found to be larger than suggested in previous studies. Accounting for these deviations in PTHA is important, as it leads to a pronounced increase in predicted tsunami run-up for a given exceedance rate.

Large tsunami disasters are low-frequency events and as such may have no regional historical precedent. At any particular coastal location, relatively small, non-destructive tsunamis tend to greatly outnumber large, destructive events, while inundation and damage tend to grow rapidly with increasing tsunami flow depth (Geist & Parsons 2006; Gonzalez *et al.* 2009; Valencia *et al.* 2011). Therefore, rare tsunami events are by far the greatest contributor to tsunami risk: globally, the historical record suggests that approximately 3% of tsunamis have caused approximately 97% of tsunami fatalities, with approximately 80% of these events generated by earthquakes (Løvholt *et al.* 2014*a*; NGDC 2015). Given the absence of well-known historical precedents and our incomplete understanding of the frequency of large subduction zone earthquakes, the two most destructive tsunami in recent decades (the 2004 Andaman–Sumatra tsunami and the 2011 Tohoku tsunami) came as a surprise to many scientists, and comparable scenarios were not considered in disaster management planning (Satake & Atwater 2007; McCaffrey 2008, 2009; Synolakis 2011; Kagan & Jackson 2013; Satake 2014; Løvholt *et al.* 2014*b*; Lorito *et al.* 2015; Synolakis & Kanoglu 2015).

Management of future tsunami risk requires a sound understanding of the hazard and, in general, the historical record is too limited for a purely empirical hazard assessment (Geist & Parsons 2006; Kagan & Jackson 2013; Løvholt *et al.* 2014*a*). Instead, tsunami hazard must be assessed using a mixture of modelling and data. Two main approaches exist: the scenario approach; and

From: Scourse, E. M., Chapman, N. A., Tappin, D. R. & Wallis, S. R. (eds) 2018. *Tsunamis: Geology, Hazards and Risks*. Geological Society, London, Special Publications, **456**, 219–244.
First published online February 23, 2017, updated March 2, 2017, https://doi.org/10.1144/SP456.5

probabilistic tsunami hazard assessment (PTHA). In the scenario approach, a number of credible (and often presumed worst-case) tsunami scenarios are modelled to produce maps depicting a tsunami hazard metric, such as the wave run-up, inundation footprint or offshore wave height (Burbidge & Cummins 2007; Lorito *et al.* 2008; Tiberti *et al.* 2009; Harbitz *et al.* 2012; Løvholt *et al.* 2012*a*, *b*). This approach is advantageous because it is relatively simple and easily communicated. However, the results can be highly sensitive to the particular scenarios selected, and become undesirably subjective when the credible scenario parameters are highly uncertain, which is common, for example, for the moment magnitude of low return period earthquakes (Berryman *et al.* 2015). In comparison, PTHA is relatively complex and computationally intensive, as it involves the integration of tsunami hazard from a model of all possible events to determine the exceedance rate of some tsunami hazard metric (mean number of events per year above a threshold level) (Geist & Parsons 2006; Parsons & Geist 2009). Like the scenario approach, PTHA is dependent on subjective decisions. For example, it may be necessary to specify the maximum earthquake magnitude possible on a given source zone, and doing this incorrectly could result in strong over- or underestimations of the hazard (Annaka *et al.* 2007; Burbidge *et al.* 2008; Satake 2014). However, unlike the scenario approach, there are established methodologies for accounting for known uncertainties associated with these choices in PTHA (Annaka *et al.* 2007; Burbidge *et al.* 2008; Parsons & Geist 2009; Sørensen *et al.* 2012; Selva *et al.* 2016).

With few exceptions (Løvholt *et al.* 2012*a*, 2014*a*), most tsunami hazard assessments to date have been conducted at the regional, national or city level (Annaka *et al.* 2007; Burbidge *et al.* 2008; Gonzalez *et al.* 2009; Parsons & Geist 2009; Sørensen *et al.* 2012; Power 2013; Horspool *et al.* 2014; ten Brink *et al.* 2014). Analysis at these scales is better suited to supporting national-scale disaster risk management, given the typical limitations on computational and human resources for a single study, because a reduced spatial scale allows for: (1) greater detail in the modelling of tsunami sources and propagation; (2) linking local inundation maps to an exceedance rate; (3) a more in-depth assessment of predictions using historical data and palaeotsunami evidence where available; and (4) a more in-depth assessment of the associated uncertainties. However, spatial coverage limitations and methodological differences make it difficult to assess tsunami hazard globally by combining regional-scale hazard assessments, while the global impacts of tsunamis can extend far beyond the inundation zone via their economic and political consequences (UN-ISDR 2015). At the global level, the Hyogo Framework for Action was established by the United Nations to reduce disaster losses, and this has been followed by the Sendai Framework for Disaster Risk Reduction. The implementation goals of these agreements rely on a sound understanding of global disaster risk due to different natural hazards, and have motivated several previous global-scale tsunami hazard and risk assessments (Løvholt *et al.* 2012*a*, 2014*a*, *b*). Given the computational challenges of modelling tsunamis globally, the former studies were based largely on scenario approaches for tsunami hazard assessment, although a partial probabilistic treatment was employed by Løvholt *et al.* (2014*a*) in the Indian Ocean and south Pacific.

The current study extends these works with a full global probabilistic tsunami hazard assessment. Compared with previous work (Løvholt *et al.* 2012*a*, 2014*a*), the PTHA approach allows modelling events both smaller and larger than the scenarios treated previously (nominal $1/500$ year exceedance rate events), and permits a more complete assessment of the associated uncertainties. The latter turn out to be large, driven in part by considerable epistemic uncertainties in the exceedance rates of high-magnitude tsunamigenic earthquakes (Berryman *et al.* 2015), and limitations in our capacity to model tsunami run-up at the global scale. Combined with appropriate exposure information, the results of the current study could also be used to support broad-scale assessments of tsunami risks (Løvholt *et al.* 2014*a*, 2015, 2016) or multi-hazard risk assessments (UN-ISDR 2015), although that step is not undertaken herein.

Given its global scale, the tsunami hazard results herein are expected to be broad-brush, and the quantified uncertainties are rather large. However, in the longer term, they could be extended with detailed analyses such as currently possible for assessments over smaller spatial scales (Gonzalez *et al.* 2009; Lorito *et al.* 2016). This would require considerable computational and human resources, necessitating a coordinated global effort which is beyond the scope of the present work.

Methodology

Overview of the applied PTHA methodology

We compute the exceedance rate (events/year) of tsunami run-up height at a globally distributed set of hazard points offshore of the coast (*c.* 100 m depth), using methods that broadly follow previous approaches (Geist & Parsons 2006; Annaka *et al.* 2007; Thio *et al.* 2007; Burbidge *et al.* 2008; Horspool *et al.* 2014). We further estimate the extra amplification that these waves undergo as they propagate onshore, following the amplification-factor approach of Løvholt *et al.* (2012*a*). The key steps are:

- Define tsunami sources to be included in the analysis. In this study only oceanic earthquakes are considered.
- Discretize each earthquake source zone into a grid of 'unit-sources'.
- For each source zone create a synthetic earthquake catalogue, with all synthetic earthquakes consisting of uniform-slip combinations of the unit-sources. Assign a mean annual rate to each synthetic event using a Gutenberg–Richter-type model for the source zone, with uncertainties in parameter values accounted for using a logic tree.
- For each unit-source, compute the sea-surface deformation produced by 1 m of slip.
- Use the linear shallow-water equations to model the wave time series produced by the 1 m unit-source ruptures at a set of nearshore coastal points (hazard points).
- Extract the maximum tsunami wave height and period from wave time series at all hazard points for each earthquake in the catalogue. Because the sea-surface deformation and tsunami propagation models are linear, the hazard point time series for a single earthquake is efficiently computed by summing the tsunami wave time series for all included unit-sources and multiplying by the earthquake slip.
- Compute the tsunami run-up from the maximum tsunami wave height and period at the offshore hazard points.
- Combine the modelled run-up and earthquake rates for all events to compute the exceedance rate of tsunami run-up near every hazard point.

Details of the above steps are presented below.

Source zone definition

As the majority of large tsunamigenic earthquakes occur on subduction zones, most earthquake sources used in this study are of this type (Fig. 1). However, we included several non-subduction zone sources in the Mediterranean, eastern Indonesia and the NE Atlantic where previous assessments indicated they are regionally significant drivers of the hazard (Sørensen *et al.* 2012; Horspool *et al.* 2014; Omira *et al.* 2015). In particular, in the NE Atlantic we include the strike-slip Gloria source, and in the Gulf of Cadiz we use a schematic source to represent the large number of thrust sources which are too small to resolve in our model (Matias *et al.* 2013; Omira *et al.* 2015).

Three-dimensional source zone fault-plane geometries are defined using the Slab 1.0 subduction interface model where available (Hayes *et al.* 2012), geometries from the SHARE project in the Mediterranean (Basili *et al.* 2013*a*, *b*), and constant dip geometries elsewhere based on Bird (2003), Burbidge *et al.* (2008), Horspool *et al.* (2014) and Berryman *et al.* (2015). The maximum seismogenic source depth is taken from the upper maximum depth suggested by Berryman *et al.* (2015). The seismogenic zone is assumed to extend to the trench (i.e. minimum depth of zero), except in the Hellenic Arc where the trench is significantly further south than the expected updip limit of the active subduction interface. Rupture to the trench is preferable for simulating tsunamis since if the minimum depth is non-zero and on the order of a few kilometres, the vertical coseismic displacement field predicted by the Okada (1985) model contains an artificial spike near the shallow rupture edge which can have a strong impact on the associated tsunami (Geist & Dmowska 1999; Goda 2015).

Source zone discretization with unit-sources

Each source is discretized in a logically rectangular grid of unit-sources with dimensions approximately 100 km along-strike and 50 km downdip (Fig. 2). The use of a logically rectangular layout (with a fixed number of unit sources along-strike and a fixed number downdip) simplifies the construction of synthetic earthquake events (see the subsection on 'Synthetic earthquake event geometries' later in this section), although it means that the unit source length and width must vary to match the typically non-uniform source zone geometry (Fig. 2). Within each unit-source, a finer grid of subunit-source points records the depth and dip. These are used to retain detailed information on the non-planar source geometry when computing the unit-source seafloor deformation. The subunit-source point spacing was set to 6 km, except for unit-sources along the trench where a finer spacing (1 km) was required to ensure accurate numerical integration of the seafloor deformation. Rupture is treated as pure thrust, except on the strike-slip Gloria source zone.

Sea-surface deformation

For every unit-source we compute the seafloor deformation from 1 m of earthquake slip by summing the deformation computed at each subunit-source point, with the individual deformation computations following Okada (1985) (Fig. 2), with both Lamé constants equal to 3×10^{10} Pa. We assume that each subunit-source point represents a small rectangular region with dip and depth determined by the source depth contours, and length and width determined by the subunit-source grid spacing. Near the unit-source boundaries, these

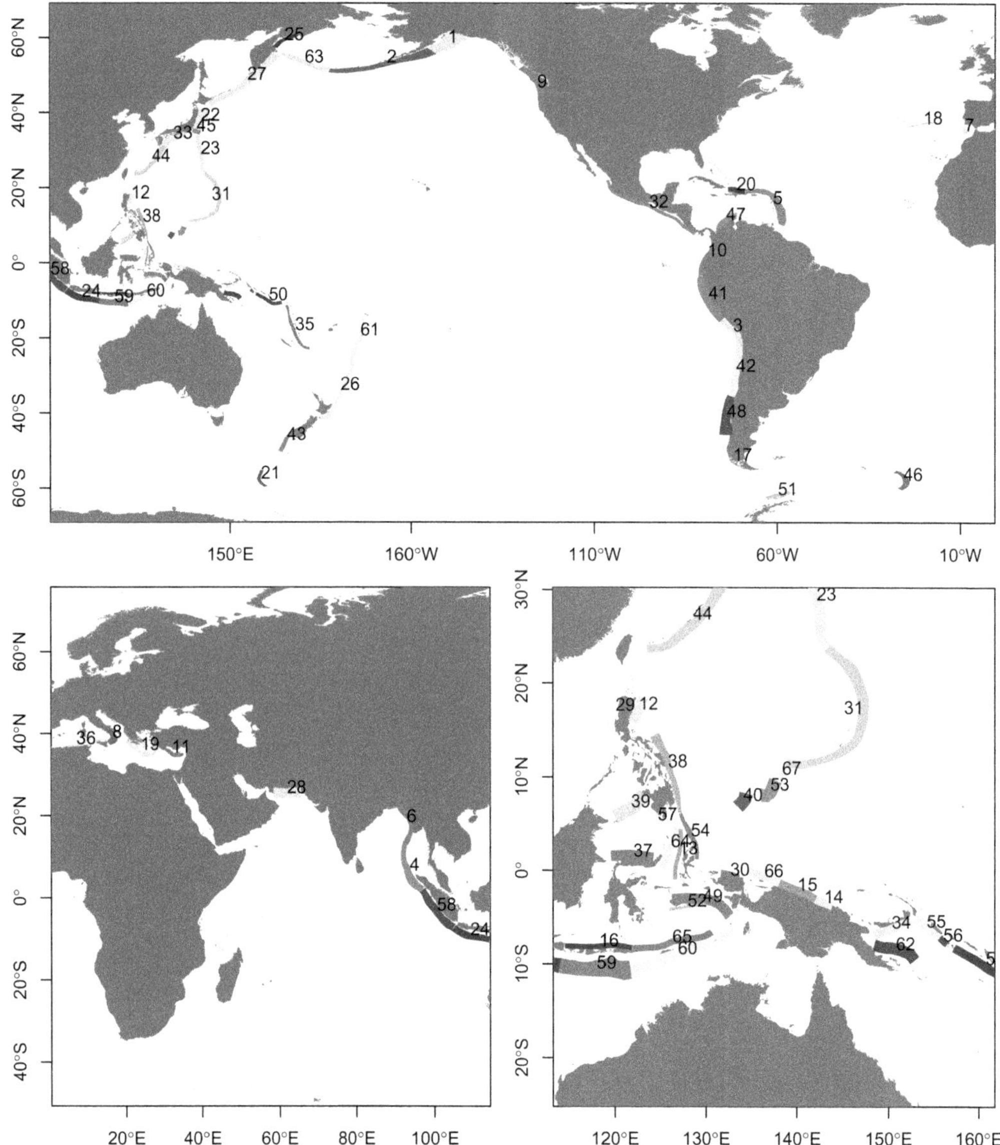

Fig. 1. Source zones used in the study: 1, Alaska; 2, Aleutians; 3, Altiplano; 4, Andaman; 5, Antilles; 6, Arakan; 7, Cadiz; 8, Calabrian; 9, Cascadia; 10, Columbia; 11, Cyprus; 12, Eastluzontrough; 13, Eastmolucca; 14, EastNewGuinea1; 15, EastNewGuinea2; 16, Flores; 17, Fuego; 18, Gloria; 19, Hellenic; 20, Hispaniola; 21, Hjort; 22, Honshu; 23, Izu; 24, Java; 25, Kamchatka; 26, Kermadec; 27, Kurils; 28, Makran; 29, Manila; 30, Manokwari; 31, Mariana; 32, Midamerica; 33, Nankai; 34, Newbritain; 35, Newhebrides; 36, North_african; 37, Northsulawesi; 38, Nphillip; 39, Nwsulu; 40, Palau; 41, Peru; 42, Puna; 43, Puysegur; 44, Ryukyu; 45, Sagami; 46, Sandwich; 47, Scaribbean; 48, Schile; 49, Seram; 50, Sesolomon; 51, Shetland; 52, Southseram; 53, South_yap; 54, Sphillip; 55, SouthSolomon; 56, SouthSolomonwood; 57, Sulu; 58, Sumatra; 59, Sumba; 60, Timor; 61, Tonga; 62, Trobriand; 63, Waleutians; 64, Westmolucca; 65, Wetar; 66, WestNewGuinea; 67, Yap.

rectangular subunit-source regions are reduced in size (if they extend outside the original unit-source boundary) to ensure that the subunit-source area sum equals the total unit-source interface area. The unit-source sea-surface deformation is computed by smoothing the seafloor deformation using a filter based on full potential linear wave theory (Kajiura 1963; Glimsdal *et al.* 2013). Compared

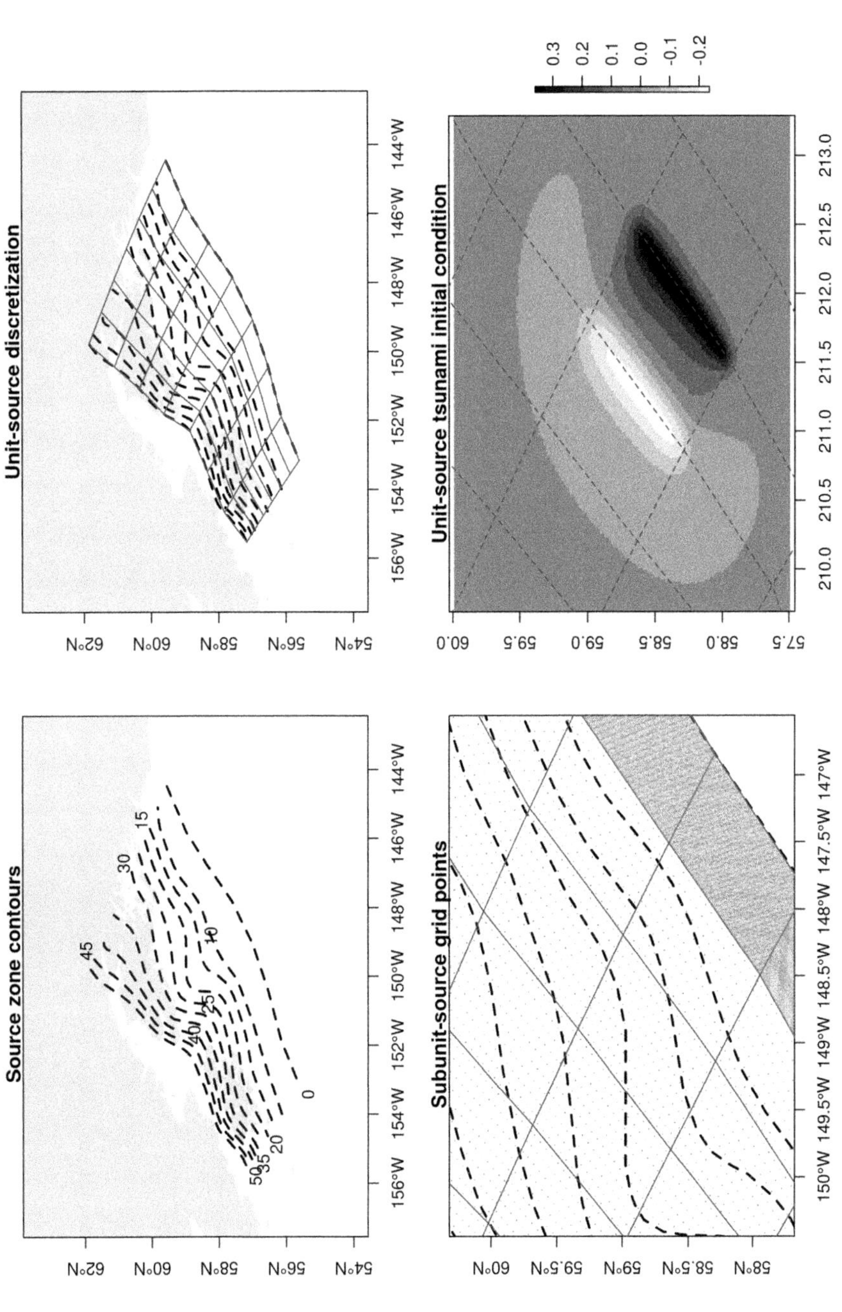

Fig. 2. Unit source definition. (Top left) Subduction interface contours for the Alaska source zone. Labels give the depth in kilometres. (Top right) The logically rectangular unit-source grid, overlain on the depth contours. (Bottom left) Zoom depicting the subunit-source grid points inside each unit-source. A finer spacing is used for unit-sources along the trench, to allow accurate numerical integration when computing the vertical coseismic displacement field. (Bottom right) The tsunami initial condition from one unit-source, with the scale in metres.

with directly applying the seafloor deformation to the sea surface, the filter leads to better agreement with the sea-surface deformation computed using primitive models such as the full potential equations, and Navier–Stokes for the generation and initial propagation (Saito & Furumura 2009; Saito 2013).

Synthetic earthquake catalogue

On each source zone we generate a catalogue of synthetic earthquake events, all of which consist of uniform slip on a logically rectangular subset of the unit-sources (see the following subsection on 'Synthetic earthquake event geometries'). A mean annual rate is assigned to each synthetic earthquake event, such that when integrated over a source zone the exceedance rates (events/year) follow a Gutenberg–Richter-type distribution (see the subsection on 'Synthetic earthquake event rates' later in this section). The latter is forced to satisfy a seismic moment conservation principle (Bird & Kagan 2004). Parameters controlling the earthquake rates are assigned to each source zone using a range of literature-derived values (see 'Source zone parameters' later in this section), with parameter uncertainties associated with the maximum magnitude, Gutenberg–Richter b value and degree of seismic coupling accounted for using a logic tree (see the subsection on 'Logic-tree weights and epistemic uncertainty quantification' later in this section). Details are provided below.

Synthetic earthquake event geometries. The synthetic earthquake events have moment magnitudes m_w in the range:

$$m_\mathrm{w} \in \{m_{\mathrm{w,min}}, (m_{\mathrm{w,min}} + 0.1), (m_{\mathrm{w,min}} + 0.2), \ldots, (m_{\mathrm{w,max}} - 0.1), m_{\mathrm{w,max}}\} \quad (1)$$

where the source zone specific computation of the minimum magnitude $m_{\mathrm{w,min}}$ and the maximum magnitude $m_{\mathrm{w,max}}$ is described later (see the following subsection on 'Source zone parameters'). For every m_w we use the scaling relationships of Strasser *et al.* (2010) to compute the desired area A, length L and width W for the synthetic earthquake events. In general, the synthetic event dimensions cannot exactly conform to A, L and W because their size is constrained by the unit-source dimensions (*c.* $100 \times 50\ \mathrm{km}^2$: Fig. 2). However, as detailed below, for each m_w we select the number of subfaults along-strike (n_s) and downdip (n_d) so that the dimensions are close to the desired values. All possible $n_\mathrm{s} \times n_\mathrm{d}$ logically rectangular subsets of the unit-sources are then included in the synthetic catalogue: that is, the earthquake floats through every possible position on the source zone. For the i-th event with magnitude $m_{\mathrm{w},i}$ and area A_i (m^2), the slip S_i (m) is computed as (Hanks & Kanamori 1979; Bird & Kagan 2004):

$$\mathrm{M}_{0,i} = 10^{1.5m_{\mathrm{w},i}+9.05} \quad (2)$$

$$S_i = \frac{\mathrm{M}_{0,i}}{(\mu A_i)} \quad (3)$$

where the shear modulus μ is taken as 3×10^{10} Pa and $\mathrm{M}_{0,i}$ is the seismic moment (N m). Note that A_i will vary somewhat for events with the same magnitude because of variations in the unit-source dimensions (Fig. 2).

The detailed methods used to select the number of unit-sources along-strike (n_s) and downdip (n_d) for a given magnitude are provided here. From m_w we compute A, L and W (Strasser *et al.* 2010), and check if A exceeds the total source zone area, in which case the synthetic event includes all subfaults. Otherwise define n'_s as L divided by the average unit-source length $\bar{l}_\mathrm{s}$ (*c.* 100 km), without rounding. If n'_s exceeds the available number of subfaults along-strike, then n_s is assumed to cover the full source zone length and n_d is computed to best fit A. Otherwise define n'_d as W divided by the average unit-source width $\bar{l}_\mathrm{d}$ (*c.* 50 km) without rounding. If n'_d exceeds the available number of subfaults downdip, then n_d is assumed to cover the full source zone width, and n_s computed to best fit A. Otherwise, we have the typical case where both n'_s and n'_d can be accommodated within the source zone. Four candidate (n_s, n_d) pairs are then investigated: two pairs come from assigning n_s to n'_s rounded up and down (with a lower bound of 1 and upper bound from the source zone geometry), and setting n_d to best match A; the other two pairs are produced analogously by rounding n'_d and finding n_s to best match A. From the four candidates (n_s, n_d) pairs we choose one which best matches the desired event aspect ratio: that is, which minimizes:

$$\left[\log_{10}\left(\frac{(n_\mathrm{s}\bar{l}_\mathrm{s})}{(n_\mathrm{d}\bar{l}_\mathrm{d})}\right) - \log_{10}\left(\frac{L}{W}\right)\right]^2 \quad (4)$$

This measures the aspect ratio error in relative terms, so that, for example, a (n_s, n_d) pair producing only 50% of the desired aspect ratio is weighted equally as one producing 200% of the desired aspect ratio.

Source zone parameters. To generate mean annual rates for events in the synthetic earthquake catalogue we require source zone specific parameter values for the minimum magnitude $m_{w,min}$, maximum magnitude $m_{w,max}$, the Gutenberg–Richter b value and the seismically coupled plate motion rate $c\dot{s}$ (the latter notation reflects that $c\dot{s}$ is the product of the seismic coupling rate c with the plate motion rate $\dot{s}$). Below, we describe how those parameters were determined, and the approach used to quantify uncertainties. The methods used to derive the associated synthetic earthquake event rates are described in following subsections on 'Synthetic earthquake event rates' and 'Logic-tree weights and epistemic uncertainty quantification'.

The minimum magnitude $m_{w,min}$ is set to 7.5 everywhere, consistent with the area of an earthquake containing one $100 \times 50\ km^2$ unit-source (Strasser *et al.* 2010). This implies that we ignore all tsunamis generated by smaller earthquakes. While smaller earthquake can generate locally significant tsunamis (especially when combined with landslides – e.g. the 1998 Papua New Guinea tsunami: Tappin *et al.* 2008; Heidarzadeh & Satake 2015), a further reduction of $m_{w,min}$ would require using a smaller unit-source size which would increase the already heavy computational demands of the study.

Parameters $m_{w,max}$, b and $c\dot{s}$ are subject to considerable epistemic uncertainty which is accounted for using a logic tree (Annaka *et al.* 2007; Thio *et al.* 2007; Horspool *et al.* 2014). Each parameter is assigned three different values (Table 1). The parameter b controls the relative frequencies of small and large earthquakes, and is assigned the values 0.7, 0.95 and 1.2 on all source zones, similar to source zone specific uncertainty ranges suggested by Berryman *et al.* (2015). This range covers a number of b values suggested from globally integrated analyses (Bird & Kagan 2004; Schorlemmer *et al.* 2005; Kagan 2010). The seismically coupled trench-normal plate motion rate $c\dot{s}$ is computed using three separate seismic coupling values c which are based on Berryman *et al.* (2015). Where our source zones are not included in the latter study, default values of 0.5 ± 0.2 are used. A single average plate motion rate $\dot{s}$ is assigned to each source zone, where the trench-normal component of plate motion is used to reflect that we only model thrust earthquakes (Burbidge *et al.* 2008), except on the strike-slip Gloria source zone (Omira *et al.* 2015). Convergence angles are based on Bird (2003), and absolute plate motion rates are primarily based on Bird (2003) and Berryman *et al.* (2015); although we refer to Cummins (2007) and Socquet *et al.* (2006) for Arakan, and Horspool *et al.* (2014) for South-Seram and Timor. In the Gulf of Cadiz, we use a single schematic source zone but artificially increase $c\dot{s}$ by a factor of 4 to reflect the moment rate on numerous sources in this complex region (Matias *et al.* 2013).

For $m_{w,max}$ an overall upper limit of 9.6 is applied following Berryman *et al.* (2015). Within that constraint, the largest $m_{w,max}$ value is computed with an empirical relationship for predicting m_w from rupture-area at the +1 SD limit (Strasser *et al.* 2010) assuming that the largest possible earthquake ruptures the entire source zone. This gives conservative $m_{w,max}$ values which are usually similar to those derived with the conservative $m_{w,max}$ approach of McCaffrey (2008) and Berryman *et al.* (2015), while still being consistent with our use of the Strasser *et al.* (2010) magnitude–area scaling relationships elsewhere. A second non-conservative estimate of $m_{w,max}$ is employed following Berryman *et al.* (2015), which is based on the largest earthquake thought to have occurred on each source zone. Where our source zones are not included in Berryman *et al.* (2015), we derive values from Burbidge *et al.* (2008), Storchak *et al.* (2012), Matias *et al.* (2013), Horspool *et al.* (2014) and Omira *et al.* (2015). Our third intermediate estimate of $m_{w,max}$ is derived by averaging the conservative and non-conservative estimates, following Berryman *et al.* (2015). For computational consistency with $m_{w,min} = 7.5$, we finally apply a lower limit of 7.6 to the maximum magnitudes and round all $m_{w,max}$ values to one decimal place.

Synthetic earthquake event rates. We assume that earthquake timings on each source zone follow a Poisson distribution: that is, the event rate is stationary and inter-event times are exponentially distributed. The mean annual rate assigned to each event in the synthetic catalogue is derived from source zone specific values for $m_{w,min}$, $m_{w,max}$, b and $c\dot{s}$. Because of uncertainty in these parameters, it is ultimately necessary to integrate over multiple parameter combinations, combining their corresponding rates as explained in the following subsection on 'Logic-tree weights and epistemic uncertainty quantification'. However, this subsection presents the rate computation assuming fixed source zone parameters, which is applied to each parameter combination separately prior to their integration.

Two different parametric forms for the source zone integrated earthquake magnitude–frequency relationship are considered (giving the mean annual number of events exceeding any magnitude on the source zone): first, a Gutenberg–Richter distribution with exceedance rate function truncated between $m_{w,min}$ and $m_{w,max}$:

$$
\begin{aligned}
\mathrm{GR}(m_w) &= 10^{a-bm_w} && \text{for } m_{w,min} \le m_w \le m_{w,max} \\
&= 0 && \text{for } m_w > m_{w,max} \\
&= 10^{a-bm_{w,min}} && \text{for } m_w < m_{w,min}
\end{aligned}
\quad (5)
$$

Table 1. *Values for the maximum moment magnitude* $m_{w,max}$ *and trench-normal coupled slip rate* $c\dot{s}$ *(mm* a^{-1}*) used on all source zones (trench-parallel slip is applied on the strike-slip Gloria source zone)*

Source name	$m^1_{w,max}$	$m^2_{w,max}$	$m^3_{w,max}$	$c\dot{s}^1$	$c\dot{s}^2$	$c\dot{s}^3$
Alaska	9.30	9.30	9.40	40.00	45.00	50.00
Aleutians	9.10	9.30	9.60	19.00	29.00	43.00
Altiplano	8.80	9.10	9.40	40.00	46.00	52.00
Andaman	9.00	9.30	9.60	15.00	18.00	20.00
Antilles	8.00	8.80	9.60	3.30	5.50	7.70
Arakan	7.60	8.40	9.40	4.90	8.10	11.00
Cadiz	8.20	8.40	8.60	4.80	7.90	11.00
Calabrian	7.60	8.10	9.00	0.53	0.88	1.20
Cascadia	8.80	9.20	9.50	26.00	30.00	34.00
Columbia	8.80	9.20	9.50	34.00	39.00	43.00
Cyprus	7.70	8.30	9.00	2.00	3.40	4.80
Eastluzontrough	7.60	8.10	8.80	3.00	5.00	6.90
Eastmollucca	8.50	8.60	8.70	25.00	42.00	59.00
Eenewguinea	7.60	7.90	8.50	39.00	46.00	53.00
Enewguinea	7.60	8.20	8.90	38.00	44.00	50.00
Flores	8.10	8.50	8.90	8.10	14.00	19.00
Fuego	8.00	8.80	9.60	3.80	6.40	8.90
Gloria	8.30	8.60	8.80	1.20	2.00	2.80
Hellenic	8.00	8.60	9.10	2.00	6.00	10.00
Hispaniola	7.60	8.20	9.00	1.40	2.40	3.30
Hjort	7.60	8.00	8.70	2.60	4.30	6.00
Honshu	9.00	9.20	9.40	53.00	62.00	80.00
Izu	7.60	8.20	9.30	4.70	9.40	33.00
Java	7.80	8.60	9.40	5.80	12.00	41.00
Kamchatka	7.60	8.20	8.80	0.71	1.20	1.60
Kermadec	8.20	8.90	9.60	14.00	21.00	44.00
Kurils	9.00	9.30	9.60	56.00	64.00	72.00
Makran	8.10	8.80	9.50	5.40	9.00	13.00
Manila	7.80	8.50	9.20	4.40	13.00	61.00
Manokwari	7.80	8.10	8.50	11.00	14.00	17.00
Mariana	7.60	8.40	9.60	4.90	9.80	34.00
Midamerica	8.20	8.90	9.60	16.00	26.00	36.00
Nankai	8.50	8.80	9.20	38.00	43.00	47.00
Newbritain	8.00	8.40	8.90	55.00	64.00	73.00
Newhebrides	8.30	8.80	9.30	17.00	23.00	45.00
North_african	7.60	8.00	9.10	0.47	0.79	1.10
Northsulawesi	7.90	8.40	9.00	9.30	16.00	22.00
Nphillip	7.60	8.30	9.10	2.80	6.90	21.00
Nwsulu	8.00	8.50	9.00	2.30	3.80	5.40
Palau	7.60	7.80	8.70	0.37	0.73	2.60
Peru	8.80	9.20	9.60	43.00	49.00	55.00
Puna	8.60	9.10	9.60	51.00	59.00	66.00
Puysegur	7.80	8.40	9.00	7.40	10.00	12.00
Ryukyu	8.00	8.70	9.30	8.60	17.00	61.00
Sagami	8.00	8.20	8.50	14.00	16.00	18.00
Sandwich	7.60	8.20	9.00	6.20	12.00	44.00
Scaribbean	7.60	7.80	9.30	2.70	4.50	6.30
Schile	9.50	9.60	9.60	53.00	61.00	68.00
Seram	8.00	8.60	9.20	14.00	23.00	32.00
Sesolomon	8.10	8.70	9.20	29.00	34.00	39.00
Shetland	7.60	8.20	8.80	2.90	4.90	6.90
South_yap	7.60	8.00	8.80	0.48	0.96	3.40
Southseram	7.80	8.10	8.40	0.88	1.50	2.00
Sphillip	7.60	8.10	8.80	2.90	7.20	22.00
Ssolomon	8.00	8.40	8.80	40.00	46.00	53.00
Ssolomonwood	7.70	8.00	8.30	57.00	67.00	76.00
Sulu	8.20	8.40	8.70	2.50	4.20	5.80

(*Continued*)

Table 1. (*Continued*)

Source name	$m^1_{\text{w,max}}$	$m^2_{\text{w,max}}$	$m^3_{\text{w,max}}$	$c\dot{s}^1$	$c\dot{s}^2$	$c\dot{s}^3$
Sumatra	9.00	9.30	9.60	25.00	29.00	32.00
Sumba	8.30	8.80	9.40	6.70	13.00	47.00
Timor	7.60	8.50	9.40	1.70	2.80	3.90
Tonga	8.00	8.70	9.30	16.00	33.00	120.00
Trobriand	7.60	7.80	9.10	3.20	5.40	7.60
Waleutians	8.90	9.10	9.30	5.50	9.20	13.00
Westmollucca	8.50	8.70	8.90	1.60	2.70	3.80
Wetar	7.60	8.30	9.10	7.80	13.00	18.00
Wnewguinea	8.10	8.50	8.80	14.00	17.00	19.00
Yap	7.60	7.70	8.30	0.59	1.20	4.10

Three values are applied for each, and additionally all sources are assigned $b = 0.7$, 0.95 and 1.2, and $m_{\text{w,min}} = 7.5$.

and, secondly, a Gutenberg–Richter distribution with the density function truncated between the same range:

$$\begin{aligned}\mathrm{GR}(m_\mathrm{w}) &= 10^{a-bm_\mathrm{w}} - 10^{a-bm_{\mathrm{w,max}}} \quad \text{for } m_{\mathrm{w,min}} \le m_\mathrm{w} \le m_{\mathrm{w,max}} \\ &= 0 \quad \text{for } m_\mathrm{w} > m_{\mathrm{w,max}} \\ &= 10^{a-bm_{\mathrm{w,min}}} - 10^{a-bm_{\mathrm{w,max}}} \quad \text{for } m_\mathrm{w} < m_{\mathrm{w,min}} \qquad (6)\end{aligned}$$

Here $\mathrm{GR}(m_\mathrm{w})$ is the exceedance rate for magnitude m_w (events/year) and a is a parameter controlling the overall frequency of earthquake events. Equation (5) was termed a characteristic distribution by Kagan (2002*a*, *b*) as it assigns a finite rate to events with magnitude exactly $m_{\mathrm{w,max}}$, although it differs from the characteristic distribution of Youngs & Coppersmith (1985). Equation (6) is often termed a truncated Gutenberg–Richter distribution or a truncated Pareto distribution (Kagan 2002*a*, *b*; Kagan & Jackson 2013), although Youngs & Coppersmith (1985) refer to it as an exponential magnitude distribution. For the same source zone parameters, equation (5) implies a greater frequency of high-magnitude events than equation (6). Both appear as branches in the logic tree.

The parameter a in equations (5) and (6) is adjusted so that the synthetic earthquake catalogue reproduces the seismically coupled fraction of the long-term moment rate (Bird & Kagan 2004):

$$\sum_{i \in E} \mu A_i S_i r_i = \xi \mu A_\mathrm{T} c\dot{s} / \cos(\delta) \qquad (7)$$

where the value of a affects r_i as explained below. In equation (7), i represents the i-th synthetic earthquake event on the source zone among the finite set of all such synthetic events E, μ is the shear modulus (3×10^{10} Pa), A_i (m^2) is the area of the i-th synthetic earthquake event with slip S_i (m) and mean annual rate r_i, A_T is the total interface area of the source zone (m^2), δ is the mean dip of the source zone in radians (for the strike-slip Gloria source zone, the factor $\cos(\delta)$ is ignored in equation 7 since the motion is transverse), and ξ is the fraction of seismic moment released by earthquakes with $m_\mathrm{w} \ge m_{\mathrm{w,min}}$ on the source zone, which will be close to unity unless $m_{\mathrm{w,max}}$ is close to $m_{\mathrm{w,min}}$ (Kagan & Jackson 2013):

$$\xi = \frac{\left(\int_{m_{\mathrm{w,min}}}^{m_{\mathrm{w,max}}} (\mathrm{GR})' M_0(m_\mathrm{w})\, \mathrm{d}m_\mathrm{w}\right)}{\left(\int_{-\infty}^{m_{\mathrm{w,max}}} (\mathrm{GR})' M_0(m_\mathrm{w})\, \mathrm{d}m_\mathrm{w}\right)} \qquad (8)$$

Here $(\mathrm{GR})'$ is the derivative of GR (equations 5 and 6) with respect to m_w (ignoring the lower magnitude bound: i.e. $m_{\mathrm{w,min}} = -\infty$ for the derivative computation only), and $\mathrm{M}_0(m_\mathrm{w})$ is the seismic moment for an earthquake of magnitude m_w (equation 3). The value of ξ is independent of a for both our modified Gutenberg–Richter relations (equations 5 and 6).

To compute r_i when solving for a (equation 7), note that:

$$r_i = \Pr(i | m_{\mathrm{w},i}) r(m_{\mathrm{w},i}) \qquad (9)$$

where $\Pr(i|m_{\mathrm{w},i})$ is the conditional probability that the i-th earthquake event occurs given that an earthquake with the same magnitude has occurred, and $r(m_{\mathrm{w},i})$ is the mean annual rate of earthquakes of size $m_{\mathrm{w},i}$ occurring anywhere on the source zone. We assume that $\Pr(i|m_{\mathrm{w},i})$ is proportional to the area of the i-th event A_i (equivalently, inversely proportional to its slip: equation 3) and to the average long-term slip rate along the trench where the i-th event occurs (denoted $\dot{s}_i$):

$$\Pr(i|m_{\mathrm{w},i}) = \frac{A_i \dot{s}_i}{\left(\sum_{j \in E(m_{\mathrm{w},i})} A_j \dot{s}_j\right)} \qquad (10)$$

where the denominator is just a normalizing constant, and $E(m_{\mathrm{w},i})$ is the set of all events in the synthetic catalogue with magnitude $m_{w,i}$. For each event, $\dot{s}_i$ is computed as the area-weighted-average of the local long-term slip rates on each unit-source which ruptures in the i-th event, with the unit-source long-term slip rates assigned from the nearest spatially varying trench-normal slip rates of Bird (2003) where available, and otherwise set to a constant. The use of equation (10) is further motivated below.

To model the mean annual rate of earthquakes of size m_w in equation (9), we assume:

$$r(m_w) = \mathrm{GR}(m_w - 0.05) - \mathrm{GR}(m_w + 0.05) \quad (11)$$

where the factor 0.05 results from the synthetic earthquake catalogue containing events with m_w in increments of 0.1. Equations (5) and (6) imply that equation (11) will contain a factor of 10^a, which can be used to solve for a using equations (7–11). Once a is known, the mean annual rate of all events in the synthetic catalogue is determined directly with equations (9–11).

The use of equation (10) to assign non-constant conditional probabilities to events with the same magnitude contrasts with previous studies which employ a constant conditional probability (Horspool *et al.* 2014; Løvholt *et al.* 2014*a*; Lorito *et al.* 2016). As shown by the following idealized examples, equation (10) allows improved treatment of source zones with spatially variable width and/or long-term slip rate. First, consider a rectangular source zone where all unit-sources have the same area, but the long-term slip rate $\dot{s}_i$ varies along the trench. Equation (10) implies that an earthquake event on a fast-moving part of the source zone is proportionately more likely than an event with the same magnitude on a slower region, which we consider preferable to assuming they occur at the same rate. Secondly, consider a source zone where $\dot{s}_i$ is constant (i.e. uniform long-term slip rate) but the down-dip width is non-uniform because of variations in the source zone dip (as is typical: e.g. Fig. 2). Unit-sources then have greater area in wider parts than in narrower parts of the source zone, because of their logically rectangular layout. For a given magnitude, earthquake events with a smaller area will have higher slip S_i than those with a greater area (since slip is inversely related to area: equation 3). Therefore, a constant conditional probability would imply greater long-term slip contributions from events in narrower parts of the source zone (where our synthetic earthquakes have lower area), which seems undesirable if $\dot{s}_i$ is actually constant along-strike. In contrast, equation (10) ensures that if $\dot{s}_i$ is constant, all events with the same magnitude induce the same long-term slip rate where they occur (i.e. $r_i S_i$ is constant for fixed $m_{w,i}$).

Logic-tree weights and epistemic uncertainty quantification. Owing to parameter uncertainties (see the earlier subsection on 'Source zone parameters'), it is necessary to compute the synthetic earthquake event rates for all combinations of $m_{w,max}$, b and $c\dot{s}$ (three values each), and for the two different Gutenberg–Richter relationships (equations 5 and 6), leading to $2 \times 3^3 = 54$ different rate equations for each source zone (Fig. 3). These are combined with a weighted average to produce a mean magnitude exceedance rate equation on each source zone. The procedure for determining the weights is detailed below, and involves deriving a preliminary set of weights for each parameter combination, which are then updated based on the probability that the parameter combination would produce the observed events in the global CMT catalogue (Ekstrom *et al.* 2012).

On each source zone each parameter combination is assigned a preliminary weight equal to the product of its individual parameter weights. The individual parameter weights for each $c\dot{s}$ value are set to 1/3; similarly, all b values are given weight 1/3; each variant of the Gutenberg–Richter model is given weight 0.5; and for $m_{w,max}$ we assign a weight of 0.45 to the upper value, 0.45 to the middle value and 0.1 to the smallest value, reflecting that it is unlikely $m_{w,max}$ has been observed on most source zones.

The weights defined with this heuristic procedure require further revision, because on some source zones particular parameter combinations lead to unrealistic earthquake rate predictions. For example, the Kermadec source zone is assigned $m_{w,max} \in \{8.2, 8.9, 9.6\}$, $b \in \{0.7, 0.95, 1.2\}$ and $c\dot{s} \in \{14, 21, 44\}$ mm a^{-1}. In order to conserve seismic moment, logic-tree branches with low $m_{w,max}$ and high $c\dot{s}$ need to have events with $m_w \geq m_{w,min}(=7.5)$ every few years on average (e.g. 0.42 a^{-1} for the top curve in Fig. 3). In contrast, only one such event was observed in the 38 year global CMT catalogue 1976–2013 (Fig. 3: details on data selection are presented below). If the true rate of $m_w \geq 7.5$ events was 0.42, then on average we should observe about 16 events in 38 years, and we are extremely unlikely to observe only one or zero events (probability of 2×10^{-6} for Poisson-distributed event times). Therefore, this particular branch of the logic tree seems unlikely to be correct. Although this conclusion is obvious from the absence of many events exceeding $m_{w,min}$ in the observational record, our method for assigning preliminary weights cannot detect such unrealistic logic-tree branches.

To address this issue, we update the preliminary weights based on the observed number of $m_w \geq m_{w,min}$ thrust earthquakes on the source zone. The observed events are counted using the global CMT

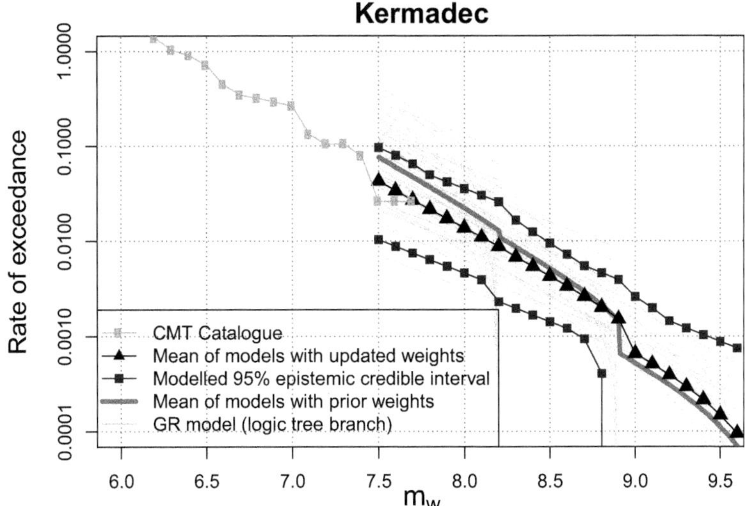

Fig. 3. Empirical and modelled m_w exceedance rates (events/year) for the Kermadec source zone. The global CMT catalogue events include those within 0.2° of our Kermadec source zone, with depth <70 km and rake within $\pi/4$ of pure thrust. By updating the weights of the logic-tree branches based on the observed number of $m_w \geq m_{w,min}$ events, unrealistic parameter combinations (e.g. corresponding to the upper lines in this case) are given minimal weight.

catalogue (1976–2013) (Ekstrom *et al.* 2012) if they are within 0.2° of our source zones, have depth $\leq$70 km, $m_w \geq 7.5$ and rake within $\pi/4$ of pure thrust. The weight update is equivalent to treating the preliminary weights as prior probabilities for each parameter combination, and using Bayes' theorem to calculate posterior probabilities using the data (Kruschke 2011):

$$w_{ijkl} = \frac{w'_{ijkl}\Pr(r_e(m_w \geq 7.5)|ijkl)}{\Pr(r_e(m_w \geq 7.5))} \quad (12)$$

where w_{ijkl} is the updated weight for the parameter combination with the i-th $m_{w,max}$ value, the j-th b value, the k-th $c\dot{s}$ value and the l-th variant of the Gutenberg–Richter equation (equation 5 or 6); w'_{ijkl} is the corresponding preliminary weight; $r_e(m_w \geq 7.5)$ is the empirically observed rate of thrust earthquakes in the global CMT catalogue; $\Pr(r_e(m_w \geq 7.5)|ijkl)$ is the probability that the empirical rate of earthquakes with $m_w \geq 7.5$ would be observed if the $ijkl$ rate equation were correct, and is computed assuming the event timings behave like a Poisson process with mean rate of $m_w \geq 7.5$ events computed from the $ijkl$ rate equation; and the denominator is a normalizing constant:

$$\Pr(r_e(m_w \geq 7.5)) = \sum_i \sum_j \sum_k \sum_l w'_{ijkl}\Pr(r_e(m_w \geq 7.5)|ijkl) \quad (13)$$

In the case of Kermadec, this leads to a mean rate curve with less frequent small earthquakes and more frequent large earthquakes, compared with the mean rate curve derived from the preliminary weights (Fig. 3). This reflects the down-weighting of unrealistic logic-tree branches which predict frequent m_w events of 7.5 magnitude, as discussed above.

Although it is clearly desirable to down-weight logic-tree branches which are unlikely to be consistent with observations, in the case of Kermadec it is not visually obvious whether the 'updated-weights' mean rate curve is an improvement on the 'prior-weights' mean rate curve (Fig. 3). Because of natural variations in the observed number of earthquakes, we cannot be sure of improvements every time weight updating is applied, but improvements are expected on average when applied globally to many source zones. For illustration, consider a large number of hypothetical scenarios in which the true source zone rate curve is unknown, but corresponds to one particular branch in Figure 3 (with probability corresponding to the preliminary weights). Without weight updating, the difference between the true rate of $m_w \geq 7.5$ events and the modelled mean rate has a mean of zero and standard deviation of 0.066. However, if 38 years of data is randomly sampled from the true model and weight updating is applied, then the difference has a mean of zero and standard deviation of 0.037 (i.e. the typical error is almost halved). It is also worth noting that at the globally aggregated level, weight updating has little impact on the modelled magnitude exceedance rate curve, with the largest change being a small reduction (c. 10%) in the rate of events with $m_w < 8$ which slightly improves agreement with globally aggregated data (see the subsection on 'Earthquake magnitude–frequency relationships' later in this paper). Although not pursued here, more extensive use of historical and palaeoseismic data may further improve the logic-tree weights. There is also potential to modify the approach to allow for possible non-Poisson event timings, and more fully exploit details of the earthquake magnitude observations (rather than just counting the observed events above $m_{w,min}$), although an advantage of ignoring such details is robustness to occasional large earthquakes.

The revised weights are also used to compute credible intervals representing our epistemic uncertainty: for example, the 95% credible interval is a region containing 95% of the total logic-tree weight for each m_w, with $\leq 2.5\%$ of the total logic-tree weight above the upper bound and $\leq 2.5\%$ below the lower bound (Fig. 3). Note that in this paper we use the term 'credible interval' to refer to intervals representing our epistemic uncertainty computed using the logic-tree weights, whereas the term 'confidence interval' is used for standard frequentist intervals (Kruschke 2011). The advantage of quantifying epistemic uncertainties is: (1) it highlights regions where the mean hazard is more likely to change in future hazard updates, as new knowledge may decrease epistemic uncertainties (Marzocchi *et al.* 2015); (2) the use of the weighted mean rate curve alone is insufficient testing PTHA predictions from the full logic tree (Marzocchi & Jordan 2014) and; (3) it allows the effects of uncertainty on risk assessment to be handled in an explicit manner (McGuire *et al.* 2005). A limitation of the credible interval approach used herein is that it implicitly assumes the model and logic tree cover all possible true models, which will not be correct. Although not pursued here, 'ensemble modelling' approaches have been proposed to address this issue by treating the logic-tree branches as a sample of the epistemic uncertainty which is subject to further statistical modelling (Marzocchi *et al.* 2015; Selva *et al.* 2016).

Wave propagation to hazard points

For each unit-source, we model the tsunami caused by an earthquake with 1 m of slip using a numerical solution of the linear shallow-water equations in spherical coordinates on a grid with cell size $1' \times 1'$. We use the same finite-difference code as Thio *et al.* (2007), Burbidge *et al.* (2008) and Horspool *et al.* (2014), based on Satake (1995), with a fixed time step of 2 s. Global elevation values are set with the GEBCO $30'' \times 30''$ bathymetric grid. The model is evolved for 26 h of simulated time, as this is longer than required for trans-Pacific tsunami propagation from, for example, Chile to Japan (Wessel 2009), and also short enough for our simulations to complete within the time limits of our computing facility.

Because of computer storage limitations, we cannot save full time series for the unit-source tsunami model runs and so, instead, store the tsunami wave height time series at 20 s intervals at a subset of points near the coastline (termed hazard points). The hazard points are placed approximately along the 100 m depth contour at a spacing of 25 km, subject to the constraints that they are not too near (<1.5 arc-min) or too far (>22 arc-min) from the coastline. If this constraint is not met on the 100 m depth contour, then the hazard point is placed at the point closest to 100 m depth which satisfies the constraint.

Because of the linearity of the propagation model, the hazard point time series associated with any synthetic earthquake event in our catalogue can be produced by summing the time series for each contributing unit-source and multiplying by the earthquake slip (Thio *et al.* 2007). While very computationally efficient, a drawback is that we cannot model wave propagation in shallow water (c. 50 m). This is both because of insufficient mesh resolution to resolve the shorter tsunami wavelengths, and because non-linear processes associated with finite flow velocities (friction and momentum advection) are ignored in the linear model. Hence, our hazard points are situated offshore in depths of approximately 100 m, and other methods are required to estimate amplification and run-up in shallower water.

Wave amplification onshore

Our tsunami propagation model cannot directly simulate inundation. While the usual practice in tsunami science is to use local inundation models for run-up estimation, in the current study this was not possible due to the large number of simulations, the global extent of the study, and the need for high-resolution topographic and bathymetric data. An alternative approach, which is less accurate but practically feasible, is required to estimate wave run-up (above mean sea level) from the hazard point water-surface elevation time series. Two different approaches are applied in this study.

First, we apply the amp-factor method of Løvholt *et al.* (2012*a*). This estimates the wave run-up from the modelled wave height, period and depth at the nearest hazard point, assuming that the near-shore topography can be approximated with one of seven different idealized bathymetric profile types (subjectively assigned to hazard points for this study). For each bathymetric profile type, the amplification factors are derived by interpolation of the amplification predicted by numerical models for plane waves. Details are provided in Løvholt *et al.* (2012*a*), who found the run-up predictions compared fairly well with run-up predictions from detailed inundation models in a number of test cases.

Secondly, we estimate the tsunami run-up using Green's law to amplify the maximum water-surface elevation at the nearest hazard point, with the water depth at the coastline set to 0.5 m (Kamigaichi 2009; Sørensen *et al.* 2012; Brizuela *et al.* 2014; Horspool *et al.* 2014; Hébert & Schindelé 2015). Kamigaichi (2009) used Green's law to estimate nearshore peak tsunami wave height from the modelled water-surface elevation at offshore sites (*c.* 50 m depth), and found the estimates compared well with those from a fine mesh model for two separate earthquake–tsunamis with $m_w = 8.0$ and 6.8. Hébert & Schindelé (2015) found the approach gives a useful indication of zones of maximum run-up compared with field data for the Indian Ocean tsunami, although the errors were often of similar magnitude as the observed run-up, and no particular value of the coastline depth would consistently give the best predictions.

It is stressed that for plane non-breaking waves, the amplification factor method of Løvholt *et al.* (2012*a*) reproduces the tsunami run-up obtained from non-linear shallow-water models (Carrier & Greenspan 1958). Using Green's law for run-up estimation requires a subjective choice of the water depth at the coastline, and will produce infinite run-up heights as the coastline depth converges to zero (i.e. the shoreline). Regardless, we examine both methods here as Green's law is straightforward to apply once the coastline depth is selected, and has been employed in several past applications of PTHA.

Run-up exceedance rate computation

At every hazard point, the methodology above provides the amplified tsunami run-up (see the previous subsection on 'Wave amplification onshore') and a corresponding mean annual rate for every earthquake event in the synthetic catalogue (equation 9). From this, we can compute the mean annual rate of events, with amplified wave height h exceeding a threshold h_T at any particular hazard point p, denoted $r_p(h \geq h_T)$:

$$r_p(h \geq h_T) = \sum_{i \in E} r_i \Pr(h \geq h_T | h_{i,p}) \quad (14)$$

where i is an event from the full synthetic earthquake catalogue E with rate r_i, and $\Pr(h \geq h_T | h_{i,p})$ gives the probability that the real run-up exceeds h_T, given that the modelled run-up is $h_{i,p}$. The latter term accounts for the run-up variability associated with aleatory uncertainties (e.g. spatially variable slip) which are not directly simulated in the model, as well as model structural errors caused by, for example, errors in the subduction interface geometry and our simplified treatment of the hydrodynamics. Although the latter factors are actually epistemic uncertainties which might theoretically be treated in a logic tree, their characterization would be computationally expensive and so for simplicity herein they are lumped in with the aleatory uncertainties (Thio *et al.* 2010; Thio 2012; Horspool *et al.* 2014; Løvholt *et al.* 2016). We assume that h follows a lognormal distribution with median equal to the predicted run-up at the nearest hazard point, plus some bias β. The probability of h exceeding a threshold h_T is:

$$\Pr(h \geq h_T | h_{i,p}) = 1 - \Phi\big(\ln(h_T) | [\ln(h_{i,p}) + \beta], \sigma\big) \quad (15)$$

where, on the right-hand-side, Φ is the cumulative distribution function for a normal distribution with mean $[\ln(h_{i,p}) + \beta]$ and standard deviation σ, evaluated at $\ln(h_T)$ (Annaka *et al.* 2007; Thio 2012; Horspool *et al.* 2014).

Values for the logarithmic bias β and standard deviation σ in equation (15) were determined by comparing predictions of uniform slip scenarios in our PTHA database with observations from the NGDC/WDS global historical tsunami database (NGDC 2015) for four large earthquake–tsunami events: the 1960 Chile m_w 9.5; the 1964 Alaska m_w 9.2; the 2004 Andaman–Sumatra m_w 9.2; and the 2011 Tohoku m_w 9.0. The bias β was estimated as the mean of $[\ln(h_{obs}) - \ln(h_p)]$, where h_{obs} is the observed run-up height and h_p is the predicted run-up height at the hazard point nearest to the observation. The standard deviation of the latter log-difference provided an estimate of σ.

Results

Earthquake magnitude–frequency relationships

For each source zone, magnitude–frequency curves similar to Figure 3 were derived and, in some instances, these have been compared with independently estimated exceedance rates based on longer-term historical or palaeoseismic data, or seismic moment conservation. For the Alaska source zone, the modelled $m_w = 9.2$ exceedance rate is 1/897 (1/3236, 1/538), where values in parenthesis give a credible interval covering logic-tree branches with 95% of the total weight, thus reflecting the modelled epistemic uncertainty for the source zone. This is in reasonable agreement with 1/650–1/750 reported by Wesson *et al.* (2007) based on palaeoseismic studies. On the Honshu source zone, our modelled $m_w = 9.0$ exceedance rate is 1/486 (0, 1/209). The lower bound of 0 occurs because $m_w = 9.0$ is the smallest $m_{w,max}$ in the logic tree (corresponding to the 2011 Tohoku earthquake) with zero probability of being exceeded according to the corresponding truncated Gutenberg–Richter–frequency curves (equation 6). This estimate is comparable to values of 1/400–1/300, 1/532 and 1/880–1/260 derived using various seismic moment conservation type analyses, and 1/1300–1/400 based on tsunami deposit data (McCaffrey 2008; Uchida & Matsuzawa 2011; Kagan & Jackson 2013). For the Nankai source zone, our modelled $m_w = 8.0$ exceedance rate is 1/76 (1/136, 1/37), towards the higher end of the $1/(124 \pm 96)$ rate suggested by Burbidge *et al.* (2008) based on 11 observed events in 1500 years. On Nankai, our $m_w = 8.6$ exceedance rate is 1/284 (0, 1/137), again consistent with 1/660 years in Burbidge *et al.* (2008), noting the latter is based on only two events and thus very uncertain. In the South Chile source zone, we estimate an $m_w = 8.0$ exceedance rate of 1/48 (1/90, 1/27), and for $m_w = 8.5$ we obtain 1/131 (1/202, 1/86). These are comparable to estimates of $1/(65 \pm 41)$ and $1/(128 \pm 46)$, respectively, by Burbidge *et al.* (2008) based on the historical record since 1570. Overall, the agreement with previous studies is encouraging, and suggests the model is providing reasonable estimates of the exceedance rate of large earthquakes.

It is also of interest to compare the globally integrated magnitude–frequency curve with data. This is a challenging comparison for our model because each source zones' magnitude–frequency curve is determined separately by combining the seismic moment conservation model with the logic-tree parameters and weights. By updating the logic-tree weights with the global CMT catalogue data, we ensure some consistency between the model and data at the source zone specific level, but this process does not force the model to agree with data at the globally aggregated level. Regardless, when aggregated globally, the mean modelled exceedance rate is similar to the global CMT catalogue empirical exceedance rate (Fig. 4: CMT events include only those within 0.2° of our source zones with depth ≤70 km and a rake within $\pi/4$ radians of pure thrust). The largest discrepancy occurs for m_w of approximately 8.1 where the model predicts a higher rate of events (Fig. 4). Since we expect the global CMT catalogue to be fairly complete for $m_w \geq 7.5$ (Kagan 2003), any discrepancies between the model and observed data are predominantly due to the following factors: (1) sampling variability due to the short duration (1976–2013) of the global CMT catalogue; (2) epistemic uncertainty leading to a wide range of observed rates being consistent with the model; and (3) bias in the model and inappropriate logic-tree weights.

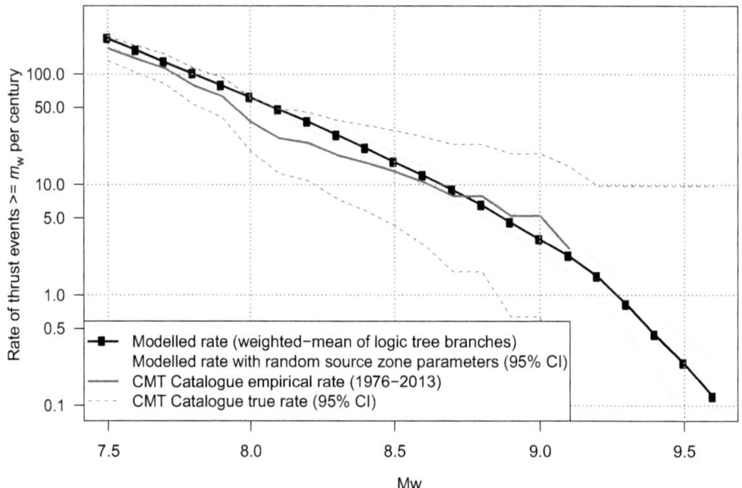

Fig. 4. Empirical and modelled m_w exceedance rates over all source zones. Events from the global CMT catalogue were included if they had a depth ≤70 km, were within 0.2° of our source zones and had (strike–rake) within 45° of the associated source zone's updip direction.

To assess the likely impacts of sampling variability on global-scale differences between the model and data, we computed 95% confidence intervals for the true exceedance rate from the global CMT catalogue using a standard exact method (Garwood 1936). This assumes globally observed thrust earthquake events have Poisson behaviour in time. 95% confidence intervals for the true exceedance rate generally contained the modelled mean exceedance rate, although for m_w of approximately 8.0 the model is close to the upper limit (Fig. 4). Based on this, we cannot confidently reject the consistency of the mean modelled rate and data at the global level. However, the uncertainties in the global rate are very large because the catalogue duration is relatively short (38 years), and there are no additional constraints imposed by models (e.g. there is no constraint from seismic moment conservation or from bounds on $m_{w,max}$ imposed in source zones with a small area). It would be possible to develop a narrower confidence interval using a longer duration database: however, this has not been pursued here since we need information on the earthquake rake to identify thrust events. To our knowledge such information is not comprehensively available in longer duration catalogues.

Even if the model is correct, epistemic uncertainties will lead to deviation between the true exceedance rate and the modelled weighted-mean rate because the latter is an average reflecting all parameter combinations in our logic tree. To assess the expected epistemic uncertainty in the globally aggregated exceedance rate, we randomly assigned each source zone to a branch of its logic tree and computed the global mean exceedance rate. The probability that each branch was assigned to its source zone was proportional to its weight w_{ijkl} (equation 12). Figure 4 shows 95% credible intervals obtained by repeating the latter procedure 1000 times and computing the 2.5 and 97.5% quantiles. Importantly, this credible interval assumes that source zone parameters are independent of each other and that the model is correct, in which case the true global exceedance rate of thrust earthquakes on our source zones is likely to occur within the envelope. The substantial overlap of this credible interval and the global CMT-based confidence interval further indicates plausible consistency between the model and data at the global scale. The credible interval is much narrower than the confidence interval predicted using the 38 year global CMT catalogue (Fig. 4). In part, this reflects the information content of our source zone parameter combinations, their prior weights and the model itself: in particular, the assumptions that the earthquake rates conserve seismic moment, and that $m_{w,max}$ is constrained by the size of the source zone (using an empirical relationship of Strasser *et al.* 2010 herein). These assumptions constrain the modelled exceedance rates more strongly than does naive inference from the global CMT earthquake catalogue. The credible interval probably underestimates the true uncertainties because it neglects the possibility of correlations between the source zone parameters, which could arise, for example, if the upper $m_{w,max}$ or $c\dot{s}$ rates were actually appropriate on most source zones. Herein, we have not tried to model these correlations because it introduces additional complexity, and our conclusion that there is reasonable consistency between the model and the data would be unaffected by a broader credible interval. However, for other applications, a more careful accounting of these correlations may be required.

While the above tests do not suggest the model is inconsistent with the data, the model is nonetheless likely to be limited in a number of ways. Realistically, the model will have structural errors, and the logic-tree branches are not expected to be exhaustive of all possibilities (Field *et al.* 2014; Marzocchi *et al.* 2015). Owing to our conservative treatment of seismic coupling, no source zone has a maximum seismic coupling of less than 0.7 (Berryman *et al.* 2015). While a reasonable reflection of uncertainty at the level of an individual source zone, when globally aggregated this will lead to over-prediction of earthquake rates if many source zones have coupling values lower than 0.7. Similarly, the inclusion of a low $m_{w,max}$ value in the logic tree can lead to high rates of low m_w earthquakes on some logic-tree branches (Fig. 3), potentially causing over-prediction of small events at the globally aggregated level. Although unrealistic magnitude–frequency curves are effectively down-weighted by updating the logic-tree weights with the empirical rate of $m_w \geq 7.5$ thrust events in the global CMT catalogue, at the individual source zone level they may appear reasonable given the limited available data.

In summary, at the source zone specific level our tests suggest reasonable agreement of several modelled magnitude–frequency relationships with data. There is typically considerable uncertainty in the modelled rates, and in some locations this uncertainty might be reduced by integrating palaeoseismic data into the model. However, even in its current form, when aggregated globally, the modelled magnitude–frequency curve is consistent with the observed rate of thrust events in the global CMT catalogue.

Wave run-up from individual events

The PTHA scenario amplified wave heights were compared against tsunami run-up observations for four large historical events in the NGDC/WDS

Table 2. *Summary statistics of the NGDC/WDS historical tsunami database run-up observations, compared with a corresponding event in the PTHA database with the same magnitude and source zone which best fits the data*

Event	Amp-factor bias β	Amp-factor σ	Green's law bias β	Green's law σ
1960 Chile $m_w = 9.5$	−0.188	0.789	−0.514	0.759
1964 Alaska $m_w = 9.2$	−0.107	1.073	−0.059	1.010
2004 Indian Ocean $m_w = 9.2$	−0.0584	0.8715	0.075	0.847
2011 Tohoku $m_w = 9.0$	−0.302	0.954	−0.079	0.924

The amp-factor method follows Løvholt *et al.* (2012*a*) to estimate the wave run-up height. Alternatively, Green's law was applied to a 0.5 m water depth. For each event, we computed $[\ln(h_{obs}) - \ln(h_p)]$, where h_p is the PTHA amplified wave height spatially closest to the observed run-up h_{obs}. The bias is the mean of the logarithmic difference and σ is its standard deviation.

historical tsunami database (Table 2; Fig. 5). For each observed event, the PTHA scenario had the same m_w and source zone, and where several such PTHA scenarios existed the best-fitting one was used. This allows estimation of the logarithmic bias β and standard deviation σ relating the predicted wave run-up height to the observations (see the earlier subsection on 'Wave amplification onshore').

Overall, the amp-factor method (Løvholt *et al.* 2012*a*) and Green's law produced a similar logarithmic bias β and standard deviation σ (Table 2). The bias varied substantially among the events, with the amp-factor method always over-predicting median run-up, while the Green's law approach did so in three-quarters of the cases (Table 2). Given the small overall difference in the performance of these methods, we hypothesize that the model error is due to the simplified uniform-slip representation of the tsunami source, and the fact that the complex onshore inundation, including local phenomena such as focusing and refraction, is not included in the amp-factor method, as well as the neglect of friction terms (Fritz *et al.* 2008; Lynett *et al.* 2012). The σ values derived here are somewhat larger than the $\sigma = 0.71$ value previously reported by comparing coarse-grid uniform slip tsunami simulations with observations, and accounting for uncertainties in dip and non-uniform slip distributions (Thio *et al.* 2010; Thio 2012). Thio *et al.* (2010) stated σ could be further reduced from 0.71 to 0.52 using a finer-grid tsunami propagation model (90 m cell size), which suggests that a substantial improvement in the hydrodynamic modelling may yield only moderate improvements in the accuracy of run-up predictions. Our results suggest that σ values even larger than used previously may be appropriate for a 1 arc-min resolution linear shallow-water model with initial conditions based on uniform-slip earthquakes (Thio *et al.* 2010; Horspool *et al.* 2014; Løvholt *et al.* 2014*a*).

While noting similar overall performance of both amplification methods trialled in this study, hereinafter we estimate run-up heights using the amp-factor method (Løvholt *et al.* 2012*a*) because it has a more firm theoretical basis for quantifying the run-up than the Green's law approach (see above). We employ a bias $\beta = -0.164$ and $\sigma = 0.927$, corresponding to the mean and root-mean-square of the respective values in Table 2. This implies that approximately 54% of run-up heights are predicted within a factor of 2 by the best corresponding PTHA scenario.

Wave run-up exceedance rates

The global pattern of 1/500 and 1/2500 exceedance rate run-up heights are shown in Figure 6. At the 1/500 exceedance rate, the model predicts high run-up around most of the Pacific Rim and eastern and southern Indonesia, with relatively low run-up heights in the Atlantic and western Indian oceans, southern Australia, and other areas which are significantly sheltered from the major subduction sources. Areas with intermediate run-up heights include many Pacific Islands, southern New Zealand, and parts of eastern and western Australia. At the 1/2500 exceedance rate, our model predicts large areas with an exceedance wave height >10 m in the Pacific and eastern Indian Oceans, as well as the Western Mediterranean near the Hellenic and Cyprus source zones, and areas adjacent the Makran source zone (Fig. 6).

The run-up heights associated with these rare events are subject to considerable epistemic uncertainty because they involve large-magnitude earthquakes. The logic-tree branches usually assign widely varying rates to large earthquakes, especially where no such events have been observed, implying that $m_{w,max}$ has a large range (Fig. 3). For example, in the Kermadec source zone, the 95% epistemic credible intervals for the rate of events with $m_w > 8.9$ always includes a lower bound of zero (Fig. 3) because our epistemic uncertainty puts significant weight on the possibility that $m_{w,max} \leq 8.9$. To quantify the impact of this on tsunami wave heights, we compute new exceedance

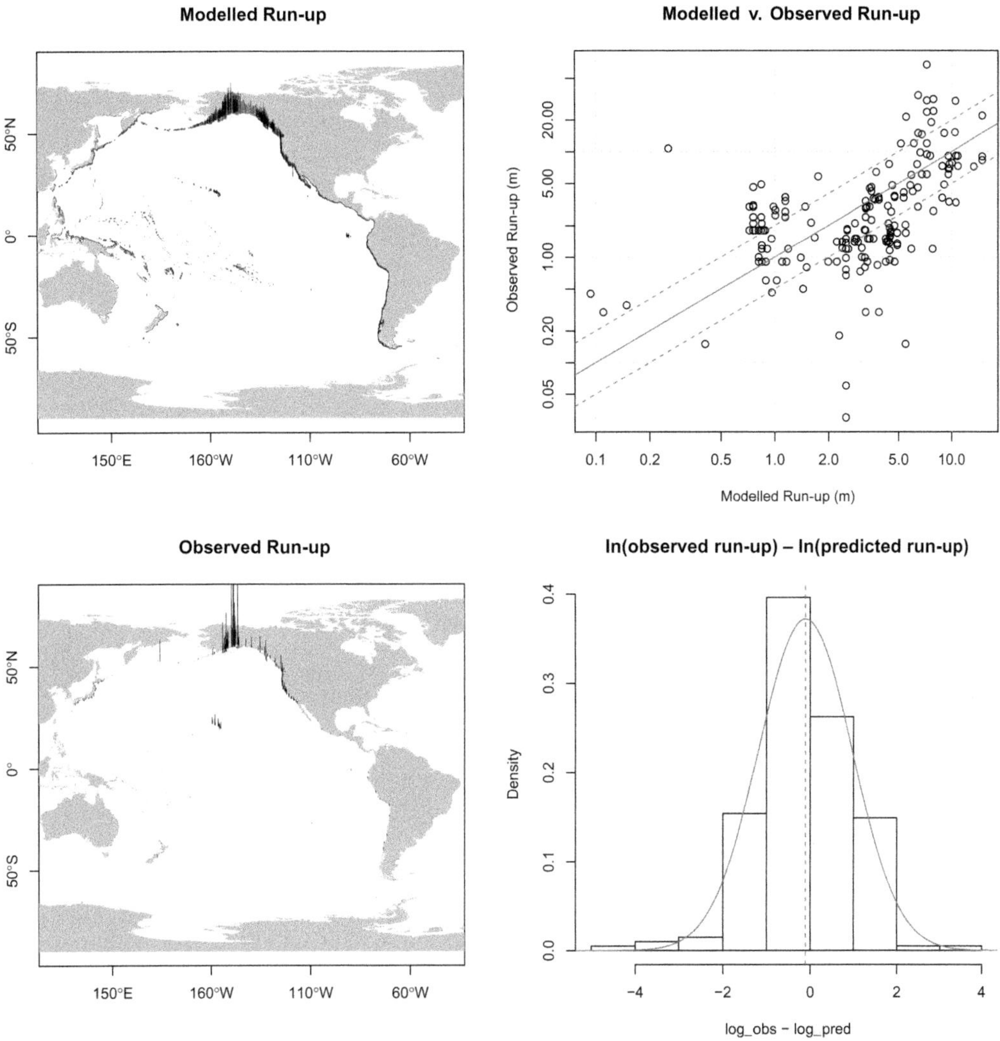

Fig. 5. Observations of the 1964 Alaska tsunami run-up in the NGDC/WDS historical tsunami database compared with the corresponding PTHA scenario. (Top left) Modelled wave run-up heights (>0.1 m). (Bottom left) Observed wave run-up heights (>0.1 m). (Top right) Modelled v. observed run-up heights. The solid line is $y = x$, and the dotted lines are $y = 2x$ and $y = 0.5x$. (Bottom right) Histogram of errors in log run-up and the corresponding normal distribution.

wave heights, assuming that the rates on all source zones matched, first, the lower 2.5% and, secondly, the upper 97.5% quantiles of our epistemic uncertainties (Fig. 7). For any coastal site, we can then investigate the range in exceedance wave heights predicted if all source zones which significantly affect it have 'high' or 'low' earthquake rates relative to our expectations. This simple method of uncertainty quantification is conservative because it ignores the potential for uncertainties from multiple source zones to average-out if their parameters are uncorrelated. However, at the site-specific level, we expect tsunami hazard is usually dominated by just a few source zones, which may have correlated parameters (e.g. we might consider the highest or lowest $m_{w,max}$ or $c\dot{s}$ value could be appropriate for all relevant sources). In the absence of detailed modelling of the inter-source zone parameter correlations, the conservative approach seems more reasonable for site-specific hazard estimation than does ignoring inter-source zone parameter correlations entirely.

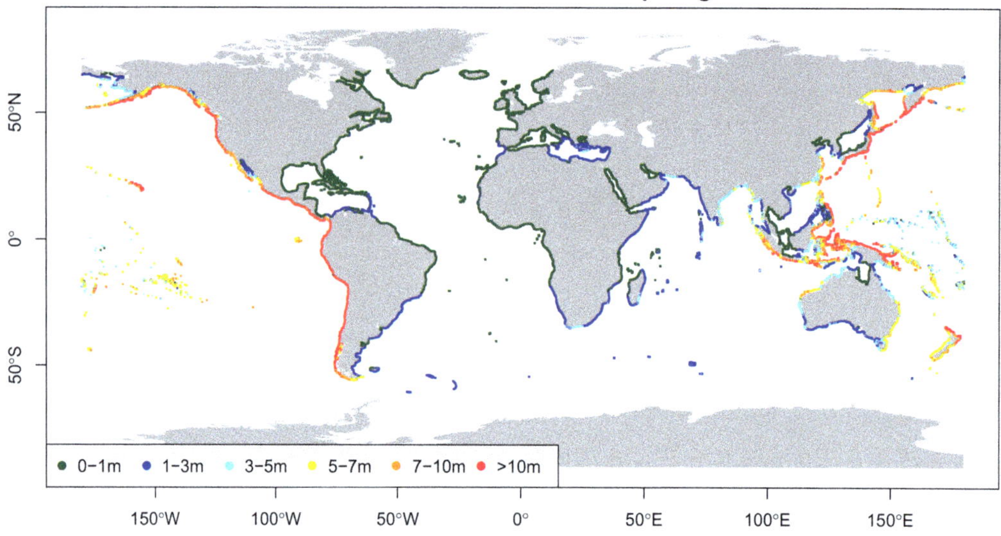

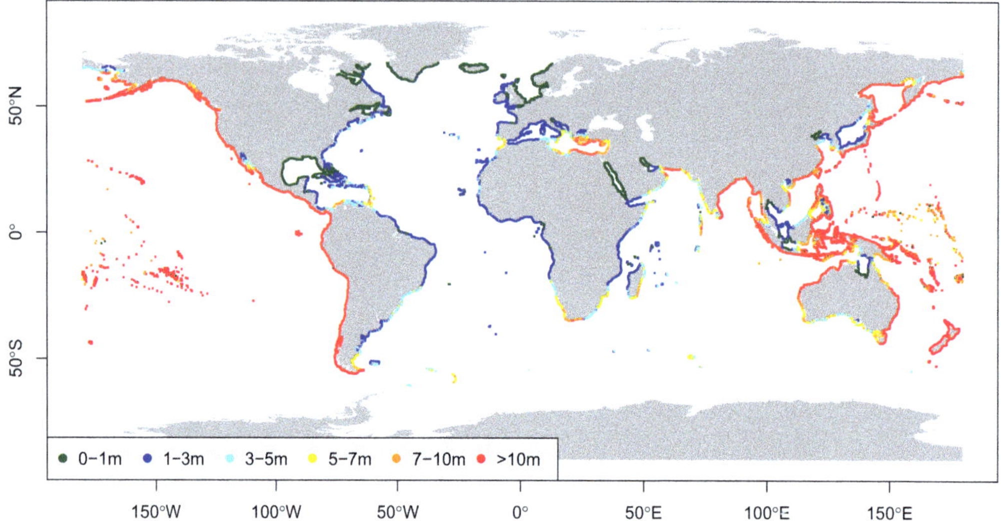

Fig. 6. Wave run-up heights associated with 1/500 (top) and 1/2500 (bottom) exceedance rates.

There is high epistemic uncertainty in the 1/500 exceedance rate run-up for many locations globally, with the 95% credible interval for run-up ranging from 1 to >10 m in much of Oceania and the eastern Indian Ocean (Fig. 7). This reflects exposure to source zones where $m_{\mathrm{w,max}}$ and the seismic coupling rate have substantial epistemic uncertainty, such as Kermadec (Fig. 3). High waves are predicted even at the lower limit of the 95% credible interval near rapidly converging source zones with high seismically coupled slip rates (i.e. the South American and Japan/Kurils source zones) (Fig. 7). Other sites with high hazard even at the lower limit of the 95% credible interval include Cascadia, southern Alaska, parts of northern and eastern New Guinea, Sumatra, and the northern Moluccas, which are exposed to source zones with high seismically coupled slip rates (Berryman *et al.* 2015). Much of the Atlantic and Mediterranean coast is predicted to have relatively low tsunami hazard (<3 m run-up) for the 1/500 exceedance rate within our 95% credible

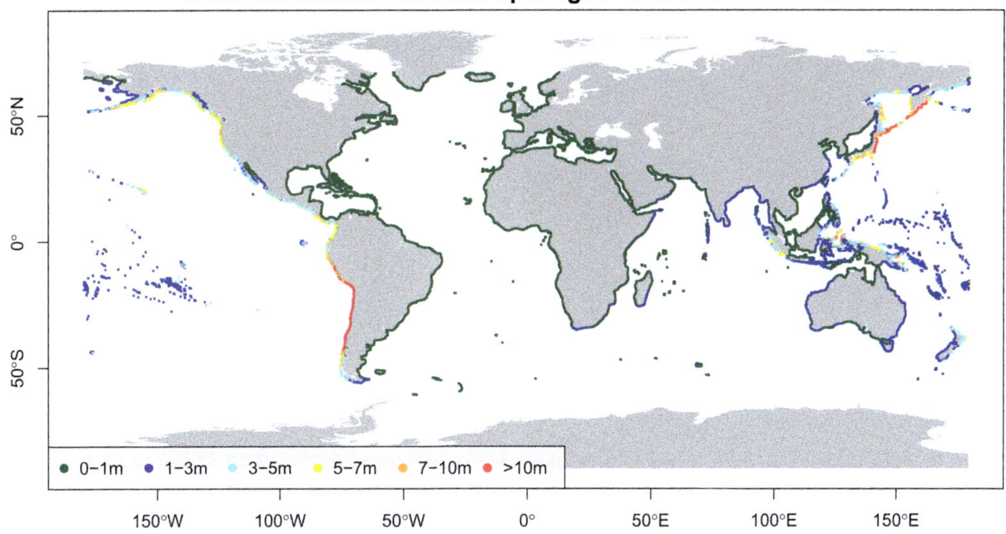

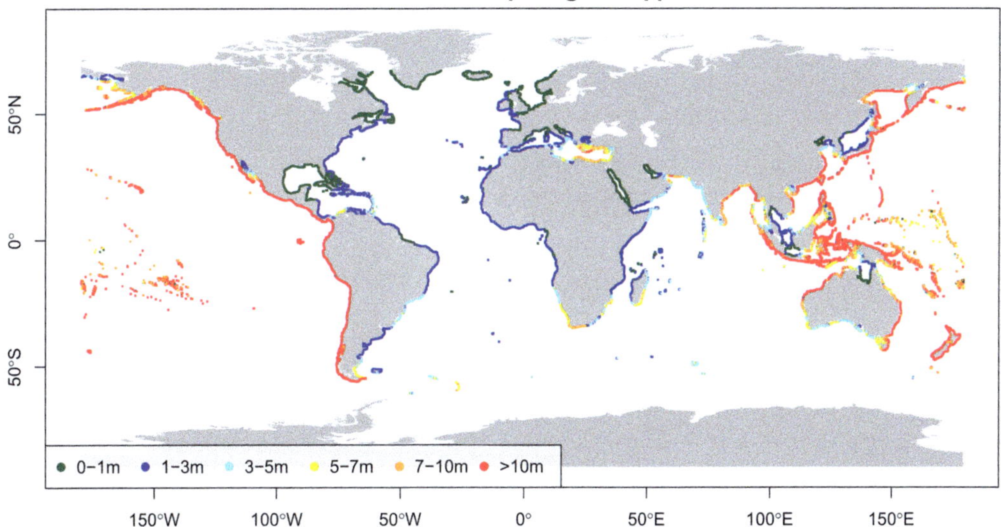

Fig. 7. 95% credible interval (CI) for the wave height with a 1/500 exceedance rate. Lower credible interval bound (top) and upper credible interval bound (bottom).

intervals, with exceptions close to source zones in the east and south Caribbean and the Eastern Mediterranean.

It is worth noting that in the northern Moluccas (NE Indonesia), the convergence rates suggested in Berryman *et al.* (2015) for the Halmahera source (90–100 mm a^{-1}, as used in this study for the Eastmolucca source zone) are significantly higher than those used by Løvholt *et al.* (2012*b*) and Horspool *et al.* (2014) (10–30 mm a^{-1}), and this contributes to our high hazard predictions in this area (Figs 6 & 7). Even assuming lower convergence rates, the area is considered high hazard (Løvholt *et al.* 2012*a*, *b*; Horspool *et al.* 2014). The higher convergence rates used here seem consistent with a geodetic velocity value reported in Bird (2003). However, the plate model of Bird (2003) suggests convergence rates vary rapidly in space. Given its complexity, this region warrants further analysis in future.

Discussion and conclusions

Comparison with previous global analyses

The current study builds on several previous global-scale analyses of tsunami hazard from earthquake sources (Løvholt *et al.* 2012*a*, 2014*a*) which largely estimated 1/500 annual exceedance rate run-up heights by direct simulation of tsunami from 1/500 earthquake scenarios. To demonstrate the accuracy of the simpler scenario-based method, the hazard from earthquake scenarios was compared with a full PTHA in the Indian and southern Pacific oceans in Løvholt *et al.* (2014*a*), with the PTHA results producing less conservative predictions than scenario methods in the majority of cases. The most significant advance herein is the global extension of the PTHA, which allows global estimation of wave heights for arbitrary exceedance rates with a uniform methodology (Fig. 6), and quantification of the associated epistemic uncertainties (Fig. 7).

There are substantial methodological differences between the current study and the PTHA component of Løvholt *et al.* (2014*a*). Rather than using constant dip source zone geometries, we use detailed subduction zone geometries where available (Hayes *et al.* 2012; Basili *et al.* 2013*a*, *b*), and allow for along-strike variations in the long-term convergent slip rate and downdip width to be accounted for when estimating the relative rates of thrust earthquakes with the same magnitude (equation 10). Our approach to defining the range of $m_{w,max}$ in the logic tree more closely follows Berryman *et al.* (2015) and often leads to a greater range of $m_{w,max}$ than the $m_{w,max} \pm 0.2$ approach applied previously. We also treat epistemic uncertainties in the Gutenberg–Richter *b* parameter, which was assumed to be unity in the earlier study; and we use a simple technique to down-weight unrealistic logic-tree branches based on the observed rate of thrust events $\geq$7.5 in the global CMT catalogue (equation 12). While our estimation of run-up from the offshore wave heights follows the same approach (Løvholt *et al.* 2012*a*), in this study uncertainties in the PTHA scenario run-up heights have been reassessed by comparison against run-up observations from four large tsunamis (Table 2). This has led to substantially larger $\sigma = 0.92$ compared to the values $\sigma = 0.52$ or 0.71 that have been used previously (Thio *et al.* 2010; Horspool *et al.* 2014). The value of σ has a particularly strong effect on the results, with higher σ associated with a major increase in the run-up height for a given exceedance rate (compare Figs 6 & 8). Using the lower $\sigma = 0.5$, the 1/500 run-up height reduces considerably, with values of less than 5 m now dominating over much of the central and SW Pacific Ocean, and in the Indian Ocean away from the Sunda Arc. If the uncertainties in the modelled tsunami run-up height are entirely neglected ($\sigma = 0$), then the 1/500 run-up height further reduces (Fig. 8).

Despite these methodological differences, there is broad similarity of our mean 1/500 run-up heights with those of Løvholt *et al.* (2014*a*), with relatively high hazard around much of the Pacific Rim and southern and NE Indonesia, and relatively low hazard in much of the Atlantic and western Indian oceans. The current study often predicts slightly higher run-up: for example, in the central south Pacific and eastern Australia, we often estimate 1/500 run-up heights of approximately 5–7 m, whereas corresponding values are typically $\leq$5 m in Løvholt *et al.* (2014*a*) (Fig. 6). This is consistent with the use of a higher σ in the current study (Figs 6 & 8). There are also regions where the current study predicts lower 1/500 tsunami run-up: for example, around northern Sumatra. One reason for this seems to be our use of Slab 1.0 to define the source zone geometry, which in this location leads to unit-sources having low dip near the shallow part of the source zone compared with the constant dip value used previously. The lower dip implies less vertical coseismic deformation for a given slip on the shallowest unit-sources. In addition, the sources used previously had a top-depth of 3 km (Horspool *et al.* 2014), compared with 0 km herein, so we expect higher slip near the trench in the former case (Goda 2015). Given the substantial methodological difference between the current study and Løvholt *et al.* (2014*a*), we are not surprised by some variations in the results: however, in general, the differences do not seem large compared with our epistemic uncertainties (Fig. 7).

Comparison with regional studies

At a number of sites, the hazard results from this study have been compared with previous estimates from regional-scale tsunami hazard assessments. Overall, the comparisons reported below are encouraging, and suggest that the current PTHA methodology is providing estimates of tsunami hazard associated with major earthquake sources that are reasonably consistent with previous work. Despite this, given the global scale of this study and associated limitations (see the following subsection on 'Limitations'), we suggest that the current study be used for global-scale evaluations of tsunami hazard and risk (Løvholt *et al.* 2014*b*), and as a reference against which the results of smaller-scale studies may be compared. Detailed tsunami hazard assessments combining inundation modelling and historical or palaeotsunami data (Gonzalez *et al.* 2009; Power 2013) provide the

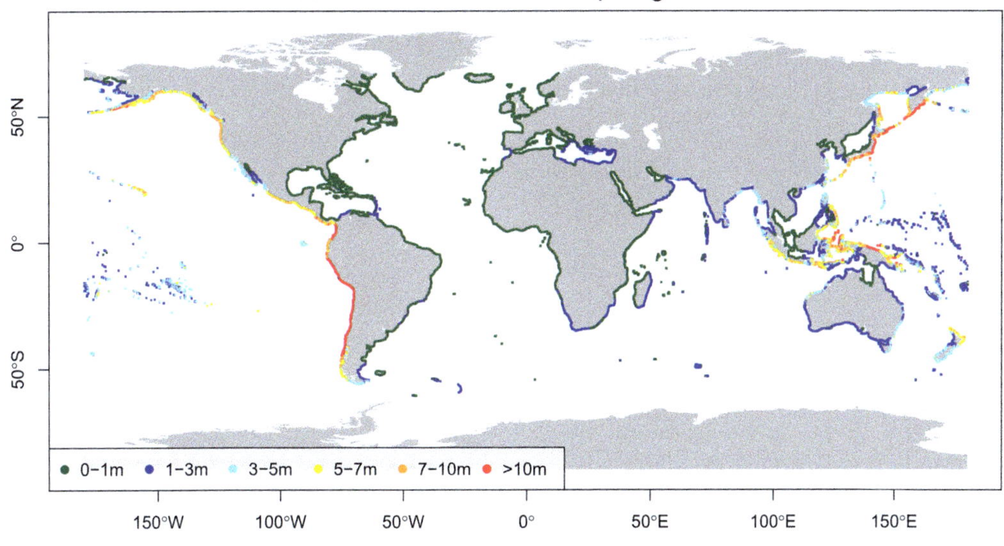

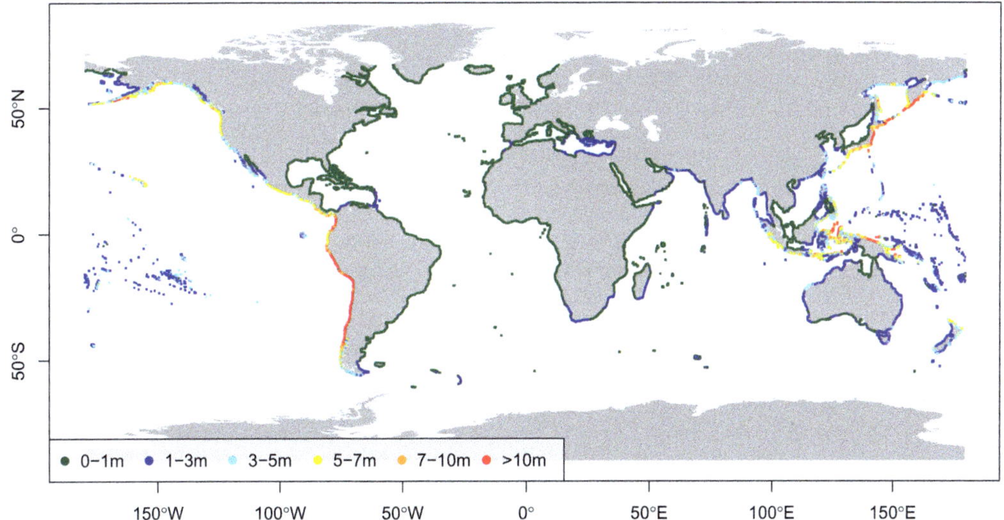

Fig. 8. The $1/500$ exceedance rate run-up height with (top) $\sigma = 0.5$ and (bottom) $\sigma = 0$.

most appropriate information to support local-scale disaster management and planning.

In Acapulco, Mexico, Geist & Parsons (2006) analysed tsunami run-up data, and suggested $1/10$ exceedance run-up of approximately 1 m, $1/100$ exceedance run-up of approximately 6.5 m and $1/500$ exceedance run-up of approximately 10 m. This compares quite well with the corresponding values for our nearest hazard point of 0.7, 4.2 and 10.7 m, respectively.

Along the coast of western Australia, Burbidge *et al.* (2008) modelled the exceedance rates of offshore wave height (in 100 m water depth). Tsunami hazard here is dominated by the eastern Sunda Arc. They predicted a spatial pattern in $1/500$ offshore wave heights similar to that estimated herein for run-up heights (Fig. 6), with a distinct peak around latitudes of approximately 20°–22° S. In the latter region, Burbidge *et al.* (2008) modelled $1/500$ wave heights of around 0.7 m in 100 m water depth (implying *c.* 3–5 m run-up height assuming amplification factors of *c.* 4–7). This is lower than the approximately 5–10 m predicted in the current study, although well within our epistemic uncertainties (Fig. 7). While there are substantial methodological differences between the current study and Burbidge *et al.* (2008), it seems likely that the differences are in part caused by the logarithmic uncertainties in predicted wave height being employed in the current study. This factor was not treated in Burbidge *et al.* (2008) but leads to an overall increase in the run-up (Figs 6 & 8).

In the Caribbean, Parsons & Geist (2009) estimated the probability of tsunami run-up exceeding 0.5 m over a 30 year time horizon using a combination of models and historical data. Their results suggesting highest hazard in the eastern Lesser Antilles, with 0.5 m run-up exceedance rates of approximately 0.0035–0.0075 events/year. Similar 0.5 m run-up exceedance rates are predicted in the current study, with 95% of eastern Lesser Antilles sites in $\in [0.0032, 0.0079]$.

In the town of Seaside, Oregon on the NW USA coast, Gonzalez *et al.* (2009) performed a detailed local-scale PTHA, and estimated a 1/100 exceedance wave height of approximately 4 m and a 1/500 exceedance wave height of approximately 10.5 m. This compares quite well with values of 5.0 and 13.2 m, respectively, estimated at our nearest hazard point.

In the Mediterranean Sea, Sørensen *et al.* (2012) conducted a detailed PTHA, including many more earthquake sources than treated in the current study. They predicted highest 1/500 run-up (*c.* 5 m) in the Eastern Mediterranean around the Hellenic source zone, with low run-up <1 m in the Western Mediterranean. Similar patterns and magnitudes for 1/500 tsunami run-up are predicted in this study (Fig. 6).

In New Zealand Power (2013) compiled detailed analyses of tsunami hazard, combining modelling with historical and palaeotsunami data: 1/500 wave heights largely ranged within 4–12 m on the north coast, 4–8 m on the east and south coast, and 2–4 m on the west coast. Our model suggests comparable values of typically 7–10+ m around the north coast, 5–10 m around the east and south coast, and 3–7 m around the west coast (Fig. 6).

In Indonesia, Horspool *et al.* (2014) modelled exceedance rates for tsunami run-up height, predicting 1/500 run-up heights of 5–10+ m along the south and NE coastlines. While similar results are suggested herein, the current study suggests that the high hazard zone in eastern Indonesia may extend further southwards to include regions adjacent to the Flores and Banda Sea (Fig. 6). This is largely because our logic tree includes greater maximum $m_{w,max}$ values for source zones in this region than Horspool *et al.* (2014). However, $m_{w,max}$ values in this region are highly uncertain, leading to large epistemic uncertainties in tsunami run-up exceedance rates (Fig. 7).

In the NE Atlantic, Omira *et al.* (2015) modelled tsunami wave height exceedance rates, suggesting highest hazard on the mainland around the Gulf of Cadiz with exceedance rates of approximately 0.007 for a 1 m wave (i.e. a 50% chance of exceedance in 100 years). The current study also predicts highest mainland hazard around the Gulf of Cadiz, although our 1 m wave height exceedance rates in this region are lower (*c.* 0.002 − 0.003), which seems to be driven by lower estimates of earthquake exceedance rates on the key regional source zones in the present study.

Limitations

The current study only considers tsunami hazard due to thrust earthquakes on sources identified in Figure 1 (aside from the strike-slip Gloria source zone), and we expect this to oversimplify the earthquake sources for some areas: for example, in the Mediterranean (Tiberti *et al.* 2009; Sørensen *et al.* 2012; Basili *et al.* 2013*b*). Uncertainties in the source zone schematization are not accounted for in the analysis but are expected to be significant, particularly away from major subduction zones. In the current methodology earthquakes are not permitted to cross source zones, and so our fault segmentation imposes potentially artificial limits on the range of events which are included in the model. We ignore earthquakes with $m_w < 7.5$, all non-thrust earthquake sources (except the strike-slip Gloria source zone) and most non-subduction sources, as well as tsunami from landslides, volcanoes and meteotsunami, although these are of significance in some contexts (Sørensen *et al.* 2012; Geist *et al.* 2014; Harbitz *et al.* 2014; ten Brink *et al.* 2014): for example, landslides are expected to contribute significantly to tsunami hazard in eastern Indonesia (Løvholt *et al.* 2012*b*), the Caribbean (Harbitz *et al.* 2012) and the Atlantic coast of the USA (ten Brink *et al.* 2014).

A significant source of uncertainty is related to the schematization of our earthquake sources. We assume a constant shear modulus μ which in reality will be spatially variable, with low near-trench values potentially leading to tsunami earthquakes with high slip and run-up for their magnitude (Geist & Bilek 2001; Newman *et al.* 2011). Although uncertainty in modelled run-up is accounted for using the lognormal distribution, parameters were estimated from high–magnitude earthquake–tsunami events and do not appropriately account for the tsunami hazard due to tsunami earthquakes. Uncertainties in the source zone geometry (e.g. dip, maximum depth) and the non-uniform slip distribution of real earthquakes are also not explicitly modelled, although to some extent they are indirectly accounted for via the lognormal run-up distribution. Even for uniform slip earthquakes, small changes in the source zone geometry can have a strong impact on models of the resulting tsunami (Burbidge *et al.* 2015), and this is probably an important factor driving the large variation in wave heights predicted by different tsunami early warning systems (Greenslade *et al.* 2014). In addition, our models of earthquake rates show substantial epistemic uncertainty (e.g. Fig. 3), and our quantification of this is influenced by the subjective choice of logic-tree parameters and their weightings. On the one hand, the possible range of source zone parameters is not fully considered (since only three values are used for $m_{w,max}$, b and $c\dot{s}$) suggesting the uncertainties may be underestimated. Conversely, it may be possible to reduce the uncertainties through better constraints on, for example, $m_{w,max}$ using the palaeo-record (Holschneider *et al.* 2014).

Given the global nature of this study, we have not been able to explicitly simulate tsunami run-up, but instead estimate run-up from offshore tsunami propagation models combined with amp-factors. This does not account for realistic nearshore topography which can have a large impact on tsunami run-up, and undoubtedly contributes to the large uncertainties in predicted run-up when PTHA scenarios are compared with observed events (Table 2). It may be possible to reduce these uncertainties using a higher-resolution nearshore tsunami-propagation model (Thio *et al.* 2010) which would have the advantage of concentrating higher run-up in areas where the nearshore bathymetry is particularly conducive to amplification. However, this would be computationally expensive on the global scale since the linear shallow-water equations will cease to be appropriate, and with non-linear models we cannot rely on superposition to efficiently produce many scenarios. Further, Thio *et al.* (2010) reported only moderate reductions in the logarithmic standard deviation σ using a high-resolution hydrodynamic model, suggesting that large run-up uncertainties are predominantly driven by the representation of earthquakes, and may persist even using a more complex (and computationally expensive) treatment of tsunami inundation.

This paper is published with the permission of the CEO, Geoscience Australia. The tsunami propagation simulations in this research were undertaken on the NCI National Facility in Canberra, Australia, which is supported by the Australian Commonwealth Government.

Correction notice: The original version was incorrect. There was an error in the legend for the bottom part of Figure 7. 'lower' had been replaced by 'upper'.

References

ANNAKA, T., SATAKE, K., SAKAKIYAMA, T., YANAGISAWA, K. & SHUTO, N. 2007. Logic-tree approach for probabilistic tsunami hazard analysis and its applications to the Japanese coasts. *Pure and Applied Geophysics*, **164**, 577–592, https://doi.org/10.1007/s00024-006-0174-3

BASILI, R., KASTELIC, V. ET AL. 2013*a*. *The European Database of Seismogenic Faults (EDSF) compiled in the framework of the Project SHARE*. SHARE (Seismic Hazard Harmonization in Europe), Swiss Seismological Service, Zurich, https://doi.org/10.6092/INGV.IT-SHARE-EDSF; http://diss.rm.ingv.it/share-edsf/

BASILI, R., TIBERTI, M.M., KASTELIC, V., ROMANO, F., PIATANESI, A., SELVA, J. & LORITO, S. 2013*b*. Integrating geologic fault data into tsunami hazard studies. *Natural Hazards and Earth System Science*, **13**, 1025–1050, https://doi.org/10.5194/nhess-13-1025-2013

BERRYMAN, K., WALLACE, L. ET AL. 2015. *The GEM Faulted Earth Subduction Interface Characterisation Project: Version 2.0 April 2015*. GEM Technical Report GEM (Global Earthquake Model) Faulted Earth Project, available from http://www.nexus.globalquakemodel.org/gem-faulted-earth/posts

BIRD, P. 2003. An updated digital model of plate boundaries. *Geochemistry, Geophysics, Geosystems*, **4**, 1–52.

BIRD, P. & KAGAN, Y.Y. 2004. Plate-tectonic analysis of shallow seismicity: apparent boundary width, beta, corner magnitude, coupled lithosphere thickness, and coupling. *Bulletin of the Seismological Society of America*, **94**, 2380–2399.

BRIZUELA, B., ARMIGLIATO, A. & TINTI, S. 2014. Assessment of tsunami hazards for the Central American Pacific coast from southern Mexico to northern Peru. *Natural Hazards and Earth System Science*, **14**, 1889–1903, https://doi.org/10.5194/nhess-14-1889-2014

BURBIDGE, D. & CUMMINS, P. 2007. Assessing the threat to Western Australia from tsunami generated by earthquakes along the Sunda Arc. *Natural Hazards*, **43**, 319–331, https://doi.org/10.1007/s11069-007-9116-3

BURBIDGE, D., CUMMINS, P., MLECZKO, R. & THIO, H. 2008. A probabilistic tsunami hazard assessment for Western Australia. *Pure and Applied Geophysics*, **165**, 2059–2088, https://doi.org/10.1007/s00024-008-0421-x

BURBIDGE, D., MUELLER, C. & POWER, W. 2015. The effect of uncertainty in earthquake fault parameters on the maximum wave height from a tsunami propagation model. *Natural Hazards and Earth System Science*, **15**, 2299–2312, https://doi.org/10.5194/nhess-15-2299-2015

CARRIER, G.F. & GREENSPAN, H.P. 1958. Water waves of finite amplitude on a sloping beach. *Journal of Fluid Mechanics*, **4**, 97–109.

CUMMINS, P.R. 2007. The potential for giant tsunamigenic earthquakes in the northern Bay of Bengal. *Nature*, **449**, 75–78, https://doi.org/10.1038/nature06088

EKSTROM, G., NETTLES, M. & DZIEWONSKI, A. 2012. The global CMT project 2004, 2010: centroid-moment tensors for 13 017 earthquakes. *Physics of the Earth and*

Planetary Interiors, **200**, 201, 1–9, https://doi.org/10.1016/j.pepi.2012.04.002

Field, E.H., Arrowsmith, R.J. *et al.* 2014. Uniform California Earthquake Rupture Forecast, Version 3 (UCERF3). *Bulletin of the Seismological Society of America*, **104**, 1122–1180, https://doi.org/10.1785/0120130164

Fritz, H.M., Kalligeris, N., Borrero, J.C., Broncano, P. & Ortega, E. 2008. The 15 August 2007 Peru tsunami runup observations and modeling. *Geophysical Research Letters*, **35**, 1–5, https://doi.org/10.1029/2008GL033494

Garwood, F. 1936. Fiducial limits for the Poisson distribution. *Biometrika*, **28**, 437–442, https://doi.org/10.1093/biomet/28.3-4.437

Geist, E. & Bilek, S. 2001. Effect of depth-dependent shear modulus on tsunami generation along subduction zones. *Geophysical Research Letters*, **28**, 1315–1318.

Geist, E.L. & Dmowska, R. 1999. Local tsunamis and distributed slip at the source. *Pure and Applied Geophysics*, **154**, 485–512.

Geist, E. & Parsons, T. 2006. Probabilistic analysis of tsunami hazards. *Natural Hazards*, **37**, 277–314, https://doi.org/10.1007/s11069-005-4646-z

Geist, E.L., ten Brink, U.S. & Gove, M. 2014. A framework for the probabilistic analysis of meteotsunamis. *Natural Hazards*, **74**, 123–142, https://doi.org/10.1007/s11069-014-1294-1

Glimsdal, S., Pedersen, G., Harbitz, C. & Løvholt, F. 2013. Dispersion of tsunamis: does it really matter? *Natural Hazards and Earth System Science*, **13**, 1507–1526, https://doi.org/10.5194/nhess-13-1507-2013

Goda, K. 2015. Effects of seabed surface rupture versus buried rupture on tsunami wave modeling: a case study for the 2011 Tohoku, Japan, Earthquake. *Bulletin of the Seismological Society of America*, **105**, https://doi.org/10.1785/0120150091

Gonzalez, F.I., Geist, E.L. *et al.* 2009. Probabilistic tsunami hazard assessment at Seaside, Oregon, for near- and far-field seismic sources. *Journal of Geophysical Research*, **114**, 1–19, https://doi.org/10.1029/2008JC005132

Greenslade, D.J., Annunziato, A. *et al.* 2014. An assessment of the diversity in scenario-based tsunami forecasts for the Indian Ocean. *Continental Shelf Research*, **79**, 36–45, https://doi.org/10.1016/j.csr.2013.06.001

Hanks, T. & Kanamori, H. 1979. A moment magnitude scale. *Journal of Geophysical Research*, **84**, 2348–2350.

Harbitz, C., Glimsdal, S. *et al.* 2012. Tsunami hazard in the Caribbean: regional exposure derived from credible worst case scenarios. *Continental Shelf Research*, **38**, 1–23, https://doi.org/10.1016/j.csr.2012.02.006

Harbitz, C.B., Løvholt, F. & Bungum, H. 2014. Submarine landslide tsunamis: how extreme and how likely? *Natural Hazards*, **72**, 1341–1374, https://doi.org/10.1007/s11069-013-0681-3

Hayes, G.P., Wald, D.J. & Johnson, R.L. 2012. Slab1, 0: a three-dimensional model of global subduction zone geometries. *Journal of Geophysical Research*, **117**, https://doi.org/10.1029/2011JB008524

Hébert, H. & Schindelé, F. 2015. Tsunami impact computed from offshore modeling and coastal amplification laws: insights from the 2004 Indian Ocean Tsunami. *Pure and Applied Geophysics*, **172**, 3385–3407, https://doi.org/10.1007/s00024-015-1136-4

Heidarzadeh, M. & Satake, K. 2015. Source properties of the 1998 July 17 Papua New Guinea tsunami based on tide gauge records. *Geophysical Journal International*, **202**, 361–369.

Holschneider, M., Zöller, G., Clements, R. & Schorlemmer, D. 2014. Can we test for the maximum possible earthquake magnitude? *Journal of Geophysical Research: Solid Earth*, **119**, 2019–2028, https://doi.org/10.1002/2013JB010319

Horspool, N., Pranantyo, I. *et al.* 2014. A probabilistic tsunami hazard assessment for Indonesia. *Natural Hazards and Earth System Science*, **14**, 3105–3122, https://doi.org/10.5194/nhessd-2-3423-2014

Kagan, Y.Y. 2002*a*. Seismic moment distribution revisited: 1. Statistical results. *Geophysical Journal International*, **148**, 520–541.

Kagan, Y.Y. 2002*b*. Seismic moment distribution revisited: 2. Moment conservation principle. *Geophysical Journal International*, **149**, 731–754.

Kagan, Y.Y. 2003. Accuracy of modern global earthquake catalogs. *Physics of the Earth and Planetary Interiors*, **135**, 173–209, https://doi.org/10.1016/S0031-9201(02)00214-5

Kagan, Y.Y. 2010. Earthquake size distribution: power-law with exponent beta = 1/2. *Tectonophysics*, **490**, 103–114.

Kagan, Y.Y. & Jackson, D.D. 2013. Tohoku earthquake: a surprise? *Bulletin of the Seismological Society of America*, **103**, 1181–1194, https://doi.org/10.1785/0120120110

Kajiura, K. 1963. The leading wave of a tsunami. *Bulletin of the Earthquake Research Institute*, **41**, 535–571.

Kamigaichi, O. 2009. Tsunami forecasting and warning. *In*: Meyers, R. (ed.) *Encyclopedia of Complexity and Systems Science*. Springer, Berlin, https://doi.org/10.1007/SpringerReference_60740

Kruschke, J.K. 2011. *Doing Bayesian Data Analysis: A Tutorial with R and BUGS*. 2nd edn. Elsevier, Amsterdam.

Lorito, S., Tiberti, M.M., Basili, R., Piatanesi, A. & Valensise, G. 2008. Earthquake-generated tsunamis in the Mediterranean Sea: scenarios of potential threats to Southern Italy. *Journal of Geophysical Research*, **113**, 1–14, https://doi.org/10.1029/2007JB004943

Lorito, S., Selva, J., Basili, R., Romano, F., Tiberti, M. & Piatanesi, A. 2015. Probabilistic hazard for seismically induced tsunamis: accuracy and feasibility of inundation maps. *Geophysical Journal International*, **200**, 574–588, https://doi.org/10.1093/gji/ggu408

Lorito, S., Romano, F. & Lay, T. 2016. Tsunamigenic earthquakes (2004–13): source processes from data inversion. *In*: Meyers, R. (ed.) *Encyclopedia of Complexity and Systems Science*. Springer Science + Business Media, New York, https://doi.org/10.1007/978-3-642-27737-5_641-1

Løvholt, F., Glimsdal, S. *et al.* 2012*a*. Tsunami hazard and exposure on the global scale. *Earth-Science Reviews*, **110**, 58–73, https://doi.org/10.1016/j.earscirev.2011.10.002

Løvholt, F., Kuhn, D., Bungum, H., Harbitz, C. & Glimsdal, S. 2012*b*. Historical tsunamis and present tsunami hazard in eastern Indonesia and southern Philippines. *Journal of Geophysical Research*, **117**, https://doi.org/10.1029/2012JB009425

Løvholt, F., Glimsdal, S., Harbitz, C., Horspool, N., Smebye, H., de Bono, A. & Nadim, F. 2014*a*. Global tsunami hazard and exposure due to large co-seismic slip. *International Journal of Disaster Risk Reduction*, **10**, 406–418, https://doi.org/10.1016/j.ijdrr.2014.04.003

Løvholt, F., Setiadi, N.J. et al. 2014*b*. Tsunami risk reduction – are we better prepared today than in 2004? *International Journal of Disaster Risk Reduction*, **10**, 127–142, https://doi.org/10.1016/j.ijdrr.2014.07.008

Løvholt, F., Glimsdal, S., Smebye, H., Griffin, J. & Davies, G. 2015. *Tsunami Methodology and Result Overview*. Technical Report, UN-ISDR Global Assessment Report 2015 Geoscience Australia, Canberra.

Løvholt, F., Griffin, J. & Salgado-Gálvez, M. 2016. Tsunami hazard and risk assessment at a global scale. *In*: Meyers, R. (ed.) *Encyclopedia of Complexity and Systems Science*. Springer Science + Business Media, New York, https://doi.org/10.1007/978-3-642-27737-5_642-1

Lynett, P., Weiss, R., Renteria, W., Morales, G.D.L.T., Son, S., Arcos, M.E.M. & MacInnes, B.T. 2012. Coastal impacts of the March 11th Tohoku, Japan tsunami in the Galapagos Islands. *Pure and Applied Geophysics*, **170**, 1189, https://doi.org/10.1007/s00024-012-0568-3

Marzocchi, W. & Jordan, T.H. 2014. Testing for ontological errors in probabilistic forecasting models of natural systems. *Proceedings of the National Academy of Sciences of the United States of America*, **111**, 11 973–11 978, https://doi.org/10.1073/pnas.1410183111

Marzocchi, W., Taroni, M. & Selva, J. 2015. Accounting for epistemic uncertainty in PSHA: logic tree and ensemble modeling. *Bulletin of the Seismological Society of America*, **105**, 2151–2159, https://doi.org/10.1785/0120140131

Matias, L.M., Cunha, T., Annunziato, A., Baptista, M.A. & Carrilho, F. 2013. Tsunamigenic earthquakes in the Gulf of Cadiz: fault model and recurrence. *Natural Hazards and Earth System Science*, **13**, 1–13, https://doi.org/10.5194/nhess-13-1-2013

McCaffrey, R. 2008. Global frequency of magnitude 9 earthquakes. *Geology*, **36**, 263–266, https://doi.org/10.1130/G24402A.1

McCaffrey, R. 2009. The tectonic framework of the sumatran subduction zone. *Annual Review of Earth and Planetary Sciences*, **37**, 345–366, https://doi.org/10.1146/annurev.earth.031208.100212

McGuire, R.K., Cornell, C.A. & Toro, G.R. 2005. The case for using mean seismic hazard. *Earthquake Spectra*, **21**, 879–886, https://doi.org/10.1193/1.1985447

Newman, A., Hayes, G., Wei, Y. & Convers, J. 2011. The 25 October 2010 Mentawai tsunami earthquake, from real-time discriminants, finite-fault rupture, and tsunami excitation. *Geophysical Research Letters*, **38**, 1–7, https://doi.org/10.1029/2010GL046498

NGDC. 2015. *National Geophysical Data Center/World Data Service Global Historical Tsunami Database*. National Geophysical Data Center (NGDC), Boulder, CO, https://doi.org/10.7289/V5PN93H7; https://http://www.ngdc.noaa.gov/hazard/tsu_db.shtml [last accessed 24 September 2015].

Okada, Y. 1985. Surface deformation due to shear and tensile faults in a half-space. *Bulletin of the Seismological Society of America*, **75**, 1135–1154.

Omira, R., Baptista, M.A. & Matias, L. 2015. Probabilistic tsunami hazard in the Northeast Atlantic from near and far field tectonic sources. *Pure and Applied Geophysics*, **172**, 901–920, https://doi.org/10.1007/s00024-014-0949-x

Parsons, T. & Geist, E.L. 2009. Tsunami probability in the Caribbean Region. *Pure and Applied Geophysics*, **165**, 2089–2116, https://doi.org/10.1007/s00024-008-0416-7

Power, W. 2013. *Review of Tsunami Hazard and Risk in New Zealand 2013 Update*. GNS Science Consultancy Report, 2013/131. GNS Science, Lower Hutt, New Zealand.

Saito, T. 2013. Dynamic tsunami generation due to sea-bottom deformation: analytical representation based on linear potential theory. *Earth, Planets and Space*, **65**, 1411–1423, https://doi.org/10.5047/eps.2013.07.004

Saito, T. & Furumura, T. 2009. Three-dimensional simulation of tsunami generation and propagation: Application to intraplate events. *Journal of Geophysical Research*, **114**, B02307, https://doi.org/10.1029/2007JB005523

Satake, K. 1995. Linear and nonlinear computations of the 1992 nicaragua earthquake tsunami. *Pure and Applied Geophysics*, **144**, 455–470.

Satake, K. 2014. Advances in earthquake and tsunami sciences and disaster risk reduction since the 2004 Indian ocean tsunami. *Geoscience Letters*, **1**, https://doi.org/10.1186/s40562-014-0015-7

Satake, K. & Atwater, B. 2007. Long-term perspectives on giant earthquakes and tsunamis at subduction zones. *Annual Reviews of Earth and Planetary Science*, **35**, 349–374, https://doi.org/10.1146/annurev.earth.35.031306.140302

Schorlemmer, D., Wiemer, S. & Wyss, M. 2005. Variations in earthquake-size distribution across different stress regimes. *Nature*, **437**, https://doi.org/10.1038/nature04094

Selva, J., Tonini, R. et al. 2016. Quantification of source uncertainties in Seismic Probabilistic Tsunami Hazard Analysis (SPTHA). *Geophysical Journal International*, **205**, 1780–1803, https://doi.org/10.1093/gji/ggw107

Socquet, A., Vigny, C., Chamot-Rooke, N., Simons, W., Rangin, C. & Ambrosius, B. 2006. India and Sunda plates motion and deformation along their boundary in Myanmar determined by GPS. *Journal of Geophysical Research*, **111**, https://doi.org/10.1029/2005JB003877

Sørensen, M.B., Spada, M., Babeyko, A., Wiemer, S. & Grünthal, G. 2012. Probabilistic tsunami hazard in the Mediterranean Sea. *Journal of Geophysical Research*, **117**, https://doi.org/10.1029/2010JB008169

STORCHAK, D., GIACOMO, D.D. ET AL. 2012. *ISC-GEM Global Instrumental Earthquake Catalogue (1900–2009)*. GEM Technical Report 2012-01. GEM (Global Earthquake Model) Faulted Earth Project.

STRASSER, F., ARANGO, M. & BOMMER, J.J. 2010. Scaling of the source dimensions of interface and intraslab subduction-zone earthquakes with moment magnitude. *Seismological Research Letters*, **81**, 941–950, https://doi.org/10.1785/gssrl.81.6.941

SYNOLAKIS, C. 2011. Tsunamis: When will we learn? *Newsweek Magazine*, http://www.newsweek.com/tsunamis-when-will-we-learn-66213

SYNOLAKIS, C. & KANOGLU, U. 2015. The Fukushima accident was preventable. *Philosophical Transactions of the Royal Society A: Mathematical, Physical and Engineering Sciences*, **373**, 20140379, https://doi.org/10.1098/rsta.2014.0379

TAPPIN, D.R., WATTS, P. & GRILLI, S.T. 2008. The Papua New Guinea tsunami of 17 July, 1998: anatomy of a catastrophic event. *Natural Hazards and Earth System Sciences*, **8**, 243–266, https://doi.org/10.5194/nhess-8-243-2008

TEN BRINK, U., CHAYTOR, J., GEIST, E.L., BROTHERS, D.S. & ANDREWS, B.D. 2014. Assessment of tsunami hazard to the U.S. Atlantic margin. *Marine Geology*, **353**, 31–54, https://doi.org/10.1016/j.margeo.2014.02.011

THIO, H. 2012. *URS Probabilistic Tsunami Hazard System: A User Manual*. Technical Report URS Corporation, San Francisco, CA.

THIO, H.K., SOMERVILLE, P. & ICHINOSE, G. 2007. Probabilistic analysis of strong ground motion and tsunami hazards in Southeast Asia. *Journal of Earthquake and Tsunami*, **1**, 119–137, https://doi.org/10.1142/S1793431107000080

THIO, H., SOMERVILLE, P. & POLET, J. 2010. *Probabilistic Tsunami Hazard in California*. Technical Report Pacific Earthquake Engineering Research Center, Berkeley, CA.

TIBERTI, M.M., LORITO, S., BASILI, R., KASTELIC, V., PIATANESI, A. & VALENSISE, G. 2009. Scenarios of earthquake-generated tsunamis for the Italian coast of the Adriatic Sea. *Pure and Applied Geophysics*, **165**, 2117–2142, https://doi.org/10.1007/s00024-008-0417-6

UCHIDA, N. & MATSUZAWA, T. 2011. Coupling coefficient, hierarchical structure, and earthquake cycle for the source area of the 2011 off the Pacific coast of Tohoku earthquake inferred from small repeating earthquake data. *Earth, Planets and Space*, **63**, 675–679, https://doi.org/10.5047/eps.2011.07.006

UN-ISDR 2015. *Global Assessment Report on Disaster Risk Reduction, Making Development Sustainable: The Future of Disaster Risk Management*. Technical Report United Nations Office for Disaster Risk Reduction, Geneva.

VALENCIA, N., GARDI, A., GAURAZ, A., LEONE, F. & GUILLANDE, R. 2011. New tsunami damage functions developed in the framework of SCHEMA project: application to European–Mediterranean coasts. *Natural Hazards and Earth System Science*, **11**, 2835–2846, https://doi.org/10.5194/nhess-11-2835-2011

WESSEL, P. 2009. Analysis of observed and predicted tsunami travel times for the Pacific and Indian Oceans. *Pure and Applied Geophysics*, **166**, 301–324, https://doi.org/10.1007/s00024-008-0437-2

WESSON, R.L., BOYD, O.S., MUELLER, C.S., BUFE, C.G., FRANKEL, A.D. & PETERSEN, M.D. 2007. *Revision of Time-Independent Probabilistic Seismic Hazard Maps for Alaska*. United States Geological Survey, Open-File Report, **2007-1043**.

YOUNGS, R.R. & COPPERSMITH, K.J. 1985. Implications of fault slip rates and earthquake recurrence models to probabilistic seismic hazard estimates. *Bulletin of the Seismological Society of America*, **75**, 939–964.

Index

Page numbers in *italics* refer to Figures. Page numbers in **bold** refer to Tables.